10 -

MW00338613

THE STRUCTURE OF COMPLEX NETWORKS

The Structure of Complex Networks

Theory and Applications

ERNESTO ESTRADA

OXFORD

UNIVERSITY PRESS

OXFORD

UNIVERSITY PRESS

Great Clarendon Street, Oxford OX2 6DP

Oxford University Press is a department of the University of Oxford.
It furthers the University's objective of excellence in research, scholarship,
and education by publishing worldwide in

Oxford New York

Auckland Cape Town Dar es Salaam Hong Kong Karachi
Kuala Lumpur Madrid Melbourne Mexico City Nairobi
New Delhi Shanghai Taipei Toronto

With offices in

Argentina Austria Brazil Chile Czech Republic France Greece
Guatemala Hungary Italy Japan Poland Portugal Singapore
South Korea Switzerland Thailand Turkey Ukraine Vietnam

Oxford is a registered trade mark of Oxford University Press
in the UK and in certain other countries

Published in the United States
by Oxford University Press Inc., New York

British Library Cataloguing in Publication Data
Data available

Library of Congress Cataloging in Publication Data
Data available

Typeset by SPI Publisher Services, Pondicherry, India
Printed in Great Britain
on acid-free paper by
CPI Group (UK) Ltd, Croydon, CR0 4YY

ISBN 978–0–19–959175–6

10 9 8 7 6 5 4 3 2 1

To Gissell, Doris, and Puri

Preface

"If you want to understand function, study structure".

F. H. C. Crick

The idea of writing this book arose in Venice in the summer of 2009, while I was attending the scientific conference NetSci09. This international workshop on network science was mainly attended by physicists who were studying the structure and dynamics of complex networks. I had previously attended a workshop in Rio de Janeiro devoted to spectral graph theory and its applications, which was mainly attended by mathematicians. Although the two areas are very much overlapped, the language, approaches, and results presented there were almost disjointed. It is frequently read that the study of complex networks is a truly interdisciplinary research field. However, it is hard to find a book that really reflects this character, unifying the languages and results used in graph theory and network science. The situation is even more complicated if we consider that many practitioners of network science work in application areas such as chemistry, ecology, molecular and cellular biology, engineering, and socio-economic sciences. The result of this *melting pot* is that many results are duplicated in different areas, usually with different denominations, and more critically there is a lack of unification of the field to form an independent scientific discipline that can be properly called 'network science'.

This book is a modest contribution to the development of a unified network theory. I have tried to express in a common language the results from *algebraic, topological, metrical*, and *extremal* graph theory, with the concepts developed in statistical mechanics and molecular physics, and enriched with many ideas and formulations developed in mathematical chemistry, biology, and social sciences. All these concepts are first mathematically formulated, and then explained and illustrated with simple examples from artificial and real-world networks. This comprises the first part of this volume, which is specifically devoted to the development of network theory. The second part is dedicated to applications in different fields, including the study of biochemical and molecular biology networks, anatomical networks, ecological networks, and socioeconomic networks. The idea of these chapters is not to describe a few results previously reported in the literature but rather to critically analyse the role of network theory in tackling real problems in these fields. Then examples of successful applications of network theory in these fields are illustrated, while erroneous applications and interpretations are critically analysed—the overall idea being to present a unified picture of the field of network science. That is, I consider these application areas as being at the intersection of the scientific disciplines and network theory where the problems arise.

Why study the structure of networks? The quote from Francis H. C. Crick, co-discoverer of the structure of DNA, and Nobel laureate, is self-explanatory.

This is the real ethos of this book. Of course, it is hard to define what we understand by 'structure', and a small section is devoted to a discussion of this in the introductory chapter. However, even without a complete understanding of this concept, architects, scientists, and artists have shaped our world as we see it today. Children usually open their toys to 'see their internal structures' in an attempt to understand how they work. Knowing the structure of complex networks does not guarantee that we automatically understand how they work, but without this knowledge it is certainly impossible to advance our understanding of their functioning. This book is devoted to the characterisation of the structure of networks based on a combination of mathematical approaches and physical analogies. This characterisation is still incomplete, and more effort is needed in the unification of apparently disconnected concepts. The reader is encouraged to take this route to advance to a definitive theory of complex networks. Finally, I have tried to correct errors and misconceptions found here and there across all fields in which networks are studied, from mathematics to biology. However, I would be pleased to be notified of anything which might require correction.

This book was written mainly in Glasgow (UK), but some parts were written in Remedios (Cuba) and Santiago de Compostela (Spain). Throughout this time I had the support of many colleagues and friends, as well as my family, although mentioning all of them would be an impossible task. However, I would like to express my gratitude to all my past and present collaborators—those who have shared datasets, software, and figures which are used in this book, or who have contributed in some way with one or other part of the process of writing: A. Allendes, U. Alon, N. A. Alves, C. Atilgan, S. R. Aylward, G. Bagler, A.-L. Barabási, V. Batagelj, M. Benzi, A. G. Bunn, D. Bassett, P. A. Bates, M. Boguñá, C. Cagatay Bilgin, P. Chebotarev, J. Crofts, J. A. Dunne, L. da Fontoura Costa, J. de los Rios, K.-I. Goh, L. H. Greene, R. Guimerà, N. Hatano, Y. He, D. J. Higham, E. Koonin, D.-S. Lee, M. Martin, R. Milo, J. Moody, M. E. J. Newman, P. Pereira, A. Perna, E. Ravasz, J. A. Rodríguez-Velazquez, R. Singh, R. V. Solé, O. Sporn, S. J. Stuart, M. Takayasu, H. Takayasu, V. van Noort, E. Vargas, and D. J. Watts. I also want to thank all the members of my family—in particular, Puri—for their infinite patience. Last but not least, I thank S. Adlung, A. Warman, and C. Charles at Oxford University Press for their effective support.

Ernesto Estrada
Glasgow, 3 February, 2011

Contents

Appendix   423

Theory

It is the theory which decides what we can observe.

Albert Einstein

Introduction

The impossibility of separating the nomenclature of a science from the science itself, is owing to this, that every branch of physical science must consist of three things; the series of facts which are the objects of the science, the ideas which represent these facts, and the words by which these ideas are expressed. Like three impressions of the same seal, the word ought to produce the idea, and the idea to be a picture of the fact.

Antoine-Laurent de Lavoisier,
Elements of Chemistry (1790)

1.1 What are networks?

The concept of networks is very intuitive to everyone in modern society. As soon as we see this word we think of a group of interconnected items. According to the *Oxford English Dictionary* (2010) the word 'network' appeared for the first time in the English language in 1560, in the Geneva Bible, Exodus, xxvii.4: 'And thou shalt make unto it a grate like networke of brass.' In this case it refers to 'a net-like arrangement of threads, wires, etc.', though it is also recorded in 1658 as referring to reticulate *patterns* in animals and plants (*OED*, 2010). New uses for the word were introduced subsequently in 1839 for rivers and canals, in 1869 for railways, in 1883 for distributions of electrical cables, in 1914 for wireless broadcasting, and in 1986 to refer to the Internet, among many others. Currently, the *OED* (2010) defines a network as an 'arrangement of intersecting horizontal and vertical lines' or 'a group or system of interconnected people or things', including the following examples: a complex system of railways, roads, or other; a group of people who exchange information for professional or social purposes; a group of broadcasting stations that connect for the simultaneous broadcast of a programme; a number of interconnected computers, machines, or operations; a system of connected electrical conductors. The generality of this concept ensures that 'network' is among the most commonly used words in current language. For instance, a search for the word via Google (on 3 August 2010) produced 884 million items. Figure 1.1 illustrates the comparative results of a search for other frequently used words, such as 'food,' 'football,' 'sex', and 'coffee'.

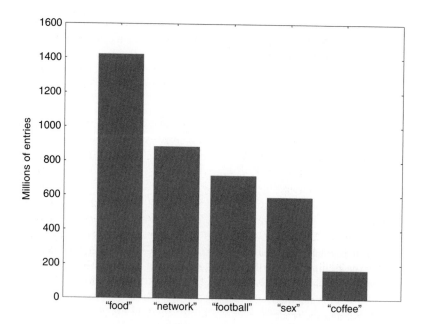

Fig. 1.1
Relevance of the word 'network' in current language. Results of a search via Google (on 3 August 2010) of some popular words in the English language.

According to the previous semantic definitions, the property of being interconnected appears as the most salient 'structural' characteristic of a network. Obviously, in order to have a system of interconnections we need a group of 'entities' and a way by which these items can be connected. For instance, in a system of railways the entities to be connected are stations, and the way they are connected to each other is with railway lines. In general, entities can represent stations, individuals, computers, cities, concepts, operations, and so on, and connections can represent railway lines, roads, cables, human relationships, hyperlinks, and so forth. Therefore, a way of generalizing the concept of networks in an abstract way is as follows.

> A *network (graph)* is a diagrammatic representation of a system. It consists of *nodes (vertices)*, which represent the entities of the system. Pairs of nodes are joined by *links (edges)*, which represent a particular kind of interconnection between those entities.

1.2 When did the story begin?

In mathematics the study of networks is known as 'graph theory'. We will use the words 'network' and 'graph' indistinctly in this book, with a preference for the first as a more intuitive and modern use of the concept. The story of network theory begins with the work published by Leonhard Euler in 1736: *Solutio problematic as geometriam situs pertinentis* (*The Solution of a Problem Relating to the Theory of Position*) (Euler, 1736). The problem in question was formulated as follows:

In Königsberg in Prussia, there is an island A, called the Kneiphof; the river which surrounds it is divided into two branches, as can be seen [in Figure 1.2]... and these branches are crossed by seven bridges, a, b, c, d, e, f, and g. Concerning these bridges, it was asked whether anyone could arrange a route in such a way that he would cross each bridge once and only once.

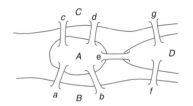

Fig. 1.2
The Königsberg bridges. Schematic illustration of the bridges in Königsberg.

Despite Euler's solving this problem without using a picture representing the network of bridges in Königsberg, he reformulated it in a way that was the equivalent of what is now referred to as a 'graph'. A representation derived from Euler's formulation of Königsberg's bridges problem is illustrated in Figure 1.3 by continuous lines which are superposed on the scheme of the islands and bridges of Figure 1.2. His paper is therefore recognised as the very first one in which the concepts and techniques of modern graph (network) theory were used. As a consequence of this paper a new branch of mathematics was born in 1736. For an account on the life and work of Euler in particular respect to the Königsberg's bridges problem, the reader is referred to the works of Assad (2007), Gribkovskaia et al. (2007), and Mallion (2007).

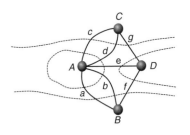

Fig. 1.3
Graph of the Königsberg bridges. Superposition of a multigraph (see further) over the schematic illustration of the bridges in Königsberg.

The history of graph theory from 1736 to 1936 is well documented by Biggs et al. (1976), and the reader is referred to this book for further details. There is one very important moral to be derived from Euler's original work, which we would like to remark on here. What Euler demonstrated is the existence of certain types of 'real-world' problems that cannot necessarily be solved by using mathematical tools such as geometry and calculus. That is, navigating through Königsberg's bridges without crossing any of them more than once depends neither on the length of the bridges nor on any of their geometric features. It depends only on the way in which the bridges are interconnected—their connectivity. This historical message is important for those who even today deny the value of topological concepts for solving important real-world problems ranging from molecular branching to the position of individuals in their social networks.

A few other historical remarks are added here to provide an idea of the interdisciplinary nature of network/graph theory since the very early days of its birth. The word 'graph' arises in a completely interdisciplinary context. In 1878 the English mathematician James J. Sylvester wrote a paper entitled 'Chemistry and Algebra', which was published in *Nature* (Sylvester, 1877–78). Here the word 'graph' derived from the word used by chemists at that time for the terminology of molecular structure. The word appeared for the first time in the following passage:

> Every invariant and covariant thus becomes expressible by a graph precisely identical with a Kekulean diagram or chemicograph.

Many examples of graph drawing have been collected by Kruja et al. (2002), including family trees in manuscripts of the Middle Ages, and 'squares of opposition' used to depict the relations between propositions and syllogisms in teaching logic, from the same period. One very early application of networks for representing complex relationships was the publication of François Quesnay's *Tableau Économique* (Quesnay, 1758), in which a circular flow of financial funds in an economy is represented with a network. In other areas, such as the study of social (Freeman, 2004) and ecological relationships, there were also important uses of network theory in early attempts at representing such complex systems. For instance, in 1912, William Dwight Pierce published a book entitled *The Insect Enemies of the Cotton Boll Weevil*, in which we find a network of the 'enemies' of the cotton plant which at the same time are attacked by their own 'enemies', forming a directed network in an ecological system (Pierce et al., 1912). Several such fields have maintained a long tradition in the use of network theory, having today entire areas of study such as chemical graph theory (Balaban, 1976; Trinajstić, 1992) social networks theory (Wasserman and Faust, 1994), and the study of food webs (Dunne, 2005).

Lastly, it should be mentioned that the transition from these pioneering works to the modern theory of networks was very much guided and influenced by Frank Harary, who is recognised as the 'father' of modern graph theory. The multidisciplinary character of his work can be considered as pioneering and inspiring because, as written in his obituary in 2005, he 'authored/co-authored more than 700 scholarly papers in areas as diverse as Anthropology, Biology, Chemistry, Computer Science, Geography, Linguistics, Music, Physics, Political Science, Psychology, Social Science, and, of course Mathematics, which brought forth the usefulness of Graph Theory in scientific thought' (Ranjan, 2005). Among his other books are *Graph Theory and Theoretical Physics* (1968), which he edited, *Structural Models in Anthropology* (1984) and *Island Networks* (1996), co-authored with Hage, and his classic *Graph Theory* (1969).

1.3 A formal definition

Although the working definition of networks presented in Section 1.1 accounts for the general concept, it fails in accounting for the different ways in which pairs of nodes can be linked. For instance, every pair of different nodes can be connected by a simple segment of a line. In this case we encounter the *simple network* as illustrated in Figure 1.4 (top left). Links in a network can also have directionality. In this case, a *directed link* begins in a given node and ends in another, and the corresponding network is known as a *directed network* (bottom right of the figure). In addition, more than one link can exist between a pair of nodes, known as multi-links, and some links can join a node to itself, known as *self-loops*. These networks are called *pseudonetworks* (pseudographs) (see Figure 1.4). For more details and examples of these types of network, the reader is referred to Harary (1969).

Finally, links can have some weights, which are real (in general positive) numbers assigned to them. In this case we have *weighted networks* (Barrat et al., 2004; Newman, 2004b). Such weights can represent a variety of concepts in the real world, such as strength of interactions, the flow of information, the number of contacts in a period of time, and so on. The combination of all these types of link is possible if we consider a weighted directed network like that illustrated in Figure 1.5 (bottom right), which is the most general representation of a network.

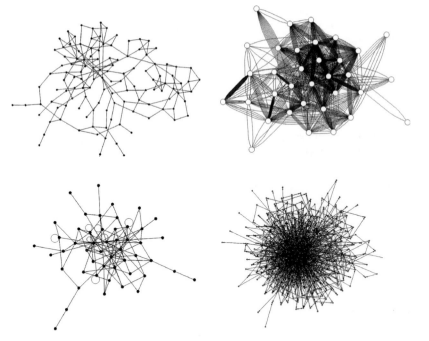

Fig. 1.4
Different kinds of network. (Top left) A simple network representing an electronic circuit. (Top right) A multi-network representing the frequency of interactions in a technical research group at a West Virginia university made by an observer every half-hour during one five-day work-week. (Bottom left) A network with self-loops representing the protein–protein interaction network of the herpes virus associated with *Kaposi sarcoma*. (Bottom right) A directed network representing the metabolic network of *Helicobacter pylori* (connected component only).

With these ingredients we will present some formal definitions of networks. Let us begin by considering a finite set $V = \{v_1, v_2, \cdots, v_n\}$ of unspecified elements, and let $V \otimes V$ be the set of all ordered pairs $[v_i, v_j]$ of the elements of V. A relation on the set V is any subset $E \subseteq V \otimes V$. The relation E is *symmetric* if $[v_i, v_j] \in E$ implies $[v_j, v_i] \in E$, and it is *reflexive* if $\forall v \in V, [v, v] \in E$. The relation E is *antireflexive* if $[v_i, v_j] \in E$ implies $v_i \neq v_j$. Then we have the following definition for simple and directed networks (Gutman and Polanski, 1987).

A *simple network* is the pair $G = (V, E)$, where V is a *finite set of nodes* and E is a *symmetric and antireflexive relation* on V. In a *directed network* the relation E is *non-symmetric*.

The previous definition does not allow the presence of multiple links and self-loops. Consequently, we introduce the following more general definition of network.

A *network* is the triple $G = (V, E, f)$, where V is a *finite set of nodes*, $E \subseteq V \otimes V = \{e_1, e_2, \cdots, e_m\}$ is a set of links, and f is a *mapping* which associates some elements of E to a pair of elements of V, such as that if $v_i \in V$ and $v_j \in V$, then $f : e_p \to [v_i, v_j]$ and $f : e_q \to [v_j, v_i]$. A *weighted network* is defined by replacing the set of links E by a set of link weights $W = \{w_1, w_2, \cdots, w_m\}$, such that $w_i \in \Re$. Then, a weighted network is defined by $G = (V, W, f)$.

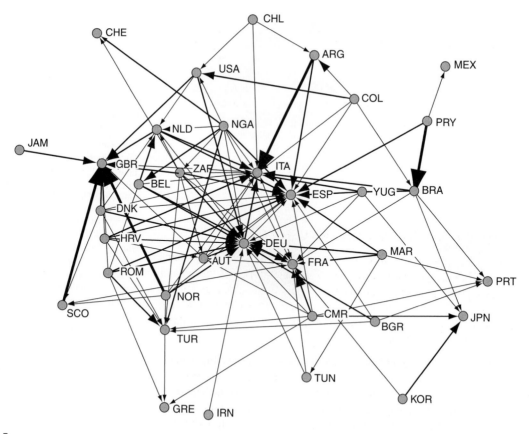

Fig. 1.5

Weighted directed network. Representation of the transfer of football players between countries after the 1998 World Cup in France. Links are drawn, with thickness proportional to the number of players transferred between two countries.

For instance, let $V = \{v_1, v_2, v_3\}$, $E = \{e_1, e_2, e_3, e_4, e_5, e_6\}$, and $f : e_1 \rightarrow [v_1, v_1]$, $f : e_2 \rightarrow [v_3, v_3]$, $f : e_3 \rightarrow [v_1, v_3]$, $f : e_4 \rightarrow [v_2, v_1]$, $f : e_5 \rightarrow [v_2, v_3]$, $f : e_6 \rightarrow [v_3, v_2]$, then we have the network illustrated in Fig. 1.3 (left). If we consider the following mapping:

$$f : 1.5 \rightarrow [v_1, v_1]$$

$$f : 0.9 \rightarrow [v_3, v_3]$$

$$f : 0.6 \rightarrow [v_1, v_3]$$

$$f : 2.1 \rightarrow [v_2, v_1]$$

$$f : 0.1 \rightarrow [v_2, v_3]$$

$$f : 1.2 \rightarrow [v_3, v_2]$$

we obtain the network illustrated in Figure 1.6 (right).

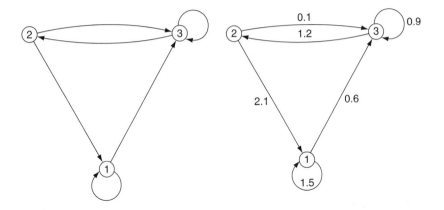

Fig. 1.6
Pseudonetwork and weighted network. The networks produced by using the general definition of networks presented in the text.

1.4 Why do we study their structure?

Complex networks are the skeletons of complex systems. Complex systems are composed of interconnected parts which display soe properties that are not obvious from the properties of the individual parts. One of the characteristics of the structure of these systems is their networked character, which justifies the use of networks in their representation in a natural way. In their essay 'Models of structure', Cottrell and Pettifor (2000) have written:

> There are three main frontiers of science today. First, the science of the very large, i.e. cosmology, the study of the universe. Second, the science of the very small, the elementary particles of matter. Third, and by far the largest, is the science of the very complex, which includes chemistry, condensed-matter physics, materials science, and principles of engineering through geology, biology, and perhaps even psychology and the social and economic sciences.

Then, the natural place to look for networks is in this vast region ranging from chemistry to socioeconomical sciences.

As we have seen in the historical account of network theory in this chapter, chemists have been well aware of the importance of the structure of networks representing molecules as a way to understand and predict their properties. To provide an idea we illustrate two polycyclic aromatic compounds in Figure 1.7. The first represents the molecule of benzo[a]pyrene, which is present in tobacco smoke and constitutes its major mutagen with metabolites which are both mutagenic and highly carcinogenic. By replacing the ring denoted as E in benzo[a]pyrene to the place below rings A and B, a new molecule is obtained. This polycyclic aromatic compound is called perylene, and it is known to be non-carcinogenic.

These two isomeric molecules are represented by networks of only 20 nodes, which reduces the 'complexity' of their analysis. However, this number can be increased to thousands when biological macromolecules, such as proteins, are represented by networks. The lesson that can be learned from this single example is that the *structure* of a network can determine many, if not all, of the properties of the complex system represented by it. It is believed that network theory can help in many important areas of molecular sciences, including the rational design of new therapeutic drugs (Csermely et al., 2005).

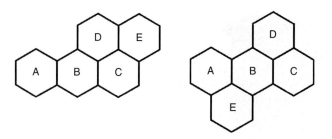

Fig. 1.7

Molecular isomeric networks. Network representation of two polycyclic aromatic compounds. The first corresponds to benzo[a]pyrene—a highly carcinogenic compound found in tobacco smoke. The second corresponds to perylene, which is a non-carcinogenic compound.

According to the *Oxford English Dictionary* (2010), 'structure' is defined as 'the arrangement of and relations between the parts or elements of something complex.' Similarly, the Cambridge Advanced Learner's Dictionary (Walter and Woodford, 2005) defines it as 'the way in which the parts of a system or object are arranged or organized, or a system arranged in this way.' The aim of our book, therefore, is to study the way in which nodes and links are arranged or organized in a network—though initially this appears to be a very easy task which probably does not merit an entire volume. In order to obtain a better idea about our subject, let us take a look at the networks which are depicted in Figure 1.8, which have 34 nodes. The structure of the first one is easily described: every pair of nodes is connected to each other. With this information we can uniquely describe this network, and can obtain precise information about all its properties. In order to describe the second network we have to say that all nodes are connected to two other nodes, except two nodes which are connected to only one node. This also uniquely describes this network, and we can know all its properties. The structure of the third network is more difficult to describe. In this case we can say that every node is connected to three other nodes. However, there are 453,090,162,062,723 networks with 34 nodes that share this property. Despite of this we can say a lot about this network with the information used to describe it. That is, there are many 'structural' properties which are shared by all networks that have the property of having every node connected to the other three. In a similar way, we can refer to the fourth and fifth networks. The fourth one has 34 nodes and 33 links—a type of structure known as a tree, which has many organizational properties in common. The fifth network has 34 nodes and 78 links, and it has been generated by a random process in which nodes are linked with certain probability. If we define such a process—the number of nodes and the linking probability—we can generate networks which asymptotically share many structural properties. We will study these types of network and some of their properties in the various chapters of this book.

Now let us try to describe, in a simple way, the last network, which represents the social interactions between individuals in a real-world system. It has 34 nodes and 78 links, but it does not appear to be as random as the previous one. It is not a tree-like, nor does it display regularity in the number of links per nodes, and it is far removed from the structures represented by the first two networks in Figure 1.8. Then, in order to describe the 'structure' of this

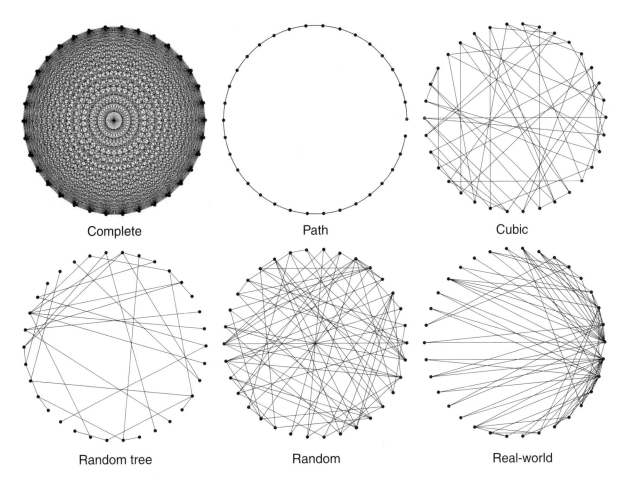

Fig. 1.8
Networks and complexity. Several types of network which can be 'defined' by using different levels of complexity.

network we need to add more and more information. For instance, we can say that the network:

(i) has *exponential degree distribution* and is strongly *disassortative* (see Chapter 2);
(ii) has small *average path length* (see Chapter 3);
(iii) has a high *clustering coefficient* (see Chapter 4);
(iv) has two nodes with high *degree*, *closeness*, and *betweenness centrality* (see Chapter 7);
(*v*) has *spectral scaling* with positive and negative deviations (see Chapter 9);
(vi) has two main *communities* (see Chapter 10);
(vii) displays small *bipartivity* (see Chapter 11), and so on.

The list of properties for this network can be considerably extended by using several of the parameters we will describe in this book, and others that are

Fig. 1.9
Road network. The international road
network in Europe.
(http://en.wikipedia.org/wiki/Image:
International_E_Road_Network.pgn.)

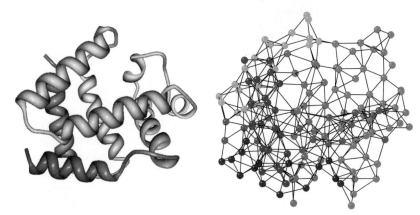

Fig. 1.10
Protein residue network. The
three-dimensional structure of a protein,
showing its secondary structure (left).
Residue interaction network built by
considering C_α atom of each amino acid
as a node of the network. Two nodes are
connected if the corresponding C_α atoms
are separated by at least a certain
distance—in this case, 7 Å.

not covered here. However, this list of properties is probably not sufficient to
account for the entire structure of this network. This situation is similar to
that described by Edgar Allan Poe when he wrote in relation to the structure
of a strange ship in his short story entitled *MS. Found in a Bottle*, written in
1833 (Poe, 1984): 'What she is not, I can easily perceive—what she is I fear
it is impossible to say.' The description of the structure of networks is not
impossible, but it requires a great deal of information to make such description
as complete as required. In other words, *the structure of some networks is in
general complex, understanding complexity as the amount of information that
we need to precisely describe this structure.*

A typical example is based on the 'nature' of the entities of the complex systems represented by these networks. In this case we can talk about
biological networks, socioeconomic networks, technological and infrastruc-
tural networks, and so forth. Every group of these networks can be further

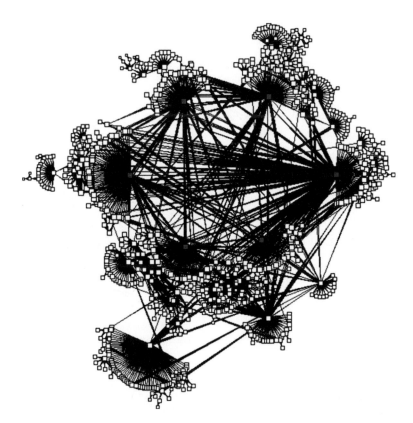

Fig. 1.11
World Wide Web. Image of
WWW-Gnutella from CAIDA.
(Obtained, with permission, from
http://www.caida.org/tools/visualization/
plankton/Images/.)

subdivided into several others. For instance, biological networks can be subdivided into interspecies networks such as food webs, intercellular networks such as neural or vascular networks; intermolecular networks such as metabolic or protein–protein interaction networks, and interatomic networks, such as protein residue networks. This classification system reveals the nature of the nodes, but not that of the interrelation between the entities of the complex systems represented by them. Another possibility of classifying complex networks is according to the nature of the links which they represent. Some of the possible classes of network in this scheme are presented below, with some examples.

- *Physical linking* (Fig. 1.9). Pairs of nodes are physically connected by a *tangible link*, such as a cable, a road, a vein, and so on. Examples are the Internet, urban street networks, road networks, vascular networks, and so on.
- *Physical interactions* (Fig. 1.10). Links between pairs of nodes represents *interactions* which are determined by a *physical force*. Examples are protein residue networks, protein–protein interaction networks, and so on.
- *'Ethereal' connections* (Fig. 1.11). Links between pairs of nodes are *intangible*, such that information sent from one node is received at another, irrespective of the 'physical' trajectory. Examples are the World Wide Web and airport networks.

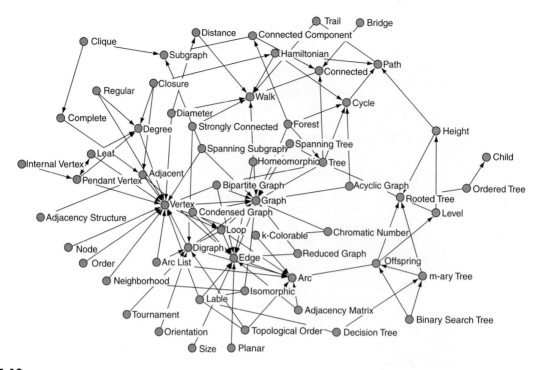

Fig. 1.12

Glossary network. The network formed by some of the terms commonly used in graph/network theory. Nodes represent concepts in graph theory, and a link is directed from one concept to another if the second appears in the definition of the first.

- *Geographic closeness.* Nodes represent regions of a surface and their connections are determined by their *geographic proximity*. Examples are countries on a map, landscape networks, and so on.
- *Mass/energy exchange.* Links connecting pairs of nodes indicate that some *energy or mass* has been *transferred* from one node to another. Examples are reaction networks, metabolic networks, food webs, trade networks, and so on.
- *Social connections.* Links represent any kind of *social relationship* between nodes. Examples are friendship, collaboration, and so on.
- *Conceptual linking* (Fig. 1.12). Links indicate *conceptual relationships* between pairs of nodes. Examples are dictionaries, citation networks, and so on.

1.5 How can we speak their 'language'?

The language of networks is the language of graph theory. We will define many of the terms used in graph and network theory in each chapter of this book, but here it is useful to present some of the most commonly used terms (Essam and Fisher, 1970; Harary, 1969; Newman, 2010).

Adjacency and incidence. Two nodes u and v are *adjacent* if they are joined by a link $e = \{u, v\}$. Nodes u and v are *incident* with the link e, and the link e

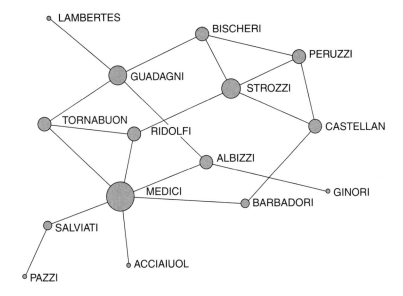

Fig. 1.13
Node degrees. Representation of the node degrees proportional to the radii of the circles of each node. The network illustrated corresponds to the marriage relations between pairs of Florentine families in fifteenth century (Breiger and Pattison, 1986).

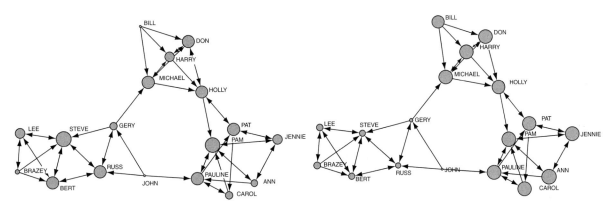

Fig. 1.14
Node degrees in a directed network. Representation of indegree (left) and outdegree (right) proportional to the radii of the nodes for a directed network of eighteen participants in a qualitative methods class. Ties signify that the ego indicated that the nominated alter was one of the three people with which s/he spent the most time during the seminar.

is *incident* with the nodes u and v. Two links $e_1 = \{u, v\}$ and $e_1 = \{v, w\}$ are *adjacent* if they are both incident with at least one node. *Node degree* is the number of links which are incident with a given node (see Fig. 1.13).

   Adjacency and incidence (directed). Node u is *adjacent* to node v if there is a directed link from u to v $e = (u, v)$.[1] A link from u to v is *incident from u* and *incident to v*; u is *incident to e*, and v is *incident from e*. The *indegree* of a node is the number of links incident to it, and its *outdegree* is the number of links incident from it (see Fig. 1.14).

---

[1] It is generally assumed that two nodes in a directed network are adjacent if they are connected in any direction by a link. This definition, however, conflicts with the definition of adjacency matrix that we will study in Chapter 2.

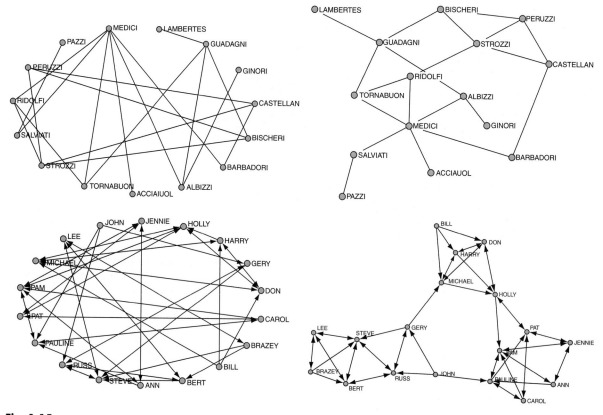

Fig. 1.15

Isomorphic networks. Different representations of the networks of Florentine families in the fifteenth century (top), and that of participants in a qualitative methods class (bottom).

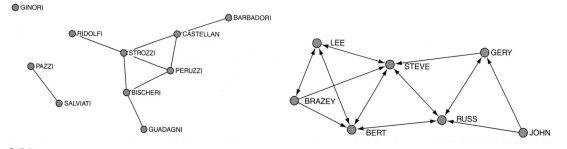

Fig. 1.16

Subgraphs in networks. Subgraphs found in the networks of Florentine families in the fifteenth century (left), and that of participants in a qualitative methods class (right).

Isomorphism (Fig. 1.15). Two networks G_1 and G_2 are isomorphic if there is a one-to-one correspondence between the nodes of G_1 and those of G_2, such as the number of links joining each pair of nodes in G_1 is equal to that joining the corresponding pair of nodes in G_2. If the networks are directed, the links must coincide not only in number but also in direction.

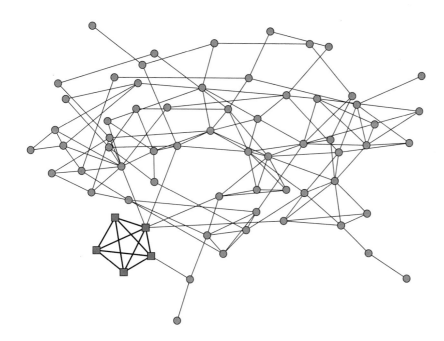

Fig. 1.17
Cliques in networks. Representation of a five-clique found in a network representing inmates in prison.

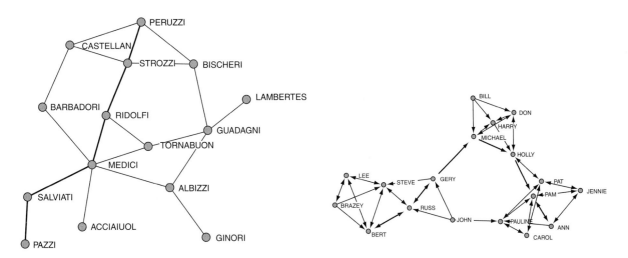

Fig. 1.18
Paths in networks. Paths in the networks of Florentine families in the fifteenth century (left), and that of participants in a qualitative methods class (right).

Subgraph (Fig. 1.16). $S = (V', E')$ is a subgraph of a network $G = (V, E)$ if and only if $V' \subseteq V$ and $E' \subseteq E$.

Clique (Fig. 1.17). A clique is a maximal complete subgraph of a network. The clique number is the size of the largest clique in a network.

Walk. A (directed) walk of length l is any sequence of (not necessarily different) nodes $v_1, v_2, \cdots, v_l, v_{l+1}$ such that for each $i = 1, 2, \cdots, l$ there is a

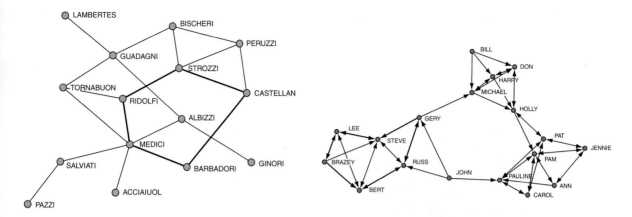

Fig. 1.19
Cycles in networks. Cycles in the networks of Florentine families in the fifteenth century (left), and that of participants in a qualitative methods class (right).

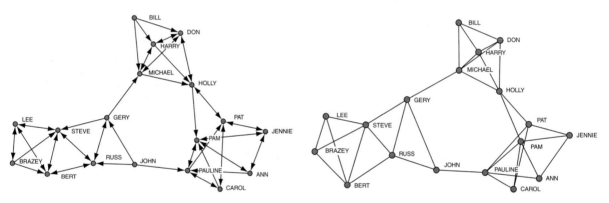

Fig. 1.20
Underlying graph of a directed network. Directed network of participants in a qualitative methods class (left) and its underlying graph (right).

link from v_i to v_{i+1}. This walk is referred to as a walk from v_1 to v_{l+1}. A *closed walk* (CW) of length l is a walk $v_1, v_2, \cdots, v_l, v_{l+1}$ in which $v_{l+1} = v_1$.

Path (Fig. 1.18). A *path* of length l is a walk of length l in which all the nodes (and all the links) are distinct. A *trial* has all the links different, but not necessarily all the nodes.

Cycle (Fig. 1.19). A cycle is closed walk in which all the links and all the nodes (except the first and last) are distinct.

Underlying network (Fig. 1.20). This is the network obtained from a directed network by removing the directionality of all links.

Connected components (Fig. 1.21). A (directed) network is *connected* if there is a path between each pair of nodes in the (underlying) network. Otherwise it is *disconnected*. Every connected subgraph is a *connected component* of the network.

Connectivity. The *node connectivity* of a connected network is the smallest number of nodes whose removal makes the network disconnected. The *edge*

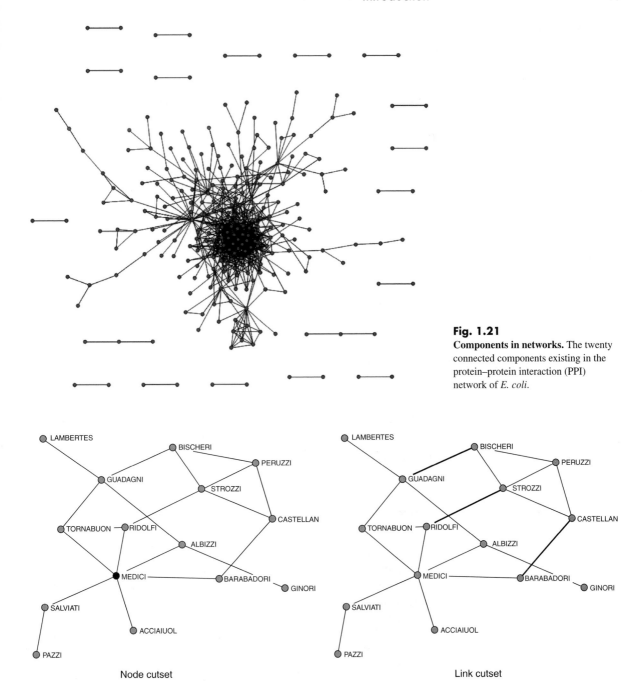

Fig. 1.21
Components in networks. The twenty connected components existing in the protein–protein interaction (PPI) network of *E. coli.*

Node cutset

Link cutset

Fig. 1.22
Cutsets in networks. A node cutset (left) and a link cutset (right) in the network representing Florentine families in the fifteenth century.

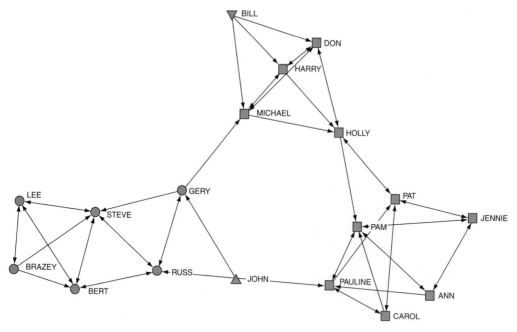

Fig. 1.23

Strongly connected components in networks. Four strongly connected components in the network of participants in a qualitative methods class. One of the strongly connected components has ten nodes and are represented by squares, another has five nodes and is represented by circles, and two others have only one node each and are represented by triangles.

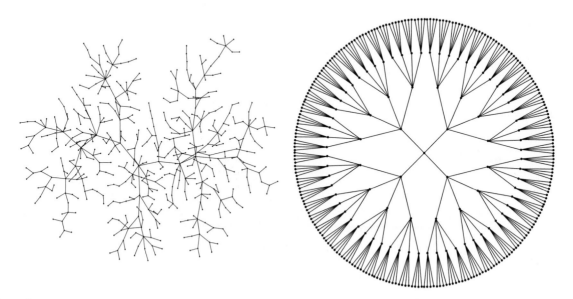

Fig. 1.24

Trees. A random tree (left) and a Cayley tree, each with 500 nodes.

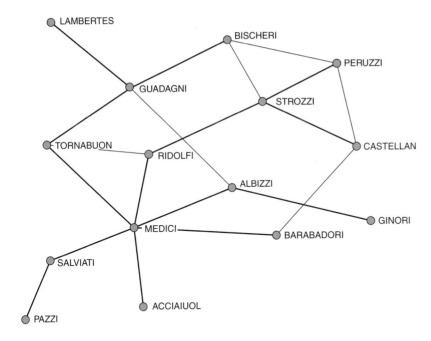

Fig. 1.25
Spanning trees in networks. A spanning tree in the network representing Florentine families in the fifteenth century.

connectivity of a connected network is the smallest number of links whose removal makes the network disconnected.

Cutsets (Fig. 1.22). A *node cutset* of a connected network is a set of nodes such as the removal of all of them disconnects the network. A *link cutset* of a connected network is a set of links such that the removal of all of them disconnects the network.

Strongly connected component (Fig. 1.23). A directed network is *strongly connected* if there is a path between each pair of nodes. The *strongly connected components* of a directed network are its maximal strongly connected subgraphs.

Tree (Fig. 1.24). A tree of *n* nodes is a network for which the following statements are equivalent:

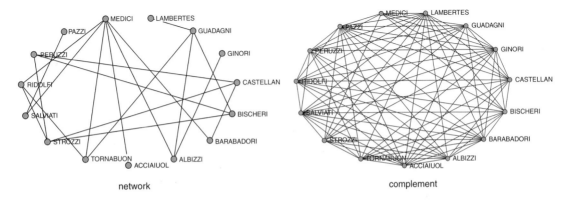

network   complement

Fig. 1.26
Network complement. Network representing Florentine families in the fifteenth century (left), and its complement (right).

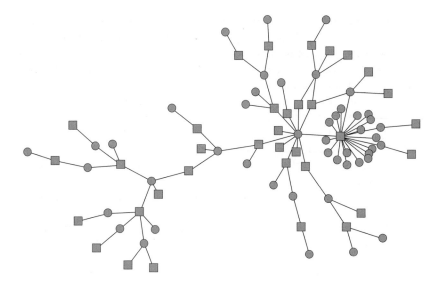

Fig. 1.27
Bipartite networks. A bipartite network representing heterosexual relationship among individuals. Individuals of each sex are represented as squares or circles, respectively.

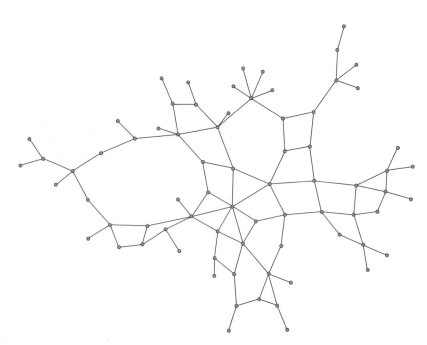

Fig. 1.28
Planar networks. A planar network representing the network of galleries made by ants.

- It is connected and has no cycles.
- It has $n - 1$ links and no cycle.
- It is connected and has $n - 1$ links.
- It is connected and becomes disconnected by removing any link.
- Any pair of nodes is connected by exactly one path.
- It has no cycles, but the addition of any new link creates a cycle.

Spanning tree (Fig. 1.25). A spanning tree of a network is a subgraph of this network that includes every node and is a tree.

Complement of a network (Fig. 1.26). The complement of a network G, denoted $\bar{G}$, is a network created by taking all nodes of G and joining two nodes whenever they are not linked in G (see Figure 1.25).

Regular network. An r-regular network is a network in which all nodes have degree r and it has $rn/2$ links.

Complete networks. A network with n nodes is complete, denoted K_n, if every pair of nodes is connected by a link. That is, there are $n(n-1)/2$ links.

Empty networks. A network that has no link. It is denoted as $\bar{K}_n$, as is the complement of the complete network.

Cycle networks. A regular network of degree 2; that is, a 2-regular network, denoted by C_n.

Network colouring. A k-colouring of a simple network is an assignment of at most k colours to the nodes of the network, such that adjacent nodes are coloured with different colours. The network is then said to be k-colourable. The chromatic number is the smallest number k for which the network is k-colourable.

Bipartite networks (Fig. 1.27). A network is bipartite if its nodes can be split into two disjoint (non-empty) subsets $V_1 \subset V$ ($V_1 \neq \phi$) and $V_2 \subset V$ ($V_2 \neq \phi$) and $V_1 \cup V_2 = V$, such that each link joins a node in V_1 and a node in V_2. Bipartite networks do not contain cycles of odd length. If all nodes in V_1 are connected to all nodes in V_2 the network is known as complete bipartite and is denoted K_{n_1,n_2}, where $n_1 = |V_1|$ and $n_2 = |V_2|$ are the number of nodes in V_1 and V_2, respectively. A network is bipartite if it is 2-colourable.

$k-$partite networks. A network is k-partite if its nodes can be split into k disjoint (non-empty) subsets $V_1 \cup V_2 \cup \cdots \cup V_k = V$, such that each link joins a node in two different subsets. If all nodes in a subset are connected to all nodes in the other subsets, the network is known as complete k-partite and is denoted $K_{n_1,n_2,\cdots,n_k}$, where $n_i = |V_i|$ is the number of nodes in V_i.

Planar networks (Fig. 1.28). A network is planar if it can be drawn in a plane in such a way that no two links intersect except at a node with which they are both incident.

Adjacency relations in networks

2

Mathematicians do not study objects, but the relations between objects; to them it is a matter of indifference if these objects are replaced by others, provided that the relations do not change.

Henri Poincaré

2.1 Node adjacency relationship

Several structural properties of networks are related to adjacency relationships between nodes. In this chapter we explore some of these properties by starting from a definition of node adjacency. We begin by considering the Hilbert space of C-valued squared-summable functions (Debnath and Mikusiski, 1990) on the set of nodes V of the network, $H := l^2(V)$. Let $\{|i\rangle, i \in V\}$ be a complete orthonormal basis of $l^2(V)$. Then, the adjacency operator of the network acting in $l^2(V)$ (Friedman and Tillich, 2008) is defined as

$$(\mathbf{A}f)(u) := \sum_{u,v \in E} f(v), \quad f \in H, \quad i \in V \tag{2.1}$$

The adjacency operator of an undirected network is a *self-adjoint operator*, which is bounded on $l^2(V)$. However, the adjacency operator of a directed network might not be self-adjoint. If the node u has k_u neighbours the adjacency operator acts on $l^2(V)$ by summing over all these neighbours. Let us represent $\mathbf{A}$ as an $|V| \times |V|$ matrix. If the network does not contain any self-loop the diagonal entries of $\mathbf{A}$ are zeroes. That is, for a network $G = (V, E)$ the entries of the adjacency matrix (Harary, 1967) are defined as

$$A_{ij} = \begin{cases} 1 & \text{if } i, j \in E \\ 0 & \text{otherwise} \end{cases}$$

Then, it is easy to see that in an undirected network the uth row or column of $\mathbf{A}$ has exactly k_u entries. The number k_u is termed the *degree* of the node u, which is the number of nearest neighbours that u has. If we denote by $\mathbf{1}$

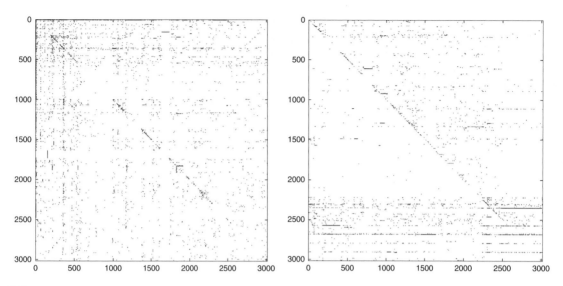

Fig. 2.1

Symmetric and unsymmetric adjacency matrices. The adjacency matrices of an undirected (left) and directed version of Internet.

a $|V| \times 1$ all-one vector, it is easy to see that for the undirected network the column vector of node degrees $\mathbf{k}$ is given by

$$\mathbf{k} = (\mathbf{1}^T \mathbf{A})^T = \mathbf{A1} \qquad (2.2)$$

However, in the directed network the adjacency matrix is not necessarily symmetric (see Figure 2.1); that is, the adjacency operator is not necessarily self-adjoint. In this case, $A_{uv} = 1$ if there is a link pointing from u to v. Then, $\mathbf{1}^T \mathbf{A}$ is not necessarily equal to $\mathbf{A1}$, and we can distinguish between two different kinds of node degrees. The *indegree* is the number of links pointing toward a given node, and the *outdegree* is the number of links departing from the corresponding node. The indegree and outdegree column vectors can be obtained as follows:

$$\mathbf{k}^{in} = (\mathbf{1}^T \mathbf{A})^T \qquad (2.3)$$

$$\mathbf{k}^{out} = \mathbf{A1} \qquad (2.4)$$

The adjacency matrix of a bipartite network having two sets of disjoint nodes V_1 and V_2, such as $|V_1| = n_1$ and $|V_2| = n_2$, can be written as

$$\mathbf{A} = \begin{pmatrix} \mathbf{0} & \mathbf{R}^T \\ \mathbf{R} & \mathbf{0} \end{pmatrix}$$

where $\mathbf{R}$ is an $n_1 \times n_2$ matrix, and $\mathbf{0}$ is an all-zeroes matrix.

In the case of weighted networks the entry A_{uv} of the weighted adjacency matrix is equal to w_{uv} if the nodes u and v are connected and the link $\{u, v\}$ has weight equal to w_{uv}. In this case the weighted degrees, respectively indegree and outdegree, are computed by using expressions like (2.2)–(2.4) in which $\mathbf{A}$ represents the weighted adjacency matrix.

When the node degrees are represented as a vector we call it the *vector degree* of the network. The quantity $\bar{k} = \frac{1}{n}\mathbf{1}^T\mathbf{k} = \frac{1}{n}\sum_{i=1}^{n}k_i$ is termed the *average node degree* of a network, and if the node degrees are arranged in non-increasing order then the corresponding array is known as the *degree sequence* of the network. Another way of representing node degrees is in the form of a diagonal matrix $\mathbf{K}$, which is known as the *degree matrix* of the network. The adjacency matrix of a network can be obtained from the incidence $\mathbf{B}$ and the degree matrices of the network. The incidence matrix is an $|V| \times |E|$ matrix whose entry B_{ue} is one if the node u is incident with the link e. Then, $\mathbf{A} = \mathbf{BB}^T - \mathbf{K}$.

Let us mention here a result that was first proved by Euler in his seminal paper of 1736 (Euler, 1736). This result is known in graph theory as the *handshaking lemma*, and will be useful in several parts of this book. The result states that in a network, the sum of all node degrees is equal to twice the number of links:

$$\mathbf{1}^T\mathbf{A}\mathbf{1} = \mathbf{K}^T\mathbf{1} = \sum_{i=1}^{n}k_i = 2|E| \tag{2.5}$$

Consequently, the average node degree is given by

$$\bar{k} = \frac{2|E|}{n}$$

2.2 Degree distributions

Using only the information provided by the node degree vector we can obtain some useful insights about the structure of the network. This elementary analysis is based on the distribution of node degrees in the network. For instance, in Figure 2.2 we illustrate the adjacency matrices of a version of the Internet as autonomous system, and the social network of the US corporate elite sorted according to their degrees, where it can be seen that they exhibit very different characteristics. While in the first the highest-degree nodes appear to be concentrated in a small region of the top-left corner of the sorted matrix, in the second they look more uniformly distributed. In the following we will analyse these distributions in more details.

Let $p(k) = n(k)/n$, where $n(k)$ is the number of nodes having degree k in the network of size n. That is, $p(k)$ represents the probability that a node selected uniformly at random has degree k. Then, a plot of $p(k)$ versus k represents the degree distribution for the network. Some of these plots are usually displayed in a log–log scale, which presents a qualitative idea about the kind of statistical distribution followed by the node degrees (Amaral et al., 2000). In Figure 2.3 we illustrate some common distributions found in complex networks. For a compendium of common probability distributions, the reader is directed to the compilations of McLaughlin (1999) and Van Hauwermeiren and Vose (2009).

Despite the apparent simplicity of determining the degree distribution of a network, in general the process is far from been straightforward. The first

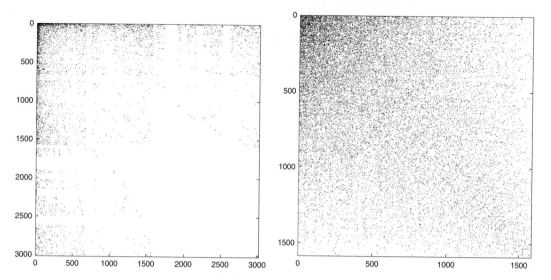

Fig. 2.2
Sorted adjacency matrices. The adjacency matrices of the Internet as an autonomous system (left) and the social network of the US corporate elite (right), sorted according to their degrees.

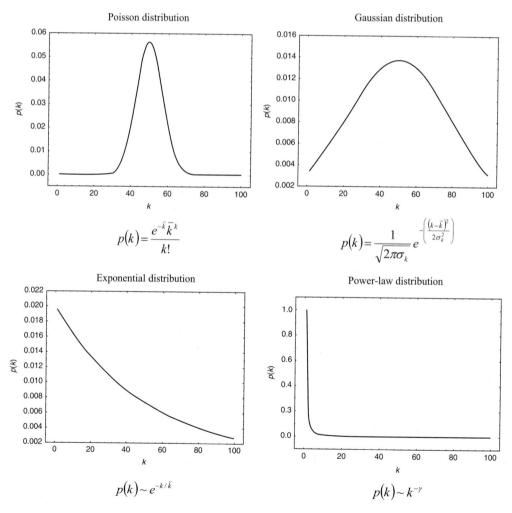

Poisson distribution

$$p(k) = \frac{e^{-\bar{k}} \bar{k}^k}{k!}$$

Gaussian distribution

$$p(k) = \frac{1}{\sqrt{2\pi}\sigma_k} e^{-\left(\frac{(k-\bar{k})^2}{2\sigma_k^2}\right)}$$

Exponential distribution

$$p(k) \sim e^{-k/\bar{k}}$$

Power-law distribution

$$p(k) \sim k^{-\gamma}$$

Fig. 2.3
Degree distributions in networks. Some examples of degree distributions found in complex networks.

difficulty encountered is that in most networks the number of data points available for the distribution is quite limited. For instance, although the network can be relatively large, the variability of node degree is not so high, which means that eventually one will have to deal with very few points to obtain the histogram of degree distributions. Another difficulty is that the number of probability distributions that can be considered to fit with the empirical data is quite large. Some of these fits are not easily discernible visually, nor even by using sophisticated statistical techniques.

From all possible degree distributions there is one that has attracted most attention in the scientific and even popular literature: the power-law degree distribution (Barabási and Albert, 1999; Barabási and Albert, 2002; Albert et al., 2000; Caldarelli, 2007; Clauset et al., 2010; Goh et al., 2002; Heyde and Kou, 2004; Keller, 2005; Kim et al., 2004; Mitzenmacher, 2003; Newman, 2005b; Solow, 2005). In this distribution (see Figure 2.3) the probability of finding a node with degree k decays as a negative power of the degree: $p(k) \sim k^{-\gamma}$. This means that the probability of finding a high-degree node is relatively small in comparison with the high probability of finding low-degree nodes. In other words, networks with power-law degree distributions resemble, in some way, star graphs.

The usual way of referring to these networks, however, is as 'scale-free' networks. The term 'scaling' describes the existence of a power-law relationship between the probability and the node degree: $p(k) = Ak^{-\gamma}$. Then, scaling the degree by a constant factor c produces only a proportionate scaling of the function:

$$p(k, c) = A(ck)^{-\gamma} = Ac^{-\gamma} \cdot p(k) \tag{2.6}$$

Power-law relations are usually represented on a logarithmic scale, by which we obtain a straight line, $\ln p(k) = -\gamma \ln k + \ln A$, where $-\gamma$ is the slope and $\ln A$ the intercept of the function. Then, scaling by a constant factor c means that only the intercept of the straight line changes but the slope is exactly the same as before: $\ln p(k, c) = -\gamma \ln k - \gamma Ac$. The existence of a scaling law reveals that the phenomenon under study reproduces itself on different time and/or space scales. That is, it has self-similarity.

In Figure 2.4 (left) we illustrate the logarithmic plot of $p(k)$ versus k for a version of the Internet. As can be seen, the tail of the distribution—the part that corresponds to high degrees—is very noisy. Two main approaches are used for reducing this noisy effect in the tail of probability distributions. One of them is the binning procedure, which consists of building a histogram with the bin sizes which increase exponentially with degree. The other approach is to consider the cumulative distribution function (CDF) (Clauset et al., 2010; Van Hauwermeiren and Vose, 2009):

$$P(k) = \sum_{k'=k}^{\infty} p(k') \tag{2.7}$$

which represents the probability of choosing at random a node with degree greater than, or equal to, k. In the case of power-law degree distributions, $P(k)$ also exhibits a power-law decay with degree

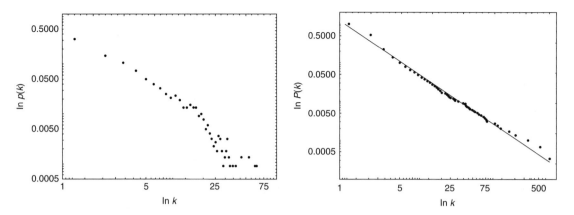

Fig. 2.4
Probability and cumulative distribution function. Plots of the probability (left) and cumulative (right) distribution function for a network displaying power-law degree distribution.

$$P(k) \sim \sum_{k'=k}^{\infty} k'^{-\gamma} \sim k^{-(\gamma-1)} \qquad (2.8)$$

which means that we will also obtain a straight line for the logarithmic plot of $P(k)$ versus k in scale-free networks. In Figure 2.4 (right) we illustrate such a plot for the case of the version of the Internet previously mentioned. As can be seen, the CDF plot is significantly less noisy than the previous one, and in this sense it should be preferred.

There is not a closed form for the CDF for a Gaussian distribution. In the case of the Poissonian and exponential distributions, the CDFs have the following forms (McLaughlin, 1999; Van Hauwermeiren and Vose, 2009), respectively:

$$P(k) = e^{-\bar{k}} \sum_{i=1}^{\lfloor k \rfloor} \frac{(\bar{k})^i}{i!} \qquad (2.9)$$

$$P(k) = 1 - e^{-k/\bar{k}} \qquad (2.10)$$

A large number of power-law degree distributions have been 'observed' in the literature. Some of these distributions have been subsequently reviewed; and in some cases, better fits with other probability distributions have been found. The more generic term of 'fat-tail' degree distribution has been proposed to globalise those distributions which include power-law, lognormal, Burr, logGamma, Pareto, and so on (Foss et al., 2011). As an example, we provide here the degree distribution of the yeast PPI. This PPI[1] has been reported as displaying a power-law degree distribution. In Figure 2.5 we plot $p(k)$ versus k for the nodes of this network, together with the exponential and power-law distributions that best fit the empirical data. Plots of Gaussian or Poissonian distributions are very far from fitting this data, and are not displayed here. A visual analysis would produce agreement in identifying this network as having

---

[1] PPI = protein-protein interaction

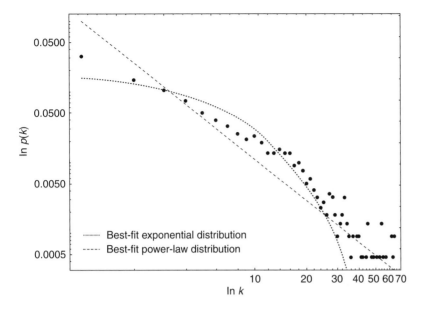

Fig. 2.5
Fitting degree distributions.
Comparison of the fit obtained for the degree distribution of a yeast PPI, using exponential and power-law functions. The best fit found for thes data is the stretched exponentially followed by a lognormal distribution.

scale-free properties. However, Stumpf and Ingram have analysed several PPI networks in order to determine the best statistical fits for their degree distributions. Accordingly, they have shown that the distribution that best fits this empirical data is the stretched exponential followed by a lognormal distribution (Stumpf and Ingram, 2005). The same distribution is observed for *H. pylori* and *E. coli*. The stretched exponential distribution is also known as the Weibull distribution, and has the following forms for the probability distribution and CDF, respectively:

$$p(k) = \frac{\alpha}{\bar{k}} \left(\frac{k}{\bar{k}} \right)^{(\alpha-1)} e^{-(k/\bar{k})^\alpha} \tag{2.11}$$

$$P(k) = 1 - e^{-(k/\bar{k})^\alpha} \tag{2.12}$$

On the other hand, the PPI network of *C. elegans* displays a power-law degree distribution, while that for *D. melanogaster* has a gamma distribution.

It has to be said that these authors (Stumpf and Ingram, 2005) have studied only six statistical distributions among the whole zoo of distributions that exist. There is therefore no guarantee that by extending the number of distributions to be analysed and the statistical criteria used to rank the fits the results will not change, and other distributions with better fits are found for these networks.

2.3 Degree–degree correlations

Another interesting characteristic of complex networks is the existence of different types of degree–degree correlation. The degree–degree correlation accounts for the way in which nodes with given degree are connected in a network. For instance, a network in which high-degree nodes tend to connect to each other displays positive degree–degree correlation and they are called assortative networks. On the other hand, networks in which high-degree nodes tend to be linked to low-degree nodes display negative degree–degree correlations and are termed 'disassortative'. An easy way of quantifying the assortativity of a network is by measuring the correlation coefficient of its degree–degree correlation; that is, by simply calculating the correlation coefficient for the degrees of the nodes existing at both sides of all links in a network. This kind of assortativity coefficient was first introduced by Newman (2002a), and is defined as follows.

Let $e(k_i k_j)$ be the fraction of links that connect a node of degree k_i to a node of degree k_j. For mathematical convenience Newman considered the degree minus one instead of the degree of the corresponding nodes, and named them 'excess degrees'. Let $p(k_j)$ be the probability that a node selected at random in the network has degree k_j. Then, the distribution of the excess degree of a node at the end of a randomly chosen link is

$$q(k_j) = \frac{(k_j + 1)p(k_j + 1)}{\sum_i k_i p(k_i)} \tag{2.13}$$

The assortativity coefficient is defined as

$$r = \frac{\sum_{k_i k_j} k_i k_j [e(k_i k_j) - q(k_i)q(k_j)]}{\sigma_q^2} \tag{2.14}$$

where σ_q^2 is the standard deviation of the distribution $q(k_j)$. Let us designate with $\mathbf{k}_i$ and $\mathbf{k}_j$ the degree vectors for nodes at both ends of the links in the network. Then, if $\mathbf{k}_i = k\mathbf{1}$ or $\mathbf{k}_j = k\mathbf{1}$, the assortativity coefficient is undefined due to the fact that there is a division by zero in (2.14). Obviously, the assortativity coefficient cannot be defined for regular networks, where (2.14) is undefined.

In the case of directed networks the assortativity coefficient is calculated by considering the distributions for the two types of ends and their respective standard deviations (Newman and Girvan, 2003). This coefficient, which is the Pearson correlation coefficient of the degrees of nodes at both ends of links, is written in the following way:

$$r = \frac{m^{-1} \sum_e k_i(e)k_j(e) - \left[m^{-1} \sum_e \frac{1}{2}[k_i(e) + k_j(e)]\right]^2}{m^{-1} \sum_e \frac{1}{2}[k_i^2(e) + k_j^2(e)] - \left[m^{-1} \sum_e \frac{1}{2}[k_i(e) + k_j(e)]\right]^2} \tag{2.15}$$

where $k_i(e)$ and $k_j(e)$ are the degrees at both ends of the link e, and $m = |E|$. We can rewrite this expression in matrix form as follows:

$$r = \frac{\langle v|\mathbf{A}|v\rangle - \frac{1}{2m}((\langle \mathbf{1}|\mathbf{E}'|\mathbf{1}\rangle)^2}{\langle \mathbf{1}|\mathbf{E}'^2|\mathbf{1}\rangle - \langle v|\mathbf{A}|v\rangle - \frac{1}{2m}((\langle \mathbf{1}|\mathbf{E}'|\mathbf{1}\rangle)^2} \tag{2.16}$$

where $\mathbf{E}'$ denotes the modified link adjacency matrix; that is, $\mathbf{E}' = \mathbf{B}^T\mathbf{B} = \mathbf{E} + 2\mathbf{I}$.

Let us explain the information provided by the assortative index by using an example. We select two food webs: St Marks and St Martin. The first represents mostly macroinvertebrates, fish, and birds associated with an estuarine sea-grass community, *Halodule wrightii*, at St Marks Refuge in Florida.

The second represents trophic interactions between birds and predators and arthropod prey of *Anolis* lizards on the island of St Martin, located in the northern Lesser Antilles. The first network has 48 nodes and 218 links, and the second has 44 nodes and 218 links. The assortativity coefficients for these two networks are: St Marks, $r = 0.118$; and St Martin, $r = -0.153$. We then use these to say that the first network displays assortative mixing and the second displays disassortative mixing. The assortative mixing of degrees means that low-degree nodes prefer to join to other low-degree nodes, while high-degree nodes are preferentially bonded to other high-degree nodes. On the other hand, disassortative mixing of degrees implies that high-degree nodes are preferentially attached to low-degree nodes. The term used for those networks having $r = 0$ is 'neutral mixing'.

In Table 2.1 we reproduce some of the results reported in the literature (Newman, 2003b) for the assortativity coefficient of several social networks. They correspond, respectively, to collaborations among scientists in physics, film actors, and business people, and a network of e-mail book addresses. As can be seen, all of them display assortative mixing of node degrees.

These results contrast with most of the values obtained for the assortativity coefficient for other classes of network. For instance, most biological, informational, and technological networks reported in the literature display disassortative mixing of node degrees. These findings have led to the conclusion that almost all networks seem to be disassortative, 'except for social networks, which are normally assortative' (Newman, 2002a). A further analysis has concluded that the presence of high clustering and community structures is among the possible causes for this characteristic mixing pattern of social networks. However, these results do not provide a clear structural interpretation about why some networks display assortative mixing of node degrees while others shown a disassortative pattern. In fact, neither all social networks are assortative, nor are all the other networks disassortative. In Table 2.2 we illustrate three social networks having disassortative mixing, and three others—one informational and two biological networks—displaying assortative mixing of node degrees.

The study of mixing patterns in networks is important beyond this classification and grouping of social networks apart from the others. For instance, it has been shown that the processes by which networks grow with assortative and disassortative mixing of node degrees display very different characteristics. The formation of a giant component, for example, appears first in a network

Table 2.1   Assortativity of social networks.
Values of the assortativity coefficient for some social networks described in the literature (Newman, 2003b).

Network	n	r
Physics coauthorship	52,909	0.363
Film actor collaborations	449,913	0.208
Company directors	7,673	0.276
E-mail address books	16,881	0.092

Table 2.2 Degree–degree correlation in networks.
The diversity of assortativity coefficient in complex networks. Some social networks are strongly disassortative,
and some non-social networks are strongly assortative.

Network	r	Network	r
Drug users	−0.118	Roget's Thesaurus	0.174
Karate club	−0.476	Protein structure	0.412
Students dating	−0.119	St Marks food web	0.118

growing with assortative mixing of node degrees, then in networks with neutral
mixing, and finally in those having disassortative mixing. This means that
the presence of assortativity in a network allows it to percolate more easily
(Newman, 2002a; Newman and Girvan, 2003). However, the giant component grows more in networks with disassortative mixing than in those with
assortative mixing (Newman, 2002a; Newman and Girvan, 2003). In general,
the degree mixing pattern has implications for network resilience to intentional attacks, epidemic spreading, synchronisation, cooperation, diffusion,
and behavioural diversity processes taking place in he network (Xulvi-Brunet
and Sokolov, 2004; Noh, 2007; Boguñá et al., 2003; Vázquez et al., 2006;
Chavez et al., 2006; Rong et al., 2007; Gallos et al., 2008; Yang et al., 2010).
In the next section we find a structural interpretation of the difference in mixing
patterns displayed by real-world networks.

2.3.1 Structural interpretation of the assortativity coefficient

The different behaviour of assortative and disassortative networks has been
explained by the intuitive reasoning that in assortatively mixed networks '*the
high-degree vertices will tend to stick together in a subnetwork or core group of
higher mean degree than the network as a whole*' (Newman, 2002a). However,
an exploration of the universe of real-world networks reveals that this is not
always the case. That is, there are assortative networks in which high-degree
nodes are not clustered together, but spread through the entire network. This
situation is analysed in detail in Chapter 8, where we study the 'clumpiness'
of central nodes in complex networks. Meanwhile, we here concentrate on
the following important questions concerning the assortativity coefficient in
complex networks: What can we learn about the structure of complex networks by analysing the assortative coefficient? That is, what kind of structural
characteristics make some networks assortative or disassortative?

We begin our analysis by rewriting the assortative coefficient in terms of
structural invariants, as well as the number of 2-paths and links, $|P_2|$ and m,
respectively:

$$r = \frac{2M_2 - \dfrac{|P_2|^2}{m} - 2m - 4|P_2|}{M_1^e + 4|P_2| + 2m - 2M_2 - \dfrac{|P_2|^2}{m}} \tag{2.17}$$

where the invariants M_2 and M_1^e—known as the second Zagreb index of the graph and the first Zagreb index of the line graph (see Chapter 8), respectively—can be expressed in terms of subgraphs, as follows:

$$M_1^e = 2|P_2| + 2|P_3| + 6|S_{1,3}| + 6|C_3| \tag{2.18}$$

$$M_2 = m + 2|P_2| + |P_3| + 3|C_3| \tag{2.19}$$

Then, the assortativity coefficient can be written in structural terms as follows:

$$r = \frac{|P_2|(|P_{3/2}| + C - |P_{2/1}|)}{3|S_{1,3}| + |P_2|(1 - |P_{2/1}|)} \tag{2.20}$$

where we have used $|P_1|$ for the number of links, and $|P_{r/s}| = |P_r|/|P_s|$. We have also designated by C the ratio of three times the number of triangles to the number of 2-paths. The reason for this will be obvious in the next chapter when we study the clustering coefficients of networks. For now, however, we will call this index the 'clustering coefficient' of the network. The denominator of (2.20) can be written as

$$\left[\frac{1}{4}\sum_{i<j} k_i k_j \left(k_i^2 + k_j^2\right) - \frac{1}{2}\sum_{i<j}(k_i k_j)^2\right] / |P_1| \tag{2.21}$$

and it is easy to show that $\frac{1}{4}\sum_{i<j} k_i k_j(k_i^2 + k_j^2) \geq \frac{1}{2}\sum_{i<j}(k_i k_j)^2$, which means that the denominator of (2.20) is always larger, or equal to, zero. The equality to zero is obtained when $k_i = k_j$, which means that for regular networks the assortativity coefficient is undefined (as explained in the previous section).

Let us now consider only the sign of the numerator of (2.20). Then, we can write the conditions for assortative and disassortative mixing as follows. A network is:

- assortative ($r > 0$) if, and only if, $|P_{2/1}| < |P_{3/2}| + C$,
- disassortative ($r < 0$) if, and only if, $|P_{2/1}| > |P_{3/2}| + C$.

Table 2.3 Structural interpretation of assortativity coefficient. Expression of assortativity coefficient in terms of path ratios and clustering coefficient in complex networks.

Network	$|P_{2/1}|$	$|P_{3/2}|$	C	r
Prison	4.253	4.089	0.288	0.103
Geom	17.416	22.089	0.224	0.168
Corporate	19.42	20.60	0.498	0.268
Roget	9.551	10.081	0.134	0.174
St. Marks	10.537	10.464	0.291	0.118
Protein3	4.406	4.45	0.417	0.412
Zackary	6.769	4.49	0.256	−0.476
Drugs	14.576	12.843	0.368	−0.118
Transc Yeast	12.509	3.007	0.016	−0.410
Bridge Brook	22.419	17.31	0.191	−0.664
PIN Yeast	15.66	14.08	0.102	−0.082
Internet	91.00	11.53	0.015	−0.229

We recall here that the clustering coefficient is bounded as $0 \leq C \leq 1$. These results explain the empirical findings about the influence of certain structural parameters on the assortative mixing of networks (Holme and Zhao, 2007; Xu et al., 2009). In Table 2.3 we illustrate the path ratios and clustering for several networks with assortative and disassortative mixing of node degrees.

As can be seen in the Table, there are some networks for which $|P_{3/2}| > |P_{2/1}|$, and they are assortative independently of the value of their clustering coefficient. In other cases, assortativity arises because $|P_{3/2}|$ is slightly smaller than $|P_{2/1}|$, and the clustering coefficient produces $|P_{3/2}| + C > |P_{2/1}|$. It is clear, then, that the role played by the clustering coefficient could be secondary in some cases and determinant in others for the assortativity of a network.

2.4 Link adjacency and beyond

The most studied adjacency relationship in networks is the node adjacency, but it is not the only one existing in networks. It is easy to extend the concept of node adjacency to that of links. Two links are adjacent if, and only if, they are incident to the same node. The link adjacency matrix $\mathbf{E}$ of undirected networks has been studied in different contexts. It is a square symmetric matrix whose entries are zeroes or ones whether or not the corresponding links are adjacent, respectively. This matrix can be obtained from the incidence matrix of the network as

$$\mathbf{E} = \mathbf{B}^T \mathbf{B} - 2\mathbf{I} \tag{2.22}$$

where $\mathbf{I}$ is the identity matrix.

In a similar way as for the node degree, we can define the degree of a link κ_i as the number of links which are adjacent to it. The link degree vector is then obtained as follows:

$$\kappa = (\mathbf{1}^T \mathbf{E})^T = \mathbf{E}\mathbf{1} \tag{2.23}$$

If the nodes u and v are the end points of the link i, $i = \{u, v\}$, then it is easy to verify that

$$\kappa_i = k_u + k_v - 2 \tag{2.24}$$

The relation between node and link adjacency relationships is evident through the use of a network transformation known as the 'line graph'. Let $G = (V, E)$ be a network. Then, its line graph $L(G) = (V', E')$ is the transformation which results from considering every link of G as a node of $L(G)$. Then, two nodes of $L(G)$ are connected if the corresponding links are adjacent in G. In Figure 2.6 we illustrate this transformation for a simple network.

By using the line graph transformation it is easy to obtain relations similar to the handshaking lemma for link degrees

$$\langle \mathbf{1}|\mathbf{E}|\mathbf{1}\rangle = \langle \kappa|\mathbf{1}\rangle = \sum_{i=1}^{n} \kappa_i = 2|P_2| \tag{2.25}$$

where $|P_2|$ denotes the number of paths of length 2 in the network.

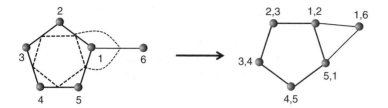

Fig. 2.6
Building a line graph. Transformation of a graph G (left) into its line graph $L(G)$(right). Nodes in $L(G)$ are the links of G, and two nodes in $L(G)$ are connected if the corresponding links are incident to the same node in G.

The idea of adjacency between other subgraphs in a network is also appealing. We can consider node adjacency as the zero-order relation, and link adjacency as the first-order relation. As an example, we will generate adjacency matrices corresponding to the participation of nodes in the cliques of a network. Let $|C_h|$ be the number of cliques of size h in a network (Bron and Kerbosch, 1973; Hubbell, 1965). Then, we can create incidence matrices $\mathbf{X}_h$ of dimensions $|V| \times |C_h|$. The adjacency between cliques of size h is then provided by the following h-clique-adjacency matrix:

$$\mathbf{C}_h = \mathbf{X}_h^T \mathbf{X}_h - h\mathbf{I} \qquad (2.26)$$

It is, however, more interesting to study the adjacency relationship between nodes in those cliques. In this case we consider that two nodes are clique-adjacent if, and only if, they are in the same clique of size h (Barber, 2008):

$$\mathbf{S}_h = \mathbf{T}_h \mathbf{T}_h^T - \mathbf{K} \qquad (2.27)$$

where $\mathbf{K} = diag(\mathbf{T}_h \mathbf{T}_h)$ is a diagonal matrix whose diagonal entries account for the number of times the corresponding node is present in a clique of size h. These diagonal entries are termed the 'clique-degree' of the corresponding node. A similar measure has been studied by Xiao et al. (2007) in real-world networks, but considering complete subgraphs which are not necessarily maximal in the graph.

2.5 Discrete Laplacian operator

Another important operator defined in network theory is the Laplacian operator (Carlson, 1998; Friedman and Tillich, 2008; Biyikoglu et al., 2007), which is defined and studied in this section. Let us begin by considering an arbitrary orientation of every link in a network. For instance, for the link $\{u, v\}$ we can consider that u is the positive and v the negative end of the oriented link. The results obtained below are independent on the orientation of the links, but we assume that once the links are oriented this orientation is not changed. Let us represent the network through an *oriented incidence matrix* $\nabla(G)$:

$$\nabla_{ij} = \begin{cases} +1 & \text{node } v_i \text{ is the positive end of link } e_j \\ -1 & \text{node } v_i \text{ is the negative end of link } e_j \\ 0 & \text{otherwise.} \end{cases}$$

This matrix represents the *gradient operator* for the network. In fact, let $f : V \to \Re$ be an arbitrary function. Then, $\nabla f : E \to \Re$ is given by

$$(\nabla f)(e) = f(u) - f(v) \qquad (2.28)$$

where u is the starting point and v the ending point of the oriented link e. Let us recall that a vector field is a function of the interval with an orientation of the interval. In this case, the interval corresponds to the link in the network, which together with its orientation forms a vector field. Then, the continuous analogous of $\nabla(G)$ is the gradient $(\nabla f) = \partial f/\partial x_1, \partial f/\partial x_2, \ldots \partial f/\partial x_n$, which produces the maximum rate of change of the function with direction.

Let us now consider the following operator:

$$\mathbf{L} f = -\nabla \cdot (\nabla f) \tag{2.29}$$

which is the network version of the Laplacian operator:

$$\nabla^2 f = \frac{\partial^2 f}{\partial x_1^2} + \frac{\partial^2 f}{\partial x_2^2} + \cdots + \frac{\partial^2 f}{\partial x_n^2} \tag{2.30}$$

Then, the Laplacian operator acting on $\mathbf{L} : \Re^{|V|} \to \Re^{|V|}$ is defined as

$$(\mathbf{L} f)(u) = \sum_{u \sim v} [f(u) - f(v)] \tag{2.31}$$

which in matrix form is represented by

$$L_{uv} = \sum_{e \in E} \nabla_{eu} \nabla_{ev} = \begin{cases} -1 & \text{if } uv \in E, \\ k_u & \text{if } u = v, \\ 0 & \text{otherwise.} \end{cases}$$

Using the degree matrix defined in Section 2.1, the Laplacian and adjacency matrices of a network are related as follows:

$$\mathbf{L} = \mathbf{K} - \mathbf{A} \tag{2.32}$$

The relation of the Laplacian operator for networks and the differential operator $\nabla^2 f$ can be explained as follows. Let us consider a node, O, which is connected to other two nodes—here designated as x_1 and x_2—which define the following two links: $e_1 = \{O, x_1\}$ and $e_2 = \{O, x_2\}$. Now, let us consider that there is particle that departs from node O, visits any of its nearest neighbours, and then returns to O. The trajectory described by this particle is known as a 'closed walk'—in this case, of length 2. (Formal definition of closed walks and their relationships, with subgraphs, is studied in Chapter 4; see also Section 2.6.) Let us designate the previously described closed walk as $O \to x_1 \to O$. Then, the first derivative evaluated in the corresponding link is given by the function evaluated at the intermediate point minus the function evaluated at the final point:

$$\frac{\partial f}{\partial x}(e_1) = f(x_1) - f(O) \tag{2.33}$$

In a similar way, we obtain

$$\frac{\partial f}{\partial x}(e_2) = f(x_2) - f(O) \tag{2.34}$$

Now we will consider the second derivatives using a similar approximation. In this case we need to consider a closed walk of the form $O \to x_1 \to O \to x_2 \to O$, which can be considered as two closed walks of length two

$(O \to x_1 \to O) + (O \to x_2 \to O)$. Applying the procedure used for the first derivatives, we obtain $f(x_1) - f(O) + f(x_2) - f(O)$. However, a similar closed walk exists in the other 'direction', $O \to x_2 \to O \to x_1 \to O$, which can be considered as two closed walks of length two $(O \to x_2 \to O) + (O \to x_1 \to O)$, producing $f(x_2) - f(O) + f(x_1) - f(O)$. Then, the second derivative evaluated in node O is given by the average of the two 'directions':

$$\frac{\partial^2 f}{\partial x^2}(O) = \frac{1}{2}[f(x_1) - f(O) + f(x_2) - f(O) + f(x_2) - f(O)$$

$$+ f(x_1) - f(O)] \tag{2.35}$$

which, after rearrangement, gives:

$$\frac{\partial^2 f}{\partial x^2}(O) = \frac{1}{2}\{[f(x_1) - f(O)] - [f(O) - f(x_1)] + [f(x_2) - f(O)]$$

$$- [f(O) - f(x_2)]\} \tag{2.36}$$

It is now easy to realise that the second derivative can be written in the usual way as

$$= \frac{1}{2}\left[\frac{\partial f}{\partial x}(e_1) + \frac{\partial f}{\partial x}(e_2)\right]$$

$$= f(x_1) + f(x_2) - 2f(O). \tag{2.37}$$

We can then write the Laplacian operator as follows:

$$(\mathbf{L}f)(O) = \sum_{i \sim O} f(x_i) - k_O f(O) \tag{2.38}$$

where k_O is the degree of node O.

In the case of connected networks it is useful to define a normalised Laplacian as

$$\mathbf{L} = \mathbf{K}^{-1/2}\mathbf{L}\mathbf{K}^{-1/2} = \mathbf{I} - \mathbf{K}^{-1/2}\mathbf{A}\mathbf{K}^{-1/2} \tag{2.39}$$

where $\mathbf{K}^{-1/2}$ is the diagonal matrix determined by taking the inverse square root of each diagonal entry of the degree matrix. Then, the entries of $\mathbf{L}$ are given by

$$\mathbf{L}_{ij} := \begin{cases} 1, & \text{if } i = j \text{ and } i \neq 0; \\ -(k_i k_j)^{-1/2}, & \text{if } i \sim j; \\ 0, & \text{otherwise.} \end{cases}$$

It has to be said that some authors (Chung, 1996) call this matrix the Laplacian of the graph; but we prefer to reserve this term for the matrix $\mathbf{L}$, and simply call $\mathbf{L}$ the normalised Laplacian.

2.6 Spectral properties of networks

The study of spectral properties of adjacency matrices (operators) represents an entire area of research in algebraic graph theory. There is a huge amount of information relating the spectra of adjacency matrices and the structure of graphs, and the reader is referred to the excellent monographs existing for a deep study of this topic (Biggs, 1993; Chung, 1996; Cvetković et al., 1995; 1997; 2010). An introduction to the literature on this topic, with references to the applications in chemistry, physics, computer science, engineering, biology, and economics, can be found in Cvetković (2009).

The spectrum of the node adjacency matrix of a network is the set of eigenvalues of $\mathbf{A}$ together with their multiplicities. Let $\lambda_1(\mathbf{A}) > \lambda_2(\mathbf{A}) > \cdots > \lambda_5(\mathbf{A})$ be the distinct eigenvalues of $\mathbf{A}$ and let $m(\lambda_1(\mathbf{A})), m(\lambda_2(\mathbf{A})), \cdots, m(\lambda_5(\mathbf{A}))$ be their multiplicities; that is, the number of times each of them appears as an eigenvalue of $\mathbf{A}$. Then, the spectrum of $\mathbf{A}$ is written as

$$Sp\mathbf{A} = \begin{pmatrix} \lambda_1(\mathbf{A}) & \lambda_2(\mathbf{A}) & \dots & \lambda_5(\mathbf{A}) \\ m(\lambda_1(\mathbf{A})) & m(\lambda_2(\mathbf{A})) & \dots & m(\lambda_5(\mathbf{A})) \end{pmatrix} \tag{2.40}$$

The eigenvalues of the adjacency matrix $\mathbf{A}$ are the zeros of the so-called characteristics polynomial of the network, $\det(\lambda\mathbf{I} - \mathbf{A})$ (Rowlinson, 2007), and the numbers λ clearly satisfy the equation $\mathbf{A}|\varphi\rangle = \lambda(\mathbf{A})|\varphi\rangle$, where each of the non-zero vectors $|\varphi\rangle$ are referred to as an eigenvector of $\mathbf{A}$.

Let us first consider some results relating the structure of networks and the spectra of their $\mathbf{A}$ matrices. The spectra of some simple networks are as follows:

$$\text{Path, } P_n : \lambda_j(\mathbf{A}) = 2\cos\left(\frac{\pi j}{n+1}\right), j = 1, \dots, n \tag{2.41}$$

$$\text{Cycle, } C_n : \lambda_j(\mathbf{A}) = 2\cos\left(\frac{2\pi j}{n}\right), j = 1, \dots, n \tag{2.42}$$

$$\text{Star, } S_n : Sp(\mathbf{A}) = \left\{ \sqrt{n}\ 0^{n-2}\ -\sqrt{n} \right\} \tag{2.43}$$

$$\text{Complete, } K_n : Sp(\mathbf{A}) = \left\{ 1\ -1^{n-1} \right\} \tag{2.44}$$

$$\text{Complete bipartite, } K_{n_1,n_2} : Sp(\mathbf{A}) = \left\{ \sqrt{n_1 n_2}\ 0^{n-2}\ -\sqrt{n_1 n_2} \right\} \tag{2.45}$$

where we have used the notation expressed as $\{\lambda_1^{m_1} \cdots \lambda_5^{m_5}\}$

The largest eigenvalue of the adjacency matrix $\mathbf{A}$ is termed the index of the network or spectral radius of $\mathbf{A}$. Then, if $\lambda_1(\mathbf{A})$ is the index of a connected undirected network it has multiplicity equal to one and has a positive eigenvector associated with it, which we call the principal eigenvector of the network (Cvetković and Rowlinson, 1988; 1990). In the case of directed networks, the previous statements are true if the network is strongly connected. (In the following part of this section we restrict ourselves to undirected networks. The property of being strongly connected will be analysed in the Chapter 3.)

Let $\bar{k}$, k_{min} and k_{max} be the average, minimum, and maximum degree, respectively, in a network. Then, the index of the network is bounded as follows:

$$k_{min} \leq \bar{k} \leq \lambda_1(\mathbf{A}) \leq k_{max} \tag{2.46}$$

where equality holds if and only if the network is regular. The index for any network satisfies the following inequality:

$$2 \cos \frac{\pi}{n+1} \leq \lambda_1(\mathbf{A}) \leq n - 1 \qquad (2.47)$$

where the lower bound is obtained for the path P_n and the upper one is obtained for the complete network K_n. In the case of bipartite networks, the spectra of $\mathbf{A}$ is symmetric with respect to the zero point, which means that the eigenvalues occur in pairs of the form $(\lambda(\mathbf{A}), -\lambda(\mathbf{A}))$. Then, by ordering the eigenvalues as, $\lambda_1 \geq \lambda_2 \geq \cdots \geq \lambda_n$, we have:

$$\lambda_i(\mathbf{A}) = -\lambda_{n-i+1}(\mathbf{A}) \qquad (2.48)$$

On the other hand, the index of a network is useful in bounding other structural properties of the network (which is discussed in several parts of this book). For instance, the size of the maximum clique in the network is at least $n/(n - \lambda_1(\mathbf{A}))$ and a network for which $\lambda_1(\mathbf{A}) \geq \sqrt{m}$ necessarily contains a triangle.

The spectrum of the link adjacency matrix $\mathbf{E}$ is defined in a similar way as for the $\mathbf{A}$ matrix. That is, the link spectrum of a network is formed by the eigenvalues ε of the matrix $\mathbf{E}$ together with their multiplicities, where the eigenvalues are the solution of $\mathbf{E}|\phi\rangle = \lambda(\mathbf{E})|\phi\rangle$ and of the characteristic equation $\det(\lambda(\mathbf{E})\mathbf{I} - \mathbf{E})$. Obviously, the link spectrum of a network is identical to the spectrum of the adjacency matrix of the line graph of the network. Then, in the mathematical literature it is used to refer to $\lambda(\mathbf{E})$ as the eigenvalues of the line graph of the graph. There is a vast literature about this topic, to which the interested reader is referred for details (Cvetković et al., 2004). The spectrum of other adjacency matrices can be defined in similar ways as for the previous ones. The results about such spectra—for example, spectra of cycle-adjacency or clique-adjacency matrices—are, however, very scarce.

The spectrum of the Laplacian matrix of a network is defined in a similar way as for the adjacency matrix. That is, the Laplacian spectrum is the set of eigenvalues of $\mathbf{L}$ together with their multiplicities (Biyikoglu et al., 2007; Jamakovic and van Mieghem, 2006). Let $\lambda_1(\mathbf{L}) < \lambda_2(\mathbf{L}) < \cdots < \lambda_5(\mathbf{L})$ be the distinct eigenvalues of $\mathbf{L}$ and let $m(\lambda_1(\mathbf{L})), m(\lambda_2(\mathbf{L})), \cdots, m(\lambda_5(\mathbf{L}))$ be their multiplicities. Then, the spectrum of $\mathbf{L}$ is written as

$$Sp\mathbf{L} = \begin{pmatrix} \lambda_1(\mathbf{L}) & \lambda_2(\mathbf{L}) & \dots & \lambda_5(\mathbf{L}) \\ m(\lambda_1(\mathbf{L})) & m(\lambda_2(\mathbf{L})) & \dots & m(\lambda_5(\mathbf{L})) \end{pmatrix} \qquad (2.49)$$

The Laplacian matrix is positive semidefinite, having eigenvalues bounded as

$$0 \leq \lambda_j(\mathbf{L}) \leq 2k_{max} \qquad (2.50)$$

and

$$\lambda_n(\mathbf{L}) \geq k_{max} \qquad (2.51)$$

The multiplicity of 0 as an eigenvalue of $\mathbf{L}$ is equal to the number of connected components in the network. Then, the second smallest eigenvalue of $\mathbf{L}$, $\lambda_2(\mathbf{L})$, is usually termed the algebraic connectivity of a network (Fiedler, 1973). (For an excellent review of the literature and some applications in complex networks, the reader is referred to the works of Abreu, (2007), Jamakovic and Mieghem (2008), and Jamakovic and Uhlig (2007)).

On the other hand, the normalised Laplacian matrix L is also positive semidefinite, having eigenvalues $0 = \lambda_1(L) \leq \lambda_2(L) \leq \cdots \leq \lambda_n(L)$, which are bounded as (Chung, 1996)

$$0 \leq \lambda_j(L) \leq 2 \tag{2.52}$$

and

$$\lambda_n(L) \geq \frac{n}{n-1} \tag{2.53}$$

The spectra of the three matrices are related by the following inequalities:

$$k_{\max} - \lambda_n(A) \leq \lambda_n(L) \leq k_{\max} - \lambda_1(A) \tag{2.54}$$

$$\lambda_j(L)k_{\min} \leq \lambda_j(L) \leq \lambda_j(L)k_{\max} \tag{2.55}$$

The following relationships exist between the spectra of adjacency, Laplacian, and normalised Laplacian of k-regular graphs:

$$\lambda_j(L) = k - \lambda_{n-j+1}(A) \tag{2.56}$$

$$\lambda_j(L) = \lambda_j(L)/k \tag{2.57}$$

$$\lambda_j(I-L) = \lambda_j(A)/k \tag{2.58}$$

Finally, some analytic expressions for the spectra of different kinds of simple network are:

Path, $P_n : \lambda_j(L) = 2 - 2\cos\left(\frac{\pi(j-1)}{n}\right),$

$$\lambda_j(L) = 1 - \cos\left(\frac{\pi(j-1)}{n-1}\right), i = 1, \ldots, n \tag{2.59}$$

Cycle, $C_n : \lambda_j(L) = 2 - 2\cos\left(\frac{2\pi j}{n}\right),$

$$\lambda_j(L) = 1 - \cos\left(\frac{2\pi j}{n}\right) i = 1, \ldots, n \tag{2.60}$$

Star, $S_n : Sp(L) = \left\{0 \ 1^{n-2} \ n\right\}, \ Sp(L) = \left\{0 \ 1^{n-2} \ 2\right\} \tag{2.61}$

Complete, $K_n : Sp(L) = \left\{0 \ n^{n-1}\right\}, \ Sp(L) = \left\{0 \ \left(\frac{n}{n-1}\right)^{n-1}\right\} \tag{2.62}$

Complete bipartite, $K_{n_1,n_2} : Sp(L) = \left\{0 \ n_1^{n_2-1} \ n_2^{n_1-1}\right\},$

$$Sp(L) = \left\{0 \ 1^{n-2} \ 2\right\} \tag{2.63}$$

Note that the normalised Laplacian does not distinguish between stars and complete bipartite networks of the same size. Two non-isomorphic networks having exactly the same spectra are termed *isospectral* or *cospectral*. There are isospectral networks for the adjacency, Laplacian, and normalised Laplacian, as well as their combinations. For instance, in Figure 2.7 a pair of

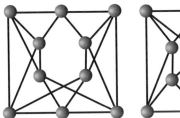

Fig. 2.7
Isospectral regular graphs. A pair of isospectral regular graphs (van Dam and Haemers, 2003). Due to regularity they are isospectral for **A**, **L**, and *L*.

non-isomorphic regular networks is illustrated (van Dam and Haemers, 2003). These two networks are isospectral for the adjacency matrix, and due to relationships (2.56)–(2.58) they are also isospectral for the Laplacian and normalised Laplacian matrices.

Another interesting area of the spectral study of networks is the analysis of network eigenvectors (Goh et al., 2001a), the study of which arises in many different situations throughout this book. Here we introduce some of the fundamental principles that guide their study in complex networks. We begin by considering the eigenvectors of the adjacency matrix. It is known that a network is completely determined by eigenvalues and a basis of corresponding eigenvectors of its adjacency matrix (Cvetković et al. 1997). Let Λ be a diagonal matrix of eigenvalues of the adjacency matrix ordered as in the previous section, $\Lambda = diag(\lambda_1(\mathbf{A}), \lambda_2(\mathbf{A}), \ldots, \lambda_n(\mathbf{A}))$, and let $\mathbf{\Phi}$ be a matrix whose columns are orthonormal eigenvectors $\vec{\varphi}_1, \vec{\varphi}_2, \ldots \vec{\varphi}_n$. Then, the spectral decomposition of the adjacency matrix is given by:

$$\mathbf{A} = \mathbf{\Phi}\Lambda\mathbf{\Phi}^T \qquad (2.64)$$

The Laplacian and normalized Laplacian matrices can also be represented by spectral formulae such as (2.64) by using diagonal matrices of eigenvalues and orthonormal matrices of associated eigenvectors.

In connected networks the eigenvector corresponding to the largest eigenvalue of the adjacency matrix, $\vec{\varphi}_1(\mathbf{A})$, is positive. We will call this eigenvector the principal or Perron–Frobenius eigenvector of the adjacency matrix (Papendieck and Recht, 2000). In the case of k-regular graphs, $\vec{\varphi}_1(\mathbf{A}) = (1/\sqrt{k} \cdots 1/\sqrt{k})$.

By using the spectral decomposition of the adjacency matrix it is easy to see that the powers of the adjacency matrix can be expressed in terms of the eigenvalues and eigenvectors of the adjacency matrix:

$$(\mathbf{A}^k)_{uv} = \sum_{j=1}^{n} \varphi_j^u(\mathbf{A})\varphi_j^v(\mathbf{A})[\lambda_j(\mathbf{A})]^k \qquad (2.65)$$

where $\varphi_j^u(\mathbf{A})$ is the uth entry of the eigenvector associated with the jth eigenvalue. This is important, because the term $(\mathbf{A}^k)_{uv}$ counts the number of walks of length k between the nodes u and v. We can then relate (as in Chapter 4) the counting of subgraphs and the spectra of the network (see also Section 2.6.1).

On the other hand, the eigenvector of the Laplacian and normalized Laplacian matrices associated with the 0 eigenvalue in a connected network are, respectively (Merris, 1998):

$$\varphi_1(\mathbf{L}) = \frac{1}{\sqrt{n}}\mathbf{1} \tag{2.66}$$

$$\varphi_1(\mathbf{L}) = \frac{1}{\sqrt{2m}}(k_1^{0.5}\cdots k_n^{0.5}) \tag{2.67}$$

In addition, if $0 \neq \lambda < n$ is an eigenvalue of the Laplacian, then the eigenvector associated with λ takes the value 0 on every node of degree $n-1$.

The eigenvectors of the adjacency, Laplacian, and normalized Laplacian matrices have been extensively used for analyzing and visualizing networks (Seary and Richards, 2003). (The use of eigenvectors in network clustering is discussed below, in Chapter 10.) All these matrices previously analysed are, in general, singular. That is, they are not invertible. For non-singular matrices $\mathbf{M}$, the entries of the inverse can be obtained as follows:

$$(\mathbf{M}^{-1})_{ij} = \sum_{k=1}^{n} \frac{1}{\tau_k} c_k(i)c_k(j) \tag{2.68}$$

where τ_k are the eigenvalue of $\mathbf{M}$, and $c_k(i)$ is the ith entry of the eigenvector associated with τ_k. In the case of a matrix being singular—that is, some of its eigenvalues are equal to zero—then a *generalised or Moore-Penrose pseudoinverse* $\mathbf{M}^+$ can be defined instead:

$$(\mathbf{M}^+)_{ij} = \sum_{k=1}^{n} \frac{1}{f(\tau_k)} c_k(i)c_k(j) \tag{2.69}$$

where

$$f(\tau_k) = \begin{cases} \tau_k & \text{if } \tau_k \neq 0 \\ 0 & \text{otherwise} \end{cases} \tag{2.70}$$

In particular, the Moore–Penrose pseudoinverse of the Laplacian matrix $\mathbf{L}^+$ has found several applications in network theory—as we will see, for instance, in Chapter 3, when we study the resistance distance between a pair of nodes in a network (Gutman and Xiao, 2004). For a connected network the matrix $\mathbf{L}^+$ is a real and symmetric matrix, which can be written as

$$\mathbf{U}^T\mathbf{L}^+\mathbf{U} = diag(1/\mu_1, 1/\mu_2, \cdots 1/\mu_{n-1}, 0) \tag{2.71}$$

where $\mathbf{U}$ is a matrix of orthonormal eigenvectors of $\mathbf{L}^+$.

2.6.1 Spectral density of the adjacency matrix

If we designate the jth eigenvalue of the adjacency matrix of a network simply as λ_j, the network spectral density can be expressed as a sum of Dirac δ functions:

$$\rho(\lambda) = \frac{1}{n}\sum_{j=1}^{n} \delta(\lambda - \lambda_j) \tag{2.72}$$

which is the density of the eigenvalues of the adjacency matrix.

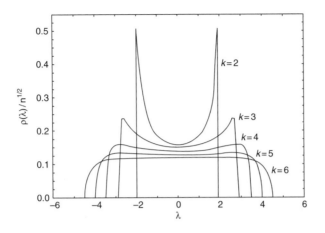

Fig. 2.8
Spectral density of regular networks.
Plot of normalised spectral density for
k-regular networks.

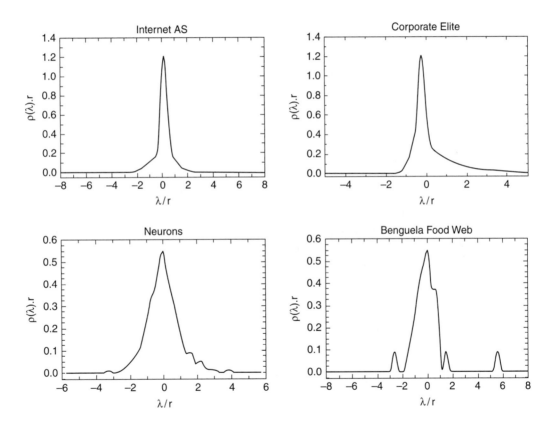

Fig. 2.9
Spectral density of complex networks. Spectral density of several real-world networks.

In the case of k-regular graphs, the spectral density can be written in the following form:

$$\rho(\lambda) = \frac{k}{2\pi} \sum_{j=1}^{n} \frac{\sqrt{4(k-1) - \lambda_j^2}}{k^2 - \lambda_j^2} \tag{2.73}$$

which is plotted in Figure 2.8 for different values of k.

The spectral density analysis of real-world networks reveals the existence of several patterns (de Aguiar and Bar-Yam, 2005; Farkas et al., 2001; 2002; Malarz, 2008; Kim and Kahng, 2007; McGraw and Menzinger, 2008; Zhan et al., 2010). For instance, the Internet as an autonomous system displays a clearly symmetric triangular distribution characteristic of power-law degree distributions, while the social network formed by the US corporate elite displays a non-symmetric triangular distribution of eigenvalues, and the neural network of *C. elegans* displays a less regular triangular-like distribution. Other networks, such as the Benguela food web, display more complex eigenvalue distributions with several sharp peaks, which have been found in certain 'small-world' networks. In Figure 2.9 we illustrate some of these distributions in complex networks from different environments.

Structural information can be recovered from the spectral density of a network via the study of the kth moment of $\rho(\lambda)$. The kth moment of $\rho(\lambda)$ can be written as

$$\mu_k = \frac{1}{n} \sum_{j=1}^{n} \lambda_j^k = \frac{1}{n} Tr(\mathbf{A}^k) \tag{2.74}$$

An important result in this field relates the moments of the spectral density and the number of walks in a network. That is, the number of walks of length k arising at node u and ending at node v is given by the entry $(\mathbf{A}^k)_{uv}$ of the kth power of the matrix $\mathbf{A}$. A closed walk or self-returning walk of length k is the one which starts and ends at the same node $(\mathbf{A}^k)_{uu}$. Then the total number of closed walks of length k is given by $Tr(\mathbf{A}^k)$ (Harary and Schwenk, 1979). The relationship between spectral moments and subgraphs is analysed in Chapter 4.

Metric and topological structure of networks

3

> Our ultimate analysis of space leads us not to a 'here' and 'there', but to an extension such that which relates 'here' and 'there'. To put the conclusion rather crudely—space is not a lot of points close together; it is a lot of distances interlocked.
>
> Arthur Stanley Eddington

The notion of how far apart two objects are in a physical system intuitively motivates the concept of through-space distance. In a discrete object, however, this concept involves the through-links separation of two nodes in the network. Then, a network metric is a function which defines a distance between the nodes of the network, such that $l : V(G) \times V(G) \to \Re$, where $\Re$ is the set of real numbers. This function needs to satisfy the following four properties:

1. $l(u, v) \geq 0$ for all $u \in V(G), v \in V(G)$
2. $l(u, v) = 0$ if and only if $u = v$
3. $l(u, v) = l(v, u)$ for all $u \in V(G), v \in V(G)$
4. $l(u, w) \leq l(u, v) + l(v, w)$ for all $u \in V(G), v \in V(G), w \in V(G)$

The last two properties are known as the symmetric and the triangle inequality properties, respectively. Although there are some alternative ways of defining the distance between two nodes in a network, the notion of shortest path distance is widely recognised as the standard definition of network distance.

3.1 Shortest path distance

In an undirected network, the shortest path distance $d(u, v)$ is the number of links in the shortest path between the nodes u and v in a network (Buckley and Harary, 1990; Chartrand and Zhang, 2003). In the case when u and v are in different connected components of the network, the distance between them is set to infinite, $d(u, v) := \infty$. In graph theory the *directed distance* $\vec{d}(u, v)$ between a pair of nodes u and v in a directed network is considered to be the length of the directed shortest path from u to v. In the case of there not being a directed path connecting two nodes, the corresponding distance is considered to be infinite. It is straightforward to realise that in general,

$\vec{d}(u, v) \neq \vec{d}(v, u)$, which violates the symmetry property of metric functions. Consequently, $\vec{d}(u, v)$ is not a distance but a *pseudo-distance* or *pseudo-metric*. A directed network is referred to as strongly connected if there is a directed path from u to v and a directed path from v to u for every pair of distinct nodes in the network (Chartrand and Zhang, 2003). An analysis of different algorithms used for finding the shortest path distance has been carried out by Zhan and Noon (1998).

In an undirected connected network the distances $d(u, v)$ among all pairs of nodes in a network can be arranged in a matrix. This square symmetric matrix is known as the *distance matrix* of the network **D** (Buckley and Harary, 1990). The distances between a node v and any other node in the network are given at the vth row or column of **D**. In the case of a directed network, the distance matrix is not necessarily symmetric and can contain entries equal to infinite.

The maximum entry for a given row/column of the distance matrix of an undirected (strongly connected directed) network is known as the *eccentricity* $e(v)$ of the node v:

$$e(v) = \max_{x \in V(G)} \{d(v, x)\} \tag{3.1}$$

The maximum eccentricity among the nodes of a network is known as the *diameter* of the network, which is $diam(G) = \max_{x, y \in V(G)} \{d(x, y)\}$. The radius of the network is the minimum eccentricity of the nodes, and a node is called central if its eccentricity is equal to the radius of the network. Then, the centre of the graph is the set of all central nodes (Hage and Harary, 1995).

The sum of all entries of a row/column of the distance matrix is known as the *distance sum* $s(u)$ of the corresponding node:

$$s(u) = \sum_{v \in V(G)} d(u, v) \tag{3.2}$$

This term has also been referred to as the total distance or status of the node. The semi-sum of all entries of the distance matrix was introduced by Wiener as early as 1947, to account for the variations of molecular branching in hydrocarbons (Wiener, 1947).[1] The connection between this index and the distance matrix, as well as the whole of graph theory, was proposed in a seminal paper by Hosoya (1971). Nowadays it is known as the Wiener index $W(G)$, and several of its mathematical properties are known (Mohar and Pisanski, 1988; Cohen et al., 2010):

$$W(G) = \frac{1}{2} \langle \mathbf{1} | \mathbf{D} | \mathbf{1} \rangle = \sum_{u=1}^{n} s(u) = \frac{1}{2} \sum_{u} \sum_{v} d(u, v) \tag{3.3}$$

An important characteristic descriptor of the topology of a network is its *average path length* $\bar{l}$, which can be expressed in terms of the Wiener index as follows:

$$\bar{l} = \frac{2W(G)}{n(n-1)} \tag{3.4}$$

---

[1] Now on we refer to undirected networks if not stated explicity.

where the average is taken by considering only those pairs of nodes for which a path connecting them exists. It has been recommended to use the 'harmonic mean' instead of $\bar{l}$, in such a way that nodes in different components for which the distance is taken to be infinite contribute nothing to the average. When analysing the average shortest path distance in networks, some authors consider the average with the inclusion of the distance between nodes to themselves, which means that they are considering $\bar{l}(n-1)/(n+1)$.

The average path length is bounded as $1 \leq \bar{l}(G) \leq \frac{n+1}{3}$, where the lower bound is obtained for the complete network and the upper one is reached for the path of n nodes. Now, let us consider all undirected trees having the same number of nodes. Then the following bounds for the average path length are obtained:

$$\frac{2(n-1)}{n} \leq \bar{l}(T) \leq \frac{n+1}{3} \tag{3.5}$$

where the lower bound is reached for the star graph with n nodes, and the upper bound is obtained for the path graph of the same size. Let us suppose that we select at random a tree among the 14,830,871,802 existing trees with 30 nodes. The probability that this tree is either the path or the star is very low indeed. Then, the selected tree will look neither as 'elongated' as a path nor as 'compact' as a star. In fact, its average path length will be close to that of a randomly generated tree, which is asymptotic to $Kn^{3/2}/(n-1)$, where $K \approx 0.568279...$ (Dobrynin and Gutman, 1999; Janson, 2003).

A similar phenomenon has been observed in the case of real-world complex networks. That is, the average path length is relatively small in comparison with the size of the network. This phenomenon is known as the *small-world effect* (Milgram, 1967), which has been popularised in many different contexts (Buchanan, 2003; Watts, 1999). In this case, the average path length is significantly smaller than that of a regular lattice, but not so small as in the complete network with the same number of nodes. In general, it is accepted that a network displays small-world effect if $\bar{l}$ scales logarithmically with the number

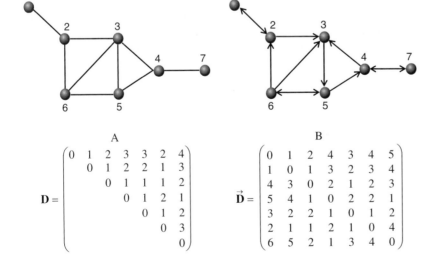

Fig. 3.1
Distances in directed and undirected networks. Undirected and strongly connected directed networks used to illustrate the shortest path distance between nodes.

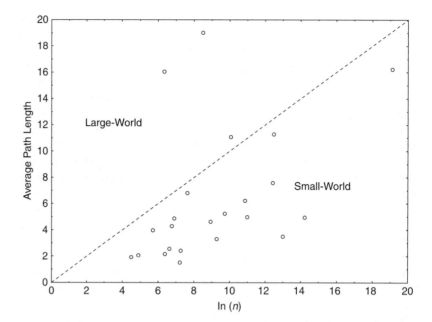

Fig. 3.2

Average distance in small-worlds. Plot of the average shortest path distance versus the logarithm of the number of nodes for several real-world networks. Small-world networks lie below the dotted line.

of nodes (Barrat and Weigt, 2000). Then, if we select a network at random from the real-world, it is expected that its average path length looks similar to that of a random network, which is, in fact, the case as $\bar{l} \sim \ln n$ for an Erdös-Rényi random network, and $\bar{l} \sim \ln n / \ln \ln n$ for a random one with power-law degree distribution (see Chapter 12). Then, the small-world phenomenon 'is not the signature of any special organising principle' in complex networks (Barrat et al., 2008). This, however, has implications for the dynamics of processes taking place on networks, such as processes in which 'something' is spreading through the network (Barrat et al., 2008). Several approaches to generating networks with 'small-world' property have been proposed in the literature (Watts and Strogatz, 1998; Kleinberg, 2000a; 2000b)—some of which will be studied in this book (see, for instance, Chapter 12).

In Figure 3.2 we illustrate a plot of $\bar{l}$ versus $\ln n$ for 23 real networks (Newman 2003b) representing social, informational, technological, and biological systems. We also plot the line $\bar{l} = \ln n$ to indicate those networks for which the average path length is smaller than or equal to $\ln n$—small-world networks. As can be seen, almost all the networks analysed so far display the small-world phenomenon. In several parts of this book—particularly in the second part—a combined definition of 'small-worldness' is used which takes into account not only the average path length of the network but also its clustering coefficient (Watts and Strogatz, 1998). (The clustering coefficient is studied below, in Chapter 4.)

3.2 Resistance distance

Despite most of the current work on metric properties of networks have been carried out by considering the shortest path distance, it is interesting to show here another network metric. This metric is inspired by physical concepts, and has important relationships with other network measures which will be analysed in this book. This network distance was introduced a few years ago in a seminal paper by Klein and Randić (1993).

In order to define this distance between a pair of nodes u and v, let us suppose that we place a fixed electrical resistor on each link of the network (Klein and Randić, 1993; Klein and Ivanciuc, 2001), and connect a battery across the nodes u and v. Then, let us calculate the effective resistance $\Omega(u, v)$ between them by using the Kirchhoff and Ohm laws. For the sake of simplicity, here we always consider resistors of $1\ Ohm$. In the simple case of a tree, the resistance distance is simply the sum of the resistances along the path connecting u and v; that is, for a tree, $\Omega(u, v) = d(u, v)$. However, in the case of two nodes for which multiple routes connecting them exist, the effective resistance $\Omega(u, v)$ can be obtained by using Kirchhoff's laws. A characteristic of the effective resistance $\Omega(u, v)$ is that it decreases with an increase in the number of routes connecting u and v. Then, in general, $\Omega(u, v) \leq d(u, v)$. Klein and Randić (1993) proved that the effective resistance is a distance function, which they termed the *resistance distance*. As for the shortest path distance, $\Omega(u, v) := \infty$ if u and v are in different connected components of the network.

The resistance distance $\Omega(u, v)$ between a pair of nodes in a connected component of a network can be calculated by using the Moore–Penrose generalised inverse $\mathbf{L}^+$ of the graph Laplacian $\mathbf{L}$. Then, the resistance distance between the nodes u and v is calculated as (Babić et al., 2002; Ghosh et al., 2008; Gutman and Xiao, 2004; Yang and Zhang, 2008):

$$\Omega(u, v) = \mathbf{L}^+(u, u) + \mathbf{L}^+(v, v) - 2\mathbf{L}^+(u, v) \tag{3.6}$$

for $u \neq v$.

Another way of computing the resistance distance for a pair of nodes in a network is as follows. Let $\mathbf{L}(G - u)$ be the matrix resulting from the removal of the uth row and column of the Laplacian, and let $\mathbf{L}(G - u - v)$ be the matrix resulting from the removal of both the uth and vth rows and columns of $\mathbf{L}$. Then, it has been proved that the resistance distance can be calculated as (Bapat et al., 2003)

$$\Omega(u, v) = \frac{\det \mathbf{L}(G - u - v)}{\det \mathbf{L}(G - u)} \tag{3.7}$$

Yet another method of computing the resistance distance between a pair of nodes in the network is presented on the basis of the Laplacian spectra (Xiao and Gutman, 2003):

$$\Omega(u, v) = \sum_{k=2}^{n} \frac{1}{\mu_k} [U_k(u) - U_k(v)]^2, \tag{3.8}$$

where $U_k(u)$ is the uth entry of the kth orthonormal eigenvector associated with the Laplacian eigenvalue μ_k, which has been ordered as $0 = \mu_1 < \mu_2 \leq \cdots \leq \mu_n$.

The resistance distance between all pairs of nodes in the network can be represented in a matrix form, called the resistance matrix $\boldsymbol{\Omega}$ of the network. This matrix can be obtained as

$$\boldsymbol{\Omega} = |\mathbf{1}\rangle diag\{[\mathbf{L} + (1/n)\mathbf{J}]^{-1}\}^{T} + diag[\mathbf{L} + (1/n)\mathbf{J}]^{-1}\langle\mathbf{1}| - 2(\mathbf{L} + (1/n)\mathbf{J})^{-1}$$

(3.9)

where $\mathbf{J} = |\mathbf{1}\rangle\langle\mathbf{1}|$ is an all-ones matrix.

The analogous distance sum for the resistance distance $R(u)$ is the sum of all entries of the uth row/column of the resistance-distance matrix:

$$R(u) = \sum_{v \in V(G)} \Omega(u, v)$$

(3.10)

For instance, for the network illustrated in Figure 3.1 (left) the resistance-distance matrix is:

$$\Omega = \begin{pmatrix} 0 & 1.00 & 1.62 & 2.14 & 1.90 & 1.62 & 3.14 \\ & 0 & 0.62 & 1.14 & 0.90 & 0.62 & 2.14 \\ & & 0 & 1.00 & 0.48 & 0.48 & 1.62 \\ & & & 0 & 0.62 & 0.90 & 1.00 \\ & & & & 0 & 0.57 & 1.62 \\ & & & & & 0 & 1.90 \\ & & & & & & 0 \end{pmatrix}$$

In Figure 3.3 we illustrate the shortest path distance and resistance-distance matrices for a social network of Doubtful Sound community of bottlenose dolphins, with nodes labelled from 1 to 62.

For the case of connected networks, the resistance-distance matrix can be related to the Moore–Penrose inverse of the Laplacian, as shown by Gutman and Xiao (2004):

$$\mathbf{L}^{+} = -\frac{1}{2}\left[\boldsymbol{\Omega} - \frac{1}{n}(\boldsymbol{\Omega}\mathbf{J} + \mathbf{J}\boldsymbol{\Omega}) + \frac{1}{n^2}\mathbf{J}\boldsymbol{\Omega}\boldsymbol{\Omega}\right]$$

(3.11)

where $\mathbf{J}$ is an all-one matrix.

The resistance-distance matrix is a *Euclidean distance matrix*. A matrix $\mathbf{M} \in \mathfrak{R}^{n \times n}$ is said to be Euclidean if there is a set of vectors $x_1, \ldots, x_n$ such that $M_{ij} = \|x_i - x_j\|^2$. Because it is easy to build vectors such that $\Omega_{ij} = \|x_i - x_j\|^2$, the resistance-distance matrix is Euclidean, and the resistance distance satisfies the weak triangle inequality

$$\Omega_{ik}^{1/2} \leq \Omega_{ij}^{1/2} + \Omega_{jk}^{1/2}$$

(3.12)

for every pair of nodes in the network.

It is important to remark that the resistance distance is proportional to the expected commute time between two nodes for a Markov chain defined by a weighted network (Ghosh et al., 2008; Doyle and Snell, 1984). Let w_{uv} be the weight of the link $\{u, v\}$, then the transition probabilities in a Markov chain defined on the network are given by

$$P_{uv} = \frac{w_{uv}}{\sum_{u,v \in E} w_{uv}}$$

(3.13)

The commute time is the time taken by 'information' starting at node u to return to it after passing through node v. The expected commuting time $\hat{C}_{uv}$ is related to the resistance distance through the following relation (Ghosh et al., 2008; Doyle and Snell, 1984):

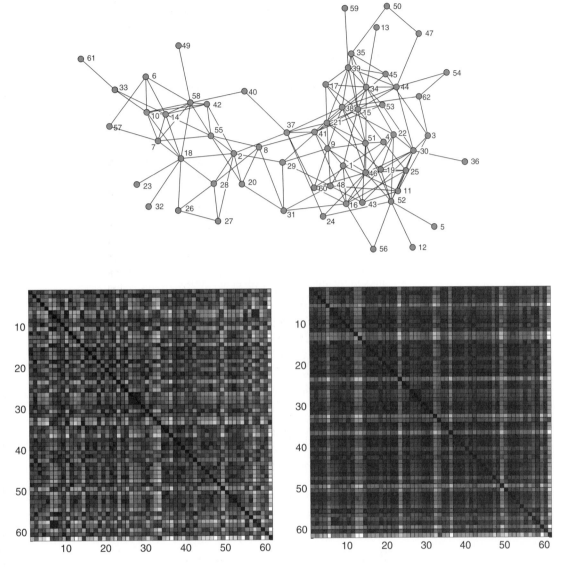

Fig. 3.3

Distance matrices of dolphin network. A social network of a Doubtful Sound community of bottlenose dolphins (top), its shortest path distance matrix (bottom left), and the resistance-distance matrix (bottom right).

$$\hat{C}_{uv} = 2\langle \mathbf{1}|w\rangle \Omega(u, v) \qquad (3.14)$$

where $\mathbf{1}$ is an all-one vector, and w is the vector of link weights. Note that if the network is unweighted, $\hat{C}_{uv} = 2m\Omega(u, v)$. Sometimes the resistance distance is also called the commute distance, for obvious reasons (von Luxburg et al. 2011).

The analogy of the Wiener index for the resistance distance is called the *Kirchhoff index* of the network (Klein and Randić, 1993),

$$Kf(G) = \frac{1}{2}\langle \mathbf{1}|\mathbf{\Omega}|\mathbf{1}\rangle = \sum_u R(u) = \frac{1}{2}\sum_u \sum_v \Omega(u, v) \qquad (3.15)$$

The Kirchhoff index of a network can be written by using the following integral (Ghosh et al., 2008):

$$Kf(G) = nTr \int_0^\infty \left(e^{-t\mathbf{L}} - \frac{1}{n}\mathbf{1}\mathbf{1}^T \right) dt \qquad (3.16)$$

If we designate by $\hat{C}$ the expected commute time averaged over all pair of links in the network, it can be seen that the Kirchhoff index is proportional to $\hat{C}$ as (Ghosh et al., 2008)

$$\hat{C} = \frac{4\langle\mathbf{1}|w\rangle}{n(n-1)}Kf(G) \qquad (3.17)$$

It is also proportional to the expected hitting time averaged over all pair of links in the network (Ghosh et al., 2008):

$$\hat{H} = \frac{2\langle\mathbf{1}|w\rangle}{n(n-1)}Kf(G) \qquad (3.18)$$

where the hitting time H_{uv} is defined as the time that takes the 'information' to arrive for the first time at node v after it departs from node u.

A couple of network invariants that are related to the Kirchhoff index are the gradient and the Hessian of $Kf(G)$, which are defined as follows (Ghosh et al., 2008):

$$\nabla Kf = -n\text{diag}\left(\mathbf{B}^T \left(\mathbf{L} + \frac{1}{n}\mathbf{1}\mathbf{1}^T \right)^{-2} \mathbf{B} \right) \qquad (3.19)$$

$$\nabla^2 Kf = 2n \left(\mathbf{B}^T \left(\mathbf{L} + \frac{1}{n}\mathbf{1}\mathbf{1}^T \right)^{-2} \mathbf{B} \right) \circ \left(\mathbf{B}^T \left(\mathbf{L} + \frac{1}{n}\mathbf{1}\mathbf{1}^T \right)^{-1} \mathbf{B} \right) \qquad (3.20)$$

where $\circ$ denotes the Hadamard product and $\mathbf{B}$ is the incidence matrix of the network.

It is worth mentioning that von Luxburg et al. (2011) have found that for very large networks, the random walk 'gets lost' in the graph. It then takes a very long time to travel through a substantial part of the network in such a way that by the time it comes close to its target node it has already 'forgotten' where it started. As a consequence, the hitting time is approximate to the inverse of the degree of the node at which the random walk starts. Similarly, the commuting time is approximated by the sum of the inverses of the degree of the nodes representing the initial and final step of the walk. Accordingly, the resistance distance in a very large network does not take into account the structure of the network. One possible way of amending this situation is by using some variations of the generalised network distances, which we analyse in the next section.

3.3 Generalized network distances

In the previous sections of this chapter we studied two different metrics for networks: the *shortest path*, and the *resistance distance*. Except for the case of acyclic networks—trees—where they coincide, both distances appear in general unrelated. Then, the question that arises is whether a sort of generalised network distance can be built, in such a way that the shortest path and the resistance distance are the two extremes of such a distance, with infinitely many others between. The answer to this question is the purpose of this section, in which we introduce this kind of *generalised network distance*.

First, let us begin with some necessary network theory definitions. A *rooted tree* is a connected tree with a marked node, called the root. Then, a *rooted forest* is a disconnected graph in which all its connected components are rooted trees. The roots of the rooted forest are the same as those of the rooted trees of which it is composed. The *global weight* of a weighted network is the product of all the weights of its links, and is designated $w(G)$. In the case of a set of graphs, $S = \{G_1, G_2, \cdots, G_t\}$, and as in a forest, the weight of the set is simply $w(S) = \sum_{i \in S} w(G_i)$. Of course, the weight of a set of unweighted networks—for example, an unweighted forest—is the number of connected components in that graph: the cardinality of the set, $|S|$. Let the matrix $\mathbf{F}$ be called the matrix of forest of a given network G, whose entries f_{ij} represent the weight of the set of all spanning rooted forests of G that have a node i belonging to a tree rooted at j. A spanning forest is a forest that contains all nodes of the graph (see Figure 3.4 for an example). The weight of the set of all spanning rooted forests in the network is denoted by f.

Chebotarev and Shamis (2002) have defined a measure of *relative forest accessibility* as a way of quantifying node proximity. This measure is defined by selecting a given parameter $\alpha > 0$ to define the following family of matrices:

$$\mathbf{Q}_\alpha = (\mathbf{I} + \alpha \mathbf{L})^{-1} \tag{3.21}$$

where $\mathbf{I}$ is the identity matrix, and $\mathbf{L} = \mathbf{K}\text{–}\mathbf{A}$ is the Laplacian. In the case of weighted networks the Laplacian is defined in the usual way: $\mathbf{L} = \mathbf{W}\text{–}\mathbf{A}$. We will study similar matrices in Chapter 6 (the resolvent of the adjacency matrix). The matrices Q_α are symmetric doubly stochastic, which means that $q_{ij}^\alpha = q_{ji}^\alpha$ and $\sum_{j=1}^n q_{ij}^\alpha = \sum_{i=1}^n q_{ij}^\alpha = 1$. The entries q_{ij}^α have been interpreted as the relative share of the connections between i and j in the totality of all i's connections with the nodes in the network, with the parameter $\alpha > 0$ determining the weight assigned to the different routes according to their lengths (see Chapters 5 and 6). Chebotarev and Shamis (2002) have shown that for a given parameter $\alpha > 0$, the measures defined below are network metrics:

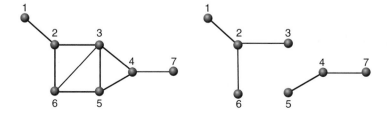

Fig. 3.4
Spanning forest. A simple graph (left) and one of its spanning forests (right).

$$d_{ij}^{\alpha} = \frac{1}{2}(q_{ii}^{\alpha} + q_{jj}^{\alpha} - q_{ij}^{\alpha} - q_{ji}^{\alpha}) \qquad (3.22)$$

$$\rho_{ij}^{\alpha} = \alpha(q_{ii}^{\alpha} + q_{jj}^{\alpha} - q_{ij}^{\alpha} - q_{ji}^{\alpha}) \qquad (3.23)$$

They were called the *forest distance* and the *adjusted forest distance* of a network, respectively. In a more recent work, Chebotarev (2011) introduced the logarithmic forest distances as follows. Let

$$\mathbf{H}_{\alpha} = \gamma(\alpha - 1)\log_{\alpha}\mathbf{Q}_{\alpha} \qquad (3.24)$$

where $\log_{\alpha}\mathbf{M}$ represents the entrywise logarithm of the matrix, $\alpha \neq 1$ and $\gamma > 0$. Then, let

$$\mathbf{X}_{\alpha} = \frac{1}{2}(|h_{\alpha}\rangle\langle 1| + |1\rangle\langle h_{\alpha}|) - \mathbf{H}_{\alpha} \qquad (3.25)$$

where $|h_{\alpha}\rangle$ is a column vector containing the diagonal entries of $\mathbf{H}_{\alpha}$. Then, because $\lim_{\alpha \to 1}[(\alpha - 1)/\ln\alpha] = 1$, Chebotarev (2011) extended

$$\mathbf{X}_{\alpha} = \frac{1}{2}(|h_{\alpha}\rangle\langle 1| + |1\rangle\langle h_{\alpha}|) - \mathbf{H}_{\alpha} \qquad (3.25)$$

to $\alpha = 1$ as

$$\mathbf{H}_1 = \gamma \log^{\circ}\mathbf{Q} \qquad (3.26)$$

Chebotarev (2011) has proved that for any connected multigraph and any $\alpha, \gamma > 0$, the entries of the matrix $\mathbf{X}_{\alpha}$ defined by (3.24) and using the extension (3.26) is a network distance. Let us multiply the weight of every link in a weighted multigraph by $\alpha > 0$. Then, we obtain the weight of the set of all weighted spanning rooted forests of G that have a node i belonging to a tree rooted at j as $\chi_{ij}(\alpha)$. Chebotarev (2011) has proved that for any connected multigraph and any $\alpha, \gamma > 0$, the matrix $\mathbf{X}_{\alpha}$ defined by (3.25) and using the extension (3.26) exists, and its entries are given by:

$$\chi_{ij}^{\alpha} = \begin{cases} \gamma(\alpha - 1)\log_{\alpha}\dfrac{\sqrt{f_{ii}(\alpha)f_{jj}(\alpha)}}{f_{ij}(\alpha)}, & \alpha \neq 1 \\[2ex] \gamma\ln\dfrac{\sqrt{f_{ii}f_{jj}}}{f_{ij}}, & \alpha = 1. \end{cases} \qquad (3.27)$$

and then the name of logarithmic forest distance proposed so far (Chebotarev, 2011) for this measure. For obvious reasons, here we will use the designation of *Chebotarev distances* to χ_{ij}^{α}. Chebotarev distances have several nice properties, including being a *network-geodetic*. A function $d : V(G) \times V(G) \to \Re$ is network-geodetic whenever for all $i, j, k \in V, d_{ij} + d_{jk} = d_{ik}$ holds, if and only if every path from i to k passes through j. However, the most important property of Chebotarev distance is related to their asymptotic properties, which show that they are a generalisation of both shortest path and resistance distance.

Let

$$\gamma = \ln(e + \alpha^{2/n}) \qquad (3.28)$$

Then, for any connected multigraph and every pair of nodes, the distance χ_{ij}^{α} with scaling parameter (3.28):

i) converges to the shortest path distance d_{ij} as $\alpha \to 0^+$,
ii) converges to the shortest path distance Ω_{ij} as $\alpha \to \infty$.

In the case of using any arbitrary parameter γ, the Chobotarev distances become proportional to the shortest path and resistance distance at the two limits given before, respectively. Below, we give the Chebotarev distance matrix for $\alpha = 2.0$ and γ given by (3.28) for the graph illustrated in Figure 3.4:

$$X_2 = \begin{pmatrix} 0 & 1.293 & 2.534 & 3.606 & 3.166 & 2.501 & 4.899 \\ & 0 & 1.240 & 2.313 & 1.872 & 1.208 & 3.606 \\ & & 0 & 1.240 & 1.039 & 1.039 & 2.534 \\ & & & 0 & 1.207 & 1.872 & 1.294 \\ & & & & 0 & 1.159 & 2.501 \\ & & & & & 0 & 3.165 \\ & & & & & & 0 \end{pmatrix}$$

The Chebotarev distance between a pair of nodes in a network provides the probability of choosing a forest partition separating both nodes in the model of random forest partitions introduced by Chebotarev and Shamis (2002). As the parameter α approaches to zero—$\alpha \to 0^+$—there is only a preservation of those partitions that connect both nodes by a shortest path and separate all other nodes. However, when $\alpha \to +\infty$, only those partitions determined by two disjoint trees are preserved. In the first case we have the shortest path distance, and in the second the resistance distance. Consequently, Chebotarev distance generalises, in a beautiful way, the most relevant metrics in networks.

3.4 Topological structure of networks

> Topology is the property of something that doesn't change when you bend it or stretch it, as long as you don't break anything.
> Edward Witten

3.4.1 Network planarity

Maps are very familiar to everyone. Then, if every country in the map is represented as a node and two nodes are connected if the corresponding countries have a common borderline, we have a network representation of the political map. If we examine, for instance, a map of Europe, we can see that Luxembourg has borders with three other countries: Belgium, France, and Germany, which also have borders among them. In terms of the network representation, this corresponds to a complete graph of four nodes K_4 (see Figure 3.5). This graph can be drawn in a plane without any crossing of its links. Consequently, it is said to be a *planar graph* (Pisanski and Potočnik, 2003).

Let us now reformulate an old puzzle set, around 1840, by Möbius for his students (Biggs et al., 1976):

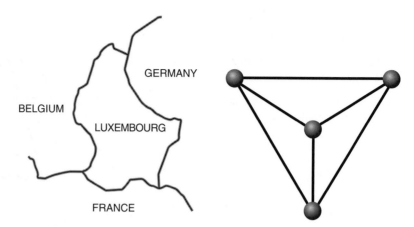

Fig. 3.5
Maps and planar networks. A section
of a map of Europe around Luxembourg,
and the planar network representing
'adjacency' between countries.

Is it possible to draw a map having five countries in such a way that every
country has a borderline in common with the other four?

In terms of our network representation, this map corresponds to the complete
graph of five nodes, K_5, and the question formulated is whether or not K_5 is
planar.

The answer to this puzzle came 90 years after its original formulation, when
K. Kuratowski (1930) formulated his celebrated theorem about graph planarity.
Accordingly, a network is planar if and only if it contains no subgraph homeomorphic to K_5 or $K_{3,3}$. Two subgraphs are homeomorphic if they can be
transformed to each other by subdividing some of their edges. In Figure 3.6
we illustrate some graphs homeomorphic to K_5 or $K_{3,3}$.

One area of the study of complex networks in which planarity plays an
important role is the study of urban street networks, in which links represent
streets and nodes represent street's intersections and *cul-de-sacs*. Although we
can find some bridges and tunnels in a city, most links in these networks
intersect basically only at nodes, and the corresponding network is planar
(Barthélemy and Flammini, 2008; Cardillo et al., 2006; Jiang, 2007). This
situation has been described by Buhl et al. (2006) by stating that 'cities
are essentially planar structures.' Other examples of planar or almost-planar
networks are interstate road maps, power-grid networks, and delivery and distribution networks, such as the US Natural Gas Pipeline Compressor Stations
illustrated in Figure 3.7.

3.4.2 Network imbedding

Planar networks like the previous ones can be imbedded in a plane. An imbedding of a network in an orientable surface S is a continuous one-to-one function
from a topological representation of the network into the surface S. A non-
orientable surface is a 2-manyfold that contains a subspace homeomorphic to
the Möbius band. Otherwise the surface is orientable (Gross and Tucker, 1987;
Pisanski and Potočnik, 2003). Closed surfaces refer to those compact surfaces
without boundary. Orientable surfaces include the sphere, the torus, the double
torus, and so on, which are commonly denoted S_0, S_1, S_2, and so on. The

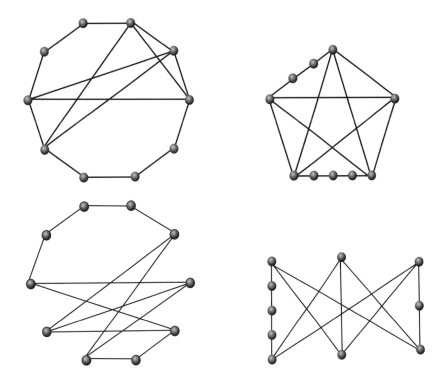

Fig. 3.6

Network homeomorphism. The two networks at left are homeomorphic to K_5 and $K_{3,3}$ respectively. The networks at right are the same as those at left, but are drawn in such a way that link subdivisions of K_5 (top) and $K_{3,3}$ (bottom) are observed.

sub-index g in S_g is known as the genus of the surface. The genus of a closed orientable surface is the number of handles that the surface has. The sphere has no handle, which means that $g = 0$. By adding a handle to the sphere we obtain a torus, which has $g = 1$, and by adding another handle a surface with $g = 2$ is obtained (Gross and Tucker, 1987). Some of these surfaces are illustrated in Figure 3.8.

If a connected network is imbedded in the sphere S_0, then the number of nodes $|V|$ minus the number of links $|E|$ is equal to 2 minus the number of faces $|F|$ of the network, which is known as the Euler's equation for the sphere (Gross and Tucker, 1987):

$$|V| - |E| = 2 - |F| \qquad (3.29)$$

For generalisations of this formula for general surfaces, the reader is referred to Gross and Tucker (1987).

The *genus of the network* $\gamma(G)$ is the minimum integer g such that there is an imbedding of the network in the orientable surface S_g (Gross and Tucker, 1987; Pisanski and Potočnik, 2003). Then, a planar network has genus equal to zero, as it can be imbedded in the sphere. In general, the determination of the genus of a graph is an NP-complete problem. However, some analytical expressions exist for determining the genus of some special kinds of network. For instance, the complete network K_n with $n \geq 3$, and the complete bipartite

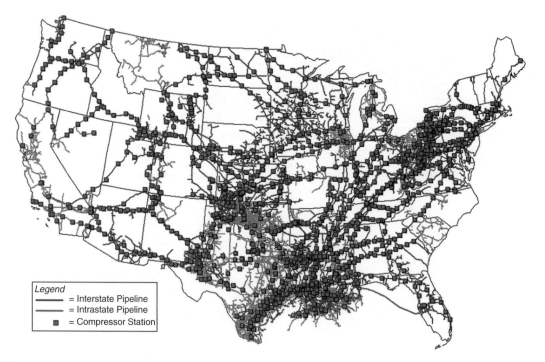

Fig. 3.7
US Natural Gas Pipeline Compressor Stations. In this figure, every node represents a compressor station, and links represent pipelines for natural gas transportation in the USA. (Image acquired from http://www.eia.doe.gov/pub/oil_gas/natural_gas/analysis_publications/ngpipeline/ngpipeline_maps.html.)

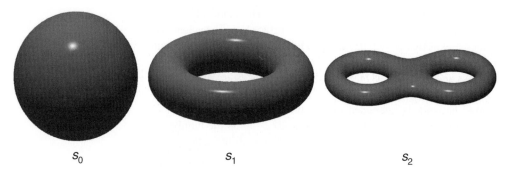

Fig. 3.8
Orientable surfaces. Three different closed surfaces corresponding to the sphere, the torus, and the bitorus.

network $K_{n,m}$ ($m, n \geq 2$), have genus (Pisanski and Potočnik, 2003):

$$\gamma(K_n) = \left\lceil \frac{(n-3)(n-4)}{12} \right\rceil \qquad (3.30)$$

and

$$\gamma(K_{n,m}) = \left\lceil \frac{(n-2)(m-2)}{4} \right\rceil \qquad (3.31)$$

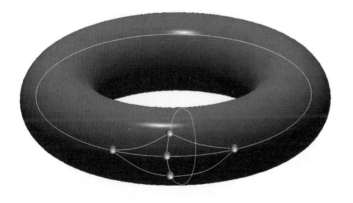

Fig. 3.9
Network imbedding. Imbedding of the complete network with five nodes K_5 on the torus.

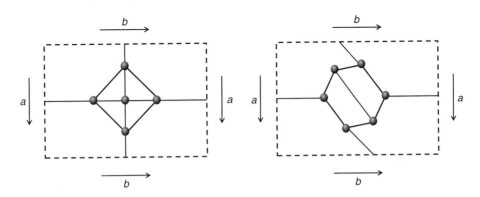

Fig. 3.10
Plane representation of imbeddings. Representation of the imbedding of K_5 (left) and $K_{3,3}$ (right) on the torus.

respectively, where $\lceil x \rceil$ denotes the ceiling function. Then, the graphs K_5 and $K_{3,3}$, which were previously shown to be non-planar, has genus equal to 1. In Figure 3.9 we illustrate an imbedding of K_5 on the torus.

A torus can be constructed from a rectangle—first by forming a tube by gluing together the two longer sides of the rectangle, and then pasting both extremes of the tube. A torus can therefore be represented as a rectangle, allowing us to draw imbeddings on a plane. In Figure 3.10 we illustrate the imbeddings of K_5 and $K_{3,3}$ on the torus represented as a rectangle (Gross and Tucker, 1987).

There are, however, surfaces which are not orientable. The non-orientable surface with k-crosscaps is denoted by N_k, and is defined as the connected sum of k copies of the projective plane N_1. One example of such surfaces is the Möbius band, which is neither closed nor orientable. It is non-orientable in the sense that if we place it in a coordinate system specifying, for instance, forward and right directions, and we move forward along the surface, then the orientation of the right direction is reversed. That is, we can paint the whole band without lifting the brush. The Möbius band is illustrated in Figure 3.11, together with the Klein bottle (usually designated as N_2), which is a closed non-orientable surface (Gross and Tucker, 1987).

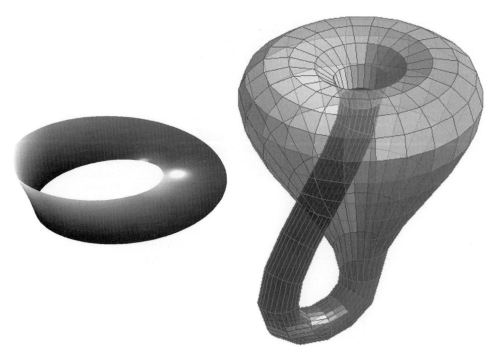

Fig. 3.11
Non-orientable surfaces. The Möbius band (left) and the Klein bottle (right).

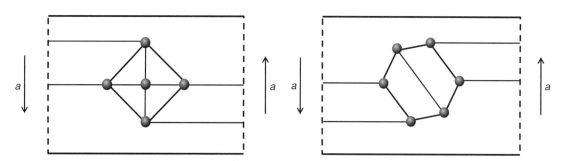

Fig. 3.12
Imbedding in a non-orientable surface. Representation of the imbedding of K_5 (left) and $K_{3,3}$ (right) in the Möbius band.

In a similar way as for the torus, imbeddings in the Möbius band and the Klein bottle can be represented in a plane. For instance, in Figure 3.12 we illustrate the imbedding of K_5 and $K_{3,3}$ in the Möbius band (Gross and Tucker, 1987).

It is easy to realise that by eliminating only one link, K_5 or $K_{3,3}$ can be imbedded in S_0. For a given network $G = (V, E)$, the number of links that need to be removed to planarise the network is known as the *skewness* of the network, $s(G)$ (Liebers, 2001). Then, by removing $s(G)$ links in a non-planar network we determine the *maximum planar subgraph* (MPS) $G' = (V', E')$

existing in such network, where $V' \subseteq V$ and $E' \subseteq E$(Liebers, 2001; Boyer et al., 2004). Obviously, $s(G) = |E| - |E'|$. Using these concepts we can here introduce the following quantitative measure for the planarity of a network:

$$\pi(G) = \frac{|E'|}{|E|} \qquad (3.32)$$

The *planarity index* $\pi(G)$ takes the value of 1 for planar graphs, where the MPG has the size of the whole network and tends to zero for very large non-planar networks with very small MPGs.

3.4.3 Topological structure of real networks

An example of planar real-world networks are the ant galleries produced in a 2-D experimental setup (Buhl et al., 2004). In this experiment, ants are dispersed around a sand disk in such a way that they can begin to dig from the periphery only. At the end of the excavation process of 72 hours, a network of galleries has been built, which, due to the spatial constraint imposed, necessarily needs to be a planar network. In Figure 3.13 we illustrate this process (Buhl et al., 2004). Obviously, a simple calculation shows that the ant gallery network has $\pi(G) = 0$.

What is the situation with other networks which are spatially constrained in three-dimensional space? Let us consider two naturally evolved networks taken from different real-world scenarios which have had to grow while constrained by the 3-D space. The first network represents the 3-D galleries and chambers which are formed inside termite nests (Perna et al., 2008). These galleries grow while constrained by the space determined by the mound (see Figure 3.14).

The second network studied here is the neural network of the worm *C. elegans*. This network—in which every node represents a neuron, and links represent synaptic connections between pairs of neurons—has evolved while constrained by the anatomical space of this organism. We can use a technological analogy to understand how this neural network has probably evolved. The analogy we will use is the one with the design of Very Large Scale Integration (VSLI) circuits (Mead and Conway, 1980). In this process, thousands of transistor-based circuits are combined into a single chip to create integrated circuits. In the network representation of VLSI circuits, the nodes represent transistors, and links represent the wires of the circuit. Then the problem consists in placing the transistors on a printed circuit board in such a way that the wires connecting them do not cross, since crossings lead to undesired signals. In order to place more and more elements in the integrated circuit it is necessary to decompose the network representing the VSLI circuit into planar subnetworks, with each of them completely embedded in one layer. The minimum number of layers required to avoid any crossing between wires is known as the *thickness* of the network (Mutzel et al., 1998). In general, it is necessary to use multilayer VSLI in order to design integrated circuits in small space. Therefore, we can think that the neural network of *C. elegans* could

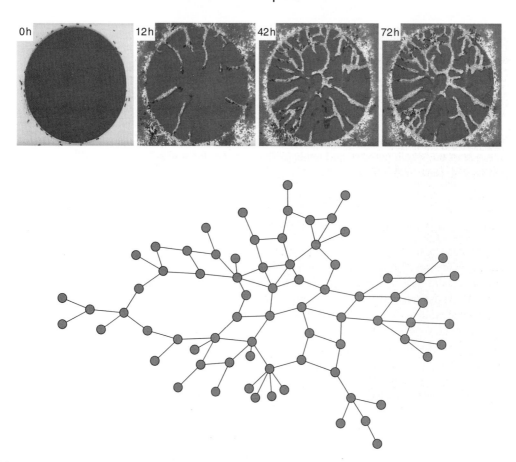

Fig. 3.13

Ant galleries. The galleries produced by 200 ants during three days (top), and a representation of one of the networks of ant galleries (bottom). From Buhl, J., Gautrais, J., Solé, R. V., Kuntz, P., Valverde, S., Deneubourg, J. L., and Theraulaz, G. (2004). 'Efficiency and robustness in ant networks of galleries', *Eur. Phys. J. B.*, **42**, 123–9. (The top part of this figure, and any original (first) copyright notice displayed with material, is reproduced with the permission of Springer Science+Business Media.)

evolve in a way that would locate as many neurons as needed in a constrained 'small' volume, by simply using a multilayer system.

We have thus calculated the planarity index $\pi(G)$ for these two networks, and the results are illustrated in Table 3.1. Surprisingly, the network of galleries in a termite nest is almost planar. In fact, by removing only 13 out of 342 links, the network becomes planar. We have explored other networks of galleries in termite nests, and the results are quite similar. Some of the networks are planar, and the planarity index is bounded between 0.865 and 1.000 for them. This situation is probably due to the fact that the nests are not built by considering an optimisation of the space. In fact, mounds are usually 6 metres high for insets which are less than 1 cm long. However, these mounds are also used for the storage of plant material, as well as for evacuating excess carbon dioxide and heat. Consequently, technically speaking these nests can be built on a plane, but termites prefer to build them in 3-D to guarantee other ecological conditions as those previously mentioned.

Fig. 3.14
Termite nest. A nest built by the *Cubitermes* termite (left), and a network of chambers (nodes) and galleries (links) in this nest. The image of the nest is from Perna, A., Valverde, S., Gautrais, J., Jost, C., Solé, R. V., Kuntz, P., and Theraulaz, G., 'Topological efficiency in the three-dimensional gallery networks of termite nests', reprinted from *Physica A*, **387**, 6235–44. (Reproduced courtesy of A. Perna, and with the permission of Elsevier, ©2008.)

Table 3.1 **Planarity in real networks**. Comparison of the planarity of two naturally evolved networks constrained by physical 3-D space. 'Termites' refers to a network of galleries in *Cubitermes* termite nests, after Perna et al. (2008). 'Neurons' refers to the neural network of *C. elegans*.

Network	$|N|$	$|E|$	$|E'|$	$\pi(G)$
Termites	287	342	329	0.962
Neurons	280	1973	592	0.300

On the other hand, the neural network of *C. elegans* is highly non-planar, and it is necessary to remove 1,381 out of the 1,973 links in order to planarise the network. In this particular example, nature teaches us that the lack of planarity can be a tool for efficiently placing many nodes in a constrained volume. The thickness of any network having more than two nodes is bounded as (Mutzel et al., 1998)

$$\theta(G) \geq \left\lceil \frac{m}{3n - 6} \right\rceil \tag{3.33}$$

which means that the neural network of *C. elegans* needs to be built in three or more layers in order to avoid 'short-circuits.' The maximum number of layers necessary can be estimated by any of the following three bounds, given that the network has more than ten nodes and maximum degree k_{max}:

$$\theta(G) \leq \left\lfloor \sqrt{\frac{m}{3}} + \frac{3}{2} \right\rfloor \tag{3.34}$$

$$\theta(G) \leq \left\lfloor \frac{k_{max}}{2} \right\rfloor \tag{3.35}$$

Then, the neural network of *C. elegans* will have less than 27 layers formed by planar subnetworks.

With these lessons learned from nature we can return to the networks from which we began our discussion: urban street networks. In Figure 3.15 we illustrate a network representation of the streets in the old part of the city of Córdoba—a moderately-sized modern city in the south of Spain, which at the end of the tenth century was the most populated city in Europe and probably the world.

The city of Córdoba has grown from the old central town founded in ancient Roman times by extending through the periphery, as many moderns cities have done. The analysis of the street network illustrated in Figure 3.15 demonstrates that the network is planar: $\pi(G) \equiv 1$. The situation can be compared to that of the termite nests, where sufficient space allows the expansion of planar or almost-planar networks to fulfil the functional requirements of the systems.

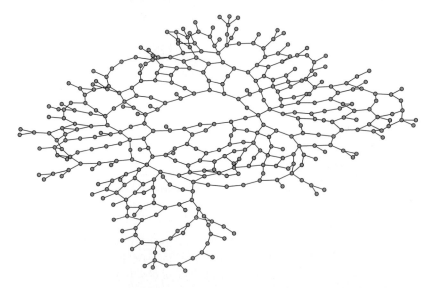

Fig. 3.15
Urban street network of Córdoba. Nodes represent intersections between streets, which are represented by links, in the old part of Córdoba, Spain.

However, what happens when the existing surface space does not permit extension of the borders of a city beyond its current limits? Let us consider the city of Tokyo, which currently has 8 million inhabitants, a total of more than 12 million in the prefecture, and more than 35 million in the Greater Tokyo area—the most populated area of the world. The Tokyo prefecture has an area of $1,457.3 \, \mathrm{km}^2$, and its current population density exceeds 8,000 inhabitants per square kilometre. Together with this, the problems of increasing land-pricing and environmental pressure has led to the proposal of some radically new approaches in the future development of the city.

One of these solutions was proposed by K. Sugizaki (1992) of Shimizu Corporation. It takes the form of a mega-pyramid, TRY2004, with a total area of 800 hectares for the foundations, and rising to a height of more than 2,000 metres. According to Sugizaki (1992), 'A pyramid shape was selected because it would not be so destructive toward the environment, enabling human beings to live in harmony with nature both inside the city itself and its vicinity.' A picture of how this city of the future will look like is provided in Figure 3.16.

The street network of this city will no longer be planar. Instead, there will be streets going up and down through the eight levels of TRY2004, as well as to one side or other of the octahedral units forming the pyramid. The strategy here is similar to that followed by the neural network of *C. elegans*; that is, to place as many elements as possible in a small space by using a system of multiple planar layers placed one over the other. The optimal use of space is obtained by considering complete networks K_n, for which the number of layers is determined by (Mutzel et al., 1998):

$$\theta(K_n) = \left\lceil \frac{m}{3n - 6} \right\rceil \tag{3.36}$$

We have seen in this chapter that metric and topological properties of networks are important elements in understanding the organisational principles of

Fig. 3.16
Pyramid city. The TRY2004 pyramid city, which according to Sugizaki, (1992) will be 2,000 metres high. From Sugizaki, (1992). Mega-City Pyramid TRY2004, *Structural Engineering International* **2**, no. 4, 287–9. Zurich, Switzerland, IABSE.

complex systems. Such principles are universal, in the sense that they can be found in natural and man-made systems which have evolved under different spatial constraints.

3.4.4 Beyond planarity: networks in hyperbolic space

An interesting question that can be posted in relation to the imbedding of complex networks on surfaces is the following. Is there an underlying geometry in which properties of imbedded complex networks arises 'naturally'? Let us take, for instance, the 'fat-tail' degree distribution discussed in Chapter 2, which is displayed by many real-world networks. The question is whether this property can 'emerge' as a consequence of the imbedding of such networks in a given space. This question has a beautiful analogy in the emergence of spacetime properties from the consideration of gravitation as a curved geometry, which is the basis of the General Relativity theory. Some approaches to answering this question have been proposed in the scientific literature by showing that many of the topological properties that produce such unique complex networks emerge naturally by imbedding them on *hyperbolic spaces* (Aste et al., 2005; Boguñá et al., 2010; Krioukov et al., 2009a; 2009b; 2010; Serrano et al., 2008).

A hyperbolic space has non-Euclidean geometry, which has a negative curvature. This means that every point in the hyperbolic space is a saddle point, such that triangles are 'thinner' than in the Euclidean space, and parallel lines diverge. By 'thinner', we mean that the sum of the angles of this triangle is smaller than π. Hyperbolic n-dimensional spaces are denoted by H^n (Anderson, 2005). A way of representing these spaces is by realising them as hyperboloids in $\Re^{n+1}$, or by means of its two different projections known as the Poincaré and Klein unit disk models (see Figure 3.17).

By considering the hyperbolic plane H^n, which has curvature $K = -1$, it is possible to place $n \gg 1$ nodes randomly distributed over a disk of radius $R \gg 1$. Then, by connecting them with some probability, which depends only on hyperbolic distances between nodes, it has been shown (Krioukov et al., 2010) that

$$\bar{k} \approx \frac{8}{\pi} n e^{-R/2} \tag{3.37}$$

which indicates that for generating a network of size n with average degree $\bar{k}$ it is necessary to select a disk of radius (Krioukov et al., 2010)

$$R = 2 \ln \frac{8n}{\pi \bar{k}} \tag{3.38}$$

More interestingly, it has been shown, using a *hidden variable approach*, that for sparse networks the degree distribution follows a power-law (Krioukov et al., 2010):

$$P(k) = 2 \left(\frac{\bar{k}}{2} \right)^2 \frac{\Gamma\left(k - 2, \bar{k}/2\right)}{k!} \sim k^{-3} \tag{3.39}$$

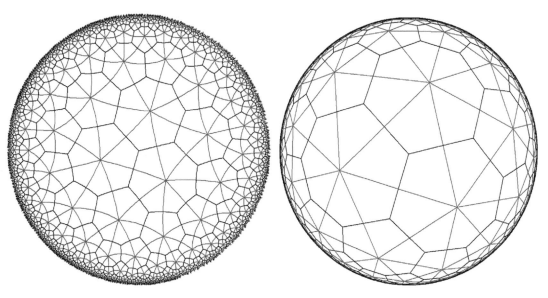

Fig. 3.17
Hyperbolic tesselations. A {7, 3}-tesselation of the hyperbolic plane with 1,000 nodes. The primal tessellation is made by heptagons (black lines), and the dual tessellation is made by triangles (grey lines). The figure illustrates both the Poincaré (left) and Klein (right) disk models.

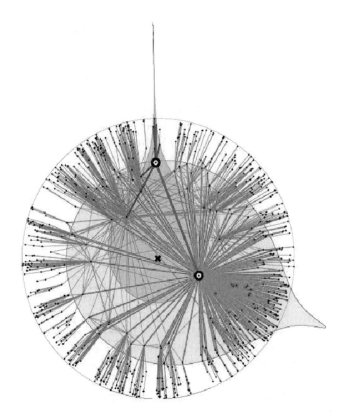

Fig. 3.18
Network embedded in the hyperbolic disk. A network with $n > 740$ nodes, power-law degree distribution with exponent $\gamma = 2.2$, and average degree $\bar{k} = 5$, embedded in the hyperbolic disk of curvature $K = -1$ and radius $R = 15.5$ centred at the origin, which is marked by a cross. (Figure courtesy M. Boguñá.)

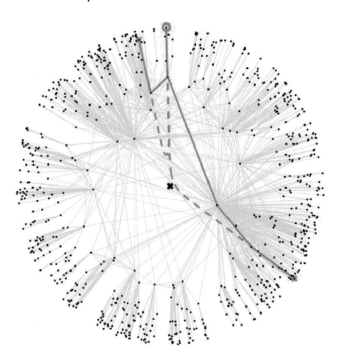

Fig. 3.19

Greedy forwarding in a network in hyperbolic space. The two dashed curves correspond to two hyperbolic straight lines and the solid line corresponds to the shortest path between the source (top circle node) and the destination (cross at the bottom). It is observed that both straight line and shortest paths follow the same pattern. (Figure courtesy M. Boguñá.)

where $\Gamma(a, b)$ is the so-called upper incomplete gamma function:

$$\Gamma(a, b) = \int_b^\infty t^{a-1} e^{-t} dt \qquad (3.40)$$

If, instead of a uniform distribution of the nodes in the hyperbolic plane, an exponential distribution with exponent $\alpha > 0$ of such nodes in a hyperbolic surface of curvature $K = -\varsigma^2$ with $\varsigma > 0$ is used, then (Krioukov et al., 2010)

$$\bar{k} = \nu \frac{2}{\pi} \left(\frac{\gamma - 1}{\gamma - 2} \right)^2 \qquad (3.41)$$

where $\nu = \pi \bar{k}/8$ and γ is the exponent of $P(k) \sim k^{-\gamma}$ with

$$\gamma = \begin{cases} 2\frac{\alpha}{\varsigma} + 1 & \text{if } \frac{\alpha}{\varsigma} \geq \frac{1}{2} \\ 2 & \text{if } \frac{\alpha}{\varsigma} \geq \frac{1}{2} \end{cases} \qquad (3.42)$$

A sample network with power-law degree distribution embedded in the hyperbolic disk of curvature $K = -1$ is illustrated in Figure 3.18 (Krioukov et al., 2010).

The model of networks imbedded in hyperbolic spaces allows us to study the efficiency of finding paths in the network without global information about the network—a situation which resembles the transmission of information in real life. It has been shown that by using a strategy that sends packets at each hop to the neighbour closest to the destination in hyperbolic space (greedy forwarding), it is possible to find almost 100% of shortest paths that reach their destinations. This efficiency is observed in both static and dynamic networks. As can be seen in Figure 3.19, this efficiency is possible due to the fact that greedy paths and hyperbolic straight lines between the same sources and destination follow the same pattern. That is, sending information through the shortest paths in a network is equivalent to sending such messages through a straight line in hyperbolic space (Krioukov et al., 2010).

Fragments (subgraphs) in complex networks

4

Beauty: the adjustment of all parts proportionately so that one cannot add or subtract or change without impairing the harmony of the whole.

L. Battista Alberti

Informally, a subgraph is a part of a network. The total number of subgraphs of a given type is an appropriate measure used to characterise a network as a whole. For instance, the number of nodes and links, which are the simplest subgraphs, are global topological characteristics of complex networks.

Formally, a subgraph is a graph $G' = (V', E')$ of the network $G = (V, E)$ if, and only if, $V' \subseteq V$ and $E' \subseteq E$. If the graph G is a digraph, the corresponding subgraphs are directed subgraphs. There are different types of subgraph, as illustrated in Figure 4.1. In general, a subgraph can be formed by one (connected subgraph) or more (disconnected subgraphs) connected components. They can be acyclic or cyclic, depending on the presence or absence of cycles in their structures.

The use of subgraphs to characterise the structure of networks dates back to 1964, when the Russian chemist E. A. Smolenskii (1964) developed a substructural molecular approach to quantitatively describe structure–property relationships. Smolenskii's procedure is based on the decomposition of the molecular graph into different fragments or subgraphs. The subgraphs of a given type then become a variable in a multiple regression model to describe a physical property of molecules. This idea was further developed in the 1980s as the *graph-theoretic cluster expansion ansatz* (Klein, 1986; McHughes and Poshusta, 1990). In general, these methods express a molecular property P as a sum of contributions of all connected subgraphs of the network G:

$$P(G) = \sum_k \eta_k(G) \cdot N_k \tag{4.1}$$

where η is the contribution of the k th subgraph to the molecular property, and the *embedding frequencies* N_k are the number of times a given subgraph appears in the molecular graph (Klein, 1986). The sum in (4.1) extends over all connected subgraphs in the network. A variation of the substructure descriptors used in chemistry is *molecular fingerprints* (Willett, 2006)—string representations of the chemical structure in which every entry of the string represents

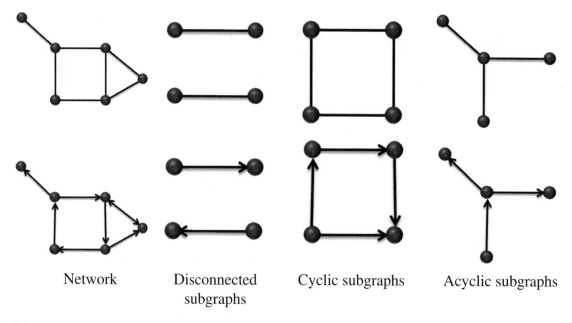

Fig. 4.1
Subgraphs. Subgraphs in undirected (top) and directed (bottom) networks.

a subgraph descriptor associated with a specific molecular feature. Molecular fingerprints are widely used in similarity searching, in which a series of molecules in a database is compared with a target molecular structure—usually with a desired property—allowing a ranking of decreasing similarity with the target for all the molecules. The molecular fingerprints are used as a vector characterizing every structure, and are compared to others by using an appropriate distance function, such as the Euclidean distance:

$$d_{ij} = \sqrt{\sum_{r=1}^{p}(f_{ir} - f_{jr})^2} \qquad (4.2)$$

where f_{ir} is rth entry of the molecular fingerprint for the target molecule, and f_{jr} is the same entry for the jth molecule in the database.

4.1 Network 'graphlets'

The idea of molecular fingerprints has been applied to protein–protein interaction networks under the name of 'graphlets' (Pržulj et al., 2004; Pržulj, 2007). A *graphlet* is a rooted subgraph in which nodes are distinguished by their topological equivalence. For instance, in a path of length 2 there are two kinds of node from a topological point of view: the node in the middle and the two (equivalent) end nodes. Pržulj has defined 73 non-equivalent types of node which are present in the 30 subgraphs having from 2 to 5 nodes (Pržulj et al., 2004; Pržulj, 2007). In Figure 4.2 we illustrate these graphlets

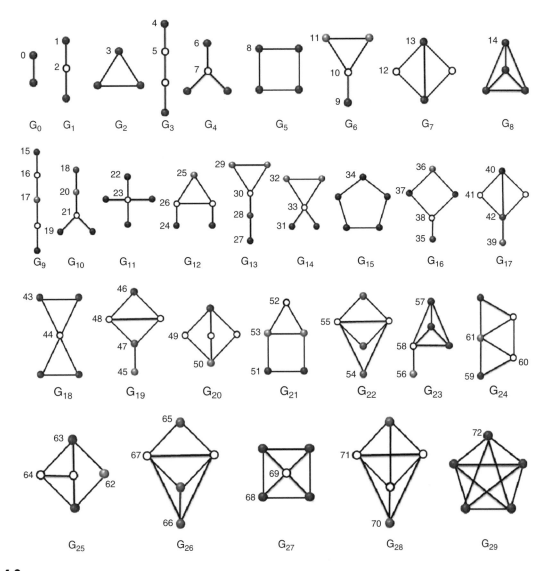

Fig. 4.2
Graphlets. Graphlets are rooted subgraphs having from three to five nodes. Here, nodes which are in the same automorphism orbit in a graphlet are drawn identically: dark grey, light grey, or empty circles.

in which non-equivalent nodes are numbered from 0 to 72. Then, the graphlet is a network fingerprint formed by a vector having 73 entries. Every entry of the graphlet represents the number of nodes of the corresponding type which are present in a given network.

Fragments, fingerprints, and graphlets can be used in different ways to study complex networks. Fragments have been the basis of quantitative structure–property relationships, such as the cluster expansion method. In this method, some properties are correlated with the number of fragments of different types in a group of networks. Another area of intensive research is the study of (molecular) similarity (Willett, 1987). Here the vector of (molecular) finger-prints of a series of objects is used to define a similarity function between the corresponding objects (molecules, networks), which are classified in different groups. Graphlets, on the other hand, have been used to extend the idea of degree distributions to subgraphs, as well as for finding similarities between proteins in a protein–protein interaction network. Some of these examples of applications will be analyzed in the second part of this book.

4.2 Network motifs

Another application of the number of subgraphs in complex networks is the concept of *network motif* (Milo et al., 2002; 2004a; 2004b; Artzy-Randrup et al., 2004). This concept was introduced by Milo et al. (2002) in order to characterize recurring significant patterns in real-world networks. A network motif is a subgraph that appears more frequently in a real network than could be expected if the network were built by a random process. This means that in order to find a motif in a real-world network we need to compare the number of occurrences of a given subgraph in this network with the number of times this subgraph appears in the ensemble of random networks with the same degree sequence. The appearance of a motif in a network then depends on the random model we select for comparison. Consequently, a large number of random networks is necessary to compute the frequency of appearance of the subgraph under study. A subgraph is then considered a network motif if the probability P of appearing in a random network an equal or greater number of times than in the real-world network is lower than a certain cut-off value, which is generally taken to be $P = 0.01$.

The statistical significance of a given motif is calculated by using the *Z-score*, which is defined as follows for a given subgraph i:

$$Z_i = \frac{N_i^{real} - \langle N_i^{random} \rangle}{\sigma_i^{random}} \tag{4.3}$$

where N_i^{real} is the number of times the subgraph i appears in the real network, $\langle N_i^{random} \rangle$, and σ_i^{random} are the average and standard deviation of the number of times that i appears in the ensemble of random networks, respectively. In Figure 4.3 we illustrate some of the motifs found by Milo et al. (2002) in different real-world directed networks.

A characteristic feature of network motifs is that they are network-specific. That is, a motif for a given network is not necessarily a motif for another network. However, a family of networks can be characterized by having the same series of motifs. In order to characterize such families, Milo et al. (2004a) have proposed using a vector whose ith entry produces the importance of the ith motif with respect to the other motifs in the network. Mathematically, the *significance profile* vector is expressed as

$$SP_i = \frac{Z_i}{\sum_j Z_j^2} \tag{4.4}$$

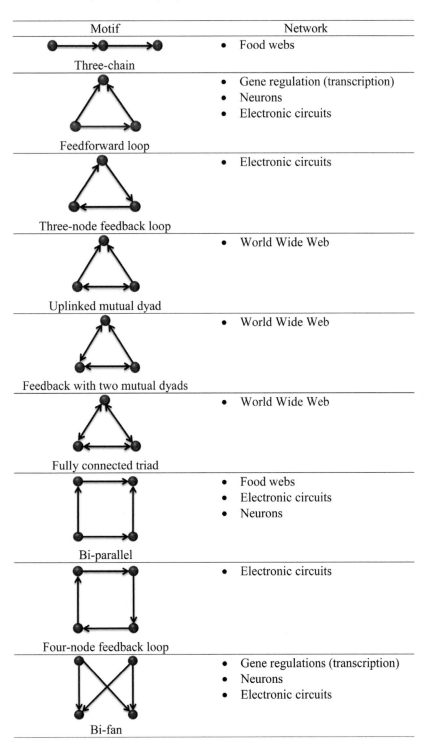

Fig. 4.3
Network motifs. Some of the motifs found in real-world networks, according to Milo et al. (2002).

4.3 Closed walks and subgraphs

Closed walks are very ubiquitous in our everyday life. If we begin our daily journey from home to our office and then return home, we have completed a closed walk of length 2. Sometimes after working we visit the supermarket and end the day by returning home. On those days we complete a closed walk of length 3. In general, a closed walk is a sequence of (not necessarily different) nodes and links in the network, beginning and ending at the same node.[1] By continuing with this analogy we can represent the places and streets we have visited at least once during a closed walk as nodes and links, respectively. In doing so, we are associating a subgraph with every closed walk. For instance, a closed walk of length 2 is associated with a subgraph formed by two nodes connected by a link, and a closed walk of length 3 is associated with a triangle. A closed walk of length 4 can visit two, three or four nodes. In Figure 4.4 the closed walks of lengths 2 to 4 are represented with their associated subgraphs.

The equivalence between closed walks and subgraphs means, in practical terms, that by counting the first ones we are able to count the second ones. Closed walks are related to the spectral moments of the adjacency matrix of the network, as we have seen in Chapter 2. Then we can relate spectral moments and the number of subgraphs of different kinds in a network. Expressions of the first spectral moments in term of subgraphs for simple networks without self-loops are:

$$\mu_0 = n \tag{4.5}$$

$$\mu_1 = 0 \tag{4.6}$$

$$\mu_2 = 2|K_2| \tag{4.7}$$

$$\mu_3 = 6|K_3| \tag{4.8}$$

$$\mu_4 = 2|K_2| + 4|P_2| + 8|C_4| \tag{4.9}$$

$$\mu_k = \sum_{i=1}^{s} c_i |S_i| \tag{4.10}$$

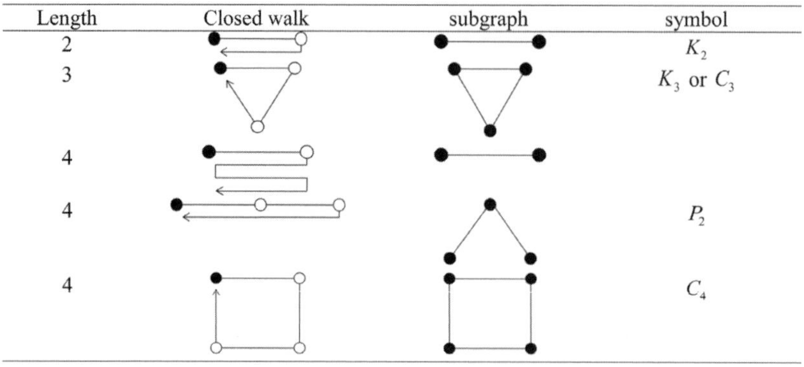

Length	Closed walk	subgraph	symbol
2			K_2
3			K_3 or C_3
4			
4			P_2
4			C_4

Fig. 4.4
Closed walks and subgraphs. Representation of closed walks of lengths 2 to 4, and the subgraphs associated with them.

[1] The nodes $v_1, v_2, \ldots v_l, v_{l+1}$ should be selected such as there is a link from v_i to v_{i+1} for each $i = 1, 2, \ldots, l$."

where $|S_i|$ represents the number of subgraphs of the corresponding type S_i—complete graphs K_n, paths P_n, and cycles C_n of size n, and so on. Here, C_i is an integer number representing the number of the corresponding fragment S_i. The first use of spectral moments and its 'decomposition' in terms of subgraphs dates back to the work of Jiang et al. (1984), in which an application in quantum molecular science was advanced.

4.4 Counting small undirected subgraphs analytically

Most of the methods used today for the analysis of network subgraphs are based on searching algorithms. However, it is possible to calculate analytically the number of certain small subgraphs present in a network (Alon et al., 1997). These calculations are based on the idea of spectral moments analysed in the previous section. In this section we present formulae for calculating the number of 17 small subgraphs in networks based on spectral moments of the adjacency matrix. We use the term $|S_i|$ to designate the number of subgraphs of type S_i, where the corresponding subgraph is illustrated in Figure 4.5. It is also worth noting here that some of these subgraphs correspond to known classes of graph, which are designated in other parts of this book with a different notation. For instance, the following subgraphs are cycles: $S_1 \equiv C_3$, $S_4 \equiv C_4$, $S_7 \equiv C_5$, $S_{14} \equiv C_6$; subgraph S_2 and S_3 are the path of length 3 P_3 and the star of 4 nodes $S_{1,4}$, respectively; the following subgraphs are tadpole graphs: $S_5 \equiv T_{3,1}$, $S_9 \equiv T_{4,1}$, $S_{1,5} \equiv T_{5,1}$, and $S \equiv T_{3,2}$. Here we also use the notation $M^{(D)}$ for a diagonal matrix with diagonal entries M_{ii}. The numbers of these 17 subgraphs are represented by the following expressions:

$$|S_1| = \frac{1}{6}\mu_3 \tag{4.11}$$

$$|S_2| = \frac{1}{2}\langle \mathbf{k'}|\mathbf{A}|\mathbf{k'}\rangle - 3|S_1| \tag{4.12}$$

where $\mathbf{k'} = \mathbf{k} - 1$,

$$|S_3| = \frac{1}{6}\langle \mathbf{k} \circ \mathbf{k'}|\mathbf{k''}\rangle \tag{4.13}$$

where $\circ$ denotes the Hadamard product, and $\mathbf{k''} = \mathbf{k} - 2$,

$$|S_4| = \frac{1}{8}[\mu_4 - 4|P_2| - 2|P_1|] \tag{4.14}$$

$$|S_5| = \frac{1}{2}\langle \mathbf{t}|\mathbf{k''}\rangle \tag{4.15}$$

where $\mathbf{t} = diag(\mathbf{A}^3)$,

$$|S_6| = \frac{1}{4}\langle \mathbf{u}^T|\mathbf{Q} \circ (\mathbf{Q} - \mathbf{A})|\mathbf{1}\rangle \tag{4.16}$$

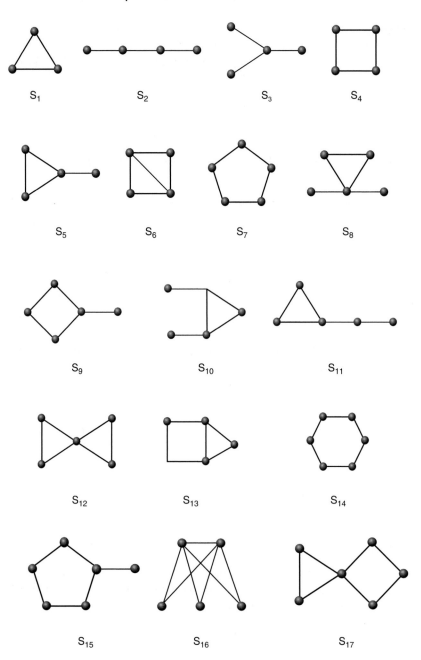

Fig. 4.5
Small subgraphs. Representation of subgraphs given by equations (4.11)–(4.27).

where $\mathbf{Q} = \mathbf{A}^2 \circ \mathbf{A}$,

$$|S_7| = \frac{1}{10}[\mu_5 - 10|S_5| - 30|S_1|] \qquad (4.17)$$

$$|S_8| = \frac{1}{4}\langle \mathbf{k}'' \circ (\mathbf{k}'' - 1)|\mathbf{t}\rangle \qquad (4.18)$$

$$|S_9| = \langle \mathbf{1}|\mathbf{P} - \mathbf{P}^{(D)}|\mathbf{k}''\rangle \qquad (4.19)$$

where $\mathbf{P} = \frac{1}{2}[\mathbf{A}^2 \circ (\mathbf{A}^2 - 1)]$,

$$|S_{10}| = \frac{1}{2}\langle \mathbf{k}''|\mathbf{Q}|\mathbf{k}''\rangle - 2|S_6| \qquad (4.20)$$

$$|S_{11}| = \frac{1}{2}\langle \mathbf{1}|\mathbf{A}^2 - \mathbf{A}^{2(D)}|\mathbf{t}\rangle \qquad (4.21)$$

$$|S_{12}| = \frac{1}{4}\left\langle \mathbf{t}\left|\frac{1}{2}\mathbf{t} - 1\right\rangle\right. - 2|S_6| \qquad (4.22)$$

$$|S_{13}| = \frac{1}{2}\langle \mathbf{1}|\mathbf{Q} \circ (\mathbf{A}^3 \circ \mathbf{A})|\mathbf{1}\rangle - 9|S_1| - 2|S_5| - 4|S_6| \qquad (4.23)$$

$$|S_{14}| = \frac{1}{12}(\mu_6 - 2|P_1| - 12|P_2| - 24|S_1| - 6|S_2| - 12|S_3| - 48|S_4| - 12|S_6| - 12|S_9| - 24|S_{12}|)$$
$$(4.24)$$

$$|S_{15}| = \langle \mathbf{k}''|\mathbf{b}\rangle - 2|S_{13}| \qquad (4.25)$$

where $\mathbf{b} = \frac{1}{2}[\mathbf{q} - 5\mathbf{t} - 2\mathbf{t} \circ \mathbf{k}'' - 2\mathbf{Q}|\mathbf{k}''\rangle - \mathbf{A}|\mathbf{t}\rangle + 2\mathbf{Q}|\mathbf{1}\rangle]$ and $\mathbf{q} = diag(\mathbf{A}^5)$,

$$|S_{16}| = \frac{1}{2}\langle \mathbf{1}|\mathbf{R} \circ \mathbf{A}|\mathbf{1}\rangle \qquad (4.26)$$

where $\mathbf{R} = \frac{1}{6}\mathbf{A}^2 \circ (\mathbf{A}^2 - 1) \circ (\mathbf{A}^2 - 2)$, and

$$|S_{17}| = \frac{1}{2}\langle \mathbf{1}|\mathbf{P} - \mathbf{P}^{(D)}|\mathbf{t}\rangle - 6|S_6| - 2|S_{13}| - 6|S_{16}| \qquad (4.27)$$

In Figure 4.6 we illustrate the relative abundance of the subgraphs shown in Figure 4.5 for six complex networks representing different systems in the real-world. The relative abundance is calculated as

$$\alpha_i = \frac{N_i^{real} - \langle N_i^{random}\rangle}{N_i^{real} + \langle N_i^{random}\rangle} \qquad (4.28)$$

where the average is taken over random networks having the same degrees for the nodes as the real ones. The algorithm used here is the same as that used by the program MFinder, developed for identifying network motifs (Kashtan et al., 2002). Fragments 6, 12, 13, and 16 are found in these networks to be highly abundant in comparison with random networks—except for the Internet, where most of the fragments, except fragments 6 and 16, show relatively low and even negative abundance.

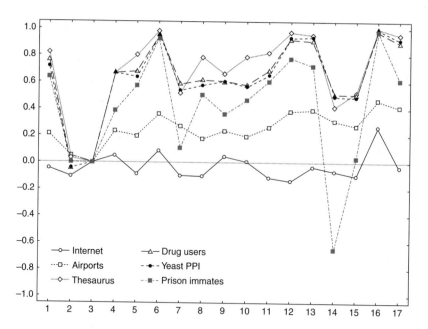

Fig. 4.6

Small subgraphs in real networks.
Relative abundance of small subgraphs
in six different real-world networks.

4.5 Fragment ratios

4.5.1 Clustering coefficients

In their seminal paper on 'small-world' networks, Watts and Strogatz (1998) introduced a way of quantifying the extent of clustering of a node. For a given node i, the clustering coefficient is the number of triangles connected to this node $|C_3(i)|$ divided by the number of triples centred on it:

$$C_i = \frac{2|C_3(i)|}{k_i(k_i - 1)} \tag{4.29}$$

where k_i is the degree of the node. The average value of the clustering for all nodes in a network $\overline{C}$ has been extensively used in the analysis of complex networks:

$$\overline{C} = \frac{1}{n}\sum_{i=1}^{n} C_i \tag{4.30}$$

Bollobás (2003) has remarked that 'this kind of "average of an average" is often not very informative.' A second clustering coefficient was introduced by Newman (2001) as a global characterisation of network cliquishness. This index takes into account that each triangle consists of three different connected triples—one with each of the nodes at the centre. This index—also known as *network transitivity*—is defined as the ratio of three times the number of triangles divided by the number of connected triples (2-paths):[2]

$$C = \frac{3|C_3|}{|P_2|} \tag{4.31}$$

---

[2] Note that C is the same appearing in the assortativity coefficient, Eq. (2.20).

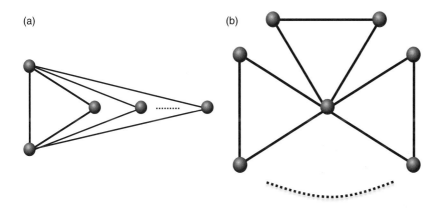

Fig. 4.7
Graphs and clustering. Two networks for which $\overline{C}$ and C diverge as the number of nodes increases.

A network with large average clustering $\overline{C}$ is not necessarily a highly clustered network, according to C. A couple of examples are provided by the networks illustrated in Figure 4.7. The first of them (A) was proposed by Bollobás (2003), and the second (B) is provided here for the first time. The graphs having the structure represented in Figure 4.7 (B) are known as 'friendship graphs' or 'Dutch-mill graphs.' The dotted lines indicate that the number of triangles in such networks can grow to infinity.

In general, we can consider that the number of triangles in these networks is designated by t. The number of nodes in these networks are $n(A) = t + 2$ and $n(B) = 2t + 1$. The average and global clustering coefficients are then represented by the following expressions:

$$\overline{C}(A) = \frac{4}{n(n-1)} + \frac{(n-2)}{n}, \quad \text{and} \quad \overline{C}(B) = \frac{1}{n(n-2)} + \frac{(n-1)}{n} \quad (4.32)$$

and

$$C(A) = C(B) = \frac{3}{n} \quad (4.33)$$

These expressions imply that as the number of triangles increases, the average clustering coefficient tends to 1 but the global clustering tends to zero. That is, $\overline{C} \to 1$ and $C \to 0$ as $n \to \infty$. This indicates that a network can be highly clustered at a local level but not on a global scale, and that taking the average of a local clustering does not produce a good representation of the global characteristics of the network. An important consequence of these differences in the two definitions of clustering is that the corresponding indices are not correlated to each other as expected. In Figure 4.8 we illustrate this lack of correlation between both indices for twenty real-world networks according to the data published by Newman (2003b). For practical applications, both indices are usually compared with the respective coefficients obtained for random graphs. Both indices tend to be significantly higher for real-world networks than for random networks of the same sizes and densities.

The extension of the concept of clustering as the ratio of a given subgraph to its maximum possible number has been carried out by Caldarelli et al. (2004),

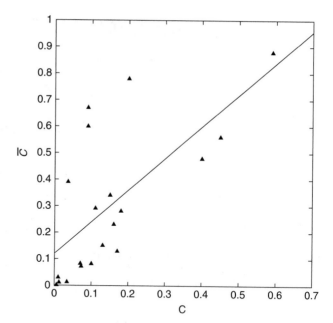

Fig. 4.8

Clustering coefficients. Correlation between the clustering coefficients defined by Watts and Strogatz (1998) and by Newman et al. (2001), $\overline{C}$ and C, respectively.

who considered the ratio of the number of different cycles in a network to its maximum possible number. In this case, the index (4.29) is just the first member of a series of analogous indices for cycles. In a similar way, Fronczak et al. (2002) have analysed the probability that there is a distance of length x between two neighbours of a node i. When the considered distance is just $x \equiv 1$, the index is reduced to the Watts–Strogatz clustering coefficient. The elimination of degree correlation biases from this coefficient has also been analysed in the literature (Soffer and Vázquez 2005).

It has been observed empirically in many real-world networks that high-degree nodes tend to display the low local clustering coefficient, C_i. That is, if we plot the average local clustering coefficient of all nodes with degree k, $C(k)$, versus degree, we obtain a clear decay, $C(k) \sim k^{-\beta}$ (Ravasz et al., 2002; Ravasz and Barabási, 2003). Such a kind of relationship is not observed for random networks with Poisson or power-law degree distributions (Ravasz and Barabási, 2003). The presence of this power-law decay of the clustering with the degree has been attributed to the existence of particular 'hierarchical' structure in complex networks. A hierarchical network according to Ravasz et al. (2002) is 'one made of numerous small, highly integrated modules, which are assembled into larger ones.' Such structure is illustrated in Figure 4.9.

As we have already seen for some graphs like those depicted in Figure 4.7, a network can display a very large average clustering coefficient, despite the possibility of the highest-degree nodes having clustering close to zero. Let us, for instance, consider a star-like node i with degree k_i. If this node has clustering coefficient C_i, the number of triangles in which it take place is represented by

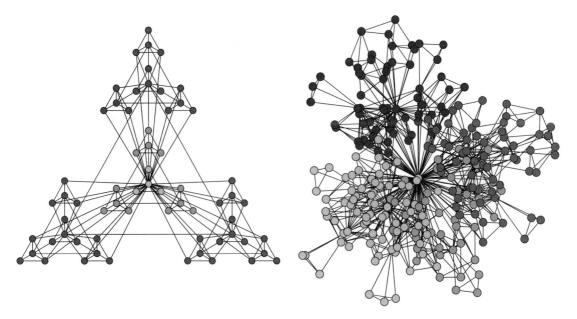

Fig. 4.9
Hierarchical networks. The hierarchical networks which have scale-free topology with embedded modularity. From Ravasz, E., Somera, A. L. Mongru, D. A., Oltvai, Z. N., and Barabási, A.-L. (2002). 'Hierarchical organization of modularity in metabolic networks'. *Science*, **297**, 1551–5. (Reproduced with the permission of AAAS.)

$$\triangle_i = \frac{1}{2} C_i k_i (k_i - 1) \tag{4.34}$$

The most efficient way of increasing the number of triangles in which node i takes place is to create cliques of size l having i as one of its nodes. For instance, in Figure 4.10 we illustrate this situation for a star graph of five nodes (A). In (B) and (C), two graphs are shown in which the central node i has clustering coefficient $C_i = 2/3$. However, in network (C), where the central node takes part in a 4-clique, the number of links is smaller than in (B). Therefore, we consider B to be a more 'efficient' way of increasing the clustering of a node, as it requires the creation of a smaller number of links in the graph.

The number of triangles in an l-clique is represented by

$$\triangle_l = \frac{1}{6} l(l^2 - 1) \tag{4.35}$$

Then, the number of l-cliques in which node i with degree k_i and clustering coefficient C_i can take place is

$$NC_l = \left[\frac{\triangle_i - \left(\sum_{j=1}^{l-3} NC_{l-j} \cdot \triangle_{l-j} \right)}{\triangle_l} \right] \tag{4.36}$$

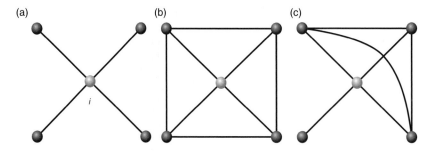

Fig. 4.10
Local clustering. A star graph of five nodes, at left, and two ways of increasing the local clustering of node i. Those at centre and right have $C_i = 2/3$, but in the latter the number of links added to the one at left is smaller than that in the centre.

For instance, let us consider a node i with degree $k_i = 8$. If $C_i = 0.357$, the number of triangles in which it takes place is equal to $\Delta_i = 10$. This node can be part of an l-clique where $3 \le l \le 8 = k_i$. Then, applying (4.36), we have

$$NC_8 = \left\lfloor \frac{10}{56} \right\rfloor = 0 \tag{4.37}$$

$$NC_7 = \left\lfloor \frac{10 - 0 \cdot 56}{35} \right\rfloor = 0 \tag{4.38}$$

$$NC_6 = \left\lfloor \frac{10 - 0 \cdot 56 - 0 \cdot 35}{20} \right\rfloor = 0 \tag{4.39}$$

$$NC_5 = \left\lfloor \frac{10 - 0 \cdot 56 - 0 \cdot 35 - 0 \cdot 20}{10} \right\rfloor = 1 \tag{4.40}$$

$$NC_4 = \left\lfloor \frac{10 - 0 \cdot 56 - 0 \cdot 35 - 0 \cdot 20 - 1 \cdot 10}{4} \right\rfloor = 0 \tag{4.41}$$

$$NC_3 = \left\lfloor \frac{10 - 0 \cdot 56 - 0 \cdot 35 - 0 \cdot 20 - 1 \cdot 10 - 0 \cdot 4}{1} \right\rfloor = 0 \tag{4.42}$$

That is, for a node i with degree $k_i = 8$ in order to have $C_i = 0.357$, it is enough to have a 5-clique. However, for the same node to have $C_i = 0.928$, the same calculations show that it is necessary to have one 6-clique, one 4-clique, and two 3-cliques. If we consider a node with degree $k_i = 20$, it is necessary to embed it into one 11-clique, one 5-clique, and one 3-clique in order to have a clustering coefficient $C_i = 0.926$. In addition, for a node of degree $k_i = 100$ to have $C_i \approx 0.9$, it is necessary to embed it into a clique of 31 nodes. This means that as the degree of a node increases, the sizes of the cliques in which it needs to take place in order to maintain the same clustering, increases dramatically. The creation of such huge cliques is very costly in terms of the number of links that need to be added around a node, which dramatically increases the density of the network. Then, it is expected that high-degree nodes display very low clustering as a consequence of their participation in small cliques only. Consequently, in a network for which the number of cliques of given sizes decays very fast with the size of these cliques, a power-law decay of the clustering with the degree is expected as a consequence of the previous

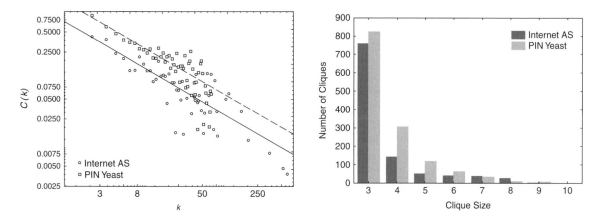

Fig. 4.11
Local clustering and node degree. Plot of the local mean clustering coefficient averaged over all nodes with the same degree (left) for two real-world networks. Histogram of the distribution of number of cliques in terms of clique size (right).

analysis. Of course, modular hierarchical networks are only one example of this kind of structure.

In Figure 4.11 we illustrate the plot of $C(k)$ versus k for two real networks representing the Internet as an autonomous system and the PIN of yeast. In both cases, the power-law correlation obtained is of the form $C(k) \sim k^{-0.69}$. That is, in both cases the nodes with the highest degree display very low clustering, and those with low degree are relatively very clustered. This version of the Internet has 17 nodes with a degree higher than 50, including nodes with degree 590, 524, 355, and so on. The PIN of yeast has 14 nodes with degree higher than 50. Thus, for these nodes to display large clustering it is necessary to have very large cliques in these networks. However, as can be seen in Fig 4.11 (B), none of these networks have cliques with more than 9 nodes, and the number of cliques decreases very quickly with the size of the clique. Thus, the existence of the relations observed in Figure 4.11 (A) is a direct consequence of the absence of large cliques in these networks more than the presence of any hierarchy in the modules of the network.

4.5.2 Network reciprocity

Another index which is based on the ratio of the number of certain subgraphs in a network is the reciprocity, defined for a directed network as

$$ r = \frac{L^{\leftrightarrow}}{L} \tag{4.43} $$

where $L^{\leftrightarrow}$ is the number of symmetric pairs of arcs, and L is the total number of directed links (Wasserman and Faust, 1994; Newman et al., 2002; Serrano and Boguñá, 2003). Thus, if there is a link pointing from A to B, the reciprocity measures the probability that there is also a link pointing from B to A.

The Structure of Complex Networks

Table 4.1 **Network reciprocity**. Some reciprocal and antireciprocal networks.

Reciprocal	ρ	Areciprocal	ρ	Antireciprocal	ρ
Metab. *E. coli*	0.764	Transc. *E. coli*	−0.0003	Skipwith	−0.280
Thesaurus	0.559	Transc. yeast	$-5.5.10^{-4}$	St. Martin	−0.130
ODLIS	0.203	Internet	$-5.6.10^{-4}$	Chesapeake	−0.057
Neurons	0.158			Little Rock	−0.025

A modification of this index was introduced by Garlaschelli and Lofredo 2004a), who considered the following normalised index:

$$\rho = \frac{r - \bar{a}}{1 - \bar{a}} \tag{4.44}$$

where $\bar{a} = L/n(n - 1)$. In this case, the number of self-loops are not considered for calculating L, such as $L = \sum_{i=1}^{n} \sum_{j=1}^{n} A_{ij} - \sum_{i=1}^{n} A_{ii}$, where A_{ij} is the corresponding entry of the adjacency matrix **A** of the directed network. It is also straightforward to realise that $L^{\leftrightarrow} = tr(\mathbf{A}^2)$. According to this index, a directed network can be reciprocal ($\rho > 0$), areciprocal ($\rho = 0$), or antireciprocal ($\rho < 0$). Some examples of network reciprocity are provided in Table 4.1.

Accounting for all parts (subgraphs)

5

> ...the different parts of each being must be co-ordinated in such a way as to render the existence of the being as a whole, not only in itself, but also in its relations with other beings, and the analysis of these conditions often leads to general laws which are as certain as those which are derived from calculation or from experiment.
>
> Georges Cuvier

We have seen in previous sections that small subgraphs form the basis of important structural characteristics of networks, such as motifs, fingerprints and clustering coefficients. The equivalence between spectral moments and subgraphs allows the definition of new topological invariants to describe the structure of a network in terms of subgraphs. For instance, let us consider a simple function based on the first spectral moments:

$$f(G, c) = \sum_{k=0}^{4} c_k M_k \qquad (5.1)$$

This invariant represents a weighted sum of the numbers of the smallest subgraphs in the network:

$$f(G, c) = c_0 n + 2(c_2 + c_4)|K_2| + 4|P_2| + 6c_3|C_3| + 8|C_4| \qquad (5.2)$$

It is straightforward to generalize this kind of invariant to account for the presence of any subgraph in the network. This kind of invariant is based on the infinite sum of all spectral moments of the adjacency matrix:

$$f(G, c) = \sum_{k=0}^{\infty} c_k M_k \qquad (5.3)$$

The key element in the definition of invariants based on the expression (5.3) is the selection of the coefficients c_k. It is known that the infinite sum of spectral moments in the network diverges: $\sum_{k=0}^{\infty} M_k = \infty$. Consequently, we need to select the coefficients c_k in such a way that the series (5.3) converges. As we require small subgraphs to make larger contributions than big ones, the coefficients c_k can be selected in such a way that the closed walks of small length are assigned more weight than the longer ones. In the following section we will introduce an index that fulfils these two conditions.

5.1 The Estrada index

Let us consider that we divide the number of closed walks of length k by the factorial of the length. In other words, we select $c_k = 1/k!$. Then, we sum all these weighted numbers of closed walks to obtain an index that accounts for all subgraphs in the network:

$$EE(G) = M_0 + M_1 + \frac{M_2}{2!} + \frac{M_3}{3!} + \cdots + \frac{M_k}{k!} + \cdots$$

Obviously, the index $EE(G)$ fulfils the condition that we assign more weight to small subgraphs than to large ones. The other condition—that the series converges—is guaranteed by considering the infinite sum:

$$EE(G) = \sum_{k=0}^{\infty} \frac{M_k}{k!} \tag{5.4}$$

By expressing the spectral moments as the traces of the different powers of the adjacency matrix of the network, and then using the property that the sum of the traces is the trace of the sum, we can see that this series converges to the trace of the exponential of the adjacency matrix:

$$EE(G) = tr(\mathbf{I}) + tr(\mathbf{A}) + \frac{tr(\mathbf{A}^2)}{2!} + \frac{tr(\mathbf{A}^3)}{3!} + \cdots + \frac{tr(\mathbf{A}^k)}{k!} + \cdots$$

$$= tr\left(\mathbf{I} + \mathbf{A} + \frac{\mathbf{A}^2}{2!} + \frac{\mathbf{A}^3}{3!} + \cdots + \frac{\mathbf{A}^k}{k!} + \cdots\right)$$

$$= tr(e^A) \tag{5.5}$$

The exponential of the adjacency matrix is a matrix function for which well-elaborated theory and techniques of calculations exist (Leonard, 1996; Moler and van Loan, 2003; Harris, Jr., et al., 2001). This index was first proposed by Estrada (2000) as a way of characterizing the degree of folding of proteins. A few years later, Estrada and Rodríguez-Velázquez (2005a) introduced this index for studying complex networks. The index was then renamed by de la Peña et al. (2007) as the 'Estrada index' of a graph, and many of its mathematical properties, as well as generalizations to other graph matrices, have appeared in the literature under this term (Bamdad et al., 2010; Benzi and Boito, 2010; Deng et al., 2009; Güngör et al., 2010).

This index can be expressed in terms of the eigenvalues of the adjacency matrix as follows:

$$EE(G) = \sum_{k=0}^{\infty} \sum_{j=1}^{n} \lambda_j^k = \sum_{j=1}^{n} e^{\lambda_j} \tag{5.6}$$

The Estrada index of a network G of size n is bounded as

$$n < EE(G) < e^{n-1} + \frac{n-1}{e} \tag{5.7}$$

where the lower bound is obtained for the graph having n nodes and no links, and the upper bound is attained for the complete graph K_n. For trees it has been proved (Deng et al., 2009) that

$$EE(S_n) > EE(T_n) > EE(P_n) \tag{5.8}$$

The Estrada indices of the star S_n and the path P_n are given by (Gutman and Graovac, 2007):

$$EE(S_n) = n - 2 + 2\cosh(\sqrt{n-1}) \tag{5.9}$$

$$EE(P_n) = \sum_{r=1}^{n} e^{2\cos(2r\pi/(n+1))} \tag{5.10}$$

The last one can be approximated as

$$EE(P_n) \approx (n+1)I_0 - \cosh(2) \tag{5.11}$$

where $I_0 = \sum_{k=0}^{\infty} \frac{1}{(k!)^2} = 2.27958530\ldots$ In the case of a cycle of n nodes the Estrada index can be calculated exactly by using the following expression:

$$EE(C_n) = \sum_{r=1}^{n} e^{2\cos(2r\pi/n)} \tag{5.12}$$

which can be approximated to

$$EE(C_n) \approx nI_0 \tag{5.13}$$

A characteristic feature of the Estrada index is the relationship between the mean value of the index and the standard deviation of the contribution of each node to it. That is, if we designate by $EE(i)$ the contribution of node i to $EE(G)$, in such a way that $EE(G) = \sum_{i=1}^{n} EE(i)$, we can define the mean and standard deviation of $EE(G)$ as follows:

$$EE_{mean} = \frac{1}{n} \sum_{i=1}^{n} EE(i) \tag{5.14}$$

$$EE_{std} = \sqrt{\frac{1}{n} \sum_{i=1}^{n} [EE(i) - EE_{mean}]^2} \tag{5.15}$$

The properties of the index $EE(i)$, which is known as the subgraph centrality, are studied in Chapter 7. A plot of the mean versus the standard deviation of the Estrada index for 11,117 networks having eight nodes is illustrated in Figure 5.1. The most important characteristic of this plot is that in many cases different networks display relatively similar values of the mean Estrada index, despite the fact that they have large variability in their local indices. This is especially important when we study networks of the same size, in which case it is recommended that there should be an analysis of not only the values of the $EE(G)$, but also its local variation.

As an example, we illustrate in Figure 5.2 the five networks marked in the previous plot as (a)–(e). While the mean Estrada index for these networks

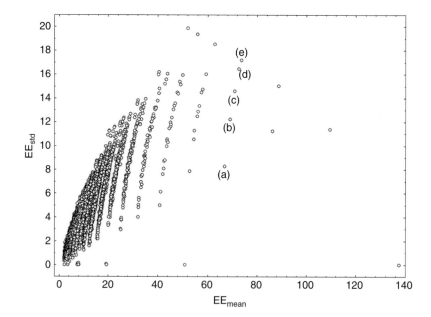

Fig. 5.1

Mean versus standard deviation of the Estrada index. Plot of the mean and standard deviation of the Estrada index of the 11,117 connected networks with eight nodes. Networks corresponding to points (a)–(e) are illustrated in Figure 7.2.

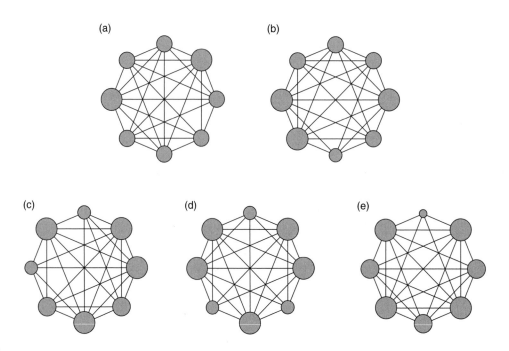

Fig. 5.2

Contributions to the Estrada index. Illustration of the networks corresponding to points (a)–(e) illustrated in Figure 5.2, where nodes are drawn with radii proportional to their contribution to the Estrada index.

changes very little—from 66.9 to 72.7—their standard deviations changes from 8.28 for network (a) to more than double for network (e). This indicates that there is a large variation in the local contribution to the Estrada index, despite the average values of the index being very close to each other for these networks.

The contributions of odd and even closed walks can also be separated in (5.6), giving rise to expressions for the Estrada index in terms of hyperbolic (matrix) functions (Estrada and Rodríguez-Velázquez, 2005b):

$$EE(G) = tr\left(\mathbf{A} + \frac{\mathbf{A}^3}{3!} + \frac{\mathbf{A}^5}{5!} + \cdots + \frac{\mathbf{A}^{2k+1}}{(2k+1)!} + \cdots\right) + tr\left(\mathbf{I} + \frac{\mathbf{A}^2}{2!} + \frac{\mathbf{A}^4}{4!} + \cdots + \frac{\mathbf{A}^{2k}}{(2k)!} + \cdots\right)$$

$$= tr[\sinh(\mathbf{A})] + tr[\cosh(\mathbf{A})] = \sum_{j=1}^{n}\sinh(\lambda_j) + \sum_{j=1}^{n}\cosh(\lambda_j) \tag{5.16}$$

$$= EE_{odd}(G) + EE_{even}(G),$$

Then, $EE_{odd}(G)$ and $EE_{even}(G)$ represent weighted sums of all odd and even closed walks in the network, respectively, giving more weight to the small ones. It is easy to see that every closed walk of odd length visits the nodes of at least one cycle of odd length. That is, the subgraph associated with an odd closed walk contains at least one odd cycle. However, closed walks of even length can be trivial in the sense that the subgraphs associated with them do not necessarily contain cycles.

In Table 5.1 we show the values of $EE_{odd}(G)$ and $EE_{even}(G)$ for several real-world networks, as well as for randomised networks having the same degree sequence.

The first three networks—a social network of drug users, yeast PPI, and Roget's Thesaurus, display significantly larger values of both indices than do their randomised analogues. This indicates that the processes giving rise to the structures of these networks generate larger numbers of small subgraphs than expected by a random evolution. However, this is not always the case, as can be seen for the version of the Internet displayed in the table. In this case there are no significant differences in the values of $EE_{odd}(G)$ and $EE_{even}(G)$ of the real network and its random version. This situation is similar to what happens when comparing the clustering coefficient of the Internet and its randomised

Table 5.1 Odd and even Estrada index of real-world networks and their random analogues. The values of the indices for random networks are the average of 100 realisations where the number of nodes, links, and degree sequences are the same as for the real network.

Network	$EE_{odd}(G)$			$EE_{even}(G)$		
	real	random	ratio	real	random	ratio
Drugs	$3.456.10^7$	$3.750.10^6$	9.21	$3.456.10^7$	$3.776.10^6$	9.15
PIN-Yeast	$9.695.10^7$	$1.538.10^7$	6.30	$9.714.10^7$	$1.272.10^7$	7.64
Roget	$1.121.10^5$	$2.529.10^4$	4.43	$1.258.10^5$	$4.593.10^4$	2.74
Internet	$3.074.10^{13}$	$4.256.10^{13}$	0.72	$3.100.10^{13}$	$4.265.10^{13}$	0.73
Trans_yeast	$1.454.10^3$	$1.492.10^5$	0.01	$3.444.10^4$	$1.613.10^5$	0.21

versions, where no significant differences are observed. However, while the clustering coefficient accounts only for triangles, Estrada indices account for all subgraphs in a global way. The analysis of network motifs carried out by Milo et al. (2002) can provide some hint about what is happening. This analysis shows that several small subgraphs are more abundant in random versions of the Internet than in the real network. Only one subgraph was found to be more abundant in the real network than in its randomised versions. Consequently, when the sum of contributions from all such subgraphs is accounted for by $EE_{odd}(G)$ and $EE_{even}(G)$, no significant difference is observed between real and random versions of the Internet. A more interesting case is the one obtained for the transcription network of yeast. In this section we are considering an undirected version of this network only. In this case, the contribution of subgraphs containing odd cycles is about 100 times larger in the random networks than in the real one. This 'anomaly' can be explained by comparing the eigenvalues of the real network and those of random graphs generated with the same degree sequence. Random graphs tend to separate the first and second eigenvalue of the adjacency matrix, which is known as 'spectral gap'. This effect is studied in more detail in Chapter 9, and for now we will say only that there are real-world networks in which this spectral gap is very small. One example is the transcription network of yeast, where the first and second largest eigenvalues are 9.976 and 8.452, respectively. This network also displays another characteristic, mentioned in Chapter 2. That is, for every positive eigenvalue there is a negative eigenvalue which has almost exactly the same modular value: $\lambda_j \approx |-\lambda_{n-j+1}|$. For instance, $\lambda_1 = 9.9761$ and $\lambda_n = 9.9648$; $\lambda_2 = 9.4518$ and $\lambda_{n-1} = 9.4301$; $\lambda_3 = 7.8083$ and $\lambda_{n-2} = 7.4113$; and so forth. We recall from Chapter 2 that the relationship $\lambda_j = |-\lambda_{n-j+1}|$ is a characteristic of bipartite networks. Then we will say that

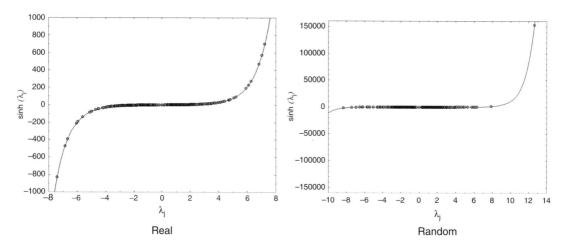

Real   Random

Fig. 5.3

Hyperbolic sine function of the eigenvalues of the transcription network of yeast. The values shown for random networks are the average of eigenvalues of 100 random realisations of the same network. The number of nodes, links, and degree sequences for random graphs are the same as for the real network.

when $\lambda_j \approx |-\lambda_{n-j+1}|$ we are in the presence of *almost-bipartite networks*. Because this concept and methods for quantifying the degree of bipartivity of a network are studied in Chapter 11, we will accept here that the transcription network of yeast is an almost-bipartite network (see Chapter 11 for a quantitative evaluation of this concept for this network). It is easy to realise that in bipartite networks the plot of λ_j versus $\sinh \lambda_j$ is a symmetric function as observed in Figure 5.3 for the transcription network of yeast. However, when the same plot is obtained for the eigenvalues of the random analogues of this network it is seen that a completely non-symmetric plot is observed. This is a consequence of the fact that random graphs do not necessarily reproduce the bipartite structure of some real networks, considerably increasing the spectral gap and producing a non-symmetric spectrum for the adjacency matrix. The number of odd cycles in these random networks is then much larger than in the real almost-bipartite networks. We recall that a bipartite network does not contain any odd cycle. This is the main difference observed in the values of $EE_{odd}(G)$ for the transcription network of yeast and its random analogues.

5.2 Parameterised Estrada index

Let us consider that every link of a network is weighted by a parameter β. Evidently, the case $\beta = 1$ corresponds to the Estrada index analysed in the previous sections. This kind of homogeneous weight for the links of a network can arise in situations where the network is submitted to a certain 'external' factor which equally affects all links in the network. In the next section we will discuss a plausible interpretation for this factor.

Let $\mathbf{W}$ be the adjacency matrix of this homogeneously weighted network. It is obvious that $\mathbf{W} = \beta \mathbf{A}$ and the spectral moments of the adjacency matrix are $M_r(\mathbf{W}) = \operatorname{Tr} \mathbf{W}^r = \beta^r \operatorname{Tr} \mathbf{A}^r = \beta^r M_r$. In this case, the Estrada index of the network can be generalised as follows (Estrada and Hatano, 2007):

$$EE(G, \beta) = \sum_{r=0}^{\infty} \frac{\beta^r M_r}{r!} = \sum_{j=1}^{N} e^{\beta \lambda_j} \qquad (5.17)$$

which can also be written as

$$EE(G, \beta) = \operatorname{Tr} \sum_{r=0}^{\infty} \frac{\beta^r \mathbf{A}^r}{r!} = \operatorname{Tr} e^{\beta \mathbf{A}} \qquad (5.18)$$

The index $EE(G, \beta)$ then accounts for the influence of the previously mentioned external factor on the subgraph structure of the network. In the next section this index is interpreted as the partition function of the network, and some statistical–mechanical parameters are introduced for a complex network.

5.2.1 β as the inverse temperature

Complex networks are continuously exposed to external 'stresses', which are independent of the organisational architecture of the network. By 'external' we mean an effect which is independent of the topology of the network. For instance, let us consider the network in which nodes represent corporations and the links represent their business relationships. In this case the external stress can represent the economical situation of the world at the moment in which the network is analysed. In 'normal' economical situations we are in the presence of a low level of external stress. In situations of economic crisis, the level of external stress is elevated. Despite the fact that these stresses are independent of the network topology, they can have a determinant role in the evolution of the structure of these systems. For instance, in a situation of high external stress such as an economic crisis, it is possible that some of the existing links between corporations are lost at the same time as new merging and strategic alliances are created. The situation is also similar for the social ties between actors in a society, which very much depend on the level of 'social agitation' existing in such a society in the specific period of time under study.

An analogy exists between the previously described situation and what happens at the molecular level of organisation in a given substance. In this case, such organisation has a strong dependence on the external temperature. For instance, the intermolecular interactions between molecules differ significantly in the gas, liquid, and solid states of matter. The agents (molecules) forming these systems are exactly the same, but their organization changes dramatically with changes of the external temperature. We will use this analogy to capture the external stresses influencing the organisation of a complex network.

Let us consider that the complex network is submerged in a thermal bath at temperature T. The thermal bath here represents the external situation which affects all the links in the network at the same time. It fulfils the requirements of being independent of the topology of the network, and of having a direct influence over it. Then, after equilibration, all links in the network will be weighted by the parameter $\beta = (k_B T)^{-1}$. The parameter β is known as the *inverse temperature*, and k_B is the Boltzmann constant.

This means that when the temperature tends to infinite, $\beta \to 0$, all links have weights equal to zero. In other words, the graph has no links. This graph is known as the 'trivial' or 'empty' graph. When $\beta = 1$ we are in the presence of the simple graph, in which every pair of connected nodes has a single link. The concept of temperature in networks can be considered as a specific case of a weighted graph in which all links have the same weights. If we consider the case when the temperature tends to zero, $\beta \to \infty$, all links have infinite weights. We then have an analogy with the different states of the matter, in which the empty graph obtained at very high temperatures $\beta \to 0$ is similar to a gas, formed by free particles with no link among them. At the other extreme, when $\beta \to \infty$, we have a graph with infinite weights in their links, which in some way resemblances a solid. The simple graph is therefore an analogy of the liquid state. An illustration of these kinds of network obtained for different temperatures is illustrated in Figure 5.4.

In extremely 'hot' situations ($\beta \to 0$) the Estrada index is simply equal to the number of nodes in the network; that is, $EE \to n$ as $\beta \to 0$. On the other hand, when the temperature is very low ($\beta \to \infty$), the principal eigenvalue dominates the kth spectral moment of the $\mathbf{A}$ matrix for large k:

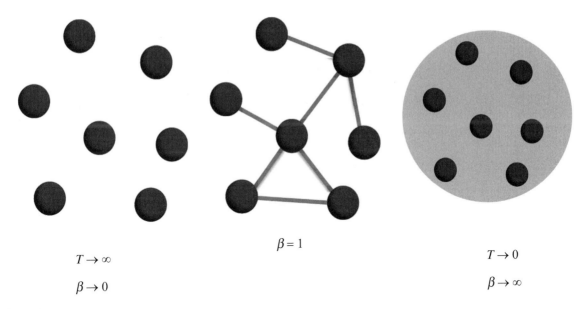

$$\beta = 1$$

$$T \to \infty \qquad\qquad\qquad\qquad\qquad\qquad T \to 0$$

$$\beta \to 0 \qquad\qquad\qquad\qquad\qquad\qquad \beta \to \infty$$

Fig. 5.4
Inverse temperature as link weights. By considering a simple graph (centre) as a link-weighted graph in which all weights are $\beta = 1$, the cases $\beta \to 0$ (right) and $\beta \to \infty$ represent analogues of a 'gas' and a 'solid'.

$$M_k \approx \lambda_1^k = e^{k \ln \lambda_1} \ (k \to \infty) \qquad\qquad (5.19)$$

which means that in the zero temperature limit, $EE \to e^{\beta \lambda_1}$ as $\beta \to \infty$.

5.3 Network entropy and free energy

It is known that the network spectrum makes partitions of the nodes which are determined by the sign pattern of the eigenvectors of the adjacency matrix (see Section 10.1.2 for more details). For instance, all entries of the principal eigenvector are positive, which can be interpreted as all nodes being grouped in one giant cluster formed by the whole network. However, the eigenvector corresponding to the second largest eigenvalue produces a bipartition of the network in which some nodes have positive and other have negative entries for this eigenvector. All these partitions of a network can be considered as 'states' in which the network exists. In Figure 5.5 we show a pictorial representation of these states of a small network. Here, the j th state corresponds to the eigenvector φ_j associated with the eigenvalue λ_j of the adjacency matrix of the network.

There are several possible physical interpretations for these states. For instance, in Chapter 6 we will study vibrations in complex networks, in which the sign patterns represented in these states correspond to different vibrational modes of the nodes. We can also consider the Schrödinger operator of a network when the potential operator is given by the node degrees, $V_i = -k_i$. In this case, the network Hamiltonian is $\mathbf{H} = -\mathbf{A}$, and then the j th configuration

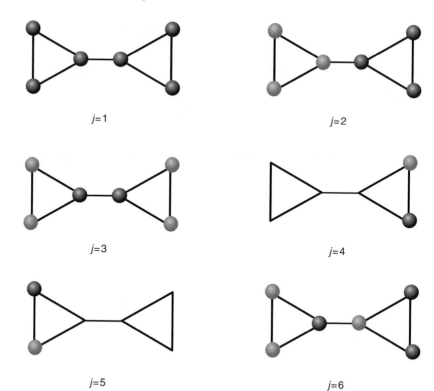

$j=1$

$j=2$

$j=3$

$j=4$

$j=5$

$j=6$

Fig. 5.5

'Energy microstates' of a simple network. Each graph corresponds to a microstate with energy $E_j = -\lambda_j$. Dark (clear) nodes represent those having positive (negative) entries of the corresponding eigenfunction, while those with zero entries are represented as the intersection of links only.

in Figure 5.5 corresponds to a state with energy $E_j = -\lambda_j$. That is, the state representing a unipartition of the network is the most stable one, followed by the one which represents a bipartition, and so forth.

The probability that the system is found in a particular state can be obtained by considering a Maxwell–Boltzmann distribution:

$$p_j = \frac{e^{-\beta E_j}}{\sum_j e^{-\beta E_j}} = \frac{e^{-\beta E_j}}{Z} \tag{5.20}$$

where the normalisation function $Z = \sum_j e^{-\beta E_j}$ is the partition function of the network. A simple substitution of the values of the energy for the different states of the network produces

$$p_j = \frac{e^{\beta \lambda_j}}{\sum_j e^{\beta \lambda_j}} = \frac{e^{\beta \lambda_j}}{EE(G, \beta)} \tag{5.21}$$

where we can easily identify the normalisation factor, which is the Estrada index, with the partition function of the complex network (Estrada and Hatano, 2007),

$$Z \equiv EE(G, \beta) \equiv \mathrm{Tr}\, e^{\beta \mathbf{A}} \tag{5.22}$$

Now we can also define the information theoretic entropy for the network, using the Shannon expression (Estrada and Hatano, 2007):

$$S(G, \beta) = -k_B \sum_j [p_j(\beta \lambda_j - \ln EE)] \tag{5.23}$$

where, for the sake of economy, we wrote $EE(G, \beta) = EE$. This expression can be written in the following equivalent way:

$$S(G, \beta) = -k_B \beta \sum_j \lambda_j p_j + k_B \ln EE \sum_j p_j \tag{5.24}$$

which, by using the standard relation $F = H - TS$, immediately suggests the expressions for the total energy $H(G)$ and Helmholtz free energy $F(G)$ of the network, respectively (Estrada and Hatano, 2007):

$$H(G, \beta) = -\frac{1}{EE} \sum_{j=1}^{n} (\lambda_j e^{\beta \lambda_j}) = -\frac{1}{EE} \mathrm{Tr}(\mathbf{A} e^{\beta \mathbf{A}}) = -\sum_{j=1}^{n} \lambda_j p_j \tag{5.25}$$

$$F(G, \beta) = -\beta^{-1} \ln EE \tag{5.26}$$

We can now analyse some of the characteristics of these statistical–mechanical functions by considering some type of special network. First, we study the complete network K_n, for which the probabilities that the network occupies the first and jth microstates ($j \geq 2$) are, respectively (Estrada and Hatano, 2007),

$$p_1 = \frac{e^{n-1}}{e^{n-1} + \frac{n-1}{e}} \quad \text{and} \quad p_j = \frac{1}{e^n + n - 1} \tag{5.27}$$

The statistical–mechanical parameters of the complete network are then given by the following expressions (Estrada and Hatano, 2007):

$$S(K_n) = -p_1[(n-1) - \ln EE] - (n-1)p_j[(-1) - \ln EE] \tag{5.28}$$

$$H(K_n) = \frac{-(n-1)(e^n - 1)}{e^n + n - 1} \tag{5.29}$$

$$F(K_n) = -\ln(e^n + n - 1) - 1 \tag{5.30}$$

We have seen in the previous section that the Estrada index of the complete network is determined by the exponential of the number of nodes minus 1 when the number of nodes tends to infinite: $EE(K_n) \to e^{n-1}$ as $n \to \infty$. Then, $p_1 \to 1$ and $p_j \to 0$ as $n \to \infty$. Consequently, the Shannon entropy tends to zero for very large complete networks, $S(K_n) \to 0$ as $n \to \infty$. Similarly, $H(K_n) \to -(n-1)$ and $F(K_n) \to -(n-1)$ as $n \to \infty$. By contrast, in the case of the null graph $\overline{K}_n$, we have $EE(\overline{K}_n) = n$ and $p_j = \frac{1}{n}$ for all j, which results in $S(\overline{K}_n) = \ln n$. Because $\lambda_j = 0 \forall j \epsilon V(\overline{K}_n)$, we have $H(\overline{K}_n) = 0$ and $F(\overline{K}_n) = -\ln n$.

Consequently, the thermodynamic functions of networks are bounded as follows:

$$0 \leq S(G, \beta) \leq \beta \ln n \tag{5.31}$$

$$-\beta(n-1) \leq H(G, \beta) \leq 0 \tag{5.32}$$

$$-\beta(n-1) \leq F(G, \beta) \leq -\beta \ln n \tag{5.33}$$

where the lower bounds are obtained for the complete graph as $n \to \infty$, and the upper bounds are reached for the null graph with n nodes (Estrada and Hatano, 2007).

Let us now analyse the thermodynamic functions of networks for extreme values of the temperature. The principal eigenvalue dominates the r th spectral moment of the A matrix for large r. That is, $\mu_r \approx \lambda_1^r = e^{r \ln \lambda_1}$ as $r \to \infty$. Then, in the zero temperature limit we approximate the value of the Estrada index as

$$EE(G) \approx \sum_{r=0}^{\infty} \frac{\beta^r e^{r \ln \lambda_1}}{r!} = e^{\beta \lambda_1} \qquad (5.34)$$

for large β. This expression indicates that in the zero temperature limit the system is 'frozen' at the ground state configuration which has the interaction energy $-\lambda_1$. In other words, $s(G, \beta \to \infty) = 0$, because the system is completely localised at the ground state with $p_1 \cong 1$. The total energy and Helmholtz free energy are then reduced to the interaction energy of the network (Estrada and Hatano, 2007):

$$H(G, \beta \to \infty) = F(G, \beta \to \infty) = -\lambda_1 \qquad (5.35)$$

At very high temperatures, $\beta \to 0$, the entropy of the system is completely determined by the Estrada index of the network, $s(G, \beta \to 0) = k_B \ln EE$ and the free energy tends to minus infinite, $F(G, \beta \to 0) \to -\infty$.

The study of real networks provides some important insights. In Table 5.2 we show the values of the statistical–mechanical functions for several real networks representing a variety of complex systems. We also show the average values of these parameters for random networks with exactly the same degrees for the nodes as the real ones. The averages are more than 100 random realisations. In general, real networks display more negative values of the total and free energies than do their random analogues. That is, real networks are more 'stable' than random networks with the same node degree sequences. The only exception concerning the networks illustrated here is the Internet, for which no significant variation is observed in comparison with random analogues. We saw in Section 5.5 that the Internet displays a small relative abundance of fragments related to random networks with the same node degree sequences. This is a plausible explanation for the small relative variation of the statistical–mechanical parameters observed for this network. That is, the partition function of a network defined by the Estrada index can be expressed in terms of subgraphs, using the following general formula:

$$Z = EE = \sum_{k=1}^{s} \eta_k |S_k| \qquad (5.36)$$

where $|S_k|$ denotes the number of fragments of type k, and the coefficients η_k can be obtained as functions of the fragments' contributions to the spectral moments discussed in Section 5.4. Then, a small number of subgraphs relative to random networks—in particular, small subgraphs—shows that the Internet has the characteristic statistical–mechanical properties observed in Table 5.2.

Table 5.2 **Statistical–mechanical parameters for networks.** Partition function (Estrada index), entropy, total energy, and free energy of real-world networks and their random analogues. Random networks are generated with the same number of nodes and links and degree distribution as the real networks, and properties are taken as averages over 100 random realisations.

	Z	S	H	F
Airports	$8.081.10^{17}$	$1.016.10^{-9}$	-41.233	-41.233
Random	$1.603.10^{16}$	$4.800.10^{-11}$	-37.298	-37.298
Variation (%)	98.0	99.8	9.5	9.5
Thesaurus	$2.379.10^{5}$	1.599	-10.781	-12.380
Random	$7.123.10^{4}$	1.950	-9.221	-11.172
Variation (%)	70.1	-18.0	14.5	9.8
Drug users	$6.912.10^{7}$	0.221	-17.830	-18.051
Random	$7.526.10^{6}$	0.026	-15.775	-15.802
Variation (%)	89.1	88.2	11.5	12.4
Yeast PPI	$1.941.10^{8}$	0.220	-18.863	-19.084
Random	$2.804.10^{7}$	0.043	-17.104	-17.148
Variation (%)	85.6	80.4	9.3	10.1
Prison inmates	707.637	2.311	-4.251	-6.562
Random	515.261	2.440	-3.804	-6.244
Variation (%)	27.2	-5.3	10.5	4.8
Internet	$6.174.10^{13}$	$1.150.10^{-4}$	-31.754	-31.754
Random	$8.521.10^{13}$	$1.245.10^{-4}$	-32.064	-32.064
Variation (%)	-27.5	-7.6	-0.97	-0.97

5.3.1 Estrada index of regular networks

An interesting observation was carried out by Ejov et al. (2007) concerning the Estrada index of regular networks. These authors plotted the mean $EE_{mean}(G, \beta)$ versus the variance $\sigma(EE)$ of the exponential of the eigenvalues of regular graphs. When $EE_{mean}(G, \beta)$ is plotted versus $\sigma(EE)$ for the regular graphs having the same number of nodes, the plot resembles a fractal of thread-like appearance. The points cluster around a line segment that has been called a *filar*.

In Figure 5.6 we illustrate the filar structure of the mean-variance plot for the cubic graphs (3-regular graphs) having ten nodes. In this figure, every filar is represented by a different set of points—for example, circles, squares, triangles, and so on. In this case, every filar is differentiated by the number of triangles in the respective graphs. That is, graphs in the first filar (circles) have no triangle, those in the second (squares) have one triangle, those in the third filar (triangles) have two triangles, and those in the fourth and fifth filars have three and four triangles, respectively. Then, the position of a graph in its corresponding filar appears to be determined by the number of other cycles. For instance, in the first filar the number of squares in every graph is 0, 2, 3, 5, 5, 6, and the two graphs with the same number of squares are differentiated by their number of pentagons. Accordingly, the membership of a graph in a filar is related to the number of cycles of various lengths that they have.

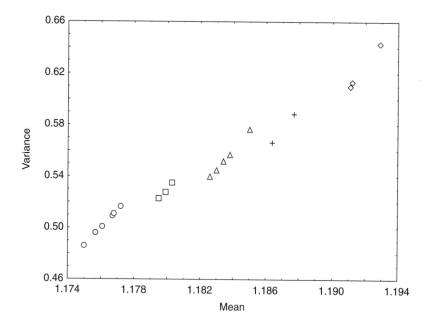

Fig. 5.6

Filars in regular graphs. A filar is obtained by plotting the mean versus the variance of the Estrada index of regular graphs. This figure illustrates the mean-variance plot for the cubic graphs having ten nodes.

Also interesting is the fact that when zooming in, a filar splits into smaller filars, which at the same time split into other filars, and so forth. This self-similarity property of the mean-variance plot of the Estrada index of regular graphs is termed a *multi-filar* structure.

The mean Estrada index $EE_{mean}(G, \beta)$ for a regular graph with n nodes of degree $d = q + 1$ is given (Ejov et al., 2007) by

$$EE_{mean}(G, \beta) = \frac{q+1}{2\pi} \int_{-2\sqrt{q}}^{2\sqrt{q}} e^{\beta s} \frac{\sqrt{4q - s^2}}{(q+1)^2 - s^2} ds + \frac{1}{n} \sum_{\gamma} \sum_{k=1}^{\infty} \frac{l(\gamma)}{2^{kl}(\gamma)/2} I_{kl}(\gamma)(2\sqrt{q}\beta)$$

(5.37)

where γ runs over all (oriented) primitive geodesics in the network, $l(\gamma)$ is the length of γ, and $I_m(z)$ is the Bessel function of the first kind:

$$I_m(z) = \sum_{r=0}^{\infty} \frac{(z/2)^{n+2r}}{r!(n+r)!}$$

(5.38)

On the other hand, the variance of the Estrada index is given by

$$\sigma(EE) = \frac{1}{n} \sum_{j=1}^{n} [e^{\beta \lambda_j} - EE_{mean}(G, \beta)]^2$$

(5.39)

By using the multiplicities of the length spectrum of the network, which are defined as the set of lengths of non-oriented primitive closed geodesics in the graph, Ejov et al. (2007) arrived at the following formulae for the mean and variance of the Estrada index for cubic graphs ($q = 2$):

$$EE_{mean}(G, \beta) = J(1/\beta) + \frac{2}{n} \sum_{l=3}^{\infty} lm_l F_l(1/\beta) \qquad (5.40)$$

$$\sigma(EE) = J(2/\beta) + \frac{2}{n} \sum_{l=3}^{\infty} lm_l F_l(2/\beta) \qquad (5.41)$$

where

$$J(\beta) = \frac{3}{2\pi} \int_{-2\sqrt{q}}^{2\sqrt{q}} e^{\beta s} \frac{\sqrt{8 - s^2}}{9 - s^2} ds \quad \text{and} \quad F_l(\beta) = \sum_{k=1}^{\infty} \frac{I_{kl}(2\sqrt{2}\beta)}{2^{kl/2}} \qquad (5.42)$$

Consequently, for cubic graphs with n nodes, all points lie above and to the right of the point given by $(J(1/\beta), J(2/\beta) - J(1/\beta)^2)$, which for the case $\beta = 1/3$ is approximately $(1.17455, 0.4217)$. Two graphs are in the same filar if they have exactly the same number of triangles. The line for this filar has been given parametrically by Ejov et al. (2007). The plot of the mean versus the variance of the Estrada index can be used as a clustering approach for regular graphs, grouping them according to the number of cycles of different lengths present in such networks. The use of clustering and partition methods to group nodes in complex networks is studied in Chapter 12.

5.4 Network returnability

When studying directed networks, use of the Estrada index can produce important hints about how much of the information departing from a given node can return to it. For instance, let us consider the partition function $Z = EE$ of a directed network written in terms of the moments $M_k(D)$ of the directed adjacency matrix:

$$Z = c_0 M_0(D) + c_2 M_2(D) + c_3 M_3(D) + \cdots + c_k M_k(D) + \cdots \qquad (5.43)$$

where $c_k = 1/k!$. The term $M_k(D)$ counts the number of walks of length k that start and end at the same node. If we consider some 'information' flowing through the network, $M_k(D)$ tells us how much of this information returns to its initial sources after k steps. In other words, it is related to the returnability of information in a network. However, the first term, $c_0 M_0(D)$, is not related to returnability, as it is only counting the number of nodes in the network. We can then exclude it from the 'returnability' partition function (Estrada and Hatano, 2009b):

$$Z' = Z - n \qquad (5.44)$$

This is illustrated in Figure 5.7, where we show three directed graphs with three nodes. In the first case, which corresponds to the feed-forward loop, there is not returnability of information departing from any of the three nodes. In the second graph—a three-node feedback loop—the information departing from a node can return to it after completing a cycle of length three. In the last

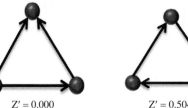

Fig. 5.7
Returnability. Values of returnability
for three directed triangles.

$Z' = 0.000$ $Z' = 0.504$ $Z' = 5.125$

case—the fully connected triad—any information departing from a node can return to it after 2, 3, 4, or more steps.

Let us now consider that under certain physical conditions the unidirectionality of a link is transformed in bidirectionality. In other words, under these physical conditions the direction of links in a network does not matter at all. For instance, in a reaction network (see Chapter 16), where the direction of the link points from a reactant to a product, under certain chemical conditions the equilibrium between both chemical species can be obtained, and the unidirectionality is transformed into bidirectionality.

Consequently, we can consider the difference of free energy between these two 'states' of a network, which is given by

$$\Delta F_r = F_D - F_U \tag{5.45}$$

where the sub-index indicates the relation with the returnability. By substituting the values of the free energy found in the previous sections (equation (5.45)), we obtain

$$\Delta F_r = -\beta^{-1} \ln \frac{Z'_D}{Z'_U} \tag{5.46}$$

The equilibrium constant is then written in terms of the change of free energy:

$$K_r = e^{-\Delta F/RT} \tag{5.47}$$

where $R = \beta^{-1}$ for single 'configuration' partition functions. Then, because Z'_D and Z'_U have no dependence on the number of configurations, the equilibrium constant is given by

$$K_r = \frac{Z'_D}{Z'_U} \tag{5.48}$$

We will call K the returnability equilibrium constant, or simply the returnability of a directed network, which is bounded as $0 \le K_r \le 1$. The lower bound is obtained for a network containing no cycles, and the upper bound is reached for symmetric directed networks. For a directed cycle of n nodes C_n, $K_r \to 0$ for $n \to \infty$. In Figure 5.8 we show the values of K_r for directed triangles.

In a similar way as for the study of chemical equilibrium, we can define $pK_r = -\log K_r$. The larger this index, the higher the influence of directionality in the network, and the lower the returnability of the information flowing through the network. In Table 5.3 we show the values of pK_r for several

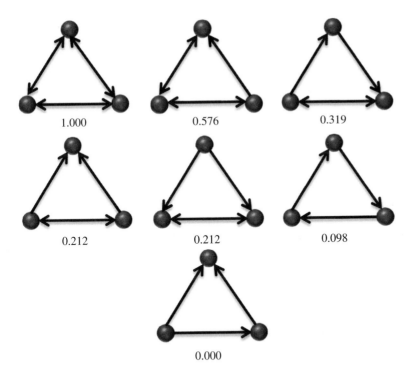

1.000   0.576   0.319

0.212   0.212   0.098

0.000

Fig. 5.8
**Returnability equilibrium constant
of directed triangles**. Values of
returnability equilibrium constants for
directed graphs of three nodes obtained
by equation (5.46).

Table 5.3 Returnability equilibrium constant of real networks. Values of $pK_r = -\log K_r$ for real networks in different scenarios.

network	pK_r	network	pK_r
Little Rock	14.66	Electronic circuit	1.84
Skipwith	7.42	Thesaurus	1.45
Neurons	6.11	Prison inmates	0.90
Trans yeast	4.33	US airports	0.00

real-world networks. The least returnable networks are the two food webs analysed (Little Rock and Skipwith), followed by the neuron network of *C. elegans*. Food webs have been found to contain large numbers of biparallel motifs as well as three chains, all of which are non-returnable motifs. In the case of the neuronal network, it contains 125 feed-forward loops, 127 bifans, and 227 biparallel motifs. These structures consist of a group of neurons acting as activators and another group acting as sink neurons, which render the structures non-returnable. It has been recognized that only 10% of synaptic couplings in this network are bidirectional. On the other hand, the US airport network is the most returnable, as all its links are bidirectional.

5.5 A brief on matrix functions

The reader not familiarised with matrix functions can be confused with the use of such terms as the exponential or the hyperbolic sine of a matrix. In fact, the functions $\exp(\mathbf{M})$ or $\sinh(\mathbf{M})$ do not refer to the entrywise operations on the matrix $\mathbf{M}$, such as $\exp(\mathbf{M}) = (\exp \mathbf{M}_{ij})$ or $\sinh(\mathbf{M}) = (\sinh \mathbf{M}_{ij})$. We therefore need to formally define what we understand—in this chapter and in the rest of this book—by a matrix function. Informally, we consider a function of the matrix $\mathbf{M}$ as the use of a scalar function f that specifies $f(\mathbf{M})$ to be a matrix of the same dimensions as $\mathbf{M}$, in a way that provides a useful generalization of the function $f(x)$ of a scalar variable. More formally, we define a matrix function as follows, by using the Cauchy integral theorem (Higham, 2008):

$$f(\mathbf{M}) := \frac{1}{2\pi i} \int_{\Gamma} f(z)(z\mathbf{I} - \mathbf{M})^{-1} dz \qquad (5.49)$$

where f is analytic on and inside a closed contour Γ that encloses the spectrum of the matrix $\mathbf{M}$.

There are other equivalent definitions of a matrix function on the basis of the Jordan canonical form as well as on Hermite interpolation, and the reader is directed to the authoritative review by Higham (2008) for an excellent account of matrix functions. Some general properties of matrix functions are the following:

i) $f(\mathbf{M})$ commutes with $\mathbf{M}$
ii) $f(\mathbf{M}^T) = f(\mathbf{M})^T$
iii) $f(\mathbf{X}\mathbf{M}\mathbf{X}^{-1}) = \mathbf{X}f(\mathbf{M})\mathbf{X}^{-1}$
iv) if λ_j is an eigenvalue of $\mathbf{M}$, then $f(\lambda_j)$ is an eigenvalue of $f(\mathbf{M})$
v) if $\mathbf{X}$ commutes with $\mathbf{M}$ then $\mathbf{X}$ commutes with $f(\mathbf{M})$

The following are some classic examples of matrix functions:

$$\exp(\mathbf{M}) = \sum_{k=0}^{\infty} \frac{\mathbf{M}^k}{k!} \qquad (5.50)$$

$$\cos(\mathbf{M}) = \sum_{k=0}^{\infty} \frac{(-1)^k}{(2k)!} \mathbf{M}^{2k} \qquad (5.51)$$

$$\sin(\mathbf{M}) = \sum_{k=0}^{\infty} \frac{(-1)^k}{(2k+1)!} \mathbf{M}^{2k+l} \qquad (5.52)$$

$$\cosh(\mathbf{M}) = \sum_{k=0}^{\infty} \frac{\mathbf{M}^{2k}}{(2k)!} \qquad (5.53)$$

$$\sinh(\mathbf{M}) = \sum_{k=0}^{\infty} \frac{\mathbf{M}^{2k+1}}{(2k+1)!} \tag{5.54}$$

$$\log(\mathbf{I} + \mathbf{M}) = \sum_{k=0}^{\infty} (-1)^{k+1} \frac{\mathbf{M}^k}{k}, \lambda_1(\mathbf{M}) < 1 \tag{5.55}$$

$$\mathrm{sgn}(\mathbf{M}) = \frac{2}{\pi} \mathbf{M} \int_0^{\infty} (t^2 \mathbf{I} + \mathbf{M}^2)^{-1} dt \tag{5.56}$$

$$\psi_k(\mathbf{M}) = \frac{1}{(k-1)!} \int_0^1 e^{(1-t)\mathbf{M}} t^{k-1} dt \tag{5.57}$$

Communicability functions in networks

6

Many incompletely understood phenomena lurk in the borderlands between physical theories—between classical and quantum, between rays and waves...

Michael Berry

6.1 Network communicability

Communication among the nodes in a network is a global process that takes places in most networks, despite their very different natures. This concept can group together a variety of processes ranging from those of interchanging thoughts, opinions, or information in social networks, the transfer of information from one molecule or cell to another by means of chemical, electrical, or other kind of signals, or the routes of transportation of any material, and so on. In many studies in complex networks it is assumed that this communication takes place through the shortest paths connecting pairs of nodes, which is the most economical way of going from one place to another in a network. However, in many real-world situations this communication takes place not only through these shortest paths but by means of any possible route that connect two nodes.

We assume here that the *communicability* between a pair of nodes in a network is assumed to depend on all routes that connect these two nodes. Among all these routes, the shortest path is the one making the most important contribution, as it is the most 'economic' way of connecting two nodes in a network. We can then use a strategy similar to that previously explained when defining the Estrada index in Chapter 5, in such a way that we make longer walks having smaller contributions to the communicability function than do shorter ones. Let $P_{rs}^{(l)}$ be the number of shortest paths between the nodes r and s having length l, and $W_{rs}^{(k)}$ the number of walks connecting r and s of length $k > l$. Let us then consider the following quantity:

$$G_{rs} = C_l P_{rs} + \sum_{k>l} C_k W_{rs}^{(k)} \tag{6.1}$$

where the coefficients C_k have to fulfil the same requirements as explained in Chapter 5.

The coefficients $C_k = 1/k!$ are used in order to derive a communicability function for pairs of nodes in a network. Then, by using the connection between the powers of the adjacency matrix and the number of walks in the network, we obtain (Estrada and Hatano, 2008):

$$G_{rs} = \sum_{k=0}^{\infty} \frac{(\mathbf{A}^k)_{rs}}{k!} = \left(e^{\mathbf{A}}\right)_{rs} \tag{6.2}$$

where $e^{\mathbf{A}}$ is the matrix exponential function. We can express the communicability function for a pair of nodes in a network by using the eigenvalues and eigenvectors of the adjacency matrix:

$$G_{rs} = \sum_{j=1}^{n} \varphi_j(r)\varphi_j(s)e^{\lambda_j} \tag{6.3}$$

By using the concept of inverse temperature introduced in Chapter 5, we can also express the communicability function in terms of this parameter (Estrada and Hatano, 2008):

$$G_{rs}(\beta) = \sum_{k=0}^{\infty} \frac{(\beta\mathbf{A}^k)_{rs}}{k!} = \left(e^{\beta\mathbf{A}}\right)_{rs} \tag{6.4}$$

In general, the communicability function $G_{rs}(\beta)$ accounts for all channels of communication between two nodes, giving more weight to the shortest paths connecting them. Intuitively, the communicability between the two nodes at the end of a linear path should tend to zero as the length of the path tends to infinite. In order to show that this is exactly the case, we can write the expression for $G_{rs}(\beta)$ for the path P_n:

$$G_{rs} = \frac{1}{n+1}\left(\sum_j \cos\frac{j\pi(r-s)}{n+1} - \cos\frac{j\pi(r+s)}{n+1}\right)e^{2\cos\left(\frac{j\pi}{n+1}\right)} \tag{6.5}$$

where $\beta \equiv 1$ is used without any loss of generality. It is then straightforward to realize by simple substitution in (6.5) that $G_{rs} \to 0$ for the nodes at the end of a linear path as $n \to \infty$. At the other extreme we find the complete network K_n, for which

$$G_{rs} = \frac{e^{n+1}}{n} + e^{-1}\sum_{j=2}^{n}\varphi_j(r)\varphi_j(s) = \frac{e^{n+1}}{n} - \frac{1}{ne} = \frac{1}{ne}(e^n - 1) \tag{6.6}$$

which means that $G_{rs} \to \infty$ as $n \to \infty$, which perfectly agrees with our intuition (Estrada and Hatano, 2008).

We have interpreted the communicability function as the thermal Green's function of the network by considering a continuous-time quantum walk on the network. Take a quantum-mechanical wave function $|\psi(t)\rangle$ at time t. It obeys Schrödinger's equation:

$$i\hbar\frac{d}{dt}|\psi(t)\rangle = -\mathbf{A}|\psi(t)\rangle \qquad (6.7)$$

where we use the adjacency matrix as the negative Hamiltonian.

Assuming from now on that $\hbar = 1$, we can write the solution of the time-dependent Schrödinger equation (6.7) in the form $|\psi(t)\rangle = e^{i\mathbf{A}t}|\psi(0)\rangle$. The final state $e^{i\mathbf{A}t}|s\rangle$ is a state of the graph that results after time t from the initial state $|s\rangle$. The 'particle' that resided on the node r at time $t = 0$ diffuses for the time t because of the quantum dynamics. Then, we can obtain the magnitude of the diffusion of this 'particle' from r to s by computing the product $\langle r|e^{i\mathbf{A}t}|s\rangle$. By continuation from the real time t to the imaginary time, we have the thermal Green's function, defined as $G_{rs} = \langle r|e^{\mathbf{A}}|s\rangle$, which is the communicability between nodes r and s n the network, as defined in this work. Consequently, the communicability between nodes r and s in the network represents the probability that a particle starting from node r ends up at node r after wandering on the complex network due to the thermal fluctuation (Estrada and Hatano, 2008). By regarding the thermal fluctuation as some forms of random noise, we can identify the particle as an information carrier in a society or a needle in a drug-user network.

6.1.1 Example

As an example we analyse a small social network previously studied by Thurman (1979). This network is the result of 16 months' observation of office politics, involving fifteen members of an overseas branch of a large international organisation. Thurman studied an informal network of friendship ties among members of the office which was not part of the official structure of the office. A pictorial representation of this network is illustrated in Figure 6.1. As can be seen, six members of the staff form a clique—a cohesive structure in which every pair of actors is connected. The members of the clique—Ann, Tina, Katy, Amy, Pete, and Lisa—are drawn in grey.

An interesting situation arose in the office during the period of Thurman's study (Thurman, 1979). Apparently, both Emma and Minna were the targets of a levelling coalition formed by the six members of staff who formed the clique. According to Thurman there was not an 'identifiable group being led by a particular individual. They were attacked by a non-group.' We can see, in the figure, that the members of the attacking coalition are among those with the largest number of ties in the network. However, Emma, who is one of the targets, has as many connections as Pete and Ann in the coalition. On the other hand, Minna has the same number of ties as Andy and Bill, who are not the targets of the coalition.

In Figure 6.2 we illustrate the communicability between all pairs of members in the office. It can be seen that the highest communicability is observed between Pete and Lisa, Pete and Ann, and Ann and Lisa. Pete has been recognised by Thurman as being at the centre of the social circle in the office, which also involves Lisa, Katy, and Amy. Pete was coming to the office from the central office, and was known to have dated both Katy and Amy. He also who arranged for Ann to be assigned to this office. Interestingly, Emma, who

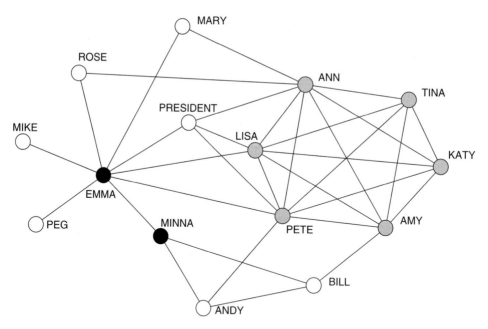

Fig. 6.1

Small office network. The social network of friendship ties in the office of an overseas branch of a large international organisation (Thurman, 1979).

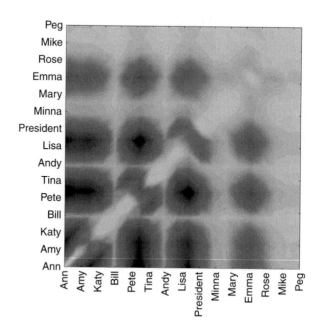

Fig. 6.2

Communicability in a small office network. Inter-node communicability among the members of a social network of friendship ties in the office of an overseas branch of a large international organisation. The darkness of the region is proportional to the communicability between the corresponding pair of nodes.

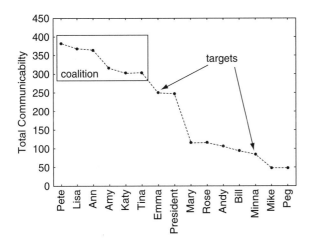

Fig. 6.3
Total communicability in a small office network. Total node communicability for every member of the social network of friendship ties in the office of an overseas branch of a large international organisation.

plays a crucial role in the office, does not show high communicability with any particular member.

If we consider the total communicability for the rth member of the office— $\sum_s G_{rs}$—we obtain the results illustrated in Figure 6.3. It can be seen that the members of the coalition are the best communicators in the office, displaying the largest total communicability among all members. Emma displays less total communicability than all of them, and a gap in the communicability is observed between the members of the coalition and Emma. More surprisingly, Minna is poorly communicated, and displays less total communicability than other members with the same or even a smaller number of friendship ties than her. Therefore, it is evident why she could not mobilise effective support to oppose attacks by members of the coalition. It is possible, however, that Emma's largest communicability makes her strong enough to repel the same series of attacks.

6.2 Communicability in directed and weighted networks

The communicability function can also be calculated for any pair of nodes in a directed network, using expression (6.1). In the general case of weighted networks, a normalisation of the adjacency matrix is recommended. Crofts and Higham (2009) have suggested the use of the following modified communicability function for weighted networks:

$$\tilde{G}_{pq} = \left[\exp(\mathbf{W}^{-1/2} \mathbf{A} \mathbf{W}^{-1/2}) \right]_{pq} \qquad (6.8)$$

where $\mathbf{A}$ is the weighted adjacency matrix and, $\mathbf{W}$ is the diagonal matrix of weighted node degrees. In Figure 6.4 we illustrate the communicability matrix for all nodes in the directed version of the neural network of the worm *C. elegans*. As can be seen, there is a clear asymmetry in the communicability between neurons in this network. Some applications of the weighted communicability function using (6.8) are provided in Chapter 17 of this book.

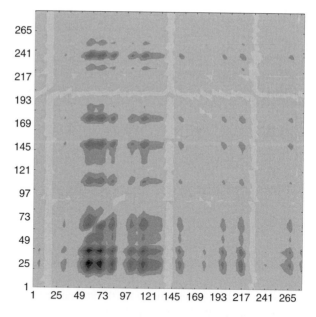

Fig. 6.4

Communicability in a neural network.
Node communicability for every pair of
neurons in the directed neural network of
the worm *C. elegans*. The darker the
region, the larger the communicability
between the corresponding pair of
neurons.

6.3 Generalised communicability functions

According to the definition of the communicability function it is possible to
use different kinds of coefficient in order to account in different ways for the
penalisation of the longer walks from one node to another in the network. If,
for instance, we are interested in using heavier penalisations for the longer
walks, we can use the following alternative definition of communicability
(Estrada, 2010c):

$$G_{rs}{}^t = \sum_{k=0}^{\infty} \frac{(\mathbf{A}^k)_{rs}}{(k+t)!} \tag{6.9}$$

In this case, instead of penalising the walks by the factorial of their length we
use $1/(k+t)!$, where $t > 1$. The generalised communicability index $G_{rs}{}^t$ can
then be written by using the following spectral formula (Estrada, 2010c):

$$G_{rs}{}^t = \sum_{j=1}^{n} \varphi_j(r)\varphi_j(s) \frac{e^{\lambda_j} - \sum_{u=1}^{t} \frac{\lambda_j^{t-u}}{(t-u)!}}{(\lambda_j)^t} \tag{6.10}$$

These indices are related to matrix functions through the formula

$$G_{rs}{}^t = [\psi_t(\mathbf{A})]_{rs} \tag{6.11}$$

where the $\psi_t(\mathbf{A})$ matrix functions (Higham, 2008) have the integral formula

$$\psi_t(\mathbf{A}) = \frac{1}{(t-1)!} \int_0^1 e^{(1-t)\mathbf{A}} x^{t-1} dx \tag{6.12}$$

obeying the following recurrence formula (Higham, 2008):

$$\psi_t(\mathbf{A}) = \mathbf{A}\,\psi_{t+1}(\mathbf{A}) + \frac{1}{t!} \tag{6.13}$$

Another alternative is to use a lighter penalization for longer walks, which can be achieved by defining the following negatively rescaled indices (Estrada, 2010c):

$$G_{rs}^{-t} = \sum_{q=0}^{t-1} (\mathbf{A}^q)_{rs} + \sum_{k=t}^{\infty} \frac{(\mathbf{A}^k)_{rs}}{(k-t)!} \tag{6.14}$$

They have the following spectral realisation:

$$G_{rs}^{-t} = \sum_{j=1}^{n} \varphi_j(r)\varphi_j(s) \left(\sum_{s=0}^{t-1} \lambda_j{}^s + \lambda_j{}^t e^{\lambda_j} \right) \tag{6.15}$$

and can be written in terms of matrix functions as follow (Estrada, 2010c):

$$G_{rs}^{-t} = \left[\mathbf{A}^t(\mathbf{I} + \mathbf{A}e^{\mathbf{A}} - e^{\mathbf{A}}) \right]_{rs} \tag{6.16}$$

A third alternative that has been explored so far is the use of the following penalising scheme:

$$G_{rs}{}^R = \sum_{k=0}^{\infty} \frac{(\mathbf{A}^k)_{rs}}{n^k} \tag{6.17}$$

where n is in this case the number of nodes in the network. Here the communicability function can be obtained explicitly from the following matrix function, known as the 'matrix resolvent' (Estrada and Higham, 2010):

$$G_{rs}{}^R = \left[\left(\mathbf{I} - \frac{\mathbf{A}}{n} \right)^{-1} \right]_{rs} \tag{6.18}$$

where $\mathbf{I}$ is the identity matrix. In terms of the eigenvalues and eigenvectors of the adjacency matrix, the resolvent communicability can be written as

$$G_{rs}{}^R = \sum_{j=1}^{n} \varphi_j(r)\varphi_j(s) \left(1 + \frac{\lambda_j}{n} \right)^{-1} \tag{6.19}$$

For the analysis of the 'multifilar' structure of the variance and mean of the resolvent-based Estrada index, the reader is referred to the work of Ejov et al. (2009). Approximations for this function using the Gauss–Radau rule have been given by Benzi and Boito (2010).

6.4 Communicability based on network vibrations

The term 'vibration' is used here in a wide context—referring not only to 'mechanical oscillations', but also to the repetitive variation of some measure about its point of equilibrium (Gregory, 2006). Such types of oscillatory process can be found in many physical, chemical, biological, and social systems. In molecular networks, for instance, where atoms are located at the vertices of a network, the study of network vibrations is a textbook example. Molecular vibrations are recognised as the 'fingerprints' of molecules, and the use of infrared (IR) spectroscopy is a very well established method for molecular identification. The idea of IR spectroscopy is that every molecule displays a unique vibrational pattern which depends on the nature of the atoms and bonds forming such a molecule (Barrow, 1962). Here we extend this idea to the study of complex networks in order to gain insights into the structural organisation of such systems. We study how some 'perturbation' occurring in one node of a network is transmitted through the whole network, using a classical mechanic analogy. This analogy considers a network of classical harmonic oscillators.

6.4.1 Network of classical harmonic oscillators

We consider that all nodes are identical and are moving only along the direction of the link. It is also considered, for the sake of simplicity, that there is no damping, and no external forces are applied to the system. The coordinates chosen to describe a configuration of the system are $x_i, i = 1, 2, \ldots, n$. We set $x_i = 0$ at the equilibrium point. Here, nodes are considered as a ball of unit mass connected by springs of force constant θ, and the network is submerged into a thermal bath of inverse temperature β, as in the scheme shown in Figure 6.5.

Let $\mathbf{x}$ be a column vector of coordinates. The equation of motion for the system is given by

$$\mathbf{M}\ddot{\mathbf{x}} + \mathbf{L}\mathbf{x} = 0 \qquad (6.20)$$

where $\mathbf{M}$ is a diagonal matrix of node masses—which in our case is the identity matrix due to unit masses selection—and $\mathbf{L}$ is the network Laplacian. The term *netons* has been proposed to refer to phonons in a complex network in order to differentiate the underlying topological structure of these systems, which is not the usual periodic lattice (Kim et al., 2003).

We are interested in considering a vibrational excitation energy from the static position of the network. The vibrational potential energy of the network can be expressed as

$$V(\vec{x}) = \frac{\theta}{2}\vec{x}^T \mathbf{L}\vec{x} \qquad (6.21)$$

where $\vec{x}$ is the vector whose i th entry x_i is the displacement of the node i from its equilibrium position. The displacement correlation can be defined as

$$\langle x_i x_j \rangle = \int x_i x_j P(\vec{x}) d\vec{x} \qquad (6.22)$$

This is the classical vibrational analogy of the communicability function studied in the previous section, and we will call it the *classical vibrational communicability* of the network. In order to calculate $\langle x_i x_j \rangle$ we need to diagonalise the Laplacian matrix $\mathbf{L}$. First, let us denote by U the matrix whose

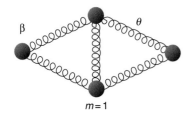

Fig. 6.5

Model of network vibrations. The network is submerged in a thermal bath at inverse temperature β. Every node is represented by a ball of unit mass, and edges represent springs of force constant θ.

columns are the orthonormal eigenvectors $\vec{\psi}_\alpha$ and $\mathbf{M}$ the diagonal matrix of eigenvalues μ_α of the Laplacian matrix. We recall from Chapter 2 that the Laplacian matrix is positive semidefinite, which means that we can write the Laplacian eigenvalues for a connected network as $0 = \mu_1 < \mu_2 \leq \cdots \leq \mu_n$. An important observation here is that the *zero eigenvalue does not contribute to the vibrational energy*. This is because the mode $\alpha = 1$ is the mode where all the nodes (balls) move coherently in the same direction, and thereby the whole network moves in such direction. In other words, this is the motion of the centre of mass, not a vibration.

In calculating equation (6.22), the integration measure is transformed as

$$d\vec{x} \equiv \prod_{i=1}^{n} dx_i = |\det U| \prod_{i=1} dy_i = d\vec{y} \qquad (6.23)$$

because the determinant of the orthogonal matrix, $\det U$, is ± 1. The partition function that appears as a normalization factor of $P(\vec{x})$ can then be written as

$$Z = \int d\vec{y} \exp\left(-\frac{\beta\theta}{2}\vec{y}^T \mathbf{M}\vec{y}\right) \qquad (6.24)$$

$$= \prod_{\alpha=1}^{n} \int_{-\infty}^{+\infty} dy_\alpha \exp\left(-\frac{\beta\theta}{2}\mu_\alpha {y_\alpha}^2\right)$$

Because we are interested in the vibrational excitation energy within the network we remove the motion of the centre of mass, which is equivalent to removing the contribution from $\mu_1 = 0$. Consequently, we use the following modified partition function:

$$\tilde{Z} = \prod_{\alpha=2}^{n} \int_{-\infty}^{+\infty} dy_\alpha \exp\left(-\frac{\beta\theta}{2}\mu_\alpha {y_\alpha}^2\right) \qquad (6.25)$$

$$= \prod_{\alpha=2}^{n} \sqrt{\frac{2\pi}{\beta\theta\mu_\alpha}}.$$

Now we calculate the vibrational communicability (6.22), beginning with the numerator of the right-hand side of this equation, which is written as

$$I_i = \int d\vec{x}\, x_i^2 \exp\left(-\frac{\beta\theta}{2}\vec{x}^T \mathbf{L}\vec{x}\right)$$

$$= \int d\vec{y}\,(U\vec{y})_i^2 \exp\left(-\frac{\beta\theta}{2}\vec{y}^T \mathbf{M}\vec{y}\right) \qquad (6.26)$$

$$= \int d\vec{y}\left(\sum_{v=1}^{n} U_{iv}y_v\right)^2 \exp\left(-\frac{\beta\theta}{2}\vec{y}^T \mathbf{M}\vec{y}\right)$$

$$\int d\vec{y}\left(\sum_{v=1}^{n}\sum_{\gamma=1}^{n} U_{iv}U_{i\gamma}\,y_v y_\gamma\right)\prod_{\alpha=1}^{n} \exp\left(-\frac{\beta\theta}{2}\mu_\alpha y_\alpha^2\right)$$

On the right-hand side, any terms with $v \neq \gamma$ will vanish after integration because the integrand is an odd function with respect to y_v and y_γ. The only possibility of a finite result is due to terms with $v = \gamma$. We therefore have

$$I_i = \int d\vec{y} \left[\sum_{v=1}^{n} (U_{iv} y_v)^2 \right] \prod_{\alpha=1}^{n} \exp\left(-\frac{\beta\theta}{2} \mu_\alpha y_\alpha^2 \right)$$

$$= \int_{-\infty}^{+\infty} dy_1 (U_{i1} y_1)^2 \times \prod_{\alpha=2}^{n} \int_{-\infty}^{+\infty} dy_\alpha \exp\left(-\frac{\beta\theta}{2} \mu_\alpha y_\alpha^2 \right) + \tag{6.27}$$

$$\sum_{v=2}^{n} \int_{-\infty}^{+\infty} dy_v (U_{iv} y_v)^2 \exp\left(-\frac{\beta\theta}{2} \mu_v y_v^2 \right) \times \prod_{\substack{\alpha=1 \\ \alpha \neq v}}^{n} \int_{-\infty}^{+\infty} dy_\alpha \exp\left(-\frac{\beta\theta}{2} \mu_\alpha y_\alpha^2 \right)$$

where we separated the contribution from the zero eigenvalue and those from the other ones. Due to the divergence introduced by the zero eigenvalue, we proceed with the calculation by redefining the quantity I_i with the zero mode removed:

$$\tilde{I}_i \equiv \sum_{v=2}^{n} \int_{-\infty}^{+\infty} dy_v (U_{iv} y_v)^2 \exp\left(-\frac{\beta\theta}{2} \mu_v y_v^2 \right) \times \prod_{\substack{\alpha=1 \\ \alpha \neq v}}^{n} \int_{-\infty}^{+\infty} dy_\alpha \exp\left(-\frac{\beta\theta}{2} \mu_\alpha y_\alpha^2 \right)$$

$$= \sum_{v=2}^{n} \frac{U_{iv}^2}{2} \sqrt{\frac{8\pi}{(\beta\theta\mu_v)^3}} \times \prod_{\substack{\alpha=1 \\ \alpha \neq v}}^{n} \sqrt{\frac{2\pi}{\beta\theta\mu_v}} \tag{6.28}$$

$$= \tilde{Z} \times \sum_{v=2}^{n} \frac{U_{iv}^2}{\beta\theta\mu_v}.$$

We then obtain the vibrational communicability as (Estrada and Hatano, 2010a; 2010b)

$$\langle x_i x_j \rangle = \sum_{v=2}^{n} \frac{U_{iv} U_{jv}}{\beta\theta\lambda_v} = \frac{1}{\beta\theta} (\mathbf{L}^+)_{ij} \tag{6.29}$$

where L^+ is the pseudoinverse or Moore–Penrose generalized inverse of the graph Laplacian (see Chapter 2).

6.4.2 Structural interpretation

We begin by recalling some concepts discussed in Chapter 3. If $\mathbf{J} = |\mathbf{1}\rangle\langle\mathbf{1}|$ and $\mathbf{O} = |\mathbf{0}\rangle\langle\mathbf{0}|$ are all-ones and all-zeroes matrices, respectively, we have seen that $\mathbf{L}^+\mathbf{J} = \mathbf{J}\mathbf{L}^+ = \mathbf{O}$, and if L and L^+ pertain to a connected network having n nodes, then

$$\mathbf{L}\mathbf{L}^+ = \mathbf{L}^+\mathbf{L} = \mathbf{I} - \frac{1}{n}\mathbf{J} \tag{6.30}$$

It has been proved by Gutman and Xiao (2004) that

$$\mathbf{L}^+\mathbf{\Omega}\mathbf{L}^+ = -2(\mathbf{L}^+)^3 \tag{6.31}$$

Then, by multiplying (6.31) by $\mathbf{L}$ from both left and right, and reordering, we obtain

$$\mathbf{L}^+ = -\frac{1}{2}\left[\mathbf{\Omega} - \frac{1}{n}\mathbf{R} + \frac{Kf}{n^2}\mathbf{I}\right] \tag{6.32}$$

where

$$\mathbf{R} = \mathbf{\Omega}\mathbf{J} + \mathbf{J}\mathbf{\Omega} \tag{6.33}$$

whose entries are

$$\mathbf{R}_{ij} = \begin{cases} 2R_i & i = j \\ R_i + R_j & i \neq j \end{cases} \tag{6.34}$$

We then have the following expression for the vibrational communicability (Estrada and Hatano, 2010a; 2010b):

$$\langle x_i x_j \rangle = (\mathbf{L}^+)_{ij} = -\frac{1}{2}\left(\Omega_{ij} - \frac{1}{n}(R_i + R_j) + \frac{2Kf}{n^2}\right) \tag{6.35}$$

The simplest interpretation of this measure can be carried out for trees. In this case we have

$$\langle x_i x_j \rangle = -\frac{1}{2}\left(d_{ij} - \frac{1}{n}(s_i + s_j) + \frac{2W}{n^2}\right) \tag{6.36}$$

where $W = \sum_{i<j} d_{ij}$ is the Wiener index, and $s_j \sum_j d_{ij}$ is the distance-sum as defined in Chapter 3. Then, the vibrational communicability for nodes p and q is given by (Estrada and Hatano, 2010a; 2010b)

$$\langle x_p x_q \rangle = \frac{1}{2n}\left[(s_p + s_q) - \left(\frac{n^2 + 2}{n}\right)d_{pq} - \left(\frac{2}{n}\right)\sum_{\substack{i\neq p \\ j\neq q \\ i<j}} d_{ij}\right] \tag{6.37}$$

Let us now interpret this expression by considering that some information is transmitted from node p in the network. In the displacement correlation function (6.37) the first term in parentheses represents the information which is transmitted simultaneously from nodes p and q. The second term is the information arising in p and reflected by q, or *vice versa*, and the last term is the information reflected from the rest of the nodes in the network. Then, $\langle x_p x_q \rangle$ represents the amount of information which is 'absorbed' by all nodes in a network when such information is transmitted simultaneously from nodes p and q.

6.4.3 Example

As an example, we return to the small social network of friendship ties in the office of an overseas branch of a large international organisation (Thurman, 1979), which was analysed in the previous section. If we consider a classical vibrational approach to the problem observed in this social network

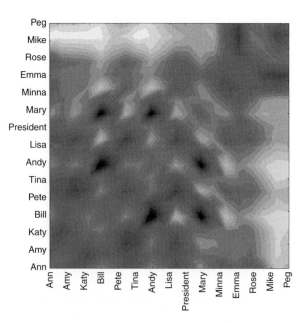

Fig. 6.6

Vibrational communicability in a small office network. The vibrational communicability function between pairs of nodes described by a network of classical harmonic oscillators in Thurman's study of a small office.

we obtain some interesting results. We can consider that the whole network is under some 'stress' due to the different disputes that have arisen among its members. That is, every member of the network is transmitting some information, which is partially 'absorbed' and partially 'reflected' by the rest of the members of the network. We can consider that the whole social network is 'vibrating', and we can see how this is reflected in the vibrational communicability function. In Figure 6.6 we illustrate the normalised matrix of vibrational communicability for all pairs of members in the office. The most interesting observations are that most of the members of the attacking faction are very well correlated among themselves, according to vibrational communicability. This means that most of them act in some kind of alignment when they are affected by some perturbation (vibration) in the network. The most interesting case is Pete, who plays a central role in the story and has positive vibrational communicability with all other members of the attacking faction. Contrarily, Minna has negative vibrational communicability with five members of the faction: Amy, Katy, Pete, Tina, and Lisa. Emma, who is also one of the targets of the attacks, is vibrationally anticorrelationated with all members of the attacking group. This means that the two targets are 'vibrating' in opposite directions as members of the attacking faction when the network is perturbed.

Centrality measures

7

> It is interesting to note how many fundamental terms which the social sciences are trying to adopt from physics have as a matter of historical fact originated in the social field.
>
> Michael R. Cohen

One of the most intuitive concepts we can think about when confronting the analysis of a network is the identification of the 'most important' nodes. The quantification of the notion of importance of actors in social networks was one of the first applications of network theory, dating back to the early 1950s. The concept is formalised in network theory under the umbrella of centrality measures (Borgatti 2005; Borgatti and Everett, 2006; Friedkin, 1991). The first attempts at quantifying such centrality of nodes where carried out by Bavelas (1948; 1950) and Leavitt (1951) by using this concept in communication networks. Freeman (2004) has pointed out that Bavelas 'believed that in any organization the degree to which a single individual dominates its communication network—the degree to which it was centralized—affected its efficiency, its morale, and the perceived influence of each individual actor.' This is still a paradigm in the use of centrality measures in complex networks: the ability of a node to communicate directly with other nodes, or its closeness to many other nodes or the quantity of pairs of nodes which need a specific node as intermediary in their communications determine many of the structural and functional properties of this node in a network. Here we account for some of the most relevant centrality measures currently in use for studying complex networks.

7.1 Degree centrality

The degree of a node was defined in Chapter 2, where we wrote it in matrix form for undirected networks as

$$k_i = (\mathbf{A}|\mathbf{1}\rangle)_i \tag{7.1}$$

where $|\mathbf{1}\rangle$ is an all-ones column vector. The use of degrees as a centrality for nodes in a network is due to Freeman (1979). In the case of directed networks we have distinguished between indegree and outdegree of a node:

$$k_i^{in} = (\langle \mathbf{1}|\mathbf{A})_i \qquad\qquad (7.2)$$

$$k_i^{out} = (\mathbf{A}|\mathbf{1}\rangle)_i \qquad\qquad (7.3)$$

Due to the simplicity and intuitive nature of this centrality measure it was proposed by several authors in the 1950s, '60s, and '70s, and then reviewed in the seminal paper of Freeman (1979). The idea behind the use of node degree as a centrality measure is that a node is more central or more influential than another in a network if the degree of the first is larger than that of the second. The degree of a node counts the number of walks of length 1 from a given node, or the number of closed walks of length 2 starting (and ending) at this particular node:

$$k_i = (\mathbf{A}^2)_{ii} \qquad\qquad (7.4)$$

This means that the degree centrality accounts for immediate effects taking place in a network. For instance, if we consider a process in which some 'information' is passed from one node to another in a network, the opportunity of receiving such information through the network is proportional to the degree of the corresponding node. This is the case of the risk of infection in a network in which some nodes are infected, and having a link to them is assumed to imply infection. For instance, in Figure 7.1 we illustrate the principal connected component of a social network of drug-users in Colorado Springs, USA,

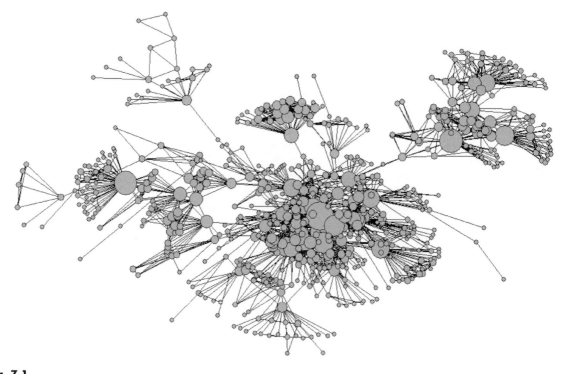

Fig. 7.1
Degree centrality. Main connected component of the social network of injecting drug-users in Colorado Springs. Nodes are drawn with radii proportional to their degree centrality.

where the links indicate whether two people have interchanged needles in the last three months. The degree centrality of every node is drawn with a size proportional to their node degree. If some of these high-degree individuals are not already infected by HIV, the risk of being infected is significantly much larger than for those non-infected nodes having low-degree centrality. That is, the individuals having a large number of connections are those displaying the largest risk of being infected and of infecting others in the network.

Degree centrality can also be seen as modelling the flow of information through the nodes of a network in an infinitely long random walk. That is, if we define the transition probability for going from one node to another as

$$p_{ij} = \frac{A_{ij}}{k_i} \tag{7.5}$$

then, if π is the stationary vector, where π_i is the stationary probability of the node i under the Markov chain with transition matrix $\mathbf{P}$, we have

$$(\pi \mathbf{P})_j = \sum_{i \in V} \pi_i p_{ij} = \frac{\sum_{i \in V} k_i p_{ij}}{\sum_{i \in V} k_i} = \frac{\sum_{i \in V} a_{ij}}{\sum_{i \in V} k_i} = \frac{k_j}{\sum_{i \in V} k_i} = \pi_j \tag{7.6}$$

In other words, the limiting probabilities for an infinitely long random walk on a network depend on the node degree,

$$\pi_i = \frac{k_i}{\sum_{i \in V} k_i} \tag{7.7}$$

and the degree centrality is an appropriate measure for walk-based transfer processes in a network.

When analysing directed networks, the use of both indegree and outdegree is necessary (see equations (7.2) and 7.3)). A representation of the food web of St Martin island is illustrated in Figure 7.2, where the radius of the nodes is proportional to the corresponding indegree (top) and outdegree (bottom).

High-indegree nodes represent species which are predated by many predators in this ecosystem, and nodes with high outdegree are predators with a large variety of prey. In general, top predators are not predated by other species, thus having high outdegree but low indegree. Highly predated species do not predate many other prey, then having high indegree but low outdegree. Then, in a food web we would expect an anticorrelation between indegree and outdegree of all nodes. This correlation between indegree and outdegree can be measured with the Pearson correlation coefficient Ω:

$$\Omega = \frac{\sum_{i=1}^{n} \left(k_i^{in} - \overline{k^{in}} \right) \left(k_i^{out} - \overline{k^{out}} \right)}{\sqrt{\sum_{i=1}^{n} \left(k_i^{in} - \overline{k^{in}} \right)^2} \sqrt{\sum_{i=1}^{n} \left(k_i^{out} - \overline{k^{out}} \right)^2}} \tag{7.8}$$

where $\overline{k^{in}}$ and $\overline{k^{out}}$ are the average indegree and outdegree for a network. According to this coefficient, networks can display positive ($\Omega > 0$), negative ($\Omega < 0$), or neutral ($\Omega = 0$) correlation between indegrees and outdegrees. In Table 7.1 we give the indegree–outdegree correlation coefficient Ω for

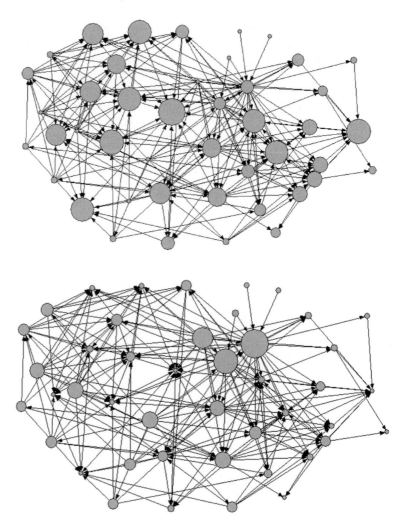

Fig. 7.2
Indegree and outdegree centrality.
Nodes represent species, and links
represent trophic links in St Martin
island. Nodes are represented with radii
proportional to their indegree (top) and
outdegree (bottom).

Table 7.1 **Indegree–outdegree correlation.** Values of indegree–outdegree
correlation coefficient Ω for several real-world networks, together with their
reciprocity coefficients as given in Table 5.1.

Network	ρ	Ω
Metab. *E. coli*	0.764	0.930
Thesaurus	0.559	0.676
Neurons	0.158	0.474
ODLIS	0.203	0.330
Internet	$-5.6.10^{-4}$	0.183
Little Rock	-0.025	-0.116
Transc. *E. coli*	-0.0003	-0.223
Transc. yeast	$-5.5.10^{-4}$	-0.242
St Martin	-0.130	-0.308
Chesapeake	-0.057	-0.309
Skipwith	-0.280	-0.682

several real networks together with their reciprocity coefficient reported in Chapter 5 (Table 5.1). As can be seen in Table 7.1, there is a good correlation between reciprocity and indegree–outdegree correlation, showing that reciprocal networks have in general $\Omega > 0$, while antireciprocal networks generally display $\Omega < 0$. The reason for the existence of this correlation is simple. If the reciprocity of a network is high it indicates that there are many nodes having reciprocal links and consequently displaying both high indegree and outdegree. In the case of antireciprocal networks there are some nodes that have more links coming in than going out, and other nodes having mainly links going out. The first group of nodes displays high indegree and low outdegree, and the second group displays high outdegree and low indegree, explaining the anticorrelation between indegree and outdegree observed; that is, $\Omega < 0$.

7.2 Centrality beyond nearest-neighbours

The most characteristic feature of the degree centrality is that it accounts only for the influence of nearest-neighbours to a given node. In many real situations it is necessary to account not only for the effects of nearest neighbours on a given node, but also for the influence of other nodes separated at a certain distance from it. For instance, in Figure 7.3 we illustrate a simple network in which nodes labelled as i and j have degree 3. The neighbours of node i have no other neighbours, so that all influence received by i comes only from these two nodes. However, every neighbour of node j has another neighbour, which can indirectly influence node j.

In a very 'artificial' way we can express the degree of a node as a sum of powers of adjacency matrices:

$$k_i = \left[\left(\mathbf{A}^0 + \mathbf{A}^1\right) - \mathbf{I}\right]|1\rangle_i \qquad (7.9)$$

where $\mathbf{I}$ is the identity matrix. $\mathbf{A}^0$ accounts for walks of length zero, which is the equivalent of remaining at the same node, and $\mathbf{A}^1$ accounts for walk of length 1, which corresponds to visiting nearest neighbours only. Note that walks of length zero are subtracted from the degree by $-\mathbf{I}$ in (7.9). One possibility of accounting for influences beyond nearest neighbours is to count

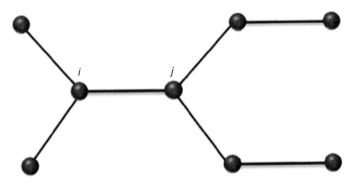

Fig. 7.3
Degree and non-nearest-neighbours.
Two nodes having the same degree but different numbers of second neighbours.

walks of length larger than 1. These are the strategies followed in the next centrality measures.

7.2.1 Katz centrality

When considering an extension of equation (7.9) on the basis of different powers of the adjacency matrix, it needs to be considered that the series $\mathbf{A}^0 + \mathbf{A}^1 + \mathbf{A}^2 + \cdots + \mathbf{A}^k + \cdots$ diverges. Then, we have to introduce some 'attenuation' factor for series convergence. Let us call this attenuation factor $\alpha > 1$, and express the new centrality measure as follows:

$$K_i = \left[\left(\alpha^0 \mathbf{A}^0 + \alpha^{-1}\mathbf{A}^1 + \alpha^{-2}\mathbf{A}^2 + \cdots + \alpha^{-k}\mathbf{A}^k + \cdots\right)|1\rangle\right]_i = \left[\left(\sum_{k=0}^{\infty} \alpha^{-k}\mathbf{A}^k\right)|1\rangle\right]_i$$

(7.10)

This series converges to

$$K_i = \left[\left(\sum_{k=0}^{\infty} \alpha^{-k}\mathbf{A}^k\right)|1\rangle\right]_i = \left[\left(\mathbf{I} - \frac{1}{\alpha}\mathbf{A}\right)^{-1}|1\rangle\right]_i \qquad (7.11)$$

This index was introduced by Katz as early as 1953 (Katz, 1953). The attenuation factor must be selected such that $\alpha \neq \lambda_1$, where λ_1 is the principal eigenvector of the adjacency matrix, in order to avoid divergence in equation (7.11). In general, the attenuation factor is selected to be smaller than λ_1 in order to obtain meaningful centrality indices. Also, the index is usually expressed by subtracting the contribution of $\mathbf{A}^0$, as was done for the degree centrality:

$$K_i = \left\{\left[\left(\mathbf{I} - \frac{1}{\alpha}\mathbf{A}\right)^{-1} - \mathbf{I}\right]|1\rangle\right\}_i \qquad (7.12)$$

For instance, the network illustrated in Figure 7.3 has $\lambda_1 = 2.1010$. Then, by using $\alpha \equiv 0.4$ we obtain $K_i = 6.847$ and $K_j = 7.839$, indicating the fact that the node j is influenced not only by its nearest neighbours but also by those separated at a distance greater than 1. Obviously, for reporting the values of the Katz centrality index for a given network it is necessary to specify the value of the attenuation factor, and in this context there can arise the question concerning 'the best' value of α for a given network. Such difficulty is solved by considering the next centrality index.

7.2.2 Eigenvector centrality

A quantum leap was accomplished in the study of centrality measures when Bonacich (1972; 1987) introduced the definition of eigenvector centrality in which actors' centrality is a function of the centralities of those actors with whom they are related. Let us consider a vector of centralities for the nodes of a network, using the following variation of the Katz index:

$$\mathbf{c} = \left(\sum_{k=1}^{\infty} \alpha^{1-k} \mathbf{A}^k \right) |\mathbf{1}\rangle \tag{7.13}$$

Then, using the spectral decomposition of the adjacency matrix

$$\mathbf{A}^k = \sum_{j=1}^{n} |\varphi_j\rangle \langle \varphi_j | \lambda_j^k \tag{7.14}$$

we can write this centrality index as:

$$\mathbf{c} = \left[\sum_{k=1}^{\infty} \alpha^{1-k} \left(\sum_{j=1}^{n} |\varphi_j\rangle \langle \varphi_j | \lambda_j^k \right) \right] |\mathbf{1}\rangle = \left\{ \sum_{k=1}^{\infty} \alpha^{1-k} \left(\sum_{j=1}^{n} |\varphi_j\rangle \langle \varphi_j | \lambda_j^k \right) \right\} |\mathbf{1}\rangle$$

$$= \alpha \left[\sum_{i=1}^{n} \sum_{k=1}^{\infty} \left(\frac{\lambda_j}{\alpha} \right)^k |\varphi_j\rangle \langle \varphi_j| \right] |\mathbf{1}\rangle, \tag{7.15}$$

which finally produces the following expression for this centrality index in vector form:

$$\mathbf{c} = \alpha \left(\sum_{j=1}^{n} \frac{\lambda_j}{\alpha - \lambda_j} |\varphi_j\rangle \langle \varphi_j| \right) |\mathbf{1}\rangle. \tag{7.16}$$

It is straightforward to realise that as $\alpha \to \lambda_1$ from below the vector $\mathbf{c}$, approaches the eigenvector associated with the largest eigenvalue of the adjacency matrix: $\lim_{\alpha \to \lambda_{\bar{1}}} \mathbf{c} = \varphi_1$, which can also be written as

$$\varphi_1 = \left\{ \lim_{k \to \infty} \left[\frac{1}{k} \left(\mathbf{A} + \lambda_1^{-1} \mathbf{A}^2 + \lambda_1^{-2} \mathbf{A}^3 + \cdots + \lambda_1^{-k} \mathbf{A}^{k+1} \right) \right] \right\} |\mathbf{1}\rangle \tag{7.17}$$

The ith entry of the principal eigenvector of the adjacency matrix is known as the eigenvector centrality of node i

$$\varphi_i(i) = \left(\frac{1}{\lambda_1} \mathbf{A} \varphi_1 \right)_i \tag{7.18}$$

Another interpretation of the eigenvector centrality that will be useful in other parts of this book is given by using a result of Wei (1952) for non-bipartite networks. Let us consider the number $N_k(i)$ of walks of length k starting at node i of a non-bipartite connected network. Let $s_k(i) = N_k(i). \left[\sum_{j=1}^{n} N_k(j) \right]^{-1}$. Then, for $k \to \infty$, the vector $[s_k(1), s_k(2), \ldots s_k(n)]^T$ tends towards the eigenvector centrality. A proof can be found in Cvetković et al. (1997). That is, the eigenvector centrality of node i can be seen as the ratio of the number of walks of length k that departs from i to the total number of walks of length k in a non-bipartite connected network when the length of these walks is sufficiently large. Other mathematical properties of the eigenvector centrality have been studied in the literature (Britta, 2000; Grassi et al., 2007; Ruhnau, 2000).

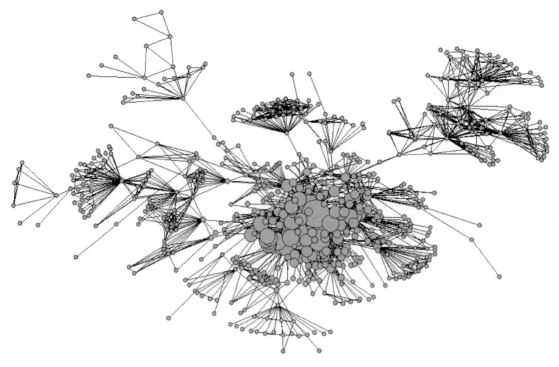

Fig. 7.4

Eigenvector centrality. Main connected component of the social network of injecting drug-users in Colorado Springs. Nodes are drawn with radii proportional to their eigenvector centrality.

For the simple network illustrated in Figure 7.3 the values of the eigenvector centrality of the two labelled nodes are, respectively, $\varphi_1(i) = 0.5000$ and $\varphi_1(i) = 0.5745$, as expected from the fact that node j is influenced not only by its nearest neighbours but also by those separated at distance longer than 1.

In Figure 7.4 we illustrate the eigenvector centrality for nodes in the main connected component of the social network of drug-users in Colorado Springs. In this case, the node with the highest eigenvector centrality coincides with that having the largest degree (see Figure 7.1). However, the general correlation between degree and eigenvector centralities for this network is very poor, having a correlation coefficient of only 0.37 (see the discussion in Chapter 19). This indicates that in this network most of the highest-degree nodes are connected to others which in general are not connected to many other nodes. In fact, many of them have degree 1.

When applied to directed networks, two types of eigenvector centrality can be defined by using the principal right and left eigenvectors of the adjacency matrix:

$$\varphi_1^R(i) = \left(\frac{1}{\lambda_1}\mathbf{A}\varphi_1^R\right)_i \tag{7.19}$$

$$\varphi_1^L(i) = \left(\frac{1}{\lambda_1}\mathbf{A}^T\varphi_1^L\right)_i \tag{7.20}$$

Right eigenvector centrality accounts for the 'importance' of a node by taking into account the 'importance' of nodes to which it points. That is, it is an extension of the outdegree concept by taking into account not only nearest neighbours. On the other hand, left eigenvector centrality measures the importance of a node by considering those nodes pointing towards the corresponding node, and is an extension of the indegree centrality. This is frequently referred to as 'prestige' in social sciences contexts. In Figure 7.5 we illustrate a directed network and the values of right and left eigenvector centralities. Node 1 is the most influential according to both measures, as it has two incoming and two outgoing links. According to right eigenvector centrality, nodes 4 and 5 are the second most central, as they have links pointing directly to node 1. However, according to the left eigenvector centrality, the second most influential nodes are 2 and 5, as they have links pointing from the most central node.

The difficulty arises when we try to apply right and left eigenvector centralities to networks where there are nodes having outdegree or indegree equal to zero, respectively. For instance, let us consider removal of the link connecting node 5 to node 1 in Figure 7.5. In this case, node 5 has one link pointing to it but none going out from it. Consequently, it has right eigenvector centrality equal to zero. The problem is that nodes pointing to node 5 do not receive any score for pointing to it, simply because its centrality is zero. Then, the right eigenvector centrality shows the same contribution for nodes 1, 2, 3, and 4, despite the fact that 1 is pointing to two nodes and the rest only to one. If we instead remove the link from 1 to 5 in Figure 7.5, then node 5 has one link going out from it but none pointing to it. Consequently, its left eigenvector centrality or prestige is equal to zero, as none point to it. As a consequence, node 1 receives a null influence from node 5 and the left eigenvector centrality ranks node 1 to 4 equally, despite node 1 being intuitively more central than the rest.

These situations can be solved by using the Katz centrality index redefined for in and out contributions. For instance,

$$K^{out} = \left[\left(\mathbf{I} - \frac{1}{\alpha} \mathbf{A} \right)^{-1} - \mathbf{I} \right] | \mathbf{1} \rangle \qquad (7.21)$$

with $\alpha \equiv 2$, produces 1.067, 1.133, 1.267, 1.067, 0.000, for the case when 1 points to 5 and no link goes out from 5, and

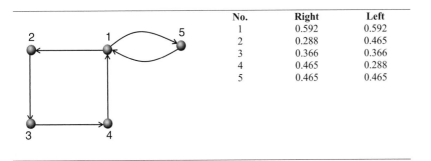

No.	Right	Left
1	0.592	0.592
2	0.288	0.465
3	0.366	0.366
4	0.465	0.288
5	0.465	0.465

Fig. 7.5
Right and left eigenvector centrality.
Right and left eigenvector centrality for a small directed network.

$$K^{in} = \langle \mathbf{1} | \left[\left(\mathbf{I} - \frac{1}{\alpha} \mathbf{A} \right)^{-1} - \mathbf{I} \right] \qquad (7.22)$$

with $\alpha \equiv 2$ produces 1.533, 1.267, 1.133, 1.067, 0.000, for the case when 5 points to 1 and no links point to it. The K^{in} is a measure of the 'prestige' of a node, as it accounts for the importance that a node has due to those that point to it.

A good characteristic of the eigenvector centrality is that it differentiates non-equivalent nodes having the same degree, such as nodes in Figure 7.3. However, this is not the case in regular networks. In these networks, all nodes have the same degree but are not necessarily equivalent. We have seen in Chapter 2 that $\lambda_1 = k$ for k-regular networks where the entries of the associated eigenvector are $\varphi_1(i) = 1/\sqrt{n}$ for all nodes in the graph. That is, in regular networks all nodes have the same eigenvector centrality, even though they can be topologically non-equivalent. The Katz centrality index does not differentiate among these nodes either, and in the next sections of this chapter we will see how to differentiate them using other centrality indices.

7.2.3   Subgraph centrality

The idea behind the subgraph centrality is that we can characterise the importance of a node by considering its participation in all closed walks starting (and ending) at it. Closed walks were introduced in Chapter 2, and used subsequently in other chapters. We recall that a closed walk of length l starting at node i is given by the ith diagonal entry of the lth power of the adjacency matrix of the network. Consequently, subgraph centralities can be defined by using the following function (Estrada and Rodríguez-Velázquez, 2005a):

$$f_i(\mathbf{A}) = \left(\sum_{l=0}^{\infty} c_l \mathbf{A}^l \right)_{ii} \qquad (7.23)$$

where coefficients c_l are selected such that the infinite series converges. The number of closed walks of length zero involving a given node is equal to 1. For any network, the number of closed walks of length $l = 2$ starting at node i is identical to the degree of this node, k_i. Additionally, if a network has no self loops, the number of closed walks of length $l = 1$ is zero, which allows us to write (7.23) as follows:

$$f_i(\mathbf{A}) - 1 = c_2 k_i + \left(\sum_{l=3}^{\infty} c_l \mathbf{A}^l \right)_{ii} \qquad (7.24)$$

We can then select the coefficients C_l such that the degree of a node has the largest contribution to $f_i(\mathbf{A})$, and as the length of the closed walk increases their contributions to the centrality becomes smaller. In other words, we can select C_l such as short closed walks involving a node make larger contributions than longer closed walks. The denomination of this index as 'subgraph' centrality derives from the following. Let us write the expression for this index in terms of the first few powers of the adjacency matrix:

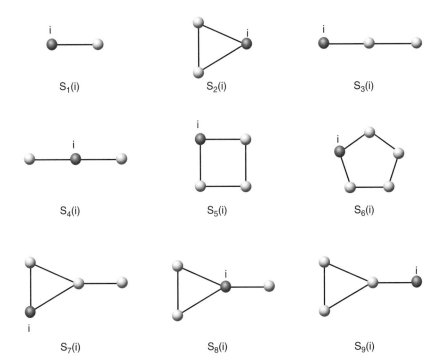

Fig. 7.6
Rooted subgraphs. Some small rooted subgraphs used in equations (7.26)–(7.30).

$$f(\mathbf{A})_i = c_2(\mathbf{A}^2)_{ii} + c_3(\mathbf{A}^3)_{ii} + c_4(\mathbf{A}^4)_{ii} + c_5(\mathbf{A}^5)_{ii} + \cdots \qquad (7.25)$$

Every diagonal entry of the powers of the adjacency matrix can be written in terms of the number of times the corresponding node participates in a rooted sugraph of given type, such as

$$(\mathbf{A}^2)_{ii} = |S_1(i)| \qquad (7.26)$$

$$(\mathbf{A}^3)_{ii} = 2|S_2(i)| \qquad (7.27)$$

$$(\mathbf{A}^4)_{ii} = |S_1(i)| + |S_3(i)| + 2|S_4(i)| + 2|S_5)(i)| \qquad (7.28)$$

$$(\mathbf{A}^5)_{ii} = 10|S_2(i)| + 2|S_6(i)| + 2|S_7|(i) + 4|S_8(i)| + 2|S_9(i)| \qquad (7.29)$$

where rooted subgraphs are illustrated in Figure 7.6. Then, $f_i(\mathbf{A})$ can be expressed as a weighted sum of the participations of node i in those rooted subgraphs:

$$f_i(\mathbf{A}) = (c_2 + c_4 + \cdots)|S_1(i)| + (2c_3 + 10c_5 + \cdots)|S_2(i)| + (c_4 + \cdots)|S_3(i)| + (c_4 + \cdots)|S_4(i)|$$

$$+ (2c_4 + \cdots)|S_5(i)| + (2c_5 + \cdots)|S_6(i)| + (2c_5 + \cdots)|S_7(i)| + (4c_5 + \cdots)|S_8(i)|$$

$$+ (2c_5 + \cdots)|S_9(i)| + \cdots \qquad (7.30)$$

There are, of course, many ways of selecting these coefficients fulfilling both conditions—the convergence and weighting scheme (Rodríguez et al., 2007). One particularly useful weighting scheme is the one used for the definition of the Estrada index discussed in Chapter 5, which eventually converges to the exponential of the adjacency matrix (Estrada and Rodríguez-Velázquez, 2005a):

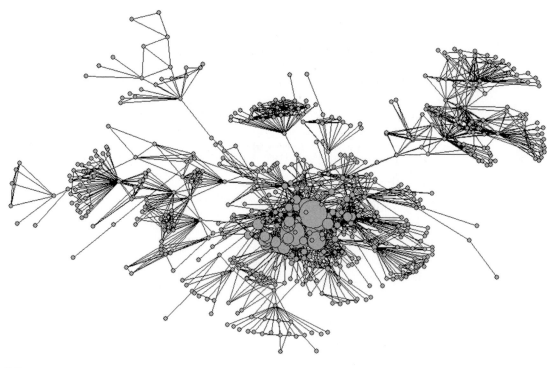

Fig. 7.7
Subgraph centrality. Main connected component of the social network of injecting drug-users in Colorado Springd. Nodes are drawn with radii proportional to their subgraph centrality.

$$EE(i) = \left(\sum_{l=0}^{\infty} \frac{\mathbf{A}^l}{l!} \right)_{ii} = \left(e^{\mathbf{A}} \right)_{ii} \qquad (7.31)$$

In a similar way as for the Estrada index, we can define subgraph centralities that take into account only contributions from odd or even closed walks in the network (Estrada and Rodríguez-Velázquez, 2005b):

$$EE_{odd}(i) = (\sinh \mathbf{A})_{ii} \qquad (7.32)$$

$$EE_{even}(i) = (\cosh \mathbf{A})_{ii} \qquad (7.33)$$

A comparison between subgraph centrality and other centrality measures, as well as the correlation between local and global network properties, has been studied by Jungsbluth et al. (2007). In Figure 7.7 we illustrate the subgraph centrality for the nodes in the social network of drug-users analysed previously. In this case it is clearly observed that the most central individuals are located in the main core of the network, indicating a large degree of inbreeding in their relationships.

A characteristic of the subgraph centrality in directed networks is that it accounts for the participation of a node in directed walks. We recall that a directed closed walk is a succession of directed links of the form $uv, vw, wx, \ldots, yz, zu$. This means that the subgraph centrality of a node in a directed network is $EE(i) > 1$ only if there is at least one closed walk that

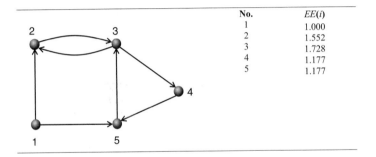

No.	EE(i)
1	1.000
2	1.552
3	1.728
4	1.177
5	1.177

Fig. 7.8
Node returnability. Subgraph centrality of a node in a directed network indicates the returnability of 'information' to this node, as illustrated in this directed graph.

starts and returns to this node. Otherwise, $EE(i) = 1$. Consequently, subgraph centrality of a node in a directed network indicates the 'returnability' of information to this node (Estrada and Hatano, 2009b). For instance, node 1 in the graph illustrated in Figure 7.8 has no returnability, as there is no directed walk starting and ending at it. Node 2 takes part in the 2-cycle 2–3–2 as well as in 2–3–4–5–3–2. However, the most central node is node 3, which takes part in 3–2–3 and in the triangle 3–4–5–3, among other longer walks.

A real-world example is provided by the Roget's Thesaurus network, where nodes represent words and there is a link from word A to word B if there is a reference in the Thesaurus to the latter among the words and phrases of the former, or if the two categories are directly related to each other by their positions in the book. Then, the returnability here has a clear meaning representing the cyclic equivalence of definitions for a given word. For instance, in Figure 7.9

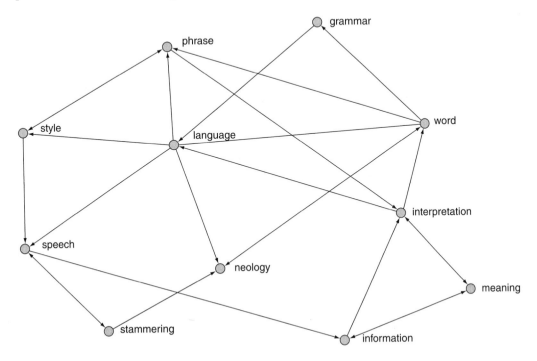

Fig. 7.9
Subnetwork of language. A small subnetwork around the word 'language' in the Roget's Thesaurus network.

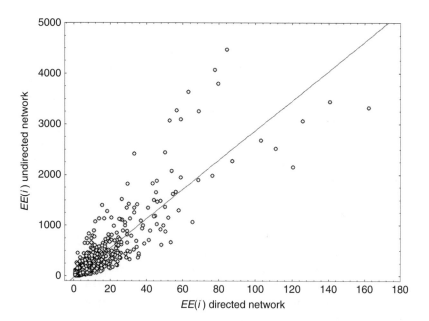

Fig. 7.10

Correlation between subgraph
centralities. Linear relationship between
the subgraph centrality obtained for
directed and undirected versions of the
Roget's Thesaurus network.

we illustrate a subnetwork for the word 'language.' In the meaning of this word, the word 'phrase' appears, from which we can go to 'interpretation' and then back to 'language'. Longer cycles also exist: for example, 'language–phrase–style–speech–information–meaning–interpretation–grammar–language.'

It is interesting to note that the information contained in the subgraph centrality differs when we use a directed and an undirected version of the same network. In Figure 7.10 we can see that the most central words, according to the subgraph centrality in the Roget's Thesaurus network do not coincide for the two versions of this network. While in the directed network the words 'information', 'error', and 'deception' are the most central, in the undirected graph the most central ones are 'inutility', 'neglect', and 'preparation,' which indicates that the participation of a node in directed cycles differs from that in undirected subgraphs.

In some situations we would be more interested in specific subgraphs than in the account of all subgraphs in which a node participates. In these cases it is possible to define subgraph-based centrality indices. Such an approach has been conducted by Koschützki and Schreiber (2008) by defining the so-called 'motif-based centrality'. In this case we need an algorithm approach that counts the number of times in which a node participates in a given subgraph. These indices have been extended by considering the different possible 'roles' of the nodes in a directed subgraph. These 'roles' are defined based simply on the topological differences of the nodes in a directed subgraph. For instance, in the so-called 'feed-forward loop' (see Chapter 4) there are three roles, because there is one node that points to the other two—a 'master regulator'—another which is pointed to by another at the same time that it points to a third—an 'intermediate regulator'—and another which only receives information and can be considered as a 'regulated node' (Koschützki and Schreiber, 2008) (see Chapter 13 for discussion).

7.2.4 PageRank centrality

One important application of centrality measures is that of ranking citations pages on the World Wide Web. The importance of this ranking is obvious when we consider that the quality of our searches through the Web depends on it. If we search the phrase 'complex network' on Google, for instance, we expect to obtain a ranking of Web pages related to the topic of complex network theory, with the most relevant ones at the top. The Google search engine which allows this ranking is based on a node centrality measure known as PageRank, which was originally proposed by Brin and Page (1998). Excellent accounts of this topic include the book by Langville and Meyer (2006), and the review papers by Bryan and Leise (2006) and Langville and Meyer (2005). A good interpretation of PageRank in the context of quantum mechanics has been presented by Perra et al. (2009).

The main idea behind PageRank is that the importance of a Web page relies on the importance of other Web pages pointing to it. In other words, the PageRank of a page is the sum of the PageRanks of all pages pointing to it. Another intuitive idea incorporated in PageRank is that the importance of the fact that one relevant Web page is pointing to another depends on the initial number of recommendations on the first page. Suppose that an important Web page A is pointing to another page B, then the PageRank of B depends on the number of links extending from the page A. If the outdegree of A is very large, the importance of being recommended by it is diluted into all the other recommended pages. However, if A is only pointing to B, the last one is receiving all the recommendations, and its importance increases. Mathematically, this intuition is captured by a modified adjacency matrix $\mathbf{H}$, whose entries are defined as follow:

$$H_{ij} = \begin{cases} 1/k_i^{out} & \text{if thre is a link from } i \text{ to } j \\ 0 & \text{otherwise} \end{cases}$$

The vector of PageRank centralities can be obtained through an iterative process:

$$\langle \pi^{k+1}| = \langle \pi^k|\mathbf{H} \tag{7.34}$$

where $\langle \pi^k|$ is the vector of PageRank at the k th iteration.

In Figure 7.11 we illustrate a simple example of six Web pages and the corresponding matrix $\mathbf{H}$. As can be seen, this matrix is 'almost' row-stochastic, except for the fact that for nodes having zero outdegree, the row sum of $\mathbf{H}$ corresponding to them is zero instead of 1 (see, for instance, node 4). These dangling nodes represent dead-end Web pages—a sort of *cul-de-sac*. If we consider a random surfer of the Web bouncing from one page to another by using the outlinks of the Web pages, he will be stacked on these dead ends without the chance of visiting any other site.

In order to solve the problem of dead-end nodes in ranking Web pages, PageRank modifies the matrix $\mathbf{H}$ by transforming it into a row-stochastic matrix $\mathbf{S}$. The new matrix is defined as follows:

$$\mathbf{S} = \mathbf{H} + \mathbf{a}\left[(l/n)\mathbf{1}^T\right] \tag{7.35}$$

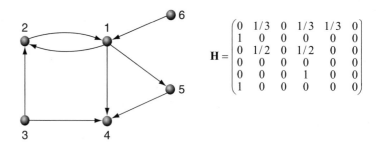

Fig. 7.11
Transformation of the adjacency matrix. The construction of the **H** matrix for the small network shown in the left part of the figure.

where the entries of the dangling vector **a** are given by

$$a_i = \begin{cases} 1 & \text{if } k_i^{out} = 0 \\ 0 & \text{otherwise} \end{cases}$$

The new matrix **S** has several attractive features, as it represents the transition probability matrix of a Markov chain, guaranteeing that a unique positive PageRank vector exists. However, it is frequently the case that a Web-surfer abandons his random approach of bouncing from one page to another and initiates a new search simply by typing a new destination in the browser's URL command line. In order to account for this 'teleportation' effect in the Web search, PageRank introduces a scaling parameter $0 \le \alpha \le 1$ in order to obtain the following matrix:

$$\mathbf{G} = \alpha \mathbf{S} + (\frac{1-\alpha}{n})\mathbf{11}^T \tag{7.36}$$

Finally, Google's PageRank is obtained by the following vector

$$\langle \pi^{k+1}| = \langle \pi^k|\mathbf{G} \tag{7.37}$$

which is the power method to obtain the eigenvectors of **G**. As the matrix **G** is row-stochastic, its largest eigenvalue is equal to 1, and the *principal left-hand eigenvector*[1] of **G** is given by

$$\langle \pi| = \langle \pi|\mathbf{G} \tag{7.38}$$

where $\langle \pi|\mathbf{1}\rangle = 1$ (Langville and Meyer, 2006).

According to PageRank, the most relevant Web page in Figure 7.11 is number 4, which has the maximum number of links pointing to it. The most interesting situation arises from analysis of nodes 1 and 2. They each have two incoming links, but PageRank gives more relevance to page 1 than to page 2. Why? The most important page which is pointing to 2 is 1. Page 1 is pointing to two other Web pages apart from 2, which means that the relevance of being recommended by 1 is diluted among three Web pages. However, the most

---

[1] Some authors define the adjacency matrix of a directed network as the transpose of the one defined in this book, which is used most. They then use the principal right-hand eigenvector of the matrix G instead. This is the case, for instance, in Newman (2010). The results are obviously the same.

Table 7.2 **PageRank of Nodes**. Values of the PageRank centrality index for the nodes of the graph in Figure 7.11.

No.	1	2	3	4	5	6
$\pi(i)$	0.2665	0.1697	0.0611	0.2900	0.1416	0.0611

important page which is pointing to 1 is 2, which recommends only this page. Thus, its recommendation is not diluted among many pages, and 1 receives the highest rank. In Table 7.2 we show the values of PageRank for all nodes in the network illustrated in Figure 7.11.

7.2.5 Vibrational centrality

In Chapter 6 we studied a communicability function based on a vibrational approach to complex networks. We can extend this idea in order to define a centrality measure for nodes based on the displacement of such nodes due to small vibrations or oscillations in the network. This centrality can be defined as the mean displacement of the corresponding node, and is defined and analysed below.

The mean displacement of a node i can be expressed by

$$\Delta x_i \equiv \sqrt{\langle x_i^2 \rangle} = \sqrt{\int x_i^2 P(\vec{x}) d\vec{x}} \tag{7.39}$$

where $\langle \cdots \rangle$ denotes the thermal average, and $P(\vec{x})$ is the probability distribution of the displacement of the nodes given by the Boltzmann distribution.

Using the same procedure as in Chapter 6, we can express the mean displacement of a node as follows:

$$\Delta x_i \equiv \sqrt{\langle x_i^2 \rangle} = \sqrt{\frac{\tilde{I}_i}{\tilde{Z}}} = \sqrt{\sum_{v=2}^{n} \frac{U_{iv}^2}{\beta \theta \mu_v}} \tag{7.40}$$

The values of the node displacements can be obtained by using the pseudoinverse or Moore–Penrose generalised inverse $\mathbf{L}^+$ of the graph Laplacian (see Chapter 2). Using $\mathbf{L}^+$, the node displacements are written as (Estrada and Hatano, 2010a; 2010b)

$$(\Delta x_i)^2 = \frac{1}{\beta \theta} (\mathbf{L}^+)_{ii} \tag{7.41}$$

It can be seen that this centrality measure is related to the resistance distance studied in Chapter 4, as follows:

$$(\Delta x_i)^2 = (\mathbf{L}^+)_{ii} = \frac{1}{n} \left(R_i - \frac{Kf}{n} \right) \tag{7.42}$$

where R_i is the sum of resistance distances for the node i, and Kf is the Kirchhoff index of the network. In the case of trees it is easy to realize that

$$(\Delta x_i)^2 = \frac{1}{n}\left(s_i - \frac{W}{n}\right) \tag{7.43}$$

and the node displacement can be expressed in terms of the shortest path distances as follows

$$(\Delta x_i)^2 = \frac{2n-1}{2n^2}\left[s_i - \left(\frac{1}{2n-1}\right)\sum_{q\neq i}s_q\right] \tag{7.44}$$

The term s_i in (7.44) represents the sum of all distances that information has to travel in order to visit all nodes in the network starting from i. After the information arrives at a particular node we can consider that some part of this information is 'absorbed' by the node and other parts are transmitted from it to the rest of the nodes in the network. We will refer to this information as being reflected from the particular node. Then, the term $-\sum_{q\neq i}s_q$ in (7.44) represents the information which is reflected from all nodes that have received it from the source i. Consequently, $(\Delta x_i)^2$ represents the difference between the total amount of information transmitted by node i minus the fraction of information which is reflected from the rest of the nodes. In other words, it is the amount of information which is absorbed by the nodes of a network after it is transmitted from a source i.

A further analysis of the relationship between node vibrations and the resistance distance is possible. For the sake of simplicity we consider the case $\beta\theta \equiv 1$. Here, the resistance distance between a pair of nodes in a network can be expressed in terms of the node displacements as follows:

$$\Omega_{ij} = \left[(\Delta x_i)^2 + (\Delta x_j)^2 - \langle x_i x_j\rangle - \langle x_i x_j\rangle\right] = \langle(x_i - x_j)^2\rangle \tag{7.45}$$

In a similar way, the Kirchhoff index of a network, introduced in Chapter 3, can be expressed as the sum of the squared node displacements multiplied by the number of nodes in the network:

$$Kf = n\sum_{i=i}^{n}(\Delta x_i)^2 = n^2\overline{(\Delta x)^2} \tag{7.46}$$

The resistance distance sum can be expressed in terms of the node displacements as

$$R_i = n(\Delta x_i)^2 + \sum_{i=1}^{n}(\Delta x_i)^2 = n\left[(\Delta x_i)^2 + \overline{(\Delta x)^2}\right] \tag{7.47}$$

due to the fact that $\sum_{j=1}^{n}L_{ij}^+ = 0$. Consequently, a plot of R_i versus $(\Delta x_i)^2$ represents a straight line with slope equal to the number of nodes and intercept equal to the sum of squared node displacements. Then, the potential energy of the vibrations in a network can be expressed as

$$\langle V(\vec{x})\rangle = \frac{1}{2n}\sum_{i=1}^{n}k_i R_i - \frac{1}{2n}\sum_{i,j\in E}(R_i + R_j - n\Omega_{ij}) \tag{7.48}$$

where k_i is the degree of node i.

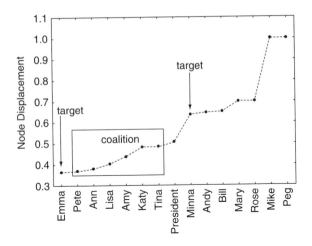

Fig. 7.12

Node displacements in a small office network. Node displacements of every member of the social network of friendship ties in the office of an overseas branch of a large international organisation.

We now return to the example of the small office (Thurman, 1979) studied in Chapter 6, and calculate the displacement of every node from its equilibrium position due to such vibrations. The results are illustrated in Figure 7.12, in which node displacements are displayed for every member of the office. It can be seen that the members of the coalition display the smallest displacements, which means that they are the least affected by all the vibrations taking place in the network. That is, most of the information they transmit is absorbed by other members of the network. In other words, they are very influential.

The two targets of their attacks correspond to the two people who display displacements immediately larger or smaller than the members of the coalition. Emma has the smallest displacement of the whole network, but she is not a member of the coalition. Minna's displacement is larger than that of the members of the coalition (note the gap between the displacement of the President and Minna in Figure 7.12). We can think that the President, which is between the coalition and Minna in terms of displacements, has other attributes due to his/her responsibilities in the organisation, which makes him/her less vulnerable to possible attacks. Then, the two targets of the attacks are those people who are similar to the members of the coalition in terms of displacements, whom members of the coalition could have considered as possible threats to their positions.

Even more interesting is the fact observed by Thurman (1979) that during this period of time 'Emma had retained and strengthened her position' by gaining more control of resources in the office. This is possibly due to the fact that most of the information that she transmits is absorbed by the others, which can enable her to change their opinions with respect to herself. However, Minna 'had failed to establish herself', as she has been unable to gain access to power in the office, and even her social relationships were dramatically deteriorated, making her feel 'frustrated, useless, and heavily criticized.' She has very little influence over the others, as the information she transmits is almost totally reflected by them. Figure 7.12 indicates the cause of what happened. Emma was very robust in her position in the social network, her 'equilibrium position' was affected very little by the 'vibrations' of the others in the network, and she was in a position of low vulnerability. However, Minna was in a more vulnerable situation with respect to the members of the coalition, who succeeded in their attack against her. Node displacement can therefore be seen as a measure of node vulnerability to external stress in a complex network.

7.3 Closeness centrality

In some situations we could be interested not in those nodes which are central due to their number of connections or relationships with other well connected nodes, but those which are relatively close to all other nodes in the network. Let us suppose that we are interested in giving some information to an individual in a social network in such a way that this information is communicated to the rest of the network in the shortest possible number of steps. If we consider, for instance, the small social network illustrated in Figure 7.13 we could think that giving such information to Steve would be a good starting point, as he is the most connected person in the network. Then, although Steve can pass information to five other people in only one step, the information will need three steps to reach Jo and Louis. However, if the information is given to Irene in the first instance, it can reach every other actor in the network in no more than two steps.

In these cases an appropriate measurement of the centrality of a node is the inverse of the sum of shortest path distances from the node in question to all other nodes in the network. This centrality measure is known as *closeness centrality* (Wasserman and Faust, 1994; Freeman, 1979), which can be expressed mathematically as follows:

$$CC(u) = \frac{n-1}{s(u)} \qquad (7.49)$$

where the distance sum is taken as defined in Chapter 3:

$$s(u) = \sum_{v \in V(G)} d(u, v) \qquad (7.50)$$

In Table 7.3 we give the values of the closeness centrality for all actors in the social network illustrated in Figure 7.13.

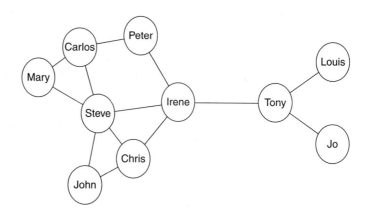

Fig. 7.13
Closeness in a small social network.
Network used to illustrate the necessity for defining closeness centrality measures for nodes in networks.

Table 7.3 **Closeness centrality**. Values of the closeness centrality of the nodes in the network illustrated in Figure 7.13.

Actor	$CC \cdot 100$	Actor	$CC \cdot 100$
Irene	64.28	Carlos	45.00
Steve	60.00	Mary	42.86
Chris	52.94	John	40.90
Tony	50.00	Jo	34.62
Peter	47.37	Louis	34.62

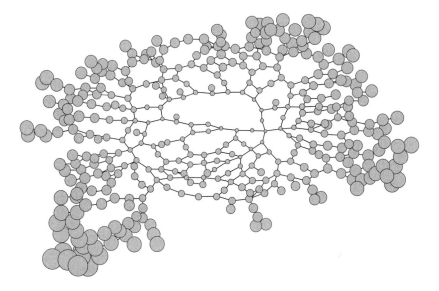

Fig. 7.14
Farness centrality in an urban network. Farness centrality for the urban street network of the old part of Cordoba.

The inverse of closeness centrality is known as *farness centrality*, which is illustrated for the urban street network of Cordoba in Figure 7.14. As can be seen in the figure, farness centrality clearly defines a core-periphery structure of the city. The city core is located at the geographical centre where most places display small farness centrality, while the periphery is characterised by locations with large farness centrality.

A possible difficulty with the closeness/farness centrality arises in situations such as that illustrated in Figure 7.15 for the nodes labelled as 1 and 2. These two nodes have the same sum of shortest path distances from them to the rest of the nodes in the network. That is, $s(1) = s(2) = 11$, and consequently both nodes have the same closeness/farness centrality. However, node 1 needs three steps to reach node 7, while all nodes are reached in only two steps from node 2. Then, intuitively, we could think that node 2 is a little more central than node 1.

This situation can be resolved if we consider an analogy of the eigenvector centrality for the shortest path distances. In Chapter 3 we introduced the distance matrix $\mathbf{D}$ for a given network. Some of the spectral properties of the distance matrix have been studied by Zhou and Trinajstic (2007). Here, we consider the i th entry of principal eigenvector $\delta_1(i)$ of the distance matrix, which can be considered as a spectral farness centrality of node i:

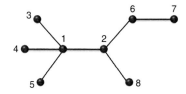

Fig. 7.15
Small network with equidistant nodes. Nodes labelled as 1 and 2 have the same distance sum and consequently the same closeness centrality, despite that intuitively node 2 appears to be more 'central' than node 1 (see text for explanation).

Table 7.4 **Farness and spectral farness of nodes**. Values of the 'classical' and spectral farness centrality for the nodes in the network represented in Figure 7.15.

Node	$s(i)$	$\delta_1(i)$	Node	$s(i)$	$\delta_1(i)$
1	11	0.2527	5	17	0.3771
2	11	0.2518	6	15	0.3278
3	17	0.3771	7	21	0.4440
4	17	0.3771	8	17	0.3763

$$\delta_1(i) = \left(\frac{1}{\eta_1} \mathbf{D}\delta_1 \right)_i \tag{7.51}$$

where η_1 is the largest eigenvalue of $\mathbf{D}$, and δ_1 is the eigenvector associated with it. In this case the nodes of the graph in Figure 7.15 have the farness and spectral farness centrality shown in Table 7.4.

In a similar way as for closeness, we can define a spectral closeness centrality simply by taking the inverse of $\delta_1(i)$. Closeness/farness and their spectral analogues are strongly correlated for most networks, which indicates that the use of the spectral measures has little effect on the ranking introduced by the non-spectral ones, although the first could be useful in differentiating nodes in which situations like the one previously analysed can arise.

A couple of measures which are related to closeness/farness have been proposed for directed networks by Valente and Foreman (1998), who called these centrality measures the 'integration' and the 'radiality' of a node. The integration of node i measures the degree to which its inward nomination integrates it into the network. On the other hand, radiality accounts for the degree to which an individual's outward nominations reach out into the network (Valente and Foreman, 1998).

7.4 Betweenness centrality

In this section we will consider the relative importance of a node in the communication between other pairs of nodes. Those nodes facilitate or inhibit the communication between other nodes in the network. These centrality measures account for the proportion of information that passes through a given node in communicating other pairs of nodes in the network, and they are generically named *betweenness centrality* (Freeman, 1979).

The most intuitive way of defining the betweenness centrality is by considering that the information going from one node to another travels only through the shortest paths connecting those nodes. Then, if $\rho(i, j)$ is the number of shortest paths from node i to node j, and $\rho(i, k, j)$ is the number of these shortest paths that pass through node k in the network, the betweenness centrality of node k is given (Wasserman and Faust, 1994; Freeman, 1979) by:

$$BC(k) = \sum_i \sum_j \frac{\rho(i, k, j)}{\rho(i, j)}, \quad i \neq j \neq k \tag{7.52}$$

Obviously, communication in a complex network does not always take place through the shortest paths connecting pairs of nodes. In many real-world situations such communication occurs by using some or even all of the available channels to go from one place to another in the network. Several measures have been proposed to account for the betweenness of a node when communication takes place by using such other alternative routes.

One of these betweenness centrality measures, accounting for the flow of information between nodes by using not only shortest paths, was introduced by Freeman, Borgatti, and White as *the flow betweenness* (Freeman et al., 1991). Apparently, this measure had been previously proposed by White and Smith in an unpublished work of 1988. The flow betweenness centrality of node k in a network is defined as follows:

$$BC_f(k) = \sum_i \sum_j \frac{f(i, k, j)}{f(i, j)}, \quad i \neq j \neq k \tag{7.53}$$

where $f(i, j)$ is the maximum flow from node i to node j, and $f(i, k, j)$ is the maximum flow between those nodes that passes through node k. The *maximum flow* between two nodes is defined as the minimum cut capacity. Let us consider two nodes i and j in a network, which are connected through at least one path. Let $E_{ij} \subseteq E$ be a subset of links containing at least one link from every path between i and j. A subset E_{ij} represents an $i-j$ *cutset*, as the removal of the links contained in it separates i and j into two isolated components. The *capacity* of E_{ij} is defined as the sum of capacities of the links contained in this set, where the capacities are simply the weights of the links or ones if the network is unweighted. If we build all $i-j$ cutsets we can obtain the *minimum cut capacity* as the smallest capacity of any of these cutsets. Then, the *maximum flow* from i to j is the minimum cut capacity of all $i-j$ cutsets.

Another variation of the betweenness centrality was introduced by Newman (2005a). In this case the *random walk betweenness centrality* is based on the number of times a random walk between i and j passes through node k. This centrality is identical when considering the betweenness based on electrical currents—for example, a current-flow betweenness centrality. It is defined as follows:

$$BC_{RW}(r) = \sum_{i<j} I_r^{pq} \tag{7.54}$$

where $I_p^{pq} = I_q^{pq} = 1$ and

$$I_r^{pq} = \frac{1}{2} \sum_j A_{rj} |T_{rp} - T_{rq} - T_{jp} + T_{jq}| \quad r \neq p, q \tag{7.55}$$

The matrix $\mathbf{T}$ is obtained as follows. First, remove any row k and the corresponding kth column of the Laplacian matrix to obtain an $(n-1) \times (n-1)$ matrix $\mathbf{L}_k$. Then obtain the matrix $\mathbf{L}_k^{-1}$, and add back in a new row and column consisting of all zeros in the position from which the row and column were previously removed.

A *communicability betweenness centrality* has also been proposed. In this case the communicability betweenness of node r is given by (Estrada et al., 2009).

$$BC_r = \frac{1}{C} \sum_p \sum_q \frac{G_{prq}}{G_{pq}}, \quad p \neq q, \ p \neq r, \ q \neq r$$

where G_{prq} is the corresponding weighted sum where we consider only walks that involve node r, and $C = (n-1)^2 - (n-1)$ is a normalization factor equal to the number of terms in the sum, so that BC_r takes values between zero and 1.

This betweenness centrality is calculated as follows. Let $\mathbf{A} + \mathbf{E}(r)$ be the adjacency matrix of the network $G(r) = (V, E')$ resulting from the removal of all edges connected to the node $r \in V$, but not the node itself. The matrix $\mathbf{E}(r)$ has non-zeros only in row and column r, and in this row and column it is -1 wherever $\mathbf{A}$ has $+1$. For simplicity, we often write $\mathbf{E}$ instead of $\mathbf{E}(r)$. Then,

$$G_{prq} = \left(e^{\mathbf{A}}\right)_{pq} - \left(e^{\mathbf{A}+\mathbf{E}(r)}\right)_{pq} \tag{7.56}$$

and the communicability betweenness is given by

$$BC_r = \frac{1}{C} \sum_p \sum_q \frac{\left(e^{\mathbf{A}}\right)_{pq} - \left(e^{\mathbf{A}+\mathbf{E}(r)}\right)_{pq}}{\left(e^{\mathbf{A}}\right)_{pq}}, p \neq q, p \neq r, q \neq r \tag{7.57}$$

The numerator of the right-hand side of (7.57) can be understood in terms of the directional or Fréchet derivative:

$$D_{\mathbf{E}(r)}(\mathbf{A}) = \lim_{h \to 0} \frac{1}{h}\left[e^{(\mathbf{A}+h\mathbf{E}(r))} - e^{\mathbf{A}}\right] \tag{7.58}$$

which accounts for the effect of introducing an infinitesimal perturbation to the adjacency matrix exponential in the direction $\mathbf{E}$. The Fréchet derivative can be expressed as a Taylor series (Al-Mohy and Higham, 2009):

$$D_{\mathbf{E}(r)}(\mathbf{A}) = \sum_{k=1}^{\infty} \frac{1}{k!} \sum_{j=1}^{k} \mathbf{A}^{j-1} \mathbf{E}(r) \mathbf{A}^{k-j} \tag{7.59}$$

which allows us to express the communicability betweenness in terms of powers of the adjacency matrix.

Analysis of the classical betweenness centrality in large complex networks has been carried out by Barthélemy (2004) and Goh et al. (2001b). In Figure 7.16 we give the values of the betweenness centrality measures previously defined for a simple network having seven nodes. As can be seen, the shortest path betweenness identifies node 2 as the most central one. However, it assigns the same 'importance' to nodes 3 and 4. When not only shortest paths are taken into account, both flow and random walk betweenness identify node 4 as the second most central one in this network, following node 2. Communicability betweenness identifies node 3 as the most central one instead of node 2, which indicates that node 3 plays a central role in distributing information among the other nodes when all kind of walk are used as communication channels.

In Figure 7.17 we illustrate the urban street network of the central part of Cordoba, with nodes of radii proportional to the betweenness centralities defined previously. In this case, shortest path and communicability between-

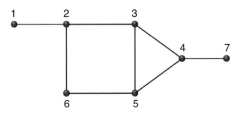

Fig. 7.16
Betweenness centralities of nodes.
Values of different betweenness
centrality measures for the nodes of a
small network.

k	BC(k)	$BC_f(k)$	$BC_{RW}(k)$	$BC_C(k)$
1	0.0	0.000	0.286	0.882
2	5.5	13.667	0.619	13.535
3	5.0	6.000	0.576	15.162
4	5.0	10.667	0.584	13.385
5	2.5	6.000	0.545	11.909
6	1.0	3.667	0.450	5.801
7	0.0	0.000	0.286	0.821

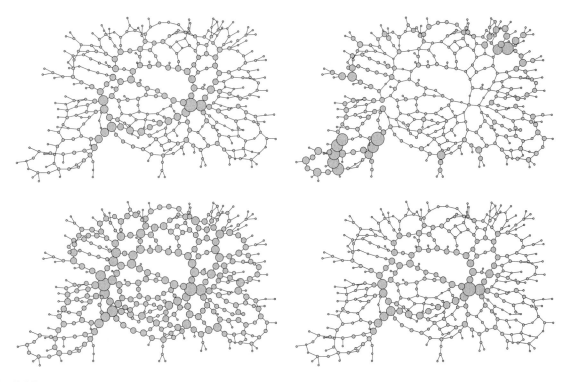

Fig. 7.17
Betweenness centrality in a urban network. Shortest path (top left), flow (top right), random walk (bottom left) and communicability
(bottom right) betweenness centrality for the urban street network of the old part of Cordoba.

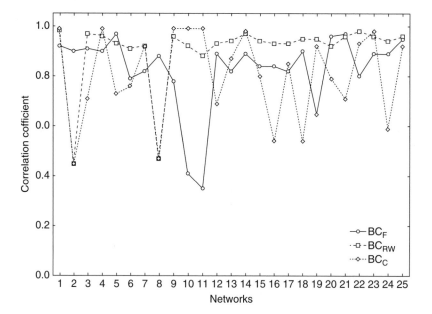

Fig. 7.18

Correlation between different betweenness centralities. Correlation coefficient between different betweenness measures for twenty-five real-world networks in different scenarios.

ness identify the most central streets at the borderline between the central core of the city and its periphery, while flow betweenness identifies several nodes in the periphery as the most central ones, and the random walk betweenness identifies nodes spread across the entire network.

Betweenness centrality measures that include not only shortest path information are not always correlated with the shortest path betweenness. For instance, in Figure 7.18 we illustrate the correlation coefficients between shortest path betweenness and the other three measures described in this section for 25 complex networks representing real-world systems. In general, the highest correlations with the shortest path betweenness are obtained for the random walk betweenness. However, these correlations depend very much on the specific networks analysed, showing that some of the measures are truly collinear or orthogonal to the shortest path betweenness for some specific cases.

7.5 Information centrality

A centrality measure based on the information that can be transmitted between any two points in a connected network has been proposed by Stephenson and Zelen (1989). If $\mathbf{A}$ is the adjacency matrix of a network, $\mathbf{D}$ a diagonal matrix of the degree of each node, and $\mathbf{J}$ a matrix with all its elements equal to 1, then IC is defined by inverting the matrix $\mathbf{B}$ defined as $\mathbf{B} = \mathbf{D} - \mathbf{A} + \mathbf{J}$ in order to obtain the matrix $\mathbf{C} = (c_{ij}) = \mathbf{B}^{-1}$, from which the information matrix is obtained as follows:

$$\mathbf{I}_{ij} = (c_{ii} + c_{jj} - c_{ij})^{-1} \tag{7.60}$$

The information centrality of the nodes i is then defined by using the harmonic average:

$$IC(i) = \left[\frac{1}{N} \sum_j \frac{1}{\mathbf{I}_{ij}} \right]^{-1} \tag{7.61}$$

Stephenson and Zelen (1989) proposed defining $\mathbf{I}_{ii}$ as infinite for computational purposes, which makes $1/\mathbf{I}_{ii} = 0$.

The information centrality is related to the resistance distance introduced in Chapter 3. In order to show this relationship, we begin by noting that $\mathbf{B} = \mathbf{L} + \mathbf{J}$. If G is a connected network, then the inverse of $\mathbf{B}$ has the same eigenvectors as the Laplacian with eigenvalues $n, \lambda_2, \lambda_3, \ldots, \lambda_n$. The inverse of $\mathbf{B}$ is given by

$$\mathbf{B}^{-1} = \mathbf{L}^+ + \frac{1}{n^2} \mathbf{J} \tag{7.62}$$

where, as before, $\mathbf{L}^+$ is the Moore–Penrose generalised inverse of $\mathbf{L}$. Now we can write the information centrality as follows:

$$IC(i) = \left[c_{ii} + \frac{(T - 2R)}{n} \right]^{-1} \tag{7.63}$$

where $\mathbf{T} = \sum_{j=1}^{n} c_{jj}$ and $\mathbf{R} = \sum_{j=1}^{n} c_{ij}$. Then,

$$c_n = (\mathbf{B}^{-1})_{ii} = \left(\mathbf{L}^+ + \frac{1}{n^2} \mathbf{J} \right)_{ii} = (\mathbf{L}^+)_{ii} + \frac{1}{n^2} = \beta k (\Delta x_i)^2 + \frac{1}{n^2} \tag{7.64}$$

$$T = \sum_{j=1}^{n} c_{jj} = \mathrm{Tr}\, \mathbf{B}^{-1} = \mathrm{Tr} \left(\mathbf{L}^+ \frac{1}{n^2} \mathbf{J} \right) = \mathrm{Tr}\, \mathbf{L}^+$$

$$+ \frac{1}{n} = \beta k \sum_{j=1}^{n} (\Delta x_i)^2 + \frac{1}{n} \tag{7.65}$$

$$R = \sum_{j=1}^{n} c_{ij} = \sum_{j=1}^{n} (\mathbf{B}^{-1})_{ij} = \frac{1}{n} \tag{7.66}$$

because $\mathrm{Tr}\, \mathbf{J} = n$ and $\sum_{j=1}^{n} (\mathbf{L}^+)_{ij} = \sum_{i=1}^{n} (\mathbf{L}^+)_{ij} = 0$. Thus, information centrality can be written as (Estrada and Hatano, 2010a)

$$IC = (i) = \left[\beta k (\Delta x_i)^2 + \frac{1}{n^2} + \frac{\left(\beta k \sum_{i=1}^{n} (\Delta x_i)^2 + \frac{1}{n} - \frac{2}{n} \right)}{n} \right]^{-1} \tag{7.67}$$

or equivalently, that

$$IC(i) = \frac{1}{\beta k \left[(\Delta x_i)^2 + \overline{(\Delta x)^2} \right]} \tag{7.68}$$

which means that the information centrality of a node is proportional to the inverse of the square node displacement plus the average displacements for all nodes in the network.

Global network invariants

8

> Among the infinite diversity of singular phenomena science can only look for invariants.
>
> Jacques L. Monod

In Chapter 7 we analysed some local invariants for characterising the nodes of networks. Here we are interested in characterising the network as a whole through the use of structural invariants. One example of such global invariants is *centralisation indices* (Wasserman and Faust, 1994). Centralisation invariants are defined as quantifying the variability of the individual node centralities. That is, a network displaying the largest variability of a given centrality measure is considered to have the largest centralisation. A network for which a given centrality measure is the same for all nodes is then expected to have the lowest centralisation. A general centralisation index C^T for a centrality of type T —for example, degree, closeness, betweenness, and so on—has been proposed by Freeman (1979), using the following expression:

$$C^T = \frac{\sum\limits_{i=1}^{n} \left(C^T_{\max} - C^T_i\right)}{\max \sum\limits_{i=1}^{n} \left(C^T_{\max} - C^T_i\right)} \tag{8.1}$$

where C^T_i is the centrality of node i in the network, and $C^T_{\max} = \max\ C^T_i$. We also analyse here some other invariants that have been used in different contexts for the analysis of networks.

8.1 Adjacency-based invariants

Let us begin by considering the node degree centralisation $C^T \equiv C^D$. It is known that in a connected network the maximum degree that a node can have is $C^T_{\max} = n - 1$, and the minimum one is $C^T_i = 1$. This difference is obtained for all nodes in a star graph excepting the central node. Consequently, the degree centralisation reaches its maximum for the star, and a normalised expression has been provided by Freeman (1979) for this case:

$$C^D = \frac{\sum\limits_{i=1}^{n}(k_{\max} - k_i)}{(n-1)(n-2)} \tag{8.2}$$

This index runs from zero to 1, where the minimum is obtained for any regular network and the maximum is obtained for the star. Using the handshaking lemma (see Chapter 2) we can transform this expression to the following one, which is easy to calculate:

$$C^D = \frac{nk_{\max} - 2m}{(n-1)(n-2)} \tag{8.3}$$

where m is the number of links.

In order to apply definition (8.1) for centralisation it is necessary to know the network for which $\sum\limits_{i=1}^{n}(C^T_{\max} - C^T_i)$ is a maximum, which is not always the star graph. Determining the graph(s) for which a given measure is a maximum is not a trivial problem. Therefore, alternative measures have been proposed to account for the centralisation of a network. One such approaches is the use of centrality variance (Wasserman and Faust, (1994), defined as follows:

$$VAR^T = \frac{1}{n}\sum\limits_{i=1}^{n}\left(C^T_i - C^T_{mean}\right)^2 \tag{8.4}$$

where C^T_{mean} is the average of the centralities of type T in the network. If we consider the particular case of the degree, we have the following expression:

$$VAR^D = \frac{1}{n}\sum\limits_{i=1}^{n}\left(k_i - \bar{k}\right)^2 \tag{8.5}$$

This index was first proposed as a measure of centralisation by Snijders (1981), as 'an index of graph heterogeneity' based on the intuition that 'centralisation is synonymous with the dispersion or heterogeneity' of a node. It takes the minimum value for any regular graph, as there is no heterogeneity at all in their degrees. However, the maximum value depends on the number of nodes in the network. In an independent work, this index was proposed in 1992 by Bell (1992) as a measure of the 'irregularity' of a network.

The study of graph irregularity can be traced back to the seminal paper of Collatz and Sinogowitz (1957). These authors raised the question of finding the most irregular graphs of a given size by using the following measure:

$$CS(G) = \lambda_1 - \bar{k} \tag{8.6}$$

where λ_1 is the principal eigenvalue of the adjacency matrix, and $\bar{k}$ is the average degree. This index takes the value of zero for regular graphs, and Collatz and Sinogowitz (1957) have conjectured that stars maximise this index. They have shown that this is, in fact, the case for graphs with up to five nodes. However, Cvetković and Rowlinson (1988) have rejected this conjecture by constructing several families of graphs having larger values of $CS(G)$ than the corresponding stars of the same size. Other measures of irregularity for graphs have been proposed in the mathematics literature (Albertson, 1997; Chartrand et al., 1988; Gutman et al., 2005; Hansen and Mélot, 2005), and the interested reader is directed to it for details.

Different approaches to generate global network invariants have arisen in different contexts. For instance, the quadratic forms $\mathbf{p}^T \mathbf{A} \mathbf{q} = \langle \mathbf{p} | \mathbf{A} | \mathbf{q} \rangle$, where $\mathbf{p}$ and $\mathbf{q}$ are vectors, are frequently used to generate network invariants in the molecular sciences (Estrada, 2001). For instance,

$$M_1 = \langle \mathbf{v} | \mathbf{A} | \mathbf{1} \rangle = \langle \mathbf{v} | \mathbf{v} \rangle = \sum_{i \in V} k_i^2 \tag{8.7}$$

$$M_2 = \langle \mathbf{v} | \mathbf{A} | \mathbf{v} \rangle = \sum_{i,j \in E} k_i k_j \tag{8.8}$$

are known as the Zagreb indices (Nikolić et al., 2003). They were originally proposed by Gutman et al. (1975), and can be expressed in terms of structural fragments of the network (Braun et al., 2005) as

$$M_1 = 2m + 2|P_2| \tag{8.9}$$

$$M_2 = m + 2|P_2| + |P_3| + 3|C_3| \tag{8.10}$$

Similarly, we can obtain invariants like Zagreb indices based on link degrees instead of node degrees. These indices were used in Chapter 2 to obtain relationships connecting subgraphs and the assortativity coefficient, and they are defined as follows:

$$M_1^e = \langle \boldsymbol{\kappa} | \mathbf{E} | \mathbf{1} \rangle = \langle \boldsymbol{\kappa} | \boldsymbol{\kappa} \rangle = \sum_{i \in V} \kappa_i^2 \tag{8.11}$$

$$M_2^e = \langle \boldsymbol{\kappa} | \mathbf{E} | \boldsymbol{\kappa} \rangle = \sum_{i,j \in E} \kappa_i \kappa_j \tag{8.12}$$

In 1975, Milan Randić proposed a branching index which is nowadays known as the Randić index of a graph (Randić, 1975). The Randić index $^1R_{-1/2}$ of a network having $|E| = m$ undirected links is given by

$$^1R_{-1/2} = \frac{1}{2} \langle \mathbf{k}^{-1/2} | \mathbf{A} | \mathbf{k}^{-1/2} \rangle = \sum_{i,j \in E}^{m} (k_i k_j)^{-1/2} \tag{8.13}$$

where $\mathbf{k}^{-1/2}$ is a vector of the inverse square roots of node degrees. This index was then generalized by Bollobás and Erdös (1998) by changing the exponent $-1/2$ in (8.13) to any real value exponent α. The connection between Randić index, branching, and spectral properties of a graph can be found in the literature (Gutman and Vidović, 2002; Michalski, 1986). The analogy of the Randić index for links was introduced by Estrada (1995) as:

$$\varepsilon = \langle \boldsymbol{\kappa}' | \mathbf{E} | \boldsymbol{\kappa}' \rangle = \sum_{i \sim j} (\kappa_i \kappa_j)^{-0.5} \tag{8.14}$$

where $\boldsymbol{\kappa}'$ is a vector of the inverse square roots of link degrees. Many of these global invariants have been proposed and applied in the literature. In particular, in the context of molecular science they are known as 'topological indices'. The reader is referred to the authoritative collection of these indices compiled by Todeschini and Consonni (2000), in which a vast list of references is also found.

Let us now consider the problem of degree heterogeneity in complex networks. In the network literature, degree heterogeneity refers to the deviations from regularity in the degrees of a network. That is, in a regular network there is no heterogeneity in their degrees, as all of them have the same value. The degree heterogeneity is usually accounted for by using the degree distribution. It is said, for instance, that a network with fat-tailed degree distribution has large heterogeneity in its degree due to the intuitive fact that it deviates very much from regularity. We can use these intuitive ideas to define a quantitative measure of degree heterogeneity. We begin by considering the intuition behind this concept. In Figure 8.1 we can see that the distribution of the degrees in a regular graph consists only of a peak at the value of the degree. That is, the probability of finding a node with degree $\bar{k} \equiv k$ in a k-regular network is exactly 1, which is trivial. In a network with Poisson degree distribution we have some 'deviations' from this regularity, as can be seen in Figure 8.1. However, we still have some regularity in this network, as most of the nodes have degree about $\bar{k}$, and only a small fraction of nodes have very small or very large degree. Therefore, we can say that a Poissonian network is almost regular, and this is what is used in network language. Now we turn to a network with a power-law degree distribution. In this case the peak of maximum probability is far from $\bar{k}$. In fact, for a network having $p(k) \sim k^{-\gamma}$ this maximum takes place at the value of $k = 1$. Then, the extreme of this deviation from the regular graph takes place for the star graph in which the probability of finding a node with degree one is $(n - 1)/n$ and that of finding a node with degree $n - 1$ is n^{-1} (see Figure 8.1).

Following previous intuition, we can account for the deviations from the regularity of a network by using the following expression:

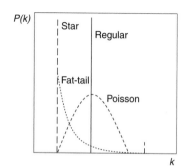

Fig. 8.1

Degree distribution and heterogeneity. The intuitive relation between degree distributions and deviations from a regular graph.

$$\frac{1}{2} \sum_{(i,j)\in E} \left(k_i^\alpha - k_j^\alpha\right)^2 \qquad (8.15)$$

Let us represent the degrees raised to the power α in a vector form: $|\mathbf{k}^\alpha\rangle = (k_1^\alpha, k_2^\alpha, \cdots, k_n^\alpha)$. The problem of minimising expression (8.15) can then be stated as a quadratic form of the Laplacian matrix (Bernstein, 2009):

$$\frac{1}{2} \sum_{(i,j)\in E} \left(k_i^\alpha - k_j^\alpha\right)^2 = \langle \mathbf{k}^\alpha | \mathbf{L} | \mathbf{k}^\alpha \rangle \qquad (8.16)$$

Then, we can write

$$\langle \mathbf{k}^\alpha | \mathbf{L} | \mathbf{k}^\alpha \rangle = \sum_{j=1}^n \left(k_j\right)^{2\alpha+1} - 2 \sum_{(i,j)\in E} \left(k_i k_j\right)^\alpha \qquad (8.17)$$

where the second term in the right-hand side is twice the generalised Randić index $^1R_\alpha$, and $^0R_\alpha = \sum_{j=1}^n (k_j)^\alpha$ is the generalised zeroth-order Randić index. Consequently, if we select $\alpha = -1/2$ for the sake of convenience, we have:

$$\sum_{i,j\epsilon E}\left(\frac{1}{\sqrt{k_i}}-\frac{1}{\sqrt{k_j}}\right)^2 = n - 2^1R_{-1/2} \qquad (8.18)$$

We can now normalise this index in such a way that it takes values in the range between zero and 1. The minimum value corresponds to regular networks, while the maximum value is reached for the networks with the largest deviation from a regular one. It is known that among networks with n nodes without isolated nodes, the star S_n attains the minimum $^1R_{-1/2}$ index (Li and Shi, 2008). On the other hand, among networks with n nodes, those without isolated nodes, in which all components are regular, have maximum $^1R_{-1/2}$ (Li and Shi, 2008). Then, among connected networks of size n, the original Randić index is bounded as

$$\sqrt{n-1} \leq {}^1R_{-1/2} \leq \frac{n}{2} \qquad (8.19)$$

where the lower bound is reached for the star S_n, and the upper bound is attained for any regular network with n nodes indistinctly of its degree. We can then use a criterion similar to that introduced by Freeman (1979) for normalising expression (8.18):

$$\rho(G) = \frac{n - 2^1R_{-1/2}}{n - 2\sqrt{n-1}} \qquad (8.20)$$

which gives a value of zero for any regular graphs, and 1 for the star in a similar way as for C^D. This index represents the normalised degree heterogeneity in a network by considering all pairs of nodes which are linked together. It can be expressed (Estrada, 2010a) as

$$\rho(G) = \frac{1}{n - 2\sqrt{n-1}} \sum_{i,j\in E}\left(\frac{1}{\sqrt{k_i}}-\frac{1}{\sqrt{k_j}}\right)^2 \qquad (8.21)$$

In Table 8.1 we illustrate the values of $\rho(G)$ for some real networks, where it can be seen that the Internet autonomous system displays the largest-degree heterogeneity or similarity to a star network. This clearly indicates the

Table 8.1 Normalised Randić index of complex networks. A measure of how similar a network is to a star graph is the normalised Randić index. Values close to 1 correspond to star-like networks, and values close to zero correspond to more regular networks.

Network	$\rho(G)$
Internet	0.548
Transcription yeast	0.448
US airports	0.369
ODLIS dictionary	0.342
PIN human	0.283
US power grid	0.099
Corporate directors	0.041

predominance of very large degree hubs in the Internet, to which many other low-degree nodes are connected. This index reflects in some quantitative and uniquely way the characteristics observed for the degree distributions studied in Chapter 2.

If we designate by φ_j an orthonormal eigenvector of the Laplacian matrix associated with the μ_j eigenvalue, we can express the generalised Randić index as follows (Estrada, 2010a):

$$^1R_\alpha = \frac{1}{2}\left[{}^0R_{2\alpha+1} - {}^0R_{2\alpha} \sum_{j=2}^{n} \mu_j \cos^2 \theta_j \right] \tag{8.22}$$

where

$$\cos\theta_j = \frac{\mathbf{k}^\alpha \cdot \vec{\varphi}_j}{\| \mathbf{k}^\alpha \|} \tag{8.23}$$

is the angle formed between the orthonormal eigenvector φ_j and the vector $\mathbf{k}^{-1/2}$, with Euclidean norm $\| \mathbf{k}^\alpha \|$, which can be written as $\| \mathbf{k}^\alpha \| = \sqrt{{}^0R_{2\alpha}}$. Then, the degree heterogeneity index $\rho(G)$ can be written as (Estrada, 2010a):

$$\rho(G) = \frac{{}^0R_{-1}}{n - 2\sqrt{n-1}} \sum_{j=1}^{n} \mu_j \cos^2 \theta_j \tag{8.24}$$

Let the eigenvalue $\mu_1 = 0$ be the origin of a Cartesian coordinate system, and let us represent $\sqrt{\mu_{j>1}}$ as a point in this system, whose coordinates are given by the angle θ_j formed by an orthonormal eigenvector associated to $\mu_{j>1}$ and the vector $\mathbf{k}^{-1/2}$. Then, the projection of $\sqrt{\mu_{j>1}}$ on the x axis is given by $CA_j = \sqrt{\mu_{j>1}} \cos \theta_j$, which is the adjacent cathetus of the triangle formed by the points $\mu_1 = 0$, $\sqrt{\mu_{j>1}}$ and b, as can be seen in Figure 8.2.

This means that the degree heterogeneity index $\rho(G)$ can be written as

$$\rho(G) = \frac{{}^0R_{-1}}{n - 2\sqrt{n-1}} \sum_{j=1}^{n} CA_j^2 \tag{8.25}$$

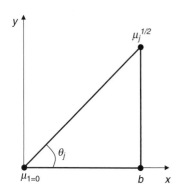

Fig. 8.2

Spectral projection. The projection of $\sqrt{\mu_{j>1}}$ in terms of the angle θ_j formed by an orthonormal eigenvector associated with $\mu_{j>1}$ and the vector $\mathbf{k}^{-1/2}$.

which is interpreted as the sum of the squares of adjacent catheti for all triangles formed by the spectral projection of Laplacian eigenvalues. Obviously, the projection of $\sqrt{\mu_{j>1}}$ on the y axis is given by $CO_j = \sqrt{\mu_{j>1}} \sin \theta_j$. This means that we can represent the degree heterogeneity of a network in a graphical form by plotting CA_j vs. CO_j for all values of j, and the heterogeneity is given by the sum of the squares of projections of such points on the abscissa. These plots are illustrated in Figure 8.3 for the Internet and for the US corporate elite, which are the two networks displaying extreme values of $\rho(G)$ in Table 8.2. We will call these plots the H-plot of a network. In the same figure we also illustrate the spectral projections of degree heterogeneity for two random networks with power-law and exponential degree distributions, which display typical shapes in their spectral projections (Estrada, 2010a).

A simple global network measure is the density, which is defined as follows:

$$\delta(G) = \frac{2m}{n(n-1)} = \frac{\bar{k}}{(n-1)} \tag{8.26}$$

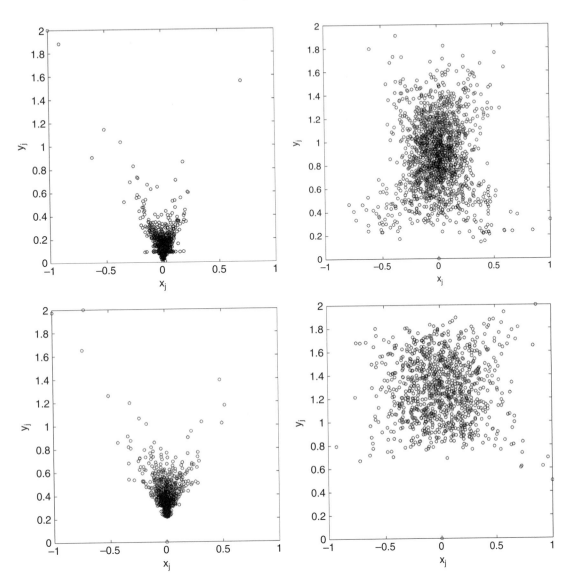

Fig. 8.3
Spectral projection of real networks. The projection of $\sqrt{\mu_{j>1}}$ in terms of the angle θ_j formed by an orthonormal eigenvector associated with $\mu_{j>1}$ and the vector $\mathbf{k}^{-1/2}$, for Internet (top left) and the USA corporate elite (top right), and for two random networks (bottom) with power-law (left) and exponential (right) degree distributions.

where $m = |E|$ is the number of links in the network, and $\bar{k}$ is the average degree.

A criterion for considering a network as *sparse* is that $\bar{k} << n$, which is equivalent to saying that $\delta(G) << 1$. Usually, plots of the matrices are given to illustrate such sparseness, as in Figure 8.4, where the adjacency matrices of two collaboration networks are illustrated. The first network corresponds to the collaboration between scientists in computational geometry and displays a

Table 8.2 **Density and global efficiency of complex networks**. Values of densities and global efficiency of some real-world networks arising in different scenarios.

Network	$\delta(G)$	E_{Global}
Small World literature	0.036	0.454
Neurons	0.051	0.422
USA Airports	0.039	0.406
ODLIS dictionary	0.004	0.338
Corporate directors	0.009	0.307
Internet	0.001	0.289
Drugs	0.011	0.229
PIN human	0.002	0.223
Transcription yeast	0.005	0.219

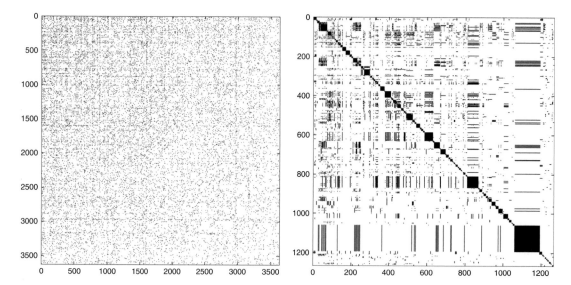

Fig. 8.4
Density in collaboration networks. Adjacency matrices of two collaboration networks: scientific collaboration in the field of computational geometry (left), and collaboration among jazz musicians (right).

density of 0.0014, while the second corresponds to collaborations among jazz musicians and displays a density equal to 0.048.

Most real-world networks analysed so far are sparse, which is particularly true for large networks. For instance, in Figure 8.5 we plot the dependence of the density with the size of the network where we observe that $\delta(G) \sim n^{-1}$, which is probably a consequence of the fact that for large networks there is a very large 'cost' associated with a large increase in the number of links.

The global density of a network can be modified to account for more local information. That is, instead of considering the network as a whole we can calculate the density of a neighbourhood around a given node. Let $N_\alpha(i) \subseteq V$

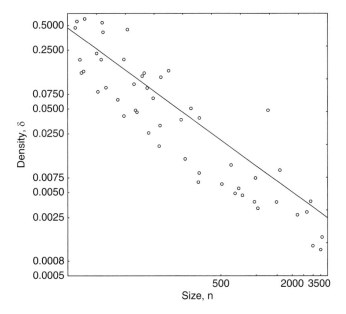

Fig. 8.5
Density versus size in real networks.
The decaying relationship between density and size in real-world complex networks. The plot is on log–log scale.

be the α-neighbourhood of the node i, defined here as the set of nodes which are at a geodesic distance smaller than or equal to α, i.e., $v_j \in N_\alpha(i)$ iff $d(v_i, v_j) \leq \alpha, \forall v_j \in V$, which also includes the node i. Let $n_i = |N_\alpha(i)|$ be the number of nodes, and m_i the number of links in $N_\alpha(i)$. The local density of node i in the graph is then defined as:

$$d_\alpha(i) = \frac{2m_i}{n_i(n_i - 1)} \tag{8.27}$$

It is obvious that when α is equal to the diameter of the network, the global and local densities are exactly the same. The average local density is then the average of the local density over all nodes in the graph:

$$\overline{d}_\alpha = \frac{1}{n} \sum_{i=1}^{n} d_\alpha(i) \tag{8.28}$$

As we have seen in previous chapters, the number of links in a network is equal to the number of walks of length 1, $W_{l=1}$. Kosub (2005) has proposed the following generalisation of the density for other lengths l of walks in a network:

$$\delta_l(G) = \frac{W_l}{n(n-1)^l} \tag{8.29}$$

where W_l accounts for all walks of length l in the network. That is,

$$W_l = \langle \mathbf{1} | \mathbf{A}^l | \mathbf{1} \rangle \tag{8.30}$$

The denominator of (8.29) arises from the fact that the maximum possible number of walks is obtained for a complete graph with n nodes:

$$\left[\mathbf{A}^l(K_n)\right]_{ii} = \frac{(n-1)^l - (-1)^l}{n} + (-1)^l \approx \frac{(n-1)^l}{n} \tag{8.31}$$

$$[\mathbf{A}^l(K_n)]_{ij} = \frac{(n-1)^l - (-1)^l}{n} \approx \frac{(n-1)^l}{n} \tag{8.32}$$

Then we have

$$\left[\mathbf{A}^l(K_n)\right]_{ii} + \left[\mathbf{A}^l(K_n)\right]_{ij} \approx n\frac{(n-1)^l}{n} + n(n-1)\frac{(n-1)^l}{n} = n(n-1)^l \tag{8.33}$$

Note that $\delta_{l=1}(G) = \delta_l(G)$, because in W_{l-1} every link is counted twice.

Now let us consider the analogy of the Estrada index (see Section 5.1), which is the sum of all walk densities in the network:

$$R = \sum_{l=0}^{\infty} \delta_l(G) = \frac{1}{n} \sum_{l=0}^{\infty} \frac{W_l}{(n-1)^l}$$

$$= \frac{1}{n} \langle \mathbf{1} | \left(\sum_{l=0}^{\infty} \frac{\mathbf{A}^l}{(n-1)^l} \right) | \mathbf{1} \rangle \tag{8.34}$$

$$= \frac{1}{n} \langle \mathbf{1} | \left(\mathbf{I} - \frac{\mathbf{A}}{n-1} \right)^{-l} | \mathbf{1} \rangle.$$

The matrix function $\left(\mathbf{I} - \frac{\mathbf{A}}{n-1} \right)^{-1}$ is the resolvent of the adjacency matrix of a network (Estrada and Higham, 2010). The communicability function based on the resolvent was introduced in Chapter 6. A global network invariant based on the adjacency matrix resolvent which is analogous to the Estrada index is then given by

$$RE = tr\left(\mathbf{I} - \frac{\mathbf{A}}{n-1} \right)^1 \tag{8.35}$$

A comparison of RE with the Estrada index EE shows that the resolvent very heavily penalises long walks—particularly in large networks—assigning much more importance to the shortest walks in the network. For instance, using the Stirling approximation $l! \approx \sqrt{2\pi l}(l/e)^l$ it can be seen that for $l < ne$ we determine that $l < (n-1)^l$. Consequently, due to the large penalisation imposed by $(n-1)^l$ in large networks, $R_r \approx 1$ and $RE \approx n$.

Another index which is related to the adjacency relationships in networks is the *rich-club coefficient*. Let $R(k_c) \subseteq V$ be the set of nodes with degree larger than certain cut-off degree k_c; that is, $R(k) = \{v \in V | k_v > k_c\}$. The rich-club coefficient has been defined by Zhou and Mondragon (2004) as:

$$\phi(k_c) = \frac{2\left|E_{R(k_c)}\right|}{|R(k_c)| (|R(k_c)| - 1)} \tag{8.36}$$

where $|E_{R(k_c)}|$ denotes the number of links in $R(k_c)$. This index measures the fraction between the number of links that exists among the nodes with degree larger than k_c to the maximum possible number of such ties. A plot of $\phi(k_c)$ versus k_c for increasing values of the cut-off degree displays a growing behaviour, indicating that high-degree nodes tend to be connected among themselves. However, this phenomenon does not reflect any structural signature of a complex network, as it is a consequence of the fact that nodes with high degree have a larger probability of being connected than that displayed by low-degree nodes. Therefore, the best way of using the rich-club coefficient is by normalising it, as proposed by Colizza et al. (2006). This normalisation is obtained as follows:

$$\phi_{norm}(k_c) = \frac{\phi(k_c)}{\phi_{ran}(k_c)} \tag{8.37}$$

where $\phi_{ran}(k_c)$ is the rich-club coefficient for a random network with the same degree distribution as the network under study.

8.2 Distance-based invariants

As in the previous section, we begin here by analysing some centralisation indices. These indices where proposed by Freeman (1979), based on node closeness and betweenness centrality, and they are expressed as follows:

$$C^c = \frac{(2n-3)\sum_{i=1}^{n}(C_{max}^c - C_i^c)}{(n-1)(n-2)} \tag{8.38}$$

$$C^B = \frac{2\sum_{i=1}^{n}(C_{max}^B - C_i^B)}{(n-1)^2(n-2)} \tag{8.39}$$

Both indices reach their maximum for the star graph. However, the minimum for the closeness centralisation is obtained for any network having the same distance sum for every node, while the betweenness centralisation is reached for a network in which all nodes are equal in betweenness. In Figure 8.6 we illustrate the intercorrelation among the centralisation measures studied in this section, together with the degree centralisation and the normalised Randić index for the 11,117 graphs having eight nodes. As can be seen, there is little correlation among these indices, despite the fact that they all are maximised for the star graphs.

In Chapter 3 we studied a distance-based invariant introduced by Wiener (1947) and formalised by Hosoya (1971), which is nowadays known as the Wiener index:

$$W(G) = \frac{1}{2}\sum_{u}\sum_{v} d(u, v) \tag{8.40}$$

where $d(u, v)$ is the shortest-path distance between nodes u and v in the network. The Wiener index for general networks is bounded as follows:

$$\frac{n(n-1)}{2} \le W(G) \le \frac{n(n^2-1)}{6} \tag{8.41}$$

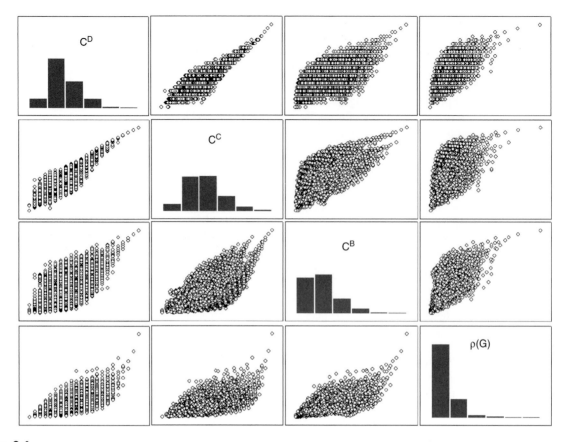

Fig. 8.6

Intercorrelation between centralisation measures. Correlation between different centralisation indices, degree, closeness, and betweenness, as well as the normalised Randić index for the 11,116 networks with eight nodes.

with the lower bound obtained for the complete network and the upper one for the path. A normalised Wiener index can then be obtained as follows:

$$\zeta(G) = \frac{n(n^2 - 1) - 6W}{n(n-1)(n-2)} \tag{8.42}$$

which takes the value of zero for a path and the value of 1 for a complete network. However, due to the small-world effect discussed in Chapter 4, real-world networks are expected to display values of $\varsigma(G)$ very close to 1. That is, due to the presence of cycles in complex networks, the Wiener index is expected to be far from that of a path. In fact, $\varsigma(G) = 0.997$ for the Internet, $\varsigma(G) = 0.998$ for the ODLIS dictionary, and $\varsigma(G) = 0.981$ for the transcription network in yeast.

The analogy of the Randić index based on the distance sum in a graph was introduced by Balaban (1982), as follows:

$$J(G) = \frac{m}{\mu + 1} \sum_{i,j \in E} (s_i s_j)^{-1/2} \tag{8.43}$$

where m, μ and s_i are, respectively, the number of links, the cyclomatic number, and the distance sum of node i. This index will not be analysed in detail here, and the reader is referred to the specialised literature for details.

A global index of network efficiency was proposed by Latora and Marchiori (2007) by using the reciprocal of the shortest path distances:

$$E_{Global} = \frac{2}{n(n-1)} \sum_{i<j} \frac{1}{d(i,j)} \qquad (8.44)$$

The index

$$H = \sum_{i<j} \frac{1}{d(i,j)} \qquad (8.45)$$

is known as the Harary index of a network, and was proposed independently by Ivanciuc et al. (1993) and Plavšić et al. (1993). Then, the efficiency index of Latora and Marchiori (2007) is simply

$$E_{Global} = \frac{2H}{n(n-1)} \qquad (8.46)$$

Also note that

$$h = (E_{Global})^{-1} \qquad (8.47)$$

is the harmonic mean of the shortest path distances in the network. In Table 8.2 we illustrate the values of the global efficiency index for some complex networks. We recall that the maximum value of this index is obtained for the complete network for which $E_{Global} = 1$. Then, as the density of a network increases, the global efficiency is also expected to increase. However, in real-world it is frequently found (see Table 8.2) that networks with relatively small density have relatively large efficiencies.

8.3 Clumpiness of central nodes

In Chapter 2 we studied the property of the nodes in a network of being grouped into two possible different ways: assortative or disassortative. This property accounts only for the statistical trend in the degree–degree correlation, which can produce some false impressions when analysing the structure of a network. For instance, in Figure 8.7 we illustrate two assortative networks which have almost equal positive assortative coefficients. The first network at the top represents trophic interactions in the St Marks food web, and the second represents a protein residue network. At the bottom we represent two networks with disassortative coefficient. The first represents the Scotch Broom food web, and the second represents a sexual network in Colorado Springs. In all cases we represent the size of the nodes proportional to their degrees.

A simple visual inspection of both pairs of networks shows that there is something about the way in which high-degree nodes are connected which escapes to the assortativity coefficient. For instance, the most central nodes in the St Marks food web are all clumped together, while those in the protein residue network are spread across the network, despite both of them being

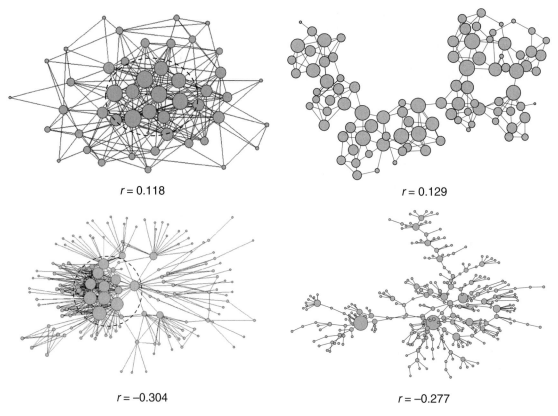

$r = 0.118$ $r = 0.129$

$r = -0.304$ $r = -0.277$

Fig. 8.7
Assortativity and grouping of hubs. Two assortative networks (top) representing the St Marks food web (left) and a protein residue network (right), and two disassortative networks (bottom) representing the Scotch Broom food web (left) and the sexual network in Colorado Springs (right). The 'clumpiness' of central nodes are different for every network in each pair. In the St Marks and Scotch Broom food webs the central nodes are clumped, while in the protein residue and the sexual network they are unclumped.

assortative networks. A similar situation is observed for both disassortative networks, where the Scotch Broom food web displays most of its central nodes concentrated in a small region, while the sexual networks has its hubs scattered through the whole network.

A way of accounting for the effects of both the degree of the pairs of nodes and the distance at which they are separated is by considering a clumpiness coefficient defined in the following way (Estrada et al., 2008a):

$$\Lambda(G, k, \alpha) = \sum_{i>j}^{n(n-1)/2} \frac{k_i k_j}{(d_{ij})^{\alpha}} \qquad (8.48)$$

where $\alpha > 0$ is a real parameter. This expression resembles those of interparticle potentials, such as the Coulombic and gravitational ones (see Chapter 19 for an example). According to the above definition, the clumpiness coefficient increases with the increase of the degrees of the nodes in the network, and decreases with the increase in the separation between these nodes. Hereafter we will use the value $\alpha \equiv 2$.

We denote by the symbol $D^{*\alpha}$ the αth entrywise power of the distance matrix $\mathbf{D}$; that is, a matrix in which every entry of $\mathbf{D}$ is raised to the power α. We can then define the following matrix:

$$\mathbf{Q} = (\mathbf{D} + \mathbf{I})^{*-2} - \mathbf{I} \tag{8.49}$$

which has entries

$$Q_{ij} = \begin{cases} (d_{ij})^{-2} & \text{for } i \neq j, \\ 0 & \text{for } i = j. \end{cases} \tag{8.50}$$

Then, we can define a clumpiness matrix as

$$\Xi_k = \mathbf{DQD} \tag{8.51}$$

whose (i, j)-entries are $\dfrac{k_i k_j}{(d_{ij})^2}$ and the diagonal entries are zeroes. Consequently, the clumpiness coefficient can be written (Estrada et al., 2008a) as

$$\Lambda(G) = \frac{1}{2} \langle \mathbf{k} | \mathbf{Q} | \mathbf{k} \rangle = \frac{1}{2} \langle \mathbf{1} | \Xi_k | \mathbf{1} \rangle \tag{8.52}$$

In Figure 8.8 we illustrate the contour plot for the clumpiness matrix of the St Marks food web (top) and protein residue network (bottom). It can be seen that the first network is formed by several groups of clumped high-degree nodes, while in the second network such groups are spread across the network.

Let us consider a connected network $G = (V, E)$ having n nodes. Then, if we remove any link $e \in E$ from G, the resulting network $G - e$ is less clumped than G:

$$\Lambda(G - e) \leq \Lambda(G) \tag{8.53}$$

This allows us to find the maximal value that the clumpiness can reach for networks of a given size. Let $G = (V, E_1)$ and $H = (V, E_2)$ be two connected networks on n vertices such that $E_1 \subseteq E_2$, and we then have $\Lambda(G) \leq \Lambda(H)$. In particular, we have $\Lambda(G) \leq \Lambda(K_n)$, which indicates that the maximum clumpiness is obtained for the complete network. The maximum clumpiness is then given (Estrada et al., 2008a) by

$$\Lambda(K_n) = \frac{n(n - 1)^3}{2} \approx n^4 \tag{8.54}$$

However, the general minimum for the clumpiness coefficient is unknown. What can be easily proved is that if the network G is Hamiltonian—that is, if it contains a path which visits every node exactly once—then $\Lambda(P_n) \leq \Lambda(G)$, because P_n can be obtained by removing links from G. It is known that

$$\Lambda(P_n) = \frac{1}{(n-1)^2} + 4 \sum_{x=1}^{n-2} \frac{1}{x^2} + \sum_{x=1}^{n-3} \frac{4(n - x - 2)}{x^2} \approx 2\pi^2(n - 2)/3 \tag{8.55}$$

The maximum value of the clumpiness of a given network having n nodes and m links is obtained by rewiring such a network to create the largest possible complete graph having m links. In other words, we can divide the n, m-graph into a complete graph K_{n_1} having m links and n_2 isolated

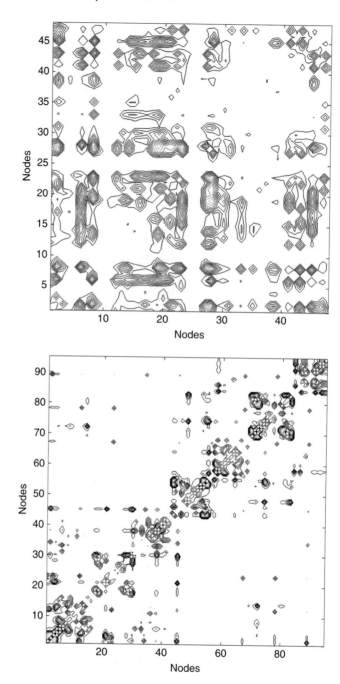

Fig. 8.8

Contour plots of degree clumpiness. Illustration of the clumpiness of nodes according to degree in the St Marks food web (top) and a protein residue network (bottom). The contour plots correspond to the normalised values of $(\Xi_k)_{ij}$ for every pair of nodes i and j in the network.

nodes with $n = n_1 + n_2$—in this case, $n_1 = \frac{1+\sqrt{1+8m}}{2}$, which can be very well approximated to $n_1 \approx \sqrt{2m}$ for large m. Consequently, the maximum clumpiness that can be obtained by rewiring an n, m-network is

$$\Lambda(K_{n_1}) = \sqrt{2m}\left(\sqrt{2m} - 1\right)^3 / 2 \qquad (8.56)$$

and we can express the clumpiness coefficient as a percentage of the maximum clumpiness that such network can attain (Estrada et al., 2008a):

$$\Phi(G) = \frac{200\Lambda(G)}{\sqrt{2m}\left(\sqrt{2m} - 1\right)^3} \qquad (8.57)$$

If we return to the networks displayed in Figure 8.7 we see that the clumpiness of the St Marks food web is $\Lambda(G) = 48.3\%$, while that of the protein residue network is only $\Lambda(G) = 11.9\%$. We recall that these two networks, illustrated in Figure 8.7, are assortative. If we calculate the clumpiness coefficient for the pair of disassortative networks displayed in Figure 8.7, we see that the Scotch Broom food web is more clumped, $\Lambda(G) = 38.6\%$, than the sexual network of Colorado Springs, $\Lambda(G) = 3.3\%$. In general, the clumpiness coefficient is not correlated with the assortativity one. In Figure 8.9 (top) we illustrate the plot of clumpiness versus assortativity for the 11,116 networks with eight nodes, where it can be seen that both properties are not correlated. Interestingly, most of these small artificial networks are disassortative and highly clumped. If we obtain the same plot for real-world networks (see Figure 8.9, bottom) we can see that most of these networks are disassortative but unclumped. Large clumpiness appears to be dissimilar in large complex systems, and only relatively small and dense networks show significant clumpiness of node degrees.

The clumpiness coefficient can be defined not only for the node degree but also for any centrality measure. Let C_i be the centrality of node i, then the clumpiness of this centrality index in a network is represented (Estrada et al., 2008a) by

$$\Lambda(G, C) = \sum_{i<j}^{n} \frac{C_i C_j}{(d_{ij})^2} \qquad (8.58)$$

For instance, for betweenness centrality $C_i = BC(i)$ we can build the following matrix:

$$\Xi_B = \beta Q \beta \qquad (8.59)$$

where β is the diagonal matrix of betweenness centrality for all nodes in the network. In Figure 8.10 we illustrate the contour plot of Ξ_B for the protein residue network for which Ξ_k is illustrated in Figure 8.8. As can be seen in Figure 8.8, most clumped pairs of nodes according to degree are distributed along the main diagonal of Ξ_k. However, those most clumped according to betweenness centrality are more homogeneously distributed through the whole network.

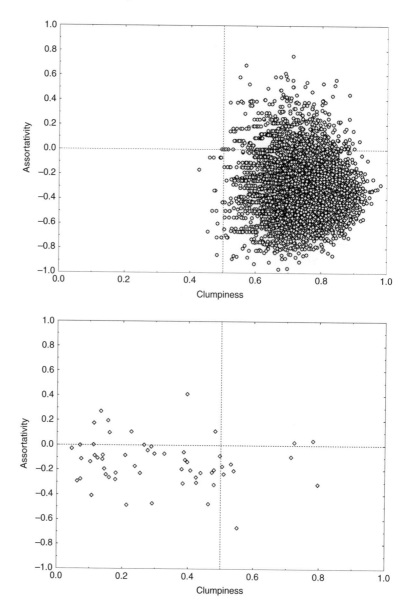

Fig. 8.9
**Degree clumpiness versus
assortativity**. Plot of the normalised
clumpiness versus assortativity
coefficient for the 11,116 connected
networks with eight nodes (top) and for
sixty real-world networks in biological,
ecological, social, and technological
systems.

8.4 Self-similarity of complex networks

One characteristic which is present in many natural systems is *self-similarity*,
which refers to the property of an object of being exactly or approximately
similar to parts of itself. This property is one of the distinctive characteristics of fractal objects, which are defined in a simple and recursive way and
display a very large heterogeneity (Mandelbrot, 1982). Examples of these
objects in nature are snowflakes, river networks, trees, systems of blood-
vessels, and many others. Mandelbrot (1982) has defined a fractal as a set

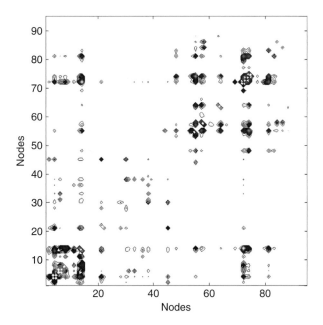

Fig. 8.10
Contour plots of betweenness clumpiness. Clumpiness of nodes according to betweenness in a protein residue network. The contour plots correspond to the normalised values of $(\Xi_B)_{ij}$ for every pair of nodes i and j in the network.

that has Hausdorff dimension greater than its topological dimension, though this excludes some highly irregular objects, such as the Hilbert curve, which are also considered as fractals. The topological dimension is an integer number that measures the size of a set. It takes values of 1, 2, and 3 for instance, for curves, surfaces, and solids, respectively. The Hausdorff dimension is defined for any subset of $\Re^n$, and in contrast with the topological dimension it is not invariant under homeomorphisms. It can be considered informally as the number of independent parameters required to describe a point in a given set; but more formally, according to the *Dictionary of Mathematics* (Borowski and Borwein, 1999), it is 'the unique positive extended real number D for which S (a set in a finite-dimensional space) has finite d-dimensional Hausdorff measure for $d < D$ and infinite measure for $d > D$.' If S is a subset of a metric space, the Hausdorff measure of S is defined by

$$H^d(S) = \inf\left\{\sum_i \gamma(d)r_i^d : C(r)\right\} \qquad (8.60)$$

where $C(r)$ is any finite cover of S by balls of radius less than r, and $\gamma(d)$ is defined in terms of gamma functions. Therefore, it is clear that the Hausdorff dimension can be a fraction, which is the case in many fractals. For example, in Figure 8.11 we illustrate some well-known fractals and their Hausdorff dimensions.

The Hausdorff dimension of many fractals can be obtained analytically. For instance, the Hausdorff dimensions for the Koch snowflake, Sierpinski gasket, Mandelbrot set, and Menger sponge are $\log(4)/\log(3)$, $\log(3)/\log(2)$, 2, and $\log(20)/\log(3)$, and, respectively. However, in many cases it is not possible to obtain such measures analytically, and some generalisations of the

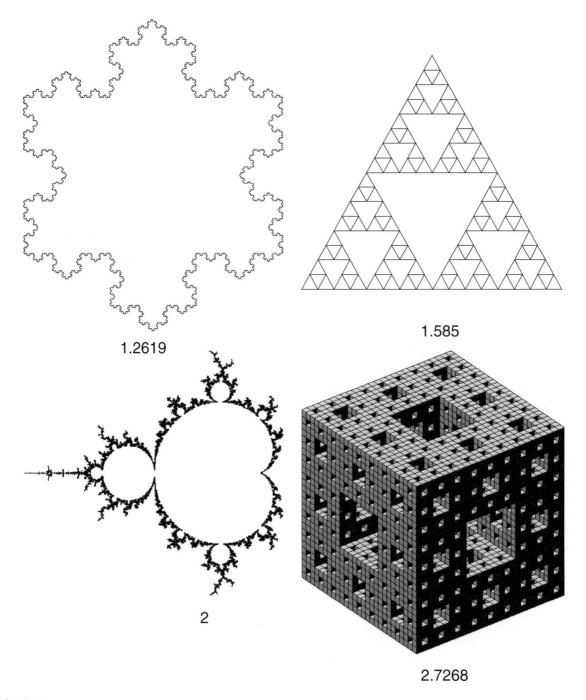

1.2619

1.585

2

2.7268

Fig. 8.11
Fractal structures and their Hausdorff dimensions. The Koch snowflake, Sierpinski gasket, Mandelbrot set, and Menger sponge, with their Hausdorf dimensions.

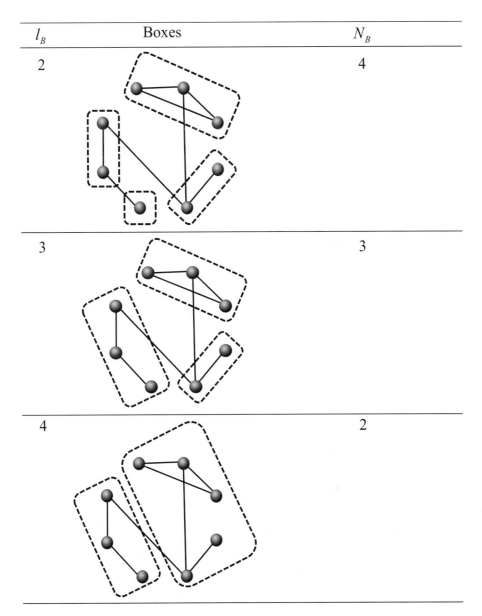

Fig. 8.12
Box counting method for networks. The box counting method used for determining fractality in complex networks.

concept of dimension are necessary. One of these measures is based on covering the fractal with boxes of size h and then counting the number $N(h)$ of boxes needed to cover the whole set. Then, the box dimension of S is defined by

$$d = \lim_{h \to 0} \frac{\ln N(h)}{\ln (1/h)} \qquad (8.61)$$

Table 8.3 **Fractal dimension of complex networks**. Values of the fractal dimension defined by the modified box counting method of some real-world networks arising in different scenarios, according to Song et al. (2005).

Network	n	d_B
WWW	325,729	4.1
Actors	392,340	6.3
PIN E. coli	429	2.3
PIN human	946	2.3
Metabolic networks[a]		3.5

[a] Average value for 43 metabolic networks from archaea, bacteria, and eukarya.

For example, three squares of edge size 1/3 are required to cover the entire Koch curve. If we reduce the size of these squares to 1/9, we need twelve such boxes to cover the fractal. Then, 48 boxes of size 1/27 are needed, and so on. Finally, the slope of the log–log plot of the values of $N(h)$ versus $1/h$ produces the value of the box dimension for this fractal.

In order to detect fractality in complex networks, an approach similar to the box counting method is used (Gallos et al., 2007; Kim et al., 2007; Kitsak et al., 2007; Song et al., 2005; 2006). Here, the cover of the network is carried out by using boxes, such as any pair of nodes contained in each box is separated at a shortest path distance $d_{ij} < l_B$, where l_B is the maximum box-diameter. As before, let N_B be the number of boxes required to cover the network (see Figure 8.12 for an example). Then, N_B and l_B scales as

$$N_B \sim l_B^{-d_B} \qquad (8.62)$$

where d_B is the fractal dimension of the network, which can be obtained as the slope of the log–log plot of N_B vs. l_B. The lack of linearity in this plot is interpreted as the lack of fractality in the network. The technical details for the algorithmic implementation of these methods can be found in the literature, and the reader is directed to it for such details.

Some results reported by Song et al., using this method for complex networks, are illustrated in Table 8.3. All these networks display fat-tail degree distribution and have been found to be self-similar. Other networks, however, which also display scale-free degree distributions, such as the Internet, have been found to display no self-similarity according to the criterion used by this method.

We end this section with some reflections about fractality, which can also be extended to other concepts such as fat-tail degree distribution and 'small-worldness' in networks. Finding that a network is fractal or self-similar is not an indication of any specific structural pattern in the network, and provides very little information about it. As Philip Ball (2009) has cleverly remarked:

> ...just saying that a structure is fractal doesn't bring you any closer to understanding how it forms. There is not a unique fractal-forming process, nor a uniquely fractal kind of pattern. The fractal dimension can be a useful measure for classifying self-similar structures, but not necessarily represent a magic key to deeper understanding.

Expansion and network classes

9

> Classes and concepts may, however, also be conceived as real objects, namely classes as 'pluralities of things' or as structures consisting of a plurality of things, and concepts as the properties and relations of things existing independently of our definitions and constructions...They are in the same sense necessary to obtain a satisfactory system of mathematics as physical bodies are necessary for a satisfactory theory of our sense perceptions.
>
> Kurt Friedrich Gödel

9.1 Network expansion

In many real-world situations it is important to understand how efficiently connected a network is with respect to subsets of nodes. A network is efficiently connected if we select a subset of nodes and the number of connections outside the subset is approximately the same as the number of nodes in this subset. Conversely, the worst case is when all the nodes are grouped together and there are relatively few links to nodes outside the subset. In this case, by removing these very few links the network is disconnected into isolated chunks.

Let us consider the subset $S \subseteq V$. The number of links that connect a node in S with a node in $V-S$ is termed the boundary of S and is denoted by ∂S. We select the subset S to be at most half the number of nodes in the network. We can then use the following measure to account for the efficiency of the connection of a network (Arora et al., 2009; Gkantsidis et al., 2003; Hoory et al., 2006):

$$\phi(G) = \inf \left\{ \frac{|\partial S|}{|S|}, S \subset V, 0 < |S| \le \frac{|V|}{2} < +\infty \right\} \qquad (9.1)$$

which is known as the *expansion constant* of the network.

In order to understand the relation between the expansion constant and the connectivity of a network, we will consider the following simple examples. First, let us consider a tree $T_n = (V, E)$, and let us consider that $S \subset V$ such that $|S| = \frac{|V|}{2} = \frac{n}{2}$. If S is selected to be the half-tree with all nodes grouped together, then $|\partial S| = 1$, because by definition there is only one path connecting two nodes in a tree. Therefore,

$$\phi(T_n) = \inf\left\{\frac{|\partial S|}{|S|}\right\} = \frac{2}{n} \tag{9.2}$$

which implies that as $n \to \infty$, $\phi(T_n) \to 0$. In other words, trees are not well-connected networks, because by removing only one single link the network can be divided into two parts, each of which contain about 50% of the nodes.

At the other extreme we have the complete networks. If we select the subset $S \subset V$ such that $|S| = p$ where $p \leq \frac{|V|}{2}$, then $|\partial S| = p(n - p)$. Thus,

$$\phi(K_n) = \inf\left\{\frac{|\partial S|}{|S|}\right\} = \frac{p(n - p)}{p} + n - p \tag{9.3}$$

It is clear that when n is even, $\phi(K_n) = \frac{n}{2}$. This means that the expansion constant increases with the number of nodes in the complete network. Consequently, the number of links that we need to remove to separate the network into two equal parts is very large. However, this high expansion is created at a very high cost—that is, by connecting all pairs of nodes to each other. The interesting question from a theoretical and practical point of view concerns the existence of sparse networks with high expansion. The expansion constant is related to the isoperimetric ratio as follows:

$$\phi(G) \approx \frac{\text{Perimeter}}{\text{Area}} \tag{9.4}$$

It is known that for a network to display good expansion, the gap between the first and second eigenvalues of the adjacency matrix ($\Delta\lambda = \lambda_2 - \lambda_1$) needs to be sufficiently large. For instance, the following, known as the Alon–Milman theorem (Dodziuk, 1984; Alon, 1986; Alon and Milman, 1985), is a well-known result in the field of expander graphs. Let G be a regular graph with spectrum $\lambda_1 \geq \lambda_2 \geq \ldots \geq \lambda_n$. The expansion factor is then bounded as

$$\frac{\lambda_1 - \lambda_2}{2} \leq \phi(G) \leq \sqrt{2\lambda_1(\lambda_1 - \lambda_2)} \tag{9.5}$$

Thus, the larger the spectral gap, the larger the expansion of the graph. Among the graphs with large spectral gap there is a family of graph named Ramanujan graphs (Ballantine, 2002; Davidoff et al., 2003; Lubotzky et al., 1988; Ram Murty, 2003) which have spectral gaps almost as large as possible. Formally, a Ramanujan graph is a d-regular graph for which

$$\lambda_2 \leq 2\sqrt{d - 1} \tag{9.6}$$

where λ_2 is the maximum of the non-trivial eigenvalues of the graph $\lambda(G) = \max_{|\lambda_i| < d} |\lambda_i|$. In a Ramanujan graph the expansion constant has a lower bound given by

$$\phi(G) \geq \frac{d - 2\sqrt{d - 1}}{2} \tag{9.7}$$

Finding graphs that are expanders is not a trivial task. In Figure 9.1 we illustrate two examples of Ramanaujan networks. The first is the Petersen graph having

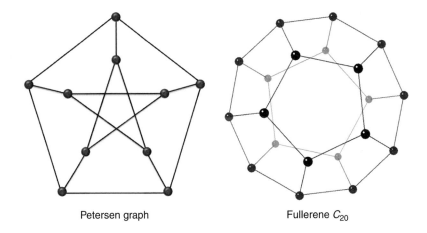

Petersen graph   Fullerene C_{20}

Fig. 9.1
Ramanujan graphs. Two small
networks with Ramanujan spectral
properties.

ten nodes, which has the characteristic polynomial $(\lambda - 3)(\lambda - 1)^5(\lambda + 2)^4$,
and consequently is easily seen to be Ramanujan. The second network represents the structure of the smallest fullerene cage C_{20}, which has the second eigenvalue equal to $\lambda_2 = 2.2360679\ldots$, and is then Ramanujan according to the definition given below. The expansion constant of the Petersen
graph is $\phi(G) = 1$, and that of the fullerene C_{20}, illustrated in Figure 9.1, is
$\phi(G) = 3/5$.

On the other hand, random d-regular networks are asymptotically Ramanujan. Friedman (1991) has shown that for these networks, $\lambda_2 \leq 2\sqrt{d-1} +$
$2 \log d + O(1)$. Consequently, a d-regular network with $d \geq 3$ is an expander
with high probability. In fact, the probability goes to 1 as $n \to \infty$. On the other
hand, Donetti et al. (2005; 2006) have found some good expansion networks
by using a deep computational search.

9.2 Universal topological classes of network

There are several classification schemes grouping networks according to their
structures. For instance, complex networks can be classified according to
the existence, or otherwise, of the 'small-world' property (see Chapter 3) or
according to their degree distribution. The last classification permits classification of networks as 'scale-free' (see Chapter 2) if their node degree distribution
decays as a power-law, 'broad-scale' networks characterised by a connectivity
distribution that has a power-law regime followed by a sharp cut-off, or 'single-
scale' networks in which degree distribution displays a fast decaying tail.
Even scale-free networks have been classified into two different subclasses
according to their exponent in the power-law distribution of the betweenness
centrality (see Chapter 7).

There are important organisational principles of complex networks which
escape the analysis of these global network characteristics. Complex networks
appear to form a complete zoo of structures ranging from very homogeneous
to very heterogeneous ones. Here we refer to heterogeneity in a network which
substantially differs from the node heterogeneity analysed in Chapter 8, and

are more interested in heterogeneities that arise by the differences in the local to global organisation of the connectivity in a network. That is, in an homogeneous network, what is seen locally is what is obtained globally from a topological point of view. In other words, we see neither 'rugosities' in the structure of the network given by some large concentration of connectivity in a given area, nor holes produced by the lack of connectivity in other regions. This means that if we study a small region of the network we can infer the global structure simply by zooming out the small region that we have characterised. The analogy with a gas is very helpful. The gas fills the whole space of the recipient containing it in an homogeneous way. Then, by analysing a small volume of this recipient we can produce a global picture of the total system. However, in heterogeneous networks we can have different organisations of the nodes in different topological regions. This situation avoids the possibility of analysing the global structure of the network by understanding one small piece of it. In an heterogeneous network the nodes can be organised in modules in such a way that the organisation of certain modules or clusters would be different from one to another and to the global characteristics of the network. Another possible situation is the one characterised by a densely connected core and a less connected periphery. In these two general cases the analysis of one small region of the network does not provide us with insights about the global architecture of the system (see Figure 9.2 for illustrations).

In Section 9.1 we saw that the existence of a power-law relationship between two variables reveals that the phenomenon under study reproduces itself on different time-scales and/or space-scales. Therefore, in an homogeneous network we should expect the existence of at least one local topological property that scales as a power-law of a global topological property. Thus, if x and y are variables representing some topological features of the network at the local and the global scale, the existence of such scaling implies that the network is topologically self-similar, and *what we see locally is what we obtain globally*. In the following section we develop an approach to account for such scaling.

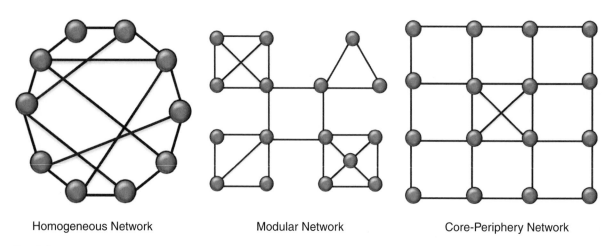

Homogeneous Network   Modular Network   Core-Periphery Network

Fig. 9.2
Network types. Models of different types of network according to their structural organisations.

9.2.1   Spectral scaling method

The aim of this method is to find a couple of appropriate topological variables for a network—one of them characterising the local environment around a node, and the other characterising the global environment. In Chapter 7 we studied several measures that characterise the topological properties of a node in the network. We have seen that the subgraph centrality index gives more weights to the participation of a node in shorter closed walks, and penalises the participation of a node in longer closed walks. Consequently, the subgraph centrality characterises the local cliquishness around a node, and it can be considered as a local spectral measure of the topological environment around a node. The subgraph centrality $EE(i)$ counts all closed walks in the network, which can be trivial when moving back and forth in acyclic subgraphs. We then use the odd-subgraph centrality $EE_{odd}(i)$, which can be considered as a topological property of local organisation in networks that characterises the odd-cyclic wiring of a typical neighbourhood.

In Chapter 7 we also showed that the eigenvector centrality EC represents the probabilities of selecting at random a walk of length L starting at node i when $L \to \infty$. Due to the infinite length of the walk which we are considering, such a walk visits all nodes and links of the network, obtaining a picture of the global topological environment around the corresponding node. Consequently, we can consider the eigenvector centrality as a global topological characterization of the environment around a node. We can now establish the mathematical relationship between the local and global spectral properties of a network.

9.2.1.1   The class of homogeneous networks

Formally, we consider that a network is homogeneous if it has good expansion (GE) properties. First we write $EE_{odd}(i)$ as follows (Estrada, 2006c; 2006d):

$$EE_{odd}(i) = \sum_{j=1} [\varphi_j(i)]^2 \sinh(\lambda_j)$$

$$= [EC(i)]^2 \sinh(\lambda_1) + \sum_{j=2} [\varphi_j(i)]^2 \sinh(\lambda_j), \qquad (9.8)$$

where $EC(i)$ denotes the principal eigenvector (eigenvector centrality) $\varphi_1(i)$, and λ_1 corresponds to the principal (Perron–Frobenius) eigenvalue of the network.

Let us assume that the network has GE properties, and, according to the Alon–Milman theorem $\lambda_1 \gg \lambda_2$, in such a way that we can consider that $[EC(i)]^2 \sinh(\lambda_1) \gg \sum_{j=2} [\varphi_j(i)]^2 \sinh(\lambda_j)$. We can then approximate the odd-subgraph centrality as

$$EE_{odd}(i) \approx [EC(i)]^2 \sinh(\lambda_1) \qquad (9.9)$$

This means that the principal eigenvector of the network is directly related to the subgraph centrality in GENs according to the following spectral power-law scaling relationship:

$$EC(i) \propto A[EE_{odd}(i)]^{\eta} \tag{9.10}$$

Here, $A \approx [\sinh(\lambda_1)]^{-0.5}$, and $\eta = 0.5$. This expression can be written in a log–log scale as

$$\log[EC(i)] = \log A + \eta \log[EE_{odd}(i)] \tag{9.11}$$

Consequently, in an homogeneous network a log–log plot of $EC(i)$ vs. $EE_{odd}(i)$ displays a perfect straight line fit with slope $\eta \approx 0.5$ and intercept $\log A$. Topologically non-homogeneous networks will display large deviations from this perfect fit.

Let us consider the homogeneous case, in which a network displays perfect spectral scaling, such that we can calculate the eigenvector centrality by using the following expression:

$$\log EC^h(i) = 0.5 \log EE_{odd}(i) - 0.5 \log[\sinh(\lambda_1)] \tag{9.12}$$

Let us now consider the deviations from this perfect fit. We can account for these deviations from *perfect homogeneity* by measuring the departure of the points from the straight line with respect to $\log EC^h(i)$ (Estrada, 2007c):

$$\Delta \log EC(i) = \log \frac{EC(i)}{EC^h(i)} = \log \left\{ \frac{[EC(i)]^2 \sinh(\lambda_1)}{EE_{odd}(i)} \right\}^{0.5} \tag{9.13}$$

We can now classify networks according to their spectral scaling. The class of homogeneous networks is designated here as Class I, and will be analysed in the following paragraphs.

Class I: homogeneous networks. These can be characterised as networks for which

$$\Delta \log EC(i) \cong 0, \forall i \in V \Rightarrow [EC(i)]^2 \sinh(\lambda_1) \cong EE_{odd}(i) \tag{9.14}$$

In Figure 9.3 we illustrate one homogeneous network and its perfect spectral scaling. In this network, what we see locally is what we obtain globally from a topological point of view.

In these networks, $[\varphi_1(i)]^2 \sinh(\lambda_1) \cong EE_{odd}(i)$, which indicates that they are formed by a tightly connected homogenous cluster which is characterised by the leading eigenvalue. This situation is very clear from the fact that

$$[\varphi_1(i)]^2 \sinh(\lambda_1) \gg \sum_{j=2}^{N} [\varphi_j(i)]^2 \sinh(\lambda_j) \tag{9.15}$$

indicating the predominance of one large highly interconnected chunk formed by almost all nodes in the network.

9.2.1.2   Classes of Non-Homogeneous Network

While homogeneous networks are classified in one single topological class, non-homogeneous networks can be classified in more than one class according to the organisation of their nodes and links (Estrada, 2007c).

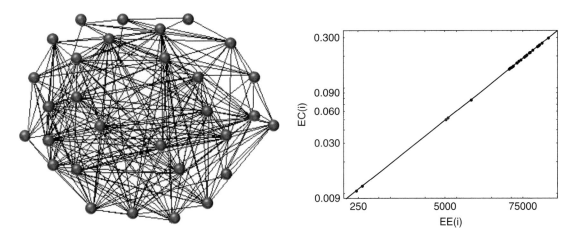

Fig. 9.3
Class I network. An homogeneous network (Class I) and its spectral scaling.

Class II networks display spectral scaling with negative deviations from the homogeneous case (Figure 9.4):

$$\Delta \log EC(i) \leq 0 \Rightarrow [EC(i)]^2 \sinh(\lambda_1) \leq EE_{odd}(i), \forall i \in V \qquad (9.16)$$

As an example of this kind of network we can consider those having modular structure in which local neighbourhoods are very densely connected but have a central hole which obviates some of these modules being highly interconnected among them. In these networks we see high local connectivity which is not reproduced globally.

The negative deviations from the homogeneity in Class II networks can be explained as follows. Let us consider a node located in one of the highly dense

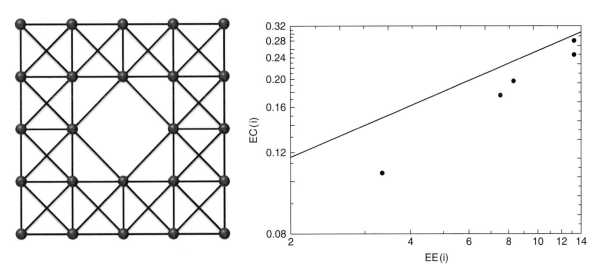

Fig. 9.4
Class II network. A network formed by clusters separated by structural holes (Class II), and its spectral scaling.

modules; for example, a node in one of the external corners of the grid in Figure 9.4. If we walk around this node we observe a high density of connections in its neighbourhood. However, if we enlarge our walk to traverse the whole network we observe that this high connectivity is not kept as soon as we arrive at the central hole in the network. Our measurement of local connectivity is the subgraph centrality, which is larger than expected from the homogeneity line of the spectral scaling. This means that the points representing the nodes of the network are placed to the right of the line representing the perfect scaling. An equivalent view is obtained by considering the eigenvector centrality, which is our global measure of connectivity. In this case the global connectivity is lower than expected from the homogeneity line placing the points below the straight line of the spectral scaling.

Mathematically, in this case need only consider the case where $[\varphi_1(i)]^2 \sinh(\lambda_1) < EE_{odd}(i)$, because $[\varphi_1(i)]^2 \sinh(\lambda_1) = EE_{odd}(i)$ produces null deviations from the perfect scaling: $\Delta \log \varphi_1(i) = 0$. Using expression (9.8) for $EE_{odd}(i)$ we obtain:

$$[\varphi_1(i)]^2 \sinh(\lambda_1) < [\varphi_1(i)]^2 \sinh(\lambda_1) + \sum_{j=2}[\varphi_j(i)]^2 \sinh(\lambda_j) \qquad (9.17)$$

which implies that

$$\sum_{j=2}[\varphi_j(i)]^2 \sinh(\lambda_j) > 0 \qquad (9.18)$$

Taking into account that we are considering networks without self-loops, the adjacency matrix has zeroes along its principal diagonal, which means that its spectrum must have both positive and negative eigenvalues. Because $\lambda_1 > 0$ we will designate by $\sum_+$ and $\sum_-$ the sums corresponding to positive and negative eigenvalues for $j \geq 2$. We can then write the left part of inequality (9.18) as follows:

$$\sum_{j=2}^{N}[\varphi_j(i)]^2 \sinh(\lambda_j) = \sum_+[\varphi_j(i)]^2 \sinh(\lambda_j) + \sum_-[\varphi_j(i)]^2 \sinh(\lambda_j) \quad (9.19)$$

We can now rewrite the inequality (9.18) in terms of the sums of positive and negative eigenvalues:

$$\sum_+[\varphi_j(i)]^2 \sinh(\lambda_j) + \sum_-[\varphi_j(i)]^2 \sinh(\lambda_j) > 0 \qquad (9.20)$$

Then, because $\sum_-[\varphi_j(i)]^2 \sinh(\lambda_j) < 0$ we immediately obtain the following new inequality:

$$\left|\sum_+[\varphi_j(i)]^2 \sinh(\lambda_j)\right| > \left|\sum_-[\varphi_j(i)]^2 \sinh(\lambda_j)\right| \qquad (9.21)$$

It is known from spectral clustering techniques that the eigenvectors corresponding to positive eigenvalues produce a partition of the network into clusters of tightly connected nodes (see Chapter 10). Conversely, the eigenvectors corresponding to negative eigenvalues produce partitions in which nodes are not close to those to which they are linked, but rather, with those to which they are not linked. In other words, the nodes will be close to other nodes which have similar patterns of connections with other sets of nodes—nodes to which they are structurally equivalent. In the case of the eigenvectors corresponding to positive eigenvalues, the nodes corresponding to larger components tend to form *quasi-cliques*; that is, clusters in which every two nodes tend to interact with each other. Conversely, for eigenvectors corresponding to negative eigenvalues, nodes tend to form *quasi-bipartites*; that is, nodes are partitioned into disjoint subsets with high connectivity between sets but with low internal connectivity.

The inequality $\left|\sum_{+}[\varphi_j(i)]^2 \sinh(\lambda_j)\right| > \left|\sum_{-}[\varphi_j(i)]^2 \sinh(\lambda_j)\right|$ means that the networks of Class II are dominated by partitions into quasi-cliques more than into quasi-bipartites. In other words, these networks are characterized by two or more clusters of highly interconnected nodes which display a low inter-cluster connectivity.

Class III: The third class of complex network is the one displaying spectral scaling with positive deviations (see Figure 9.5):

$$\Delta \log EC(i) \geq 0 \Rightarrow [EC(i)]^2 \sinh(\lambda_1) \geq EE_{odd}(i), \forall i \in V \qquad (9.22)$$

In these networks the positive deviations from the homogeneity arises from the following. If, as before, we walk around a node which is located in one of the corners of the grid, we observe a very low density of connections in its neighbourhood. However, if we enlarge our walk to traverse the whole network we observe that this low connectivity is not retained as soon as we arrive at the central core of the network. Then, the subgraph centrality is lower than expected from the homogeneity line of the spectral scaling placing the points representing the nodes of the network to the left of the perfect scaling line. The

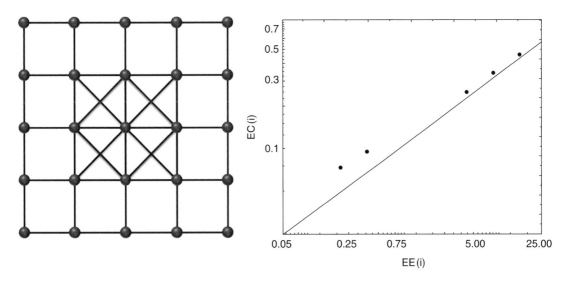

Fig. 9.5
Class III network. A network formed by a core-periphery structure (Class III), and its spectral scaling.

eigenvector centrality, however, is larger than expected from the homogeneity line, because the global density is higher than expected from the inference made at the local vicinity of the node. This places the points over the straight line of the spectral scaling, producing the pattern observed in Figure 9.5.

Mathematically, we can follow the same algebraic manipulations as for Class II networks, and we arrive at the following expression:

$$\left| \sum_{+} [\varphi_j(i)]^2 \sinh(\lambda_j) \right| < \left| \sum_{-} [\varphi_j(i)]^2 \sinh(\lambda_j) \right| \tag{9.23}$$

According to this inequality, the topological organisation of the nodes/links in Class III networks is 'dominated' by the negative eigenvalues. This means that networks in structural Class III are characterised by the dominance of quasi-bipartites more than by that of quasi-cliques. We can therefore produce a model explaining the general features of these networks by considering a central core of highly interconnected nodes surrounded by a periphery of nodes displaying low connectivity with the central core and among them (see Figure 9.5 left).

Class IV: These networks display spectral scaling with mixed deviations (see Figure 9.6):

$$\Delta \log EC(p) \le 0, \ p \in V \text{ and } \Delta \log EC(q) > 0, \ q \in V \tag{9.24}$$

The networks in Class IV display a mixture of positive and negative deviations. They are characterized by a combination of both quasi-cliques and quasi-bipartites, without the predominance of either structure over the other. This situation can be represented by the network illustrated in Figure 9.6, which is a combination of the two presented for Classes II and III. On the one hand, the central nodes connecting the highly interconnected clusters display larger

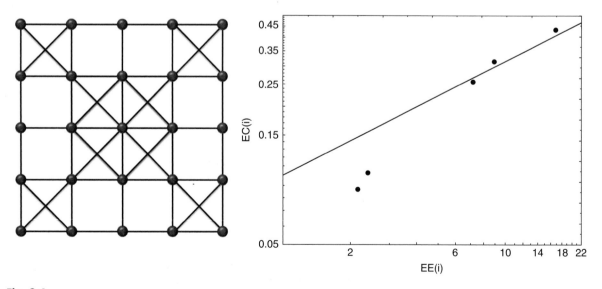

Fig. 9.6
Class IV network. A network formed by a central core and several periphery clusters separated by structural holes (Class IV), and its spectral scaling.

connectivity to all other nodes in the network than that expected from their local cliquishness; that is, they display positive deviations from the perfect scaling. On the other hand, the nodes on one side of the graph are not well connected to the nodes on the other side, despite being internally highly connected. Consequently, these nodes display negative deviations from the perfect scaling.

In summary, Class II corresponds to networks in which there are two or more communities of highly interconnected nodes, which display low inter-module connectivity. One example of this kind of network is represented by those containing holes in their structures. Class III networks can be represented by a typical 'core-periphery' structure characterized by a highly interconnected central core surrounded by a sparser periphery of nodes. Finally, Class IV networks display a combination of highly connected groups (quasi-cliques) and some groups of nodes partitioned into disjoint subsets (quasi-bipartites), without a predominance either structure.

9.3 Topological classification of real-world networks

An important question which arises from this classification of complex networks is whether these universal classes of network are represented in the real world. To extend this classification to real-world networks we need to quantify the degree of deviation of the nodes from the ideal spectral scaling. In doing so we account for the mean square error of all points with positive and negative deviations in the spectral scaling, respectively:

$$\xi^+ = \sqrt{\frac{1}{N_+} \sum_+ \left(\log \frac{EC(i)}{EC^{Homo(i)}} \right)^2} \tag{9.25}$$

$$\xi^- = \sqrt{\frac{1}{N_-} \sum_- \left(\log \frac{EC(i)}{EC^{Homo(i)}} \right)^2} \tag{9.26}$$

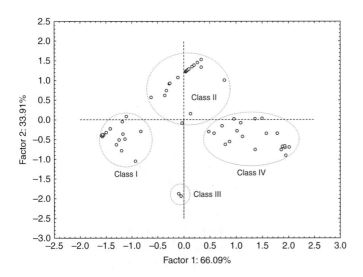

Fig. 9.7
Structural classes of real networks.
Plot of the two principal components (factors) for $x = \log(\xi^- + \varepsilon)$ and $y = \log(\xi^+ + \varepsilon)$, where $\varepsilon = 10^{-5}$ for sixty-two real-world networks. The two factors are normalised to zero mean and unit standard deviation.

where $\sum_+$ and $\sum_-$ are the sums carried out for the N_+ points having $\Delta \log EC(i) > 0$ and for the N_- having $\Delta \log EC(i) < 0$, respectively.

We calculate the values of ξ^- and ξ^+ for sixty-two real-world networks (Estrada, 2007c). Then, we define the following two new variables $x = \log(\xi^- + \varepsilon)$ and $y = \log(\xi^+ + \varepsilon)$, where $\varepsilon = 10^{-5}$ is used to avoid indeterminacy in the logarithms. Using Principal Component Analysis we transform x and y to obtain the plot presented in Figure 9.7 for all these networks.

These results represent an empirical evidence for the existence of the four structural classes of complex network in real-world systems. Classes I, II, and IV are equally represented among the sixty-two networks studied, that is, there are about 32% of networks in each of these classes. In contrast, there are only two networks in Class III, which correspond to the ecological systems

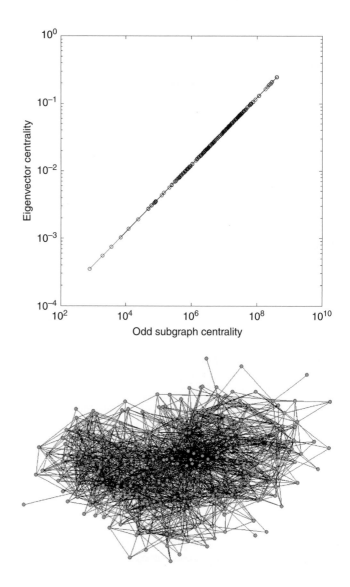

Fig. 9.8
Real-world Class I network. Spectral scaling (top) of the neural network (bottom) of the worm *C. elegans*, which shows characteristics of Class I networks.

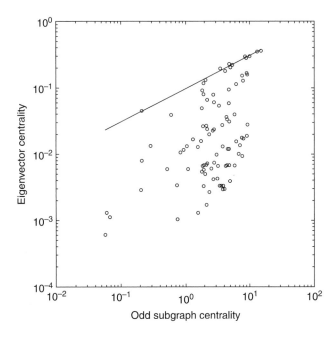

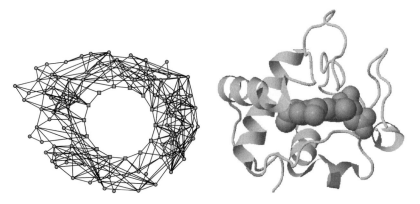

Fig. 9.9
Real-world Class II network. Spectral scaling (top) of the protein residue network of (bottom) of the rice ferricytochrome C. The network (bottom left) is illustrated by representing a central hole formed by fourteen nodes, which corresponds to the binding site of the heme group represented in the illustration (bottom right) of this protein.

of Canton Creek and Stony Stream. In general, most ecological networks correspond to Class I (70%), and they represent the only systems in which the four classes of network are represented. Most biological networks studied correspond to Class IV (67%). Informational networks are mainly classified into two classes: Class I (50%) and Class II (33.3%). On the other hand, technological networks are mainly in Class IV (64%), while 27% correspond to Class I. Social networks mainly correspond to Classes II and IV (91%).

9.3.1 Examples

Networks in Class I display large structural homogeneity. They are characterized by the lack of structural bottlenecks which separate large regions of the network by disconnecting relatively few nodes/links. An example of

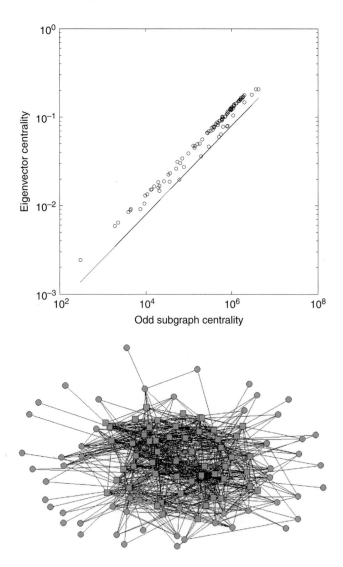

Fig. 9.10

Real-world Class III network. Spectral scaling (top) of the Canton Creek food network (bottom), which shows characteristics of Class III networks. The network is represented by a central core (squares) and a periphery (circles). The central core consists of a connected maximal induced subgraph in which every node has degree greater or equal to 11.

these networks is the neural network of the worm *C. elegans*, which has $\xi^- = 3.83 \times 10^{-4}$ and $\xi^+ = 2.32 \times 10^{-5}$. In Figure 9.8 we illustrate the spectral scaling and a picture of this network.

The prototype of Class II networks is the protein residue network (see Chapter 14). These networks are characterized by the presence of several modules of relatively highly interconnected nodes, which are separated among them by the presence of structural holes. These modules correspond to regions in the protein composed of amino acids that are close to each other in space, such as those forming structural motifs such as helices and sheets. Such modules are then spaced by certain cavities which in many cases play important roles in the protein as binding sites for other molecules. For instance, in Figure 9.9 we illustrate the spectral scaling for the protein residue network of the rice ferricytochrome C, which is an electron transport protein holding a protoporphyrin

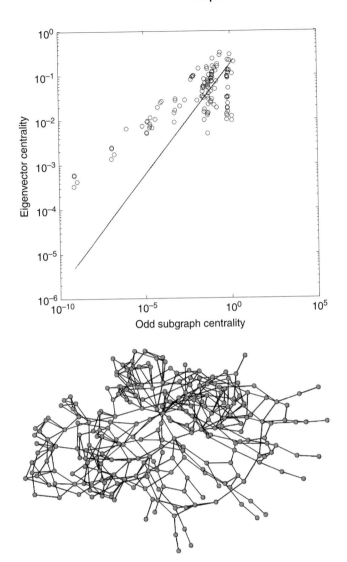

Fig. 9.11
Real-world Class IV network. Spectral scaling (top) of the flip-flop electronic circuit s420 network (bottom), which shows characteristics of Class IV networks.

X (heme) group in a structural cavity. In Figure 9.9 this cavity is represented as a hole in the network, and an illustration of the protein complexed with the heme group is also shown

Class III networks are characterized by a highly connected central core surrounded by a sparsely connected periphery. In Figure 9.10 we illustrate the spectral scaling and a picture of the Canton food web, which has $\xi^- = 0.000$ and $\xi^+ = 0.183$. We have identified a central core as the connected maximal induced subgraph in which every node has degree greater or equal to 11. The rest of the nodes are considered to be forming a periphery. The central core is formed by 50% of nodes, and displays a density $d = 0.317$, which is larger than that of the whole network, $d = 0.122$. However, the largest difference is in the densities of the core-periphery and the periphery–periphery regions, which are $d = 0.038$ and $d = 0.022$, respectively. The structure of this network is clearly dominated by the formation of quasi-bipartites over the formation of quasi-cliques, as previously analysed for the model of Class III networks. In Chapter 11 we will develop some measures of bipartivity which allow us to quantify how much bipartivity exists in a given network.

Finally, networks in Class IV are characterized by the presence of both quasi-cliques and quasi-bipartites without a predominance of any of them over another. In Figure 9.11 this situation is illustrated for a network representing a flip-flop electronic circuit. As can be seen in Fig. 9.11 (at top), the spectral scaling for this network consists of both positive and negative deviations from the straight line corresponding to homogeneous networks. This circuit is characterized by having several modules which are sparse, to avoid deviations from planarity (see Chapter 3). These modules are separated by several holes, producing a structure which resembles that of the model network in Figure 9.6.

Community structure of networks

10

The term 'community' implies a diversity but at the same time a certain
organized uniformity in the units.

J. Eugene B. Warming

In Chapters 7 and 8 we studied invariants that characterise the local and
global structure of networks. A closer inspection of the structure of many real-
world networks provides us with insights about a third order of topological
organisation in these systems. That is, in many complex networks nodes group
together forming some kind of clusters characterised by properties which are
more or less independent of the properties of individual nodes and of the
network as a whole. This kind of structure is known as 'network communities'
(Fortunato, 2010; Gulbahce and Lehmann, 2008; Porter et al., 2009). In the
mathematics and computer science literature this topic is usually termed 'graph
clustering' or 'partition' (Brandes et al., 2003; Elsner, 1997; Fjällström, 1998;
Schaeffer, 2007). For instance, the network illustrated in Figure 10.1 (top)
represents the friendship ties among individuals in a karate club in the USA. At
some point in time the members of this social network were polarised into two
different factions due to an argument between the instructor and the president,
labelled as nodes 1 and 34, respectively, in the figure. These two factions, represented at the top of the figure by circles and squares, respectively, act as cohesive groups which can be considered as independent entities in the network. In
Chapter 6 an example of this kind of collective behaviour was analysed for
the members of the attacking group in the social network of friendship ties in
the office of an overseas branch of a large international organisation. Another
example is provided in Figure 10.1 (bottom), representing sixty-two bottlenose
dolphins living in Doubtful Sound, New Zealand, where links are considered
between two animals if they are seen together more frequently than expected
at random (see Appendix). These dolphins were separated into two cohesive
groups after one of them abandoned the place for some time. Both groups are
distinguished by circles and squares at the bottom of the figure.

In all the examples provided in the previous paragraph we have some
experimental evidence that supports the split of the network into two or more
cohesive groups. The question that arises here is whether we can find such
partitions in networks without any other information than that provided by

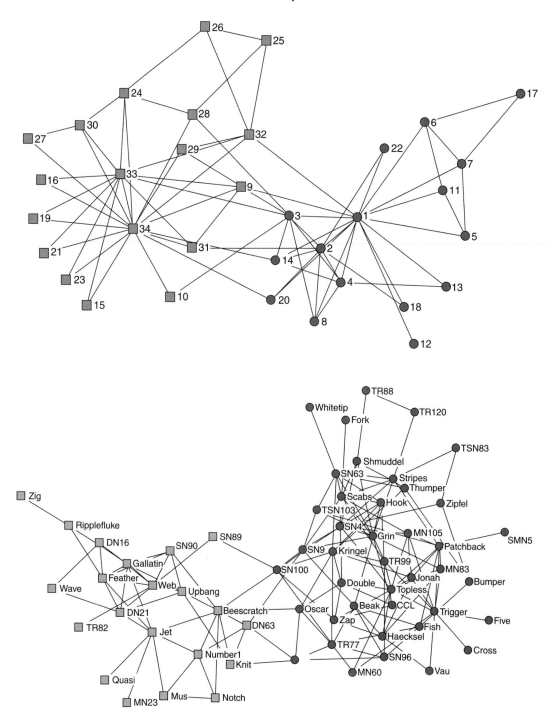

Fig. 10.1

Community structure in networks. Two social networks which are known to be divided into two communities. Zachary's karate club (top), with circles representing followers of the instructor (number 1) and squares representing followers of the president (number 34). Network of bottlenose dolphins in Doubtful Sound (New Zealand), with circles and squares representing the two groups formed after the absence of one dolphin.

the topological structure of the network. This is *grosso modo* the topic of this chapter.

Let us take a closer look at how we can detect such groups of nodes in networks. First of all, let us define the intracluster density of the subgraph $C \subset V$ having $n_C = |C|$ nodes. Let k_i^{int} and k_i^{ext} be the internal and external degree of node i, defined as follows, respectively (Fortunato, 2010):

$$k_i{}^{\text{int}} = \sum_{j \epsilon C} \mathbf{A}_{ij} \qquad (10.1)$$

$$k_i{}^{\text{ext}} = \sum_{j \epsilon \bar{C}} \mathbf{A}_{ij} \qquad (10.2)$$

where $\bar{C}$ is the complement of C. Then, the number of links which connect nodes internally in the subgraph is given by $m_C = \frac{1}{2} \sum_i k_i^{\text{int}}$, and the number of links connecting C and $\bar{C}$ is $m_{C-\bar{C}} = \frac{1}{2} \sum_i k_i^{\text{ext}}$. We recall that the number of links that connect a node in C with a node in $\bar{C} = V - C$, is termed the boundary of C, denoted by ∂C in Chapter 9. We can use the general definition of global network density studied in Chapter 8 to define the intracluster and intercluster densities as follows:

$$\delta_{\text{int}}(C) = \frac{\sum_i k_i{}^{\text{int}}}{n_C(n_C - 1)} \qquad (10.3)$$

$$\delta_{\text{ext}}(C) = \frac{\sum_i k_i{}^{\text{ext}}}{2n_C(n - n_C)} \qquad (10.4)$$

With these new tools at hand we will analyse both networks illustrated in Figure 10.1, for which we know *a priori* the community structures. In the karate club network, the values of the internal densities of the communities C_1 (followers of the instructor) and C_2 (followers of the president) are $\delta_{\text{int}}(C_1) = 0.26$ and $\delta_{\text{int}}(C_2) = 0.24$, respectively, and the intercluster density is $\delta_{\text{ext}}(C_1) = \delta_{\text{ext}}(C_2) = 0.035$. The total density of this network is $\delta(G) = 0.14$. For the network of bottlenose dolphins the cluster represented as squares has $\delta_{\text{int}}(C_1) = 0.26$, while the one represented by circles has $\delta_{\text{int}}(C_2) = 0.11$ with external density $\delta_{\text{ext}}(C_1) = \delta_{\text{ext}}(C_2) = 0.007$, while the total density of the network is $\delta(G) = 0.08$. For the clusters found 'experimentally' for these two networks, we observe the following:

(i) The internal density of every cluster is significantly larger than its external density.
(ii) The internal density of every cluster is significantly larger than the total density of the network.

Another observation concerning existing communities is that there is at least one path between every pair of nodes in a community, such that all nodes in that path are in the same community. That is, communities are *internally connected*. Thus, we can loosely define a community as follows:

> A community is formed by an internally connected set of nodes for which the internal density is significantly larger than the external one.

This principle is the basis for the construction of networks with explicit communities, which are known as benchmark graphs (Lancichinetti et al., 2008; 2010). Before starting with the details of methods and algorithms used for finding communities in complex networks, we would like to remark that communities are quite important far beyond social networks, finding important applications in biological, ecological, and technological networks.

10.1 Network partition methods

In formal terms, we are seeking to partition the network into p disjoint sets of nodes, such that:

(iii) $\bigcup_{i=1}^{p} V_i = V$ and $V_i \bigcap V_j = \phi$ for $i \neq j$

(iv) The number of links crossing between subsets ('cut size' or 'boundary') is minimized.

(v) $|V_i| \approx n/p$ for all $i = 1.2, \ldots, p$, where the vertical bars indicate the cardinality of the set.

(vi) In the case of weighted networks the cut set is defined as the sum of the weights of the links crossing subsets, and condition (iii) is taken as $W_i \approx W/p$, where W_i and W are the sum of the weights of the nodes in the i th subset and in the whole network.

When condition (iii) or its equivalent for weighted networks is fulfilled, the corresponding partition is called *balanced*. There are several algorithms that have been tested in the literature for the purpose of network partitioning, and we will analyse some of them here.

10.1.1 Local improvement methods

One of the earliest methods of graph partitioning was proposed by B. W. Kernighan and S. Lin in 1970 as a local improvement method (Kernighan and Lin, 1970). In general, these methods take a partition of the network as their input, and the most simple case is by taking a bisection. They then try to decrease the cutset by some local search approach. The generation of the initial partition can be carried out by any of the existing bisection methods, or can simply be generated at random. In the last case, several random realizations are needed, and the one with minimum cut size is selected as the input.

For the sake of simplicity we will consider that the weights of nodes are equal to 1. Then, let $\{V_1, V_2\}$ be a bisection of the network, and let us define the following quantities:

$$W_i^{\text{int}} = \sum_{\substack{i,j \in E \\ I(i)=I(j)}} w_{ij} \tag{10.5}$$

$$W_i^{ext} = \sum_{\substack{i,j \in E \\ I(i) \neq I(j)}} w_{ij} \qquad (10.6)$$

where w_{ij} is the weight of the link between nodes i and j, and $I(v)$ is the subset to which node v belongs—either V_1 or V_2. The cut size is given by

$$C(V_1, V_2) = \frac{1}{2} \sum_{i \in V} W_i^{ext} \qquad (10.7)$$

Let us now suppose that we move one node from one subset to the other. The advantage of moving such node is given by:

$$g_i = W_i^{ext} - W_i^{int} \qquad (10.8)$$

For instance, if we consider the network illustrated in Figure 10.2 in which a random partition of the nodes into subsubsets of equal size is represented by a dotted line, we can see that node v_1 has $W_{v_1}^{int} = 1$, $W_{v_1}^{ext} = 3$, and $g_{v_1} = 2$. The gains for nodes v_2 and v_3 are 1 and –3, respectively. The cut size for this partition is $C = 5$.

In the case when $W_i^{ext} > W_i^{int}$, the cut size can be reduced if we move node i from one subset to the other. The reduction obtained for the cut size is equal to the gain of the corresponding node. That is, if we move node v_1 from the set illustrated at the left of Figure 10.2 to the set illustrated at the right, the cut size is reduced from 5 to 3, as can be seen in Figure 10.3.

However, as we have started with a balanced partition of the nodes, moving one node from one subset to the other will produce an imbalance in the number of nodes in each partition. Consequently, instead of analyzing the gain obtained by moving one node from one partition to another, we need to quantify the gain produced by transposing a couple of nodes from different partitions. Let

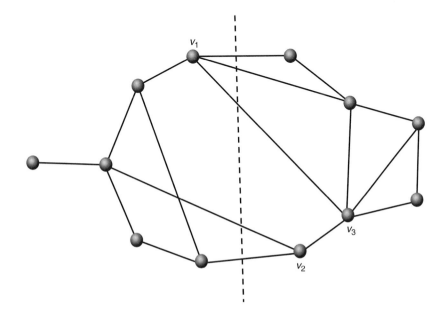

Fig. 10.2
A random balanced partition. A network having twelve nodes and used to illustrate the Kernighan–Lin algorithm. The partition illustrated by a dotted line is arbitrary.

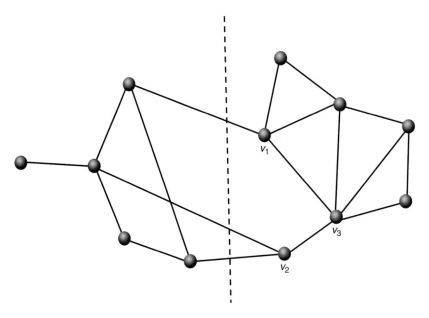

Fig. 10.3
Moving a node from partitions. The effect produced by moving the node labelled as v_1 in Figure 10.2, from one partition to another.

$v_1 \in V_1$ and $v_2 \in V_2$, and let $g(v_1, v_2)$ denote the gain of interchanging v_1 and v_2 between both subsets. Then,

$$g(v_1, v_2) = \begin{cases} g(v_1) + g(v_2) - 2w(v_1, v_2) & \text{if } v_1 \sim v_2 \\ g(v_1) + g(v_2) & \text{otherwise} \end{cases}$$

Because in the example given in Figure 10.2 nodes v_1 and v_2 are not directly linked, the gain of transposing them is $g(v_1, v_2) = 3$, which means that if we transposing them, the cut size is reduced from 5 to 2. The partition obtained so far is illustrated in Figure 10.4.

Transposing pairs of nodes between different partitions forms the basis of the Kernighan–Lin algorithm, which is sketched as follows. Given a balanced bisection $\{V_1, V_2\}$ with all nodes unmarked, we compute the cut size $C(V_1, V_2)$ of the bisection. We then repeat the following procedure from $k = 1$ to $r = \min(|V_1|, |V_2|)$. Find a pair of nodes $v_1 \in V_1$ and $v_2 \in V_2$ with the biggest value of $g(v_1, v_2)$. Note that $g(v_1, v_2)$ can be negative. Mark the pair v_1, v_2, and update the values of $g(v_j)$ for neighbours of v_1 and v_2. The result of this process is an ordered list of node pairs $\left(v_1^k, v_2^k\right)$, $k = 1, 2, \cdots, r$. We now calculate $C_k(V_1, V_2) = C_{k-1}(V_1, V_2) + g(v_1^k, v_2^k)$, which would be the cut size if $v_1^1, v_1^2, \cdots, v_1^k$ and $v_2^1, v_2^2, \cdots, v_2^k$ had been transposed. Next we find the index j such that $C_j(V_1, V_2) = \min_k C_k(V_1, V_2)$. Then, we transpose the first j node pair, such that

$$V_1 := V_1 - \left\{v_1^1, v_1^2, \ldots, v_1^j\right\} \bigcup \left\{v_2^1, v_2^2, \ldots, v_2^j\right\}$$

$$V_2 := V_2 - \left\{v_2^1, v_2^2, \ldots, v_2^j\right\} \bigcup \left\{v_1^1, v_1^2, \ldots, v_1^j\right\}$$

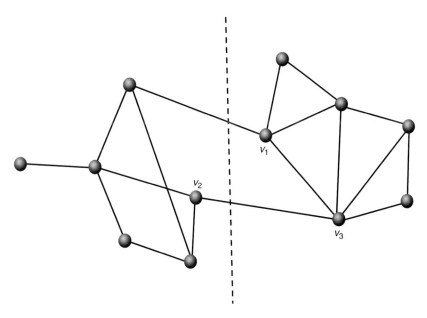

Fig. 10.4

Transposing a pair of nodes from different partitions. The effect produced by transposing nodes, labelled as v_1 and v_2 in Figure 10.2, between the two existing partitions.

We continue until no further improvement is achieved for the cut size.

The algorithm of Kernighan and Lin (1970) requires a time proportional to the third power of the number of nodes in the network, $O(n^3)$, but further implementations and modifications have reduced this time significantly. These changes are basically in the areas concerning swapping of nodes, use of a fixed number of iterations, and evaluating the gain only for those nodes which are near the boundary between the two subsets. For instance, Dutt (1993) has reduced this time to $O(|E|\max\{\log n, k_{\max}\})$. Other local improvement algorithms, using techniques such as *simulated annealing, tabu search,* or *genetic algorithms,* have also been proposed and used in the graph-partitioning literature. The reader is directed to the more specialised literature given at the beginning of this chapter to search for details of these methods.

10.1.2 Spectral partitioning

Although there are several partitioning methods that can be considered as spectral methods, we concentrate here only on those which are 'classically' considered as such. These methods are based on the use of the eigenvectors of different matrices representing the network. The roots of spectral clustering methods dates back to the early 1970s, when Fiedler (1973) first suggested that the second eigenvector of the Laplacian matrix φ_2^L separates the network into two approximately equal clusters of nodes having the fewest connections between them. This separation is carried out *grosso modo* as follows. Two nodes v_1, v_2 are considered to be in the same partition if $\operatorname{sgn} \varphi_2^L(v_1) = \operatorname{sgn} \varphi_2^L(v_2)$. Otherwise they are considered to be in two different partitions

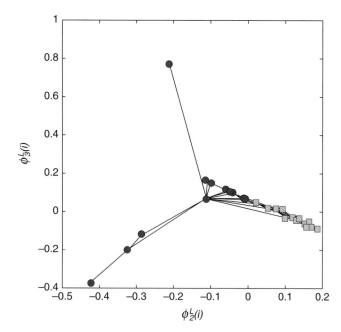

Fig. 10.5

Laplacian spectral partition. The
partition of the nodes in the karate club
network, using the two smallest
non-trivial eigenvectors of the Laplacian
matrix.

$\{V_1, V_2\}$. Such kind of separation is observed for the Zachary karate club
network illustrated in Figure 10.5, where we plot the smallest non-trivial
eigenvectors of the Laplacian. The separation made by φ_2^L for the two communities 'experimentally' found in this network is quite good, as can be seen
in Figure 10.5.

In 1988 Powers showed that eigenvectors of the adjacency matrix can also
be used to find partitions in a network (Powers, 1988). The idea behind these
methods is that the second largest eigenvector φ_2^A has both positive and negative components, allowing a partition of the network according to the sign
pattern of this eigenvector. In the case of the karate club benchmark network,
the separation provided by φ_2^A is even better than that obtained by using the
Laplacian second eigenvector (Fiedler vector), as can be seen in Figure 10.6.

In a similar way, the third largest positive eigenvector has a different pattern
of positive and negative signs, and so forth (Seary and Richards, 1995). Then,
if we arrange the rows and columns of the adjacency matrix according to the
signs of the first and second largest eigenvectors we obtain a partition of this
matrix into biants. These biants correspond to the partition of the network into
clusters of highly interconnected nodes—quasi-cliques. The use of sign patterns of further eigenvectors will produce partitions of the matrix as quadrants,
octants, and so on. At the end, the sign patterns of the last two eigenvectors—
the two largest negative ones—produce a partition of the network into n-tants,
in which nodes are closer to those with which they are not linked; that is,
quasi-bipartites. Other variants of spectral clustering use normalised Laplacian
matrix, and are based on the same fundamental principle of sign partition, as
can be seen for the karate club network in Figure 10.7.

Based on the heuristic of eigenvector partitioning, several algorithms
have been proposed and tested (Alpert et al., 1999; Blanchard and

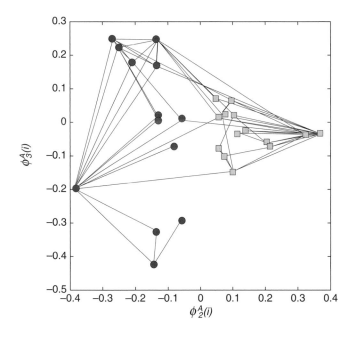

Fig. 10.6
Adjacency spectral partition. The partition of the nodes in the karate club network, using the second and third eigenvectors of the adjacency matrix.

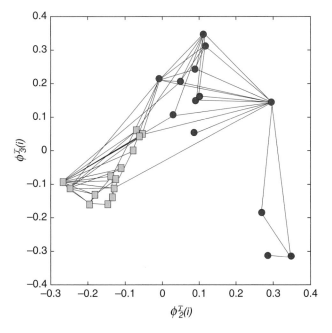

Fig. 10.7
Normalised Laplacian spectral partition. The partition of the nodes in the karate club network, using the two smallest non-trivial eigenvectors of the normalized Laplacian matrix.

Volchenkov, 2009; Capocci et al., 2005; Donetti and Muñoz, 2005; Filippone et al., 2008; Kannan et al., 2004; Shi and Malik, 2000; Qiu and Hancock, 2004; Verma and Meila, 2005; von Luxburg, 2007). In general, they consist of three main stages: (i) *preprocessing,* (ii) *spectral mapping*, and (iii) *post-processing or grouping*. One of these algorithms, introduced by Shi and Malik (2000), is based on a normalised cut criterion defined as follows. Let $Vol(V_1) = \sum_{i \in V_1} k_i$, and the normalised cut between the two sets V_1 and V_2 is

$$C_N(V_1, V_2) = C(V_1, V_2)\left(\frac{1}{Vol(V_1)} + \frac{1}{Vol(V_2)}\right) \qquad (10.9)$$

The aim of this algorithm, therefore, is to find a partition that minimises $C_N(V_1, V_2)$ over all possible bisections of the network. The method starts by calculating the following row-stochastic matrix:

$$\mathbf{P} = \mathbf{D}^{-1}\mathbf{A} \qquad (10.10)$$

It then computes the eigenvector φ_2^P associated with the second largest eigenvalue λ_2^P of $\mathbf{P}$. Next, it sorts the elements of φ_2^P in increasing order, where $\varphi_2^P(i)$ denotes the i th element in the sorted list. Now, repeat $n - 1$ times the following: compute $C_N(V_i, V_i')$, where $V_i = \{v_1, \cdots, v_i\}$ and $V_i' = \{v_{i+1}, \cdots, v_n\}$. Partition the network into two clusters $\{V_{i_0}, V_{i_0}'\}$, where $i_0 = \min_i C_N(V_i, V_i')$. The algorithm is repeated recursively on the cluster with the largest value of λ_2^P until K clusters are obtained.

A variation of this algorithm has been proposed by Kannan et al. (2004), who consider the optimal cut respect to the Cheeger conductance, which is defined as

$$\phi(V_1, V_2) = \frac{C(V_1, V_2)}{\min\{Vol(V_1), Vol(V_2)\}} \qquad (10.11)$$

Another spectral clustering algorithm m, proposed by Ng et al. (2001), uses the normalised Laplacian matrix $\mathcal{L}$ studied in Chapter 2. We recall that this matrix is defined as:

$$\mathcal{L} = \mathbf{D}^{-1/2}\mathbf{L}\mathbf{D}^{-1/2} \qquad (10.12)$$

This algorithm calculates the spectrum of $\mathcal{L}$, $0 = \lambda_1(\mathcal{L}) \leq \lambda_2(\mathcal{L}) \leq \cdots \leq \lambda_n(\mathcal{L})$, and uses the matrix $\mathbf{U} = \left[\varphi_1^{\mathcal{L}}, \varphi_2^{\mathcal{L}}, \cdots, \varphi_n^{\mathcal{L}}\right]$ in which the orthonormal eigenvectors are used as columns. Then, the matrix $\mathbf{Y}$ is formed by renormalizing each row of $\mathbf{U}$ to unit length; that is, $Y_{ij} = U_{ij}/\sqrt{\sum_j U_{ij}^2}$. Finally, the method uses the K-means algorithm to obtain the final clustering.

Meila and Shi (2001) proposed the use of the matrix $\mathbf{P}$ as in the Shi and Malik algorithm, but instead proposed the use of eigenvectors corresponding to the k largest eigenvalues, and clustering of the points in the truncated eigenvector matrix as points in a k-dimensional space. The main advantage of this last algorithm is that it uses all eigenvectors instead of only one. In fact, Alpert et al. (1999) have shown that when all eigenvectors are used, network partitioning exactly reduces to vector partitioning. They have suggested that 'as many eigenvectors as are practically possible should be used to construct a solution.'

The question that immediately arises here concerns the reason why spectral methods make these kinds of partition in networks (Brand and Huang, 2003). In order to explain this let us consider a positive–semidefinite matrix $\mathbf{M}$, such as the Laplacian or the normalised Laplacian. In case of a non-positive–semidefinite $\mathbf{N}$ matrix, such as the adjacency matrix, we consider that $\mathbf{N} = \mathbf{M} - \mathbf{C}$, where $\mathbf{M}$ and $\mathbf{C}$ are positive–semidefinite and $rank(\mathbf{N}) = rank(\mathbf{M}) + rank(\mathbf{C})$, and $\mathbf{M}$ built from the positive part of the spectrum of $\mathbf{N}$ is the best Gram approximation of $\mathbf{N}$, which offers a real-valued decomposition of the form $\mathbf{M} = \mathbf{X}^T\mathbf{X}$. We consider that the spectral decomposition of $\mathbf{M}$ is $\mathbf{M} = \mathbf{U}\mathbf{\Lambda}\mathbf{U}^T$, where $\mathbf{U}$ is a matrix with columns which are orthonormalised eigenvectors of $\mathbf{M}$, and $\mathbf{\Lambda}$ is a diagonal matrix of eigenvalues in non-increasing order (see Chapter 2). Let $\mathbf{X} = \mathbf{\Lambda}^{1/2}\mathbf{U}^2$, and let $\mathbf{X}^d$ be the top d rows of $\mathbf{X}$. It is known that $\mathbf{M}^d = (\mathbf{X}^d)^T\mathbf{X}^d$ is the best rank $-d$ approximation to $\mathbf{M}$ with respect to the Frobenius norm. Then, the angle between two columns of $\mathbf{X}^d$, $\mathbf{x}_i$, $\mathbf{x}_j$, $\in \mathbf{X}^d$, is

$$\theta_{ij} = \arccos \frac{\mathbf{x}_i^T \mathbf{x}_j}{\|\mathbf{x}_i\| \cdot \|\mathbf{x}_j\|} \tag{10.13}$$

Brand and Huang (2003) have proved the *polarization theorem*, which states that as the matrix $\mathbf{M}$ is projected to successively lower ranks using the top d eigenvectors, the sum of squared cosines of the angles between two data points $\sum_{i \neq j} \cos^2 \theta_{ij}$ is strictly increasing. This means that if we consider that the matrix $\mathbf{M}$ is imbedded in a high-dimensional space, as the dimensionality of the representation is reduced, the distribution of cosines changes from being concentrated around zero towards the two poles ± 1. The clustering is then explained, because

> truncation of the eigenbasis amplifies any unevenness in the distribution of points on the d-dimensional hypersphere by causing points of high affinity to move towards each other and others to move apart.

If we change point by node in this corollary we have an explanation of network spectral clustering.

Another interpretation can be produced by using the stochastic matrix $\mathbf{P}$ given in (10.10). This matrix can be considered as the transition matrix of a Markov random walk over the nodes of the network. Let $P_{V_1 V_2}$ be the transition probability for a random walk going from the subset V_1 to the set V_2 in one step, if the current state is V_1 and the random walk is in its stationary distribution π. Then,

$$P_{V_1 V_2} \frac{\sum\limits_{i \in V_1, j \in V_2} \pi_i P_{ij}}{\pi_A} = \frac{\sum\limits_{i \in V_1, j \in V_2} M_{ij}}{Vol(V_1)} = \frac{C(V_1, V_2)}{Vol(V_1)} \tag{10.14}$$

which by substitution in the expression for the normalised cut gives:

$$C_N(V_1, V_2) = \left(\frac{C(V_1, V_2)}{Vol(V_1)} + \frac{C(V_1, V_2)}{Vol(V_2)} \right) = P_{V_1 V_2} + P_{V_2 V_1} \tag{10.15}$$

This means that if the normalised cutset is small, a random walk which is in the set V_1 has a small probability of jumping to the set V_2 or from V_2 to V_1. This means that the random walks are mainly 'localized' in each partition, and cross-overs are scarce, which is why we observe the two clusters of nodes in the network.

10.2 Methods based on link centrality

The idea of these methods is to find those links which are central for the intercluster communication. Then, by removing them successively we can find the best partition of the network into communities. The first of these algorithms was introduced by Girvan and Newman on the basis of the edge betweenness centrality (Girvan and Newman, 2002; Newman and Girvan, 2004). They consider an extension of the idea of betweenness centrality (which we have studied in Chapter 7) to the links of the network. It is expected that links connecting nodes which are located in different communities display the highest edge betweenness. For instance, the two links with the highest betweenness in the network illustrated in Figure 10.4 are shown in Figure 10.8.

Consequently, the method provides a way of identifying inter-community links. The edge betweenness centrality is defined as the number of shortest paths between two nodes that pass through a given link, divided by the total number of such shortest paths. Then, the Girvan–Newman algorithm can be resumed in the following steps:

(vii) Calculate the edge betweenness centrality for all links in the network.
(viii) Remove the link with the largest edge betweenness, or any of them if more than one exists.
(ix) Recalculate the edge betweenness for the remaining links.
(x) Repeat until all links have been removed.
(xi) Use a dendrogram for analysing the community structure of the network.

In order to build the dendrogram we must proceed from the bottom to the top. That is, for a network consisting of n nodes the bottom line of the dendrogram consists of n points representing those nodes. Then, we find $n - 1$ communities, which will be formed by the joining of a pair of nodes and the remaining isolated nodes. Next we ask for $n - 2$ communities, and continue until only one community formed by the whole network is obtained. Then, the dendrogram is built from the bottom, as illustrated in Figure 10.9, for the small

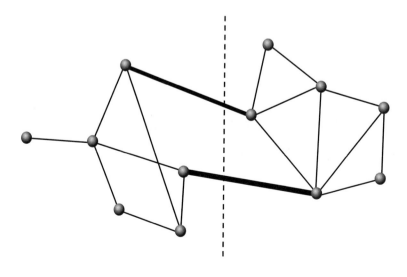

Fig. 10.8
Intercluster link betweenness. The two links represented as fat lines correspond to those having the largest link betweenness centrality among all the links of this network.

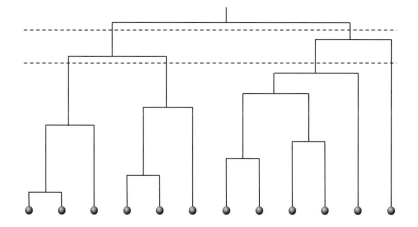

Fig. 10.9
Girvan–Newman dendrogram.
Dendrogram based on link betweenness,
using the Girvan–Newman method for
clustering the nodes of the network
shown in Figure 10.8.

network previously displayed in Figure 10.4. The top dotted line indicates the
division of the network into two communities, each formed by six nodes, and
the other dotted line indicates a division into four communities having 3, 3, 5,
and 1 nodes, respectively.

The main characteristic of this method is that it presents a range of possible
partitions of the network rather than a fixed one. Newman has remarked that
'it is up to the user to decide which of the many divisions represented is most
useful for their purposes.' However, in most real-world situations we need
some quality criterion in order to select a partition in preference to others.
This kind of quality measure is studied in the next section.

In Figure 10.10 we illustrate the bisection of the karate club network which
is obtained by the Girvan–Newman algorithm. As can be seen, it almost divides
the network into the two clusters found 'experimentally' by Zackary, except
that node 3 is classified in the 'wrong' group.

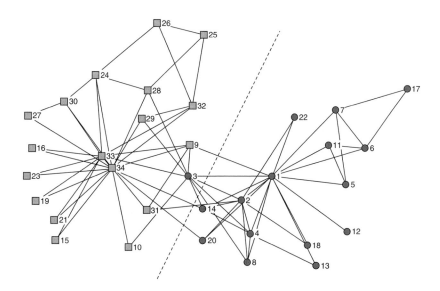

Fig. 10.10
Girvan–Newman partition. The
partition of the nodes in the karate club
network, using the Girvan–Newman
algorithm.

Another method that centres on the idea of link centrality was proposed by Radicchi et al. (2004). The intuition here is that a community is characterized by a high density of links—a situation which favours the presence of small cycles. In contrast, links connecting two communities are not expected to take part in many cycles. Consequently, Radicchi et al. (2004) have proposed a measure of the link clustering coefficient which is based on the same notion for nodes introduced by Watts and Strogatz (1998), studied in Chapter 4. The clustering coefficient for the link i, j is defined as:

$$C_{i,j}^l = \frac{z_{i,j}^l + 1}{s_{i,j}^l} \tag{10.16}$$

where $z_{i,j}^l$ is the number of cycles of length l that pass through the link i, j, and $s_{i,j}^l$ is the maximum possible value that $z_{i,j}^l$ can take. Obviously, this index is more general than the Watt–Strogatz clustering coefficient, as it accounts for cycles of length larger than 3, which can have an important influence in the structure of communities. However, for the sake of computational efficiency, usually only cycles of length 3 and 4 are used. The method then works in a similar way as the Girvan–Newman algorithm by removing those links with the smallest clustering coefficient, recalculating $C_{i,j}^l$ for the remaining links, and iterating until no link remains in the network.

Other variations of these approaches can be found in the literature, and the principle of removing links according to certain centrality measure which should differentiate intracluster and intercluster links remains the basic principle. In Chapter 7 we studied variations of the betweenness centrality for nodes which accounts for flows, random walks, and communicability, which can also be extended to links and used in Girvan–Newman-type algorithms.

10.2.1 Quality criteria

As we found in the previous section, in many situations it is necessary to decide among several different partitions of a network that have been produced by a given method. In order to evaluate the quality of these partitions we need some quantitative criteria that help us to decide on the 'best' partition. Here caution is required, as there is no 'absolute best', and the quality of a given partition depends on the criterion used to define a cluster. For instance, if we define a community as the rich club formed by the nodes with the highest degrees, the use of any criteria based on the intracluster and intercluster densities will fail to identify the clustering determined as one of 'good quality.' Having said that, we give below some of the quality criteria which are commonly used in the literature (Bolshakova and Azuaje, 2003; Boutin and Hascoët, 2004; Brandes et al., 2003; 2008; Dixon et al., 2009; Fortunato, 2007; Kovács et al., 2006).

i. Silhouette index

This index, introduced by Rousseeuw (1987), assigns a quality measure for every node in a given community. We start by considering a network that has been clustered into the following communities: $C_1, C_2, \ldots C_r$. Then, we consider that the node i has been clustered in the community C_k. Let $\bar{d}_{i,C_j}$ be the average distance between node i and all nodes in the community C_j. For $j \neq k$ we compute the minimum among all average distances $\bar{d}_{i,C_j}$, which we designate as $\bar{d}_{min}$. Then, the silhouette of the node $i \in C_k$ is given by

$$s(i) = \frac{\bar{d}_{min} - \bar{d}_{i,C_k}}{\max\{\bar{d}_{i,C_k}, \bar{d}_{min}\}} \qquad (10.17)$$

The Silhouette index is bounded between -1 and 1, where the value of 1 is reached when node i is 'well-clustered' in the community C_k and the value of -1 indicates that the node is misclassified. In addition, values close to zero indicate that the node could be also assigned to the nearest neighbouring community.

In order to consider the quality of a given cluster, the average Silhouette index is defined as characterizing the heterogeneity of the cluster:

$$s(C_k) = \frac{1}{n_{C_k}} \sum_{i=1}^{n_{C_k}} s(i) \qquad (10.18)$$

where n_{C_k} is the number of nodes in the cluster C_k. Finally, the global Silhouette index is used as an effective validity index for a clustering of a network into a series of communities $C_1, C_2, \ldots C_r$:

$$GS = \frac{1}{r} \sum_{j=1}^{r} s(C_j) \qquad (10.19)$$

ii. Performance

The performance of a clustering C has been defined by van Dongen (2000) as:

$$Per(C) = 1 - \frac{2m[1 - 2Cov(C)] + \sum_{i=j}^{c} |C_i|(|C_i| - 1)}{n(n-1)} \qquad (10.20)$$

where $Cov(C)$ is the 'coverage' of the clustering, defined as:

$$Cov(C) = \frac{1}{m} \sum_{i=j}^{c} \omega(C_i) \qquad (10.21)$$

Here ωC_i and $|C_i|$ are the number of links and nodes in the cluster C_i respectively, and m is the total number of links. According to this index a good clustering is the one having most of the links inside clusters with very few links between them. As a consequence, the larger the value of the coverage, the better the quality of the clustering C. The smaller the value of the performance of C, the better the quality of the clustering.

iii. Davies–Bouldin Validation Index

A quality criterion based on intracluster and intercluster distances has been defined by Davies and Bouldin (1979) as follows:

$$DB(C) = \frac{1}{c} \sum_{i=1}^{c} \max_{i \neq j} \left\{ \frac{\Delta(C_i) + \Delta(C_j)}{\delta(C_i, C_j)} \right\} \qquad (10.22)$$

where $\Delta(C_i)$ represents a measurement of the intracluster distance of cluster C_i, and $\delta(C_i, C_j)$ defines a distance function between clusters C_i and C_j. There are several possible definitions of these intercluster and intracluster distance functions, and the following are a few examples (Bolshakova and Azuaje, 2003):

Intracluster distances

$$\text{Complete diameter: } \Delta(C_i) = \max_{v_{ri} v_s \in C_i} \{d_{rs}\} \qquad (10.23)$$

$$\text{Average diameter: } \Delta(C_i) = \frac{1}{|C_i|(|C_i| - 1)} \sum_{\substack{r,s \in C_i \\ r \neq s}} d_{rs} \qquad (10.24)$$

$$\text{Centroid diameter: } \Delta(C_i) = 2 \left(\frac{\sum_{r \in C_i} d_{rO}}{|C_i|} \right) \qquad (10.25)$$

where O is the centroid of the cluster, which can be obtained by

$$\bar{r} = \frac{1}{|C_i|} \sum_{r,s \in C_i} d_{rs} \qquad (10.26)$$

Intercluster distances

$$\text{Single linkage: } \delta(C_i, C_j) = \min_{\substack{r \in C_i \\ s \in C_j}} \{d_{rs}\} \qquad (10.27)$$

$$\text{Complete linkage: } \delta(C_i, C_j) = \max_{\substack{r \in C_i \\ s \in C_j}} \{d_{rs}\} \qquad (10.28)$$

$$\text{Average linkage: } \delta(C_i, C_j) = \frac{1}{|C_i||C_j|} \sum_{\substack{r \in C_i \\ s \in C_j}} d_{rs} \qquad (10.29)$$

$$\text{Centroid linkage: } \delta(C_i, C_j) = d_{OP} \qquad (10.30)$$

where O and P are the centroids of the cluster C_i and C_j, respectively.

Compact clusters with centres which are far away from each other display small values of the Davies–Bouldin index. Consequently, among two partitions the one displaying the smallest value of this index is considered the one of highest quality. However, the index is biased when there are communities with only one node, as there are no internal distances in this cluster and the index tends to be too small.

Several other quality indices have been defined in the literature in different contexts, and for details the reader is directed to the works of Boutin and Hascoët (2004), Kovács et al. (2006), Brandes et al. (2008) and Dixon et al. (2009). However, there is one particular quality criterion that has been of particular attraction to practitioners of community detection methods. It is the modularity criterion defined by Newman (2006b)—the topic of the next section.

10.3 Methods based on modularity

The most popular of these quality measures is *modularity*, introduced by Newman and Girvan (2004). The intuition behind the idea of modularity is that a community is a structural element of a network that has been formed in a far from random process. If we consider the actual density of links in a community, it should be significantly larger than the density we would expect if the links in the network were formed by a random process. Consequently, modularity can be defined as (Newman and Girvan, 2004; Newman, 2006b):

$$Q = \frac{1}{4m} \sum_{ij} \left(A_{ij} - \frac{k_i k_j}{2m} \right) S_{i,r} S_{j,r} \tag{10.31}$$

where $m = |E|$, and $S_{i,r} = 1$ if $i \in V_r$ or zero otherwise. This expression can be written in the following form:

$$Q = \frac{1}{2m} tr(\mathbf{S}^T \mathbf{B} \mathbf{S}) \tag{10.32}$$

where $\mathbf{S}$ is a rectangular matrix with rows representing nodes and columns representing clusters, and where the modularity matrix is defined as

$$\mathbf{B} = \mathbf{A} - \frac{1}{2m} \mathbf{K} \mathbf{J} \mathbf{K},$$

where $\mathbf{J}$ is an all-ones matrix. Then, the entries of the modularity matrix are given by

$$B_{ij} = A_{ij} - \frac{k_i k_j}{2m} \tag{10.33}$$

When the nodes are partitioned between two groups only, the modularity can be written in quadratic form:

$$Q = \frac{1}{4m} \langle s | \mathbf{B} | s \rangle \tag{10.34}$$

where $s_i = 1$ if $i \in V_1$ or $s_i = -1$ if $i \in V_2$.

In a network consisting of n_V partitions, $V_1, V_2, \ldots, V_{n_C}$, the modularity can be interpreted as the sum over all partitions of the difference between the fraction of links inside each partition and the expected fraction, by considering a random network with the same degree for each node:

$$Q = \sum_{k=1}^{n_C} \left[\frac{|E_k|}{m} - \left(\frac{\sum_{j \in V_k} k_j}{2m} \right)^2 \right] \tag{10.35}$$

where $|E_k|$ is the number of links between nodes in the kth partition of the network. Then, if the number of intracluster links is not larger than the expected value for a random network, we obtain $Q = 0$. The maximum value for the modularity is $Q = 1$, which indicates strong community structure. The objective is then to use the values of Q in order to decide which partitions are particularly satisfactory.

There are several algorithms which are based on modularity optimisation for detecting communities in networks (Fortunato, 2010). They include agglomerative techniques such as the one originally proposed by Newman (2004a), as well as several modifications designed to improve the efficiency of the algorithm, such as simulated annealing (Guimerá et al., 2004), extremal optimisation techniques (Boettcher and Percus, 2001), and spectral methods (Newman, 2006a).

In particular, the spectral optimisation method for the modularity function is based on the following principles. The objective here is to find the value of the vector $|\mathbf{s}\rangle$ that maximises the quadratic function (10.34) for a given modularity. The procedure consists in expressing $|\mathbf{s}\rangle$ as a linear combination of the eigenvectors of the modularity matrix $\mathbf{B}$,

$$|\mathbf{s}\rangle = \sum_{i=1}^{n} \alpha_i |\varphi_i^B\rangle, \tag{10.36}$$

where $\alpha = \langle \varphi_i^B | \mathbf{s}\rangle$. Then, the modularity is written as

$$Q = \frac{1}{4m} \sum_{i=1}^{n} \left(\langle \varphi_i^B | \mathbf{s}\rangle\right)^2 \lambda_i^B \tag{10.37}$$

where λ_i^B is the eigenvalue of the modularity matrix corresponding to the eigenvector $|\varphi_i^B\rangle$. This expression indicates that modularity can be optimised on bisections in a similar way as we have done already, using other network matrices (Newman, 2006a). That is, we have to search for the eigenvector $\mathbf{B}$ corresponding to the largest positive eigenvalue, and then group the nodes according to the signs of the corresponding entries of such eigenvector. An example is shown in Figure 10.11 for the karate club network, where the second and third eigenvector of the modularity matrix are plotted for all nodes in the network. As can be seen, the second eigenvector divides the network into two clusters, which exactly corresponds to those observed experimentally.

In applying this algorithm to obtain multipartitions in networks, it is not possible to obtain first a bisection, removing the links connecting both subsets and then again applying the method to each individual part of the bisection (Newman, 2006a). The problem arises because we have degrees in the expression of modularity which will change if we remove some links from the network. The correct approach, therefore, is to consider the additional contribution to the modularity function of the entire network upon further bisecting a community C of size n_C:

$$\Delta Q = \frac{1}{4m} \langle \mathbf{s} | \mathbf{B}^C | \mathbf{s}\rangle \tag{10.38}$$

where the entries of the matrix $\mathbf{B}^C$ are given by

$$B_{ij}^C = B_{ij} - \delta_{ij} \sum_{k \in C} B_{ik} \tag{10.39}$$

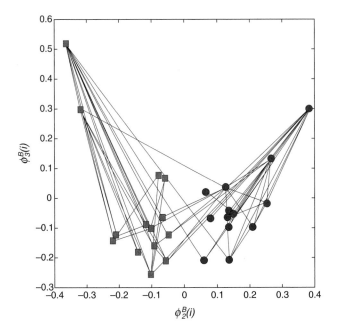

Fig. 10.11
Modularity spectral partition. The partition of the nodes in the karate club network, using the second and third largest eigenvectors of the modularity matrix.

10.3.1 The problem of resolution

When studying the detection of communities with the use of the modularity function, Fortunato and Barthélemy (2007) detected a very challenging problem which is nowadays known as the 'problem of resolution limit'. It consists in the fact that the modularity index can be fooled by some situations in which some very well defined, but small, communities are either adjacent to a bigger community or where many small communities are circularly connected (Arenas, et al., 2008; Fortunato and Barthélemy, 2007; Kumpula et al., 2007; Zhang and Zhang, 2010; Zhang et al., 2010). Fortunato and Barthélemy (2007) have found that in these cases, modularity prefers to group together pairs of the small communities instead of identifying them as independent ones. One of these 'pathological' cases is a network formed by several small cliques—say K_5—which are connected by a ring, as illustrated in Figure 10.12. In this case, modularity identifies the pairs of modules instead of the single cliques as the best partitions. This problem is not exclusive of the modularity index, as Kannan et al. (2004) have identified some pathological cases for other quality measures, such as minimal cuts, for the diameter and for the 2-median measures. Fortunato (2010) has remarked that for quality functions which are based on a null model, the problem of resolution limit has to be considered. A specific analysis of the possible causes of these pathologies has been advanced by Estrada (2011) on the basis of the local and global densities in networks, and the reader is referred to this work for details. The main lesson from the existence of such pathological cases is that in finding communities in networks it is better to use and compare more than one quality criterion. In doing so, it is necessary to use some kind of majority or consensus criterion for selecting the best partition among the many identified by a given algorithm.

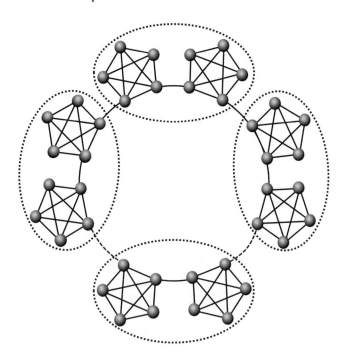

Fig. 10.12

Pathological case for modularity optimisation. An example of one of the networks found by Fortunato and Barthélemy (2007) where the modularity fails in identifying the most intuitive community partition of the network.

10.4 Clustering based on similarity

A different way of clustering objects, including the nodes of a network, is by grouping together those which are similar with respect to certain properties. To apply these methods, we first need *similarity measures* for the nodes in a network (Schaeffer, 2007). To maintain a topological context we will explore only those similarity measures that can be directly obtained from the information provided by the adjacency matrix of the network.

One popular measure of similarity is the cosine similarity, defined as the angle formed between two vectors. It can be calculated as follow:

$$\sigma_{ij} = \cos\vartheta_{ij} = \frac{\langle \mathbf{x}|\mathbf{y}\rangle}{\|\mathbf{x}\| \cdot \|\mathbf{y}\|} \tag{10.40}$$

Two perpendicular vectors are considered to be dissimilar, while two collinear ones are similar. When applying this measure to quantify the similarity between two nodes, It has been proposed considering the i th and j th columns of the adjacency matrix as two vectors representing the corresponding nodes, and then using the cosine of the angle between them to measure node–node similarity. In this case it is easy to realise (see Newman, 2010, for example) that

$$\sigma_{ij} = \frac{\eta_{ij}}{\sqrt{k_i k_j}} \tag{10.41}$$

where η_{ij} is the number of common neighbours of nodes i, and j. The intuition behind this measure is that two nodes are similar if they share a large number of common neighbours. For instance, in Figure 10.13 (top) we illustrate the adjacency matrix for a path of three nodes, P_3, and a representation of the

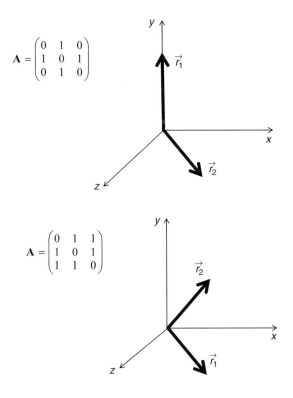

$$A = \begin{pmatrix} 0 & 1 & 0 \\ 1 & 0 & 1 \\ 0 & 1 & 0 \end{pmatrix}$$

$$A = \begin{pmatrix} 0 & 1 & 1 \\ 1 & 0 & 1 \\ 1 & 1 & 0 \end{pmatrix}$$

Fig. 10.13

Projection of adjacency matrices.
Examples of the projection of two simple adjacency matrices in which columns are represented as vectors in an n-dimensional space.

corresponding rows (or columns) corresponding to a node with degree 1, $\vec{r}_1$, and the one with degree 2, $\vec{r}_2$. As the two nodes do not share any other node, an angle exists between the two row-vectors, which is calculated by (10.40) as 90°. That is, they are perpendicular, and the two nodes are dissimilar with the lowest possible similarity between them, $\sigma_{1,2} = 0$. If we consider instead the two equivalent nodes of degree 1, it can be easily seen that the angle between the two corresponding row-vectors is equal to zero, as both nodes share exactly the same node. If, as in Figure 10.13 (bottom), we consider the nodes in a cycle of length 3, the vector representation of the adjacency matrix provides an angle of 60° between any pair of nodes, which is due to the fact that these nodes have degree 2 and share only one common node.

Another popular similarity measure is the Pearson correlation coefficient between pairs of rows (or columns) of the adjacency matrix. If $|\mathbf{a}_i\rangle$ is a vector corresponding to the ith row (column) of the matrix A, the Pearson correlation coefficient is defined as:

$$r_{ij} = \frac{n\langle \mathbf{a}_i | \mathbf{a}_j \rangle - \langle \mathbf{a}_i | \mathbf{1} \rangle \langle \mathbf{a}_j | \mathbf{1} \rangle}{\sqrt{n\langle \mathbf{a}_i | \mathbf{a}_i \rangle - (\langle \mathbf{a}_i | \mathbf{1} \rangle)^2} \sqrt{n\langle \mathbf{a}_j | \mathbf{a}_j \rangle - (\langle \mathbf{a}_j | \mathbf{1} \rangle)^2}} \tag{10.42}$$

Note that r_{ij} is identical to $\sigma_{ij} = \cos \vartheta_{ij}$ if we normalise each vector to have mean equal to zero. In computing the correlation coefficient between columns of the adjacency matrix, some authors (Schaeffer, 2007) recommend the use of $\mathbf{A} + \mathbf{I}$ instead of $\mathbf{A}$; that is, including 1s in the main diagonal of the adjacency matrix. This is done in order to force all reflexive edges to be present. However, in many circumstances this falsifies the results, and the simple adjacency matrix is preferred. For instance, if we consider $\mathbf{A} + \mathbf{I}$ instead of $\mathbf{A}$ for the case given in Figure 10.13 (top), the nodes 1 and 3 which are topologically equivalent are found to be different.

Other measures like the norms of the difference between the corresponding rows (columns) of the adjacency matrix have also been proposed. The following are a few examples.

Manhattan or taxicab norm:

$$\|\mathbf{a}_i - \mathbf{a}_j\|_1 = \sum_{k=1}^{n} |\mathbf{a}_i(k) - \mathbf{a}_j(k)| \tag{10.43}$$

Euclidean norm:

$$\|\mathbf{a}_i - \mathbf{a}_j\|_2 = \sqrt{\sum_{k=1}^{n} \left[\mathbf{a}_i(k) - \mathbf{a}_j(k)\right]^2} \tag{10.44}$$

Infinite norm:

$$\|\mathbf{a}_i - \mathbf{a}_j\|_\infty = \max_{k \in [1,n]} |\mathbf{a}_i(k) - \mathbf{a}_j(k)| \tag{10.45}$$

A comprehensive selection is outside the scope of this book, but other variations of similarity indices between nodes can be found in the more specialised literature. Let us return to the main objective of this chapter, which is that of detecting partitions in networks. When using similarity indices, the algorithms for finding network communities start by choosing a similarity index for the nodes of a network and then finding successive clusters using any of the existing linkage techniques, such as:

- *Single linkage:* the distance between two clusters is determined by the distance of the two closest nodes (nearest neighbours) in the different clusters.
- *Complete linkage:* the distances between clusters are determined by the greatest distance between any two nodes in the different clusters. It usually performs very well when the objects are clumped together, but it fails when they are somehow elongated or of a 'chain'-type nature.
- *Unweighted pair-group average:* the distance between two clusters is calculated as the average distance between all pairs of objects in the two different clusters. It is efficient both when the objects are clumped together and when they are somehow elongated.
- *Weighted pair-group average:* it is identical to the unweighted pair-group average method, except that in the computations, the size of the respective clusters is used as a weight.
- *Unweighted pair-group centroid:* the centroid of a cluster is the average point in the multi-dimensional space defined by the dimensions.
- *Weighted pair-group centroid (median):* it is identical to the previous one, except that weighting is introduced into the computations to take into consideration differences in cluster sizes.
- *Ward's method:* it is different from all other methods, because it uses an analysis of the variance approach to evaluate the distances between clusters.

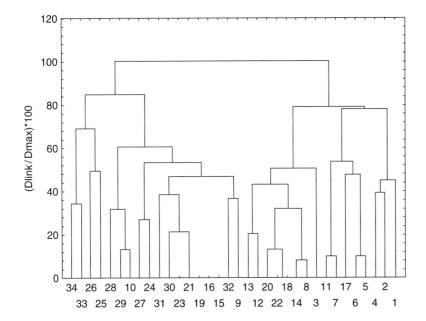

Fig. 10.14
Similarity-based clustering.
Dendrogram obtained by using the
Pearson correlation coefficient of the
matrix **A** as the similarity measure with
the complete linkage method for
grouping the clusters in the karate club
network.

In Figure 10.14 we illustrate the result of using the Pearson correlation coefficient of the matrix **A** as the similarity measure with the complete linkage method for grouping the clusters in the karate club network. As can be seen, the method perfectly identifies the two clusters observed experimentally, and allows partition into other clusters as we descend in the dendrogram. However, the use of the Euclidean norm, for instance, fails to identify the community structure of this network.

10.5 Communities based on communicability

In Chapter 6 we introduced a communicability function that accounts for how much 'information' is transmitted from one node to another in a network by using all possible routes between them. The concept of communicability introduces an intuitive way of finding the structure of communities in complex networks. We can think intuitively that a community is a group of nodes which have better communicability among themselves than with the rest of the nodes in the network.

Let us consider that a node in a network can be in one of the following three states: on–positive, on–negative, or off. Let us assume that two nodes having the same on-state can be grouped together. In a social network we can consider that an individual can have a 'positive' or 'negative' position with respect to some criterion or point of view, or he/she can be completely uninterested in it. Then, it is expected that those having similar points of view form some clusters or communities among themslves. If we think about this in terms of energy, the minimum energy configuration would be a consensus one in which all individuals share the same (positive or negative) state. The configuration of maximum energy would be the one in which individuals with

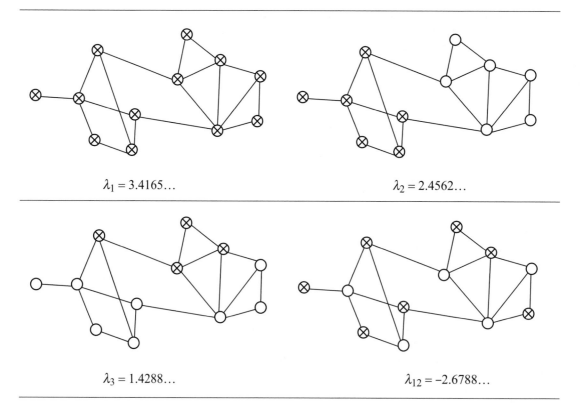

$$\lambda_1 = 3.4165\dots \qquad\qquad \lambda_2 = 2.4562\dots$$

$$\lambda_3 = 1.4288\dots \qquad\qquad \lambda_{12} = -2.6788\dots$$

Fig. 10.15

'Energy' levels in networks. Each node is represented by the sign of the corresponding eigenvector, which represents a state of the network with energy given by its corresponding eigenvalue. The first state corresponds to the most stable consensus configuration, and the last state corresponds to the most unstable configuration.

contrary opinion are close together in the network. This picture is very much analogous to the one provided by the eigenvector partition of the nodes in a network. In this case, $-\lambda_j$ would represent the energy of the jth state. The sign of the pth component of the jth eigenvector indicates the state of the node—say plus, zero, or minus. If two nodes, p and q, have the same sign for the jth eigenvector, it indicates that they share the same 'opinion.' In Figure 10.15 we illustrate this situation for the nodes in a simple network. According to the Perron–Frobenius theorem, all the components of the principal eigenvector φ_1 have the same sign. Consequently, it represents the consensus configuration of the whole network.

Let us now recall the expression for the communicability between a pair of nodes p and q in an undirected network:

$$G_{pq}(\beta) = \sum_{j=1}^{n} \varphi_j(p)\varphi_j(q)e^{\beta\lambda_j} \qquad (10.46)$$

where λ_j and φ_j are eigenvalues and eigenvectors of the adjacency matrix, respectively, and β is the inverse temperature. For the sake of simplicity we concentrate on the case $\beta = 1$. Nodes which are in an off state do not contribute to the communicability function. However, those in positive or negative on-states can contribute to it in different ways. This allows us to express the communicability function in the following way:

$$G_{pq} = \left[\varphi_1(p)\varphi_1(q)e^{\lambda_1} \right]$$

$$+ \left[\sum_{2 \leq j \leq n} {}^{++}\varphi_j(p)\varphi_j(q)e^{\lambda_j} + \sum_{2 \leq j \leq n} {}^{--}\varphi_j(p)\varphi_j(q)e^{\lambda_j} \right] \quad (10.47)$$

$$+ \left[\sum_{2 \leq j \leq n} {}^{+-}\varphi_j(p)\varphi_j(q)e^{\lambda_j} + \sum_{2 \leq j \leq n} {}^{-+}\varphi_j(p)\varphi_j(q)e^{\lambda_j} \right].$$

The first term on the right-hand side of the communicability function (10.47) represents the *consensus configuration* in which all the nodes share the same state. In the second term on the right-hand side of (10.47), the nodes p and q have the same sign of the corresponding eigenvector (positive or negative), which means that they will communicate well among themselves. In this case we can consider that p and q are in the same cluster if there is more than one cluster in the network. The last term of (10.47), on the other hand, represents a lack of consensus in the states of the nodes p and q. That is, they have different signs of the eigenvector component, which means that they do not communicate well in the network, and we can regard them as being in different clusters of the network. Consequently, we call the second term of (10.47) the *intracluster communicability*, and the third term the *intercluster communicability* between a pair of nodes (Estrada and Hatano, 2008).

The consensus configuration does not provide us with any information about the community structure of a network. Consequently, we move this term to the other member of (10.47) in such a way that we obtain the difference between the intracluster and intercluster communicability for a network as:

$$\Delta G_{pq} = + \left[\sum_{2 \leq j \leq n} {}^{++}\varphi_j(p)\varphi_j(q)e^{\lambda_j} + \sum_{2 \leq j \leq n} {}^{--}\varphi_j(p)\varphi_j(q)e^{\lambda_j} \right]$$

$$+ \left[\sum_{2 \leq j \leq n} {}^{+-}\varphi_j(p)\varphi_j(q)e^{\lambda_j} + \sum_{2 \leq j \leq n} {}^{-+}\varphi_j(p)\varphi_j(q)e^{\lambda_j} \right] \quad (10.48)$$

$$= \overset{\text{intracluster}}{\sum_{j=2}} \varphi_j(p)\varphi_j(q)e^{\lambda_j} - \left| \overset{\text{intercluster}}{\sum_{j=2}} \varphi_j(p)\varphi_j(q)e^{\lambda_j} \right|$$

where in the last line we use the fact that the intracluster communicability is positive and the intercluster communicability is negative.

We can represent the values of ΔG_{pq} as the entries of the following communicability matrix:

$$\Delta \mathbf{G} = e^{\mathbf{A}} - e^{\lambda_1} |\varphi_1\rangle\langle\varphi_1| \quad (10.49)$$

Obviously, $\Delta G_{pq} > 0$ indicates that the two nodes display larger intracluster than intercluster communicability, and their chances for being members of the same community are very large. A very restrictive way of defining a community is by saying that it is a group of nodes $C \subseteq V$ for which $\Delta G_{p,q} > 0 \quad \forall (p, q) \in C$. The method is very much restrictive, because it only allows membership to a community to a node that has $\Delta G_{pq} > 0$ with all other members of the group. Let us suppose that there is a community $C \subseteq V$ formed by n nodes, and that there is a node v which has $\Delta G_{vq} > 0$, where $q \in C$, with more than 90% of the nodes in C, but negative with the rest. According to the previous definition of community, the node v is not a member of C. As a consequence, there will be communities including v and some of the nodes in C, which will have a large degree of overlapping among them.

The main disadvantage of this definition is that due to the situation explained previously, a large number of highly overlapped communities are detected in networks. For instance, for the karate club network this approach finds five communities from which the first three are very overlapped, sharing more than 85% of nodes among them. A different approach in detecting overlapped communities is based on cliques, and has been studied by Palla et al. (2005). The communities are as follows (Estrada and Hatano, 2008; 2009a):

A :{10, 15, 16, 19, 21, 23, 24, 26, 27, 28, 29, 30, 31, 32, 33, 34}
B :{9, 10, 15, 16, 19, 21, 23, 24, 27, 28, 29, 30, 31, 32, 33, 34}
C :{10, 15, 16, 19, 21, 23, 24, 25, 26, 27, 28, 29, 30, 32, 33, 34}
D :{1, 2, 3, 4, 5, 6, 7, 8, 11, 12, 13, 14, 17, 18, 20, 22}
E :{3, 10}

In order to avoid this undesired proliferation of overlapped communities we are obliged to relax the condition $\Delta G_{p,q} > 0 \ \forall (p, q) \in C$ for defining communities in a network, by changing it to a relaxed condition (Estrada, 2011):

> A *relaxed* communicability-based community is a subset of nodes $C \subseteq V$ in the network $G = (V, E)$ for which the intracluster communicability is larger than the intercluster one *for most of the nodes* in C, which are then grouped according to a given quantitative criterion.

Then, some possible criteria for grouping nodes into communities are allowed, including:

(xii) Considering the communicability matrix $\mathbf{C}$ as the input for clustering algorithms.

(xiii) Considering the communicability matrix $\mathbf{C}$ as the input for hierarchical clustering methods to produce a dendrogram with the hierarchical structure of communities in the graph.

(xiv) Dichotomizing the matrix $\mathbf{C}$ to obtain the communicability graph, and then using the communicability graph as the input for any clustering algorithm.

An example of these methods has been tested for artificial and real-world networks (Estrada, 2011) by using communicability-based (ComBa) $K-$means clustering. The method consists of the following steps:

(xv) Normalise the adjacency matrix of the network as: $\hat{\mathbf{A}} = \mathbf{K}^{-1/2}\mathbf{A}\mathbf{K}^{-1/2}$. This avoids failures in the pathological cases found by Fortunato and Barthélemy (2007).

(xvi) Construct the communicability matrix based on the normalised adjacency matrix: $\Delta\mathbf{G}' = e^{\hat{\mathbf{A}}} - e^{\lambda_1(\hat{\mathbf{A}})}|\varphi_1(\hat{\mathbf{A}})\rangle\langle\varphi_1(\hat{\mathbf{A}})|$

(xvii) Construct the matrix: $\mathbf{C} = \Delta\mathbf{G}' - diag(\Delta\mathbf{G}')$;

(xviii) Use $\mathbf{C}$ as the input for the K-means algorithm (Liu, 2010) and select the clusters by sorting distances and taking observations at equal distances in the clusters. Other strategies, such as choosing observations that maximize the initial between-cluster distance, or choosing the first N (number of clusters) observations, are also possible. Here we will always use the first of them.

(xix) Evaluate the quality of the clusters found so far, and determine the best one according to specified criteria.

An example of the application of this algorithm and some of its advantages over other classical partition methods can be found in Chapter 19 and in Estrada (2011).

Network bipartivity

11

> To regulate something always requires two opposing factors. You cannot regulate by a single factor. To give an example, the traffic in the streets could not be controlled by a green light or a red light alone. It needs a green light and a red light as well.
>
> Albert Szent-György

Bipartite networks are also known as two-colourable graphs, as we need only two colours—say green and red—to colour their nodes in such as way that red nodes are connected only to green nodes, and vice versa. A good compendium of mathematical properties and some applications of bipartite graphs can be found in the book by Asratian et al. (1998). Bipartivity is an important topological characteristic of complex networks in which the vertex set V of the network can be partitioned into two subsets V_1 and V_2 such that all edges have one endpoint in V_1 and the other in V_2. One of these sets of nodes can be 'regulators' of the other set of 'regulated' nodes, or it can represent a group of 'sellers' that have transactions with the 'buyers' in the other set. Many natural systems can be represented as bipartite networks, such as reaction networks (see Chapter 16) or 'two-mode' networks (see Chapter 19), in which two disjointed sets of nodes are related by links representing the relationship between the elements of both classes. As an example we can mention the citation networks in which a set of nodes represents authors who cite (or are cited by) papers, which are represented as the other set of nodes. Other examples are people-institutions networks in which one set represents people that belong to institutions represented by the other set. Cities that have certain services or voting results of delegates concerning certain proposals are also represented as bipartite networks.

However, in many real-world situations this 'perfect' separation into two classes is not always possible, as some 'regulators' are also 'regulated' by other nodes, or 'sellers' usually buy products from other sellers. These situations create links that connect nodes in the same class. In the physics literature, such kinds of link are usually called *frustrated link*, due to the relation of this problem with the spin frustration in spin glasses. In a bipartite network, all links are unfrustrated links. In non-bipartite networks it is interesting to know how much bipartivity exists. Holme et al. (2003) have pointed out several areas for the potential application of a quantitative measure of bipartivity, such as network studies of sexually transmitted diseases, trade networks of buyers and sellers, 'genealogical' networks of disease outbreak, and food webs (see Chapter 18). The following scenarios are examples of the relevance of a measure of network bipartivity . Let us consider the study of a sexually transmitted disease. It is known that the transmission rates differ for homosexual and heterosexual contacts. Consequently, the transmission of the disease will depend on the degree of bipartivity of the corresponding network. In other words, by having some idea of the bipartivity of sexual networks we will be able to assess the rate of spreading of a sexually transmitted disease. Another scenario in which the analysis of network bipartivity can be of great utility is in the study of information and communication networks. In a dictionary, for instance, all entries should be related in a self-referential way showing a large transitivity between triples of words. In communication networks such as a network of airports or the Internet, the network bipartivity indicates that two separate groups exist, where direct communication is possible only between nodes in the different groups. The lack of direct communication between 'members' of the same group is an indication of the lack of efficiency of such networks.

11.1 Bipartivity measures

A simple way of defining the bipartivity of a network is by accounting for the fraction of frustrated links in the network

$$b = 1 - \frac{m_{fr}}{m} \qquad (11.1)$$

where m_{fr} is the number of links connecting nodes in the same subset, and m is the total number of links in the network (Holme et al., 2003). The problem consists in finding the best partition of the nodes in the network into two almost disjoint subsets in such a way that the smallest value of b for all possible partitions corresponds to the bipartivity of the network. Despite the simplicity of this definition, a problem arises in the computation of this measure. Because we need to find the best possible partition, the computation of this index is NP-complete, and we need some approximate algorithms for the calculation of the network bipartivity.

Holme et al. (2003) have proposed one of such approximate algorithms for calculating the bipartivity of a network. In their algorithm, O_k is the set of odd circuits of length shorter than k, and $\sum O_k$ designates the accumulated length of the circuits, for which a cut-off value of $3m$ is imposed. Let $\hat{k}$ be the smallest k such as $\sum O_k \geq 3m$, and let $\gamma(e)$ be the number of circuits in O_k passing through the link e. The number of frustrated links is then roughly estimated as the number of links that have to be marked in such a way that each circuit of length lower than $\hat{k}$ is marked at least once. The algorithm then starts by assigning $O = O_k$ and sorting the links in order of γ. Then, while $o \neq \phi$ the following three steps are repeated: i) mark the link e with highest γ, ii) remove all circuits in o containing e and, iii) recalculate γ for each link. Finally, the number of iterations m' is the estimation of m_{fr}, and the *algorithmic bipartivity* b_A is estimated as

$$b_A = 1 - \frac{m'}{m} \tag{11.2}$$

In the case of sparse networks this algorithm has a complexity of the order of the square of the number of links in the network.

Another measure of network bipartivity is based on the Estrada index of a network (see Chapter 5). This index can be expressed as the sum of two contributions—one deriving from odd and the other from even closed walks:

$$EE(G) = tr[\cosh(A)] + tr[\sinh(A)]$$

$$= \sum_{j=1}^{n} \cosh(\lambda_j) + \sum_{j=1}^{n} \sinh(\lambda_j) \tag{11.3}$$

$$= EE_{even}(G) + EE_{odd}(G)$$

It is well known that a network is bipartite if, and only if, there is no cycle of odd length. Consequently, a bipartite network is characterized by the absence of odd closed walks, so that $EE_{odd}(G) = 0$. Thus,

$$EE(G) = EE_{even}(G) = \sum_{j=1}^{n} \cosh(\lambda_j) \tag{11.4}$$

Consequently, the proportion of even closed walks to the total number of closed walks is a measure of the network bipartivity (Estrada and Rodríguez-Velázquez, 2005b):

$$b_S(G) = \frac{EE_{even}(G)}{EE(G)} = \frac{EE_{even}(G)}{EE_{even}(G) + EE_{odd}(G)} = \frac{\sum\limits_{j=1}^{n} \cosh(\lambda_j)}{\sum\limits_{j=1}^{n} e^{\lambda_j}} \tag{11.5}$$

The idea can be rooted in the finding that graphs without short odd cycles are nearly bipartite (Györi et al., 1997; Lind et al., 2005). It is evident that $b_S(G) \le 1$ and $b_S(G) = 1$ if, and only if, G is bipartite; that is, $EE_{odd} = 0$. Furthermore, as $0 \le EE_{odd}(G)$ and $\sinh(\lambda_j) \le \cosh(\lambda_j), \forall \lambda_i$, then $\frac{1}{2} < b(G)$ and $\frac{1}{2} < b_S(G) \le 1$. The lower bound is reached for the least possible bipartite graph with n nodes, which is the complete graph K_n. As the eigenvalues of K_n are $n - 1$ and -1 (with multiplicity $n - 1$), then $b_S(G) \to \frac{1}{2}$ when $n \to \infty$ in K_n. This lower bound coincides with that given by Holme et al. in the $n \to \infty$ limit for their measures (Estrada and Rodríguez-Velázquez, 2005b).

A desired property for $b_S(G)$ is that it changes monotonically as the bipartivity of the network changes. Let G be a non-complete network, and let e be an edge of the complement of G Let $G + e$ be the network obtained by adding the edge e to G. In this situation there exist real and non-negative numbers a and b, such that $EE_{even}(G + e) = EE_{even}(G) + a$ and $EE_{odd}(G + e) = EE_{odd}(G) + b$. Notice that a is the contribution of edge e to EE_{even} and b is the contribution of edge e to EE_{odd}. Thus, with the above notation, if

$b \geq a$. then $b(G) \geq b(G + e)$. That is, as $\frac{a+b}{2} \geq a$ and $EE_{even}(G) \geq \frac{EE(G)}{2}$ then $(a + b)EE_{even}(G) \geq aEE(G)$. The addition of $EE_{even}(G)EE(G)$ to both terms, and further reordering gives $EE_{even}(G)[EE(G) + a + b] \geq EE(G)[EE_{even}(G) + a]$ and, consequently,

$$b_S(G) = \frac{EE_{even}(G)}{EE(G)} \geq \frac{EE_{even}(G) + a}{EE(G) + a + b} = b_S(G + e) \qquad (11.6)$$

which proves the monotony of the change for the spectral bipartivity measure. In Figure 11.1 we illustrate the monotonic change of the bipartivity measure for some small networks, and compare the results with those obtained by using b_A. Starting with a bipartite network formed by two disjoint sets V_1 and V_2, we systematically add links connecting nodes in the same set. The spectral bipartivity measure then decreases systematically from $b = 1.000$ for the bipartite network to $b = 0.597$ for the complete network, which is the least bipartite. The algorithmic bipartivity also decreases systematically as the number of frustrated links increases. However, as can be seen in Figure 11.1 there are several pairs of networks having different position of the frustrated link that have the same algorithmic bipartivity. These networks are well differentiated by the spectral measure, which always assigns larger bipartivity to the network having the frustrated link in the largest subset.

The analysis of bipartivity in real-world networks provides some hints about the structural organisation of these networks. In Figure 11.2 we illustrate two pie diagrams of the values of bipartivity for the 11,117 networks with eight nodes (left graphic), and a similar plot for sixty real-world networks (right graphic). The number of highly bipartite ($0.9 < b_S \leq 1$) artificial networks is very small, as they account for only 6% of the total number of such graphs. In a similar way, artificial networks which are non-bipartite constitute only 10% of the total number of graphs. In other words, bipartivity of artificial networks displays a normal distribution, with most of networks displaying values of bipartivity between 0.6 and 0.8. However, in real-world scenarios the percentage of highly bipartite networks ($0.9 < b_S \leq 1$) is 15%, and the percentage of strictly non-bipartite networks ($b_S = 0$) is 28%. In addition, 32% of real-world networks are highly non-bipartite: $0.5 < b_S \leq 0.6$.

As can be seen in Figure 11.2, the real-world networks appear more polarized in the bipartite/non-bipartite spectrum. The cause for the observed high degree of non-bipartity is the presence of odd cycles in general and in particular triangles, which pervades most of complex networks. In particular, transitivity of relationships in social sciences is a known source of such non-bipartity. However, the relatively large percentage of highly bipartite real-world networks points to the presence of other mechanisms that avoid the proliferation of odd cycles in certain networks. In Table 11.1 we provide some examples of the spectral bipartivity for real-world networks.

The question that remains open is whether we can find the best bipartition of any of these networks identified as having a significant level of bipartivity. This question is answered in the next section.

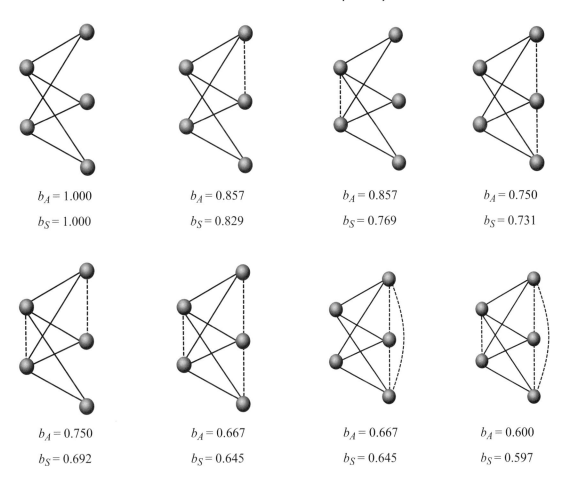

Fig. 11.1
Algorithmic and spectral bipartivity. Change of the algorithmic and spectral measures of bipartivity for some simple graphs.

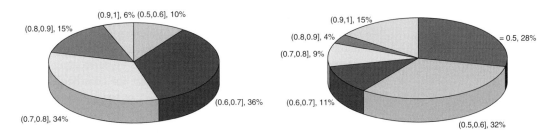

Fig. 11.2
Bipartivity in artificial and real networks. Pie plots of artificial (left) and real-world (right) networks with different percentages of spectral bipartivity.

Table 11.1 Bipartivity of real networks. Spectral bipartivity of complex networks in different real-world scenarios.

Type	Network	b_S	Type	Network	b_S
Information	ODLIS	0.500	PPIs	Yeast	0.500
	SmallWorld	0.500		Human	0.576
	SciMet	0.500		H. pylori	0.711
	Roget	0.529		A. fulgidus	0.978
Social	Drugs	0.500	Transcription	urchins	0.618
	Corporate elite	0.500		E. coli	0.831
	Karate Club	0.597		yeast	0.960
	Saw Mill	0.749			
Food webs	Coachella	0.500	Technological	USAir97	0.500
	El Verde	0.500		Internet	0.502
	Grassland	0.743		Electronic3	0.952
	Stony stream	0.815			

11.2 How do we find bipartitions in networks?

In Chapter 10 we used the communicability function previously defined in Chapter 6 to detect communities in complex networks. This method is based on the spectra of the adjacency matrix, from which it is known that the eigenvectors corresponding to positive eigenvalues produce a partition of the network into clusters of tightly connected nodes (see Chapter 10). On the other hand, the eigenvectors corresponding to negative eigenvalues produce partitions of the network in which nodes are not close to those to which they are linked, but rather with those with which they are not linked (Seary and Richards, 1995).

Let us consider, for instance, the cycle of six nodes (hexagon). The eigenvectors $\{\phi_j(p) = \mathrm{Re}e^{i\vartheta_j p}/\sqrt{6}\}$, where $\phi_j(p)$ denotes the component on the p th entry of the eigenvector with the label j, and $\vartheta_j = \pi(j-1)/3$. The corresponding eigenvalues are $\lambda_j = 2\cos\vartheta_j$. In Figure 11.3 we illustrate the values of $\phi_j(p)$ for every node in C_6 for the different values of j.

Because this network is bipartite, the lowest eigenvalue $\lambda_6 = -\lambda_1 = -2$ with eigenvector $\phi_6(p) = (-1)^p/\sqrt{6}$, which means that the nodes of the network are partitioned into two disjoint sets in which the nodes with even p have positive $\phi_6(p)$, and those with odd p have negative $\phi_6(p)$ (see Figure 11.3). From the perspective of the communicability function introduced in Chapter 6 and studied in the previous chapter, we have that a negative value of the inverse temperature β increases the contribution of the negative eigenvalues to the communicability function, which will emphasise the contribution of bipartite structures in the partition of the network. We can write the communicability function in the following way:

$$G_{pq}(\beta) = \sum_{\lambda_j < 0} \phi_j(p)\phi_j(q)e^{\beta\lambda_j} + \sum_{\lambda_j < 0} \phi_j(p)\phi_j(q)e^{\beta\lambda_j} + \sum_{\lambda_j > 0} \phi_j(p)\phi_j(q)e^{\beta\lambda_j}$$

(11.7)

which indicates that

$$G_{pq}(\beta < 0) \approx \sum_{\lambda_j < 0}^{n} \phi_j(p)\phi_j(q)e^{-|\beta||\lambda_j}$$

(11.8)

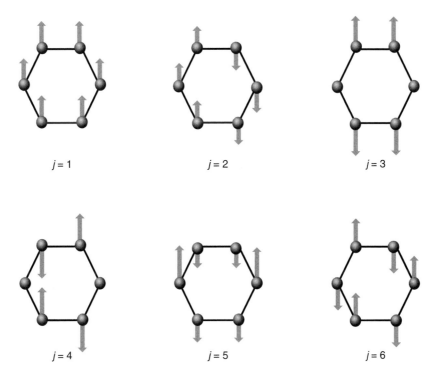

Fig. 11.3

Eigenpartitions of the hexagon. Representation of the eigenvectors $\phi_j(p)$ of the six-node cycle. The direction of the arrows indicates the sign of $\phi_j p$ —up if $\phi_j(p) > 0$, and down if $\phi_j(p) > 0$. The size of the arrow is proportional to the absolute magnitude of $\phi_j(p)$. No arrow is shown if $\phi_j(p) = 0$.

In other words, for $G_{pq}(\beta < 0)$ the network is partitioned into bipartite clusters such as the one displayed in Figure 11.3, for $j = 6$, where the nodes tend to form *quasi-bipartite clusters*; that is, nodes are partitioned into *almost* disjoint subsets with high connectivity between sets, but low internal connectivity.

In order to illustrate how the communicability function with negative values of the inverse temperature accounts for bipartite clusters, let us use a Taylor series expansion of the exponential for $\beta < 0$,

$$e^{-|\beta|A} = I - |\beta|A + \frac{(|\beta|A)^2}{2!} - \frac{(|\beta|A)^3}{3!} + \ldots \qquad (11.9)$$

which can be expressed in terms of the hyperbolic functions as

$$e^{-|\beta|A} = \cosh(|\beta|A) - \sinh(|\beta|A) \qquad (11.10)$$

where we have used the symbol $|\ldots|$ for the modulus function.

The term $[\cosh(|\beta|A)]_{pq}$ represents the weighted sum of the number of walks of even length connecting nodes p and q in the network. Similarly, $[\sinh(|\beta|A)]_{pq}$ represents the weighted sum of the number of walks of odd length connecting nodes p and q.

Let us consider a bipartite graph, and let p and q be nodes which are in two different partitions of the network. Then, there are no walks of even length starting at p and ending at q in the graph, which implies that

$$G_{pq}(\beta < 0) = [- \sinh(|\beta|A)]_{pq} < 0 \qquad (11.11)$$

If we consider that p and q are in the same partition of a bipartite network we can see that there is no walk of odd length connecting them, due to the lack of odd cycles in the bipartite graph, which produces

$$G_{pq}(\beta < 0) = [\cosh(|\beta|A)]_{pq} > 0 \qquad (11.12)$$

Consequently, the sign of $G_{pq}(\beta < 0)$ determines whether or not the corresponding pair of nodes are in the same partition:

$$G_{pq}(\beta < 0) = \begin{cases} < 0 \ p \text{ and } q \text{ are in two different partitions} \\ > 0 \ p \text{ and } q \text{ are in the same partition.} \end{cases}$$

Now we can use the same procedure developed in the Chapter 10 for detecting communities based on the communicability function, but this time using a negative value of the absolute temperature. Consequently, a bipartition of a network is defined as follows:

> Let $C \subseteq V$ be a cluster of nodes in the network. Then, C is a quasi-bipartite cluster if, and only if, $[\cosh(A)]_{pq} > [\sinh(A)]_{pq}, \forall p, q \in C$.

We start by calculating $\exp(-\mathbf{A})$ and transforming it into the adjacency matrix of a node-repulsion graph $\Theta[\exp(-A)]$, which is obtained from the element-wise application of the function $\Theta(x)$ to the matrix $\exp(-A)$, where

$$\Theta(x) = \begin{cases} 1 & \text{if } x > 0 \\ 0 & \text{if } x \leq 0 \end{cases}$$

A pair of nodes p and q in the node-repulsion graph $\Theta[\exp(-A)]$ is connected if, and only if, they have $G_{pq} > 0$. Finally, the quasi-bipartite clusters are obtained by finding all cliques in the *node-repulsion graph,* where every clique corresponds to a quasi-bipartite cluster of the network (Estrada et al., 2008b). We recall here that this approach is equivalent to the restrictive definition of communities based on communicability presented in Chapter 10. Consequently, it is possible to define relaxed ways for defining bipartitions in a similar way as was done for communities. This, however, will not be covered in this book.

11.2.1 Physical meaning of negative absolute temperatures

The reader not familiar with thermal physics can find weird the concept of negative absolute temperatures (Ramsey, 1956). A good account of this topic can be found in Baierlein (1999), which we will follow here. In order to explain this concept, let us begin with the thermodynamic definition of the temperature for a system in equilibrium at a constant volume,

$$\frac{1}{T} = \left(\frac{\partial S}{\partial U}\right)_{V,N} \qquad (11.13)$$

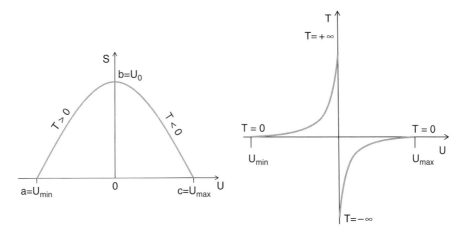

Fig. 11.4
Negative absolute temperature. Plot of the energy versus the entropy (left) and its derivative (right) for a hypothetical system in equilibrium at a constant volume.

where S and U are the entropy and the internal energy, respectively, and V and N are the volume and the number of particles in the systems, which remain constants. If we consider a plot of U versus S, then the inverse temperature is defined as the slope of this curve at a given point. In Figure 11.4 (left) we illustrate this plot for a hypothetical system. If we consider that it represents a system of N ideal paramagnets with spins $1/2\hbar$, then the point of minimum energy corresponds to the case where all spins are aligned with an external magnetic field, which is designated as (a). The point of the maximum energy (c) corresponds to an anti-alignment of all spins with respect to the external field. Both situations (a) and (c) have minimal entropy, as those systems are completely ordered; that is, $S = 0$. The point (b) corresponds to the situation of the maximum entropy. The derivative at any point between (a) and (b) is positive, which indicates that the temperature has positive values. However, at any point between (b) and (c) the slope of the curve is negative, as is the temperature. In the right-hand side of this figure we illustrate the absolute temperature calculated as the slope of the curve in the previous plot. It can be seen that points (a) and (c) correspond to temperatures of $0\ K$, because at that point the slope $(1/T)$ of the function in the first plot becomes infinite, and consequently $T = 0$. However, when the system approaches point (b), the temperature rises to plus infinite if we approach it from (a), or to minus infinite if we approach it from (c). The absolute temperature then runs from cold to hot, as $0K, \ldots, +300K, \ldots, +\infty K, \ldots, -\infty K, \ldots, -300K, \ldots, -0K$, which means that absolute negative temperatures are hotter than positive ones.

It has to be remarked that absolute zero is an unattainable value. Therefore, approaches from different sides of the temperature scale represent two very different scenarios. When we go from an infinite temperature to zero ($T \to +0$) the system goes from the point of the maximum energy to its ground state. However, when the temperature goes to zero from the negative side ($T \to -0$) it is going into its highest energy state. As we illustrate in Figure 11.5 in

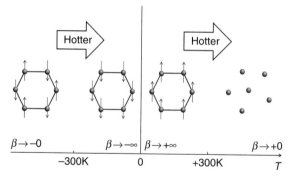

Fig. 11.5

System configuration and absolute temperature. Evolution of a system of spins as a function of the absolute temperature. The vertical line at absolute zero indicates a barrier for the 'hotter' arrow as well as an unattainable value.

the positive temperature side of the plot, as we increase T we move from a configuration in which all spins are aligned to one in which every particle is free. In contrast, in the negative side, when the temperature approaches zero, we move to the least stable configuration of the system. Consequently, the identification of bipartite structures in networks using a negative value of β corresponds to the physical metaphor in which we are moving away from the least stable configuration with all spins anti-parallel to the external field, to a configuration of the maximum entropy in which one spin is up and its neighbours are down. Then, a value of $\beta = -1$ could be a good approximation to the configuration in which the system displays its bipartite characteristics. Note that the arrangement of spin up-spin down is possible only if the network is bipartite. Otherwise, some spins are 'frustrated'.

11.3   Bipartitions in undirected networks

As an example, let us consider the complete bipartite network $K_{6,6}$, in which we have added four extra links. Because it is not possible to add more links going from one partition to the other, these four extra links reduce the bipartivity of this network. The resulting graph, illustrated in Figure 11.6 (top), displays spectral bipartivity equal to $b_S = 0.64$. The corresponding node-repulsion graph is formed by two isolated components, each having six nodes. The corresponding partition of the graph shows the two existing quasi-bipartite clusters, as illustrated in Figure 11.6 (bottom).

An example of a real-world highly bipartite network is the protein–protein interaction (PPI) network of *A. fulgidus*. We have seen that this PPI has spectral bipartivity $b_S = 0.978$. This network and its best bipartition are illustrated in Figure 11.7 (top). It can be seen that this PPI is divided into two almost disjoint sets of nodes—represented by squares and circles—and only one interaction between nodes— represented by a thick line—is observed.

Another example is provided by the social network of workers in a sawmill, where two workers are connected if they discuss work matters with relative frequency. This small company was formed by Spanish-speaking and English-speaking employees, who are divided into two main sections: the mill section and the planner section. There are also two employees working in a yard, as well as some managers and additional officials. The network displays spectral

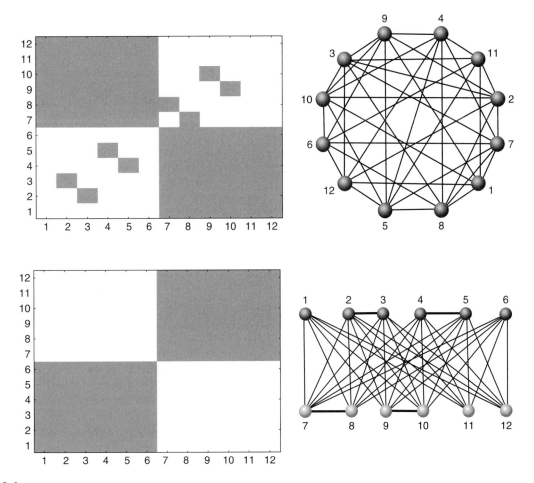

Fig. 11.6

Quasi-bipartition in an artificial graph. A small graph formed from $K_{6,6}$ by adding four extra links and its adjacency matrix (top), and its best bipartition found using $\Theta[\exp(-A)]$ (bottom).

bipartivity $b_S = 0.749$. In Figure 11.7 we illustrate both the network and its best bipartition according to the method described previously. As can be seen, this network is also partitioned into two almost disjointed clusters, here represented by squares and circles. However, in this case there are sixteen links between members of the same partition.

The causes of bipartivity in different networks might differ significantly. For instance, in the PPI of *A. fulgidus* bipartivity can be formed due to the lock-and-key nature of the protein–protein interactions (Morrison et al., 2006). That is, some proteins can act as a lock having certain cavities in which some other proteins, acting as a key, can be docked, as illustrated in Figure 11.8 (right). However, in the sawmill network the situation is quite different. Here, a bipartite structure is also observed among workers speaking the same language and working together in the same department (Michael and Massey, 1997). For instance, Spanish-speaking employees working in the planning department

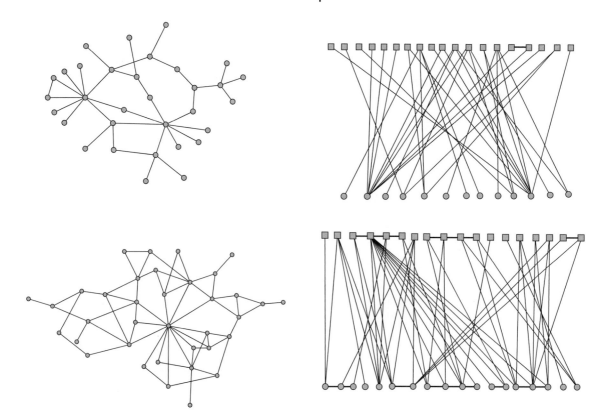

Fig. 11.7

Quasi-bipartition in real-world networks. The protein–protein interaction network of *A. fulgidus* (top left), the social network of a sawmill (bottom left), and their best bipartition found using $\Theta[\exp(-A)]$ (right).

can be split into two almost disjoint partitions having five and six nodes respectively, with only one connection between members of the same community (see Figure 11.8 left). Therefore, the conclusion that can be drawn from this network is that in contrast with other social relationships, which are characterized by a large transitivity, discussing work matters (at least in this sawmill network) is not—which means that if A frequently discuss matters of work with B and with C, there is a very small probability that B and C also discuss such matters.

Finally, a certain degree of bipartivity can also be found in networks with a clear community structure. For instance, the karate club social network studied in the previous chapter is clearly formed by two main communities: followers of the administrator, and followers of the trainer. This network, however, has spectral bipartivity $b_S = 0.579$, which indicates a certain degree of bipartition among members of the karate club. In Figure 11.9 we illustrate the best bipartition obtained by $\Theta[\exp(-A)]$ for this network. Circles represent nodes in one partition, and squares represent nodes in the other. Thick lines represent connections between members of the same partition.

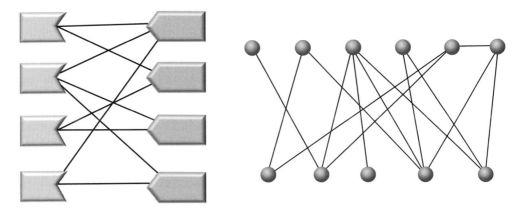

Fig. 11.8

Causes of bipartivity. A schematic of lock-and-key proteins, which can explain the emergence of bipartivity in the PPI of *A. fulgidus* (left), and a representation of the sub-network of Spanish-speaking employees in a sawmill working together in the same department (right), which might indicate that bipartivity in this network is produced neither by language nor by departmental separation of workers.

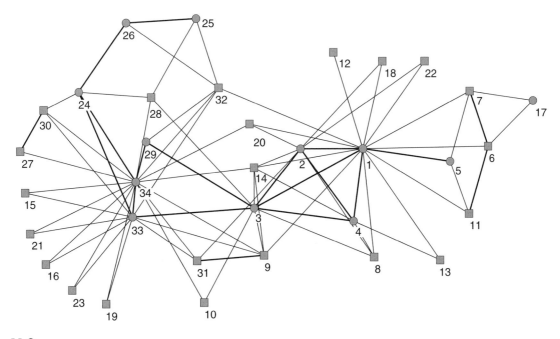

Fig. 11.9

Quasi-bipartition in the karate club network. The two partitions found by using $\Theta[\exp(-A)]$ (circles and squares) found in the karate club social network. Bold lines represent links between nodes in the same partition.

11.4 Bipartitions in directed networks

When studying directed networks we confront a situation different from the one studied previously. In a directed network we can partition the nodes into two disjoint groups in more than one way. For instance, it is possible to divide the network into two groups V_1 and V_2, such that there is no link between two members of the same group, but allowing links from both V_1 to V_2, as well as in the other direction. Another, more restrictive, approach is that links in only one of these two directions are allowed. For instance, in Figure 11.10 (left) we illustrate a directed network having two groups of nodes $V_1 = \{1, 2, 3, 4, 5\}$ and $V_2 = \{6, 7, 8, 9, 10\}$, with links only from V_1 to V_2. Therefore, while the first situation described below can be treated by considering the network as undirected, by using the methods described in the previous section, the second situation does cannot. Consequently, we define a directed bipartite network as the case in which only links from one partition to the other are allowed.

> Let $G = (V, E)$ be a directed network in which the node set is a disjoint union $V = V_1 \cup V_2$ and an arc is directed from nodes in V_1 to V_2 only. This network is here termed a *directed bipartite network*.

Here we are interested in methods which allow the identification of the maximum possible bipartite structure in such networks. For instance, in Figure 11.10 (right) we illustrate such a case in which 'perfect bipartivity' is destroyed by links going from 1 to 3 and from 9 to 5.

The strategy used here was described by Crofts et al. (2010), and is based on the fact that the $p,\ q$ -entry of a matrix obtained by

$$(\mathbf{AA}^T \mathbf{AA}^T \ldots)_{pq} \tag{11.14}$$

and having t factors, counts the number of alternating walks of length t from p to q. An alternating walk is defined as follows.

> An alternating walk of length $t - 1$ from node p_1 to node p_t is a list of nodes $p_1, p_2, \ldots, p_t$, such that $A_{i_s, i_{s+1}} \neq 0$ for S odd, and $A_{i_{s+1}, i_s} \neq 0$ for s even.

In plain words, it is a walk that successively traverses links in the forward and backward directions, such as 1–6–2–10 in Figure 11.10, but not 1–9–5–10.

Then, following the same strategy used in several parts of this book, for penalising longer walks using a division by the factorial of the length of the walk, we have the following new matrix:

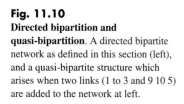

Fig. 11.10

Directed bipartition and quasi-bipartition. A directed bipartite network as defined in this section (left), and a quasi-bipartite structure which arises when two links (1 to 3 and 9 10 5) are added to the network at left.

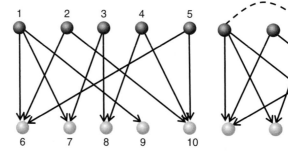

$$\kappa(A) = I - A + \frac{AA^T}{2!} - \frac{AA^T A}{3!} + \frac{AA^T AA^T}{4!} + \dots \qquad (11.15)$$

Properly speaking, this is not a matrix function according to the definition given in Chapter 5. However, using singular value decomposition (SVD) $A = U\Sigma V^T$, where U and V are orthogonal matrices and Σ is a diagonal one, we can write the new matrix as:

$$\kappa(A) = I - U\Sigma V^T + \frac{U\Sigma^2 V^T}{2!} - \frac{U\Sigma^3 V^T}{3!} + \frac{U\Sigma^4 V^T}{4!} + \dots \qquad (11.16)$$

We can group together positive and negative terms as

$$\kappa(A) = U\left(1 + \frac{\Sigma^2}{2!} + \frac{\Sigma^4}{4!}\right)U^T - U\left(\Sigma + \frac{\Sigma^3}{3!} + \frac{\Sigma^5}{5!}\right)V^T \qquad (11.17)$$

which allows us to write the new matrix as follows (Crofts et al., 2010):

$$\kappa(A) = U\ \cosh(\Sigma)U^T - U\sinh(\Sigma)V^T \qquad (11.18)$$

It is straightforward to realise that $\kappa(A) + \kappa(A^T)$ is a symmetric matrix, which is more amenable to many of the algorithms existing for clustering purposes. The first term in (11.18) accounts for the number of even alternating walks in the network, while the second term accounts for the number of odd alternating walks. If we start, for instance, at node 1 of the network in Figure 11.10 (left), any even alternating walk will finish in a node which is in the same partition as 1—such as 2, 3, 4, 5. However, any odd alternating walk will finish in a node which is in the other partition.

In order to determine which nodes are in the same directed bipartition we have to build the matrix $\kappa_{sym} = \kappa(A) + \kappa(A^T)$, then dichotomise it, and find all cliques in the network represented by the dichotomised matrix. For instance, for the network in Figure 11.10 (right) the following bipartite clusters are found: $B_1 = \{1, 2, 3, 4, 5\}$, $B_2 = \{3, 6\}$, $B_3\{3, 9\}$, and $B_4 = \{7, 8, 10\}$. These results clearly indicate that the set $B_1 = \{1, 2, 3, 4, 5\}$ can be considered as a group of 'source' nodes, $B_4 = \{7, 8, 10\}$ is formed by nodes acting as 'sinks', and $B_3 = \{3, 9\}$ is formed by nodes having both in-links and out-links. However, we can consider a less restrictive approach simply by reordering the matrix κ_{sym} and selecting those nodes having $\kappa_{sym}(p, q)$ larger than a certain predetermined cut-off value.

An excellent scenario for applying this method for detecting directed bipartite clusters is that of neuronal networks. In a relatively simple organism like the worm *C. elegans*, various questions have been posited—such as 'What is the processing depth from sensory input to motor output?' and 'To what extent is the circuitry unidirectional, progressing linearly from input to output?' These questions are specifically concerned with the existence of directed bipartitions of neurons in which neurons in one bipartition share some specific function, like sensorial or motor functions.

A directed network of 131 neurons in the frontal region of *C. elegans* and 964 synaptic connections is illustrated in Figure 11.11 (top).

A reordered version of the matrix κ_{sym} is also illustrated in Figure 11.11, where it can be seen that there are some tight clusters that can be identified by

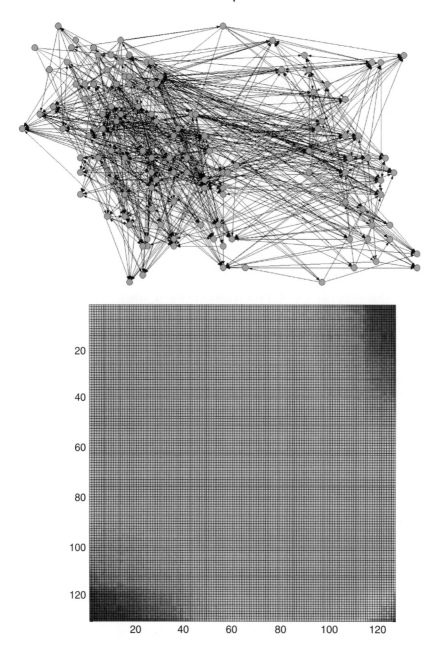

Fig. 11.11
Network of frontal neurons of *C. elegans*. A directed network representing 131 frontal neurons of the worm *C. elegans*, represented according to plane coordinates in the worm and its matrix K_{sym}. The matrix is sorted such that the largest positive values are at the bottom right and at the top left corners (Crofts et al. 2010).

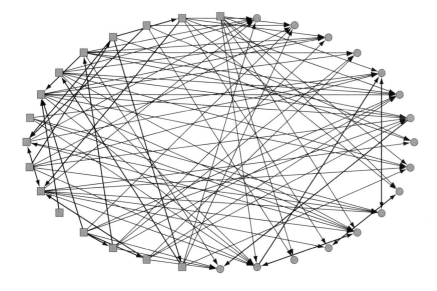

sources		sinks	
OLL	Head sensory neuron	SMD	Ring motor neuron
URY		RME	
IL2		RMD	
CEP	Ring interneuron	AVB	Command interneuron
RIH	Amphids; sensory neuron	AVA	
ADL		AVE	
ASH	Ring motor neuron	AVD	
RIM	Ring motor; interneuron		
RIV			
AVH	Interneuron		

Fig. 11.12

Directed bipartition in the network of frontal neurons of *C. elegans*. A group of thirty-two neurons identified as forming a directed bipartition in the network of frontal neurons of *C. elegans*. Neurons identified as mainly sources are represented by squares, and those identified as mainly sinks are represented by circles. Neuron labels and classes are also shown.

using appropriate methods. One such cluster is formed by thirty-two neurons in which sixteen neurons act mainly as sources, and sixteen neurons act mainly as sinks. If we consider these two subsets of nodes as V_1 and V_2, it can be seen in Figure 11.12 that there are 5, 35, and 9 times more $V_1 \rightarrow V_2$ links than $V_1 \rightarrow V_2$, $V_2 \rightarrow V_1$ and $V_2 \rightarrow V_2$ respectively. More interestingly, most neurons acting as sources have sensorial functions, while those acting as sinks are exclusively motor and command neurons—which indicates a real directionality in the functioning of this network, and a clear interpretation of the resultant directed bipartition.

Random models of networks

12

> If one were to define chance as the outcome of a random movement which interlocks with no causes, I should maintain that it does not exist at all, that it is a wholly empty term denoting nothing substantial.
>
> Herman Boerhaave

In order to fully understand the structure of complex networks we need some reference model with which we can compare them. Intuitively we can think about a model in which pairs of nodes are connected by some probability. That is, if we start with a collection of n nodes and for each of the $\frac{1}{2}n(n-1)$ possible links, we connect a pair of nodes u, v with certain probability $p_{u,v}$. Then, if we consider a set of network parameters to be fixed and allow that the links are created by a random process, we can create models that permit us to understand the influence of these parameters on the structure of networks. Let us suppose that we observe some kind of general structural characteristic in real world complex networks; for ecample, the number of triangles, degree distribution, and so on. How can we know whether these structural characteristics are present due to some functional principles of these networks, or have simply arisen as a consequence of the random linking of nodes? Using random models of network (Newman, 2002b) we can answer such questions at the same time as we dispose of models that allow us to compare one structural property in two or more different kinds of (random) network. We have placed this chapter at the end of the theory part of this book because we will need several of the concepts introduced in previous chapters to understand the structure of these networks.

12.1 'Classical' random networks

The study of random networks is not a new area of research, and its roots can be traced back to the work of Solomonoff and Rapoport in the early 1950s (Solomonoff and Rapoport, 1951). However, today the best-known model of random networks is the one introduced by Erdös and Rényi at the end of the same decade (Erdös and Rényi, 1959; 1960). This model—which is now known as the Erdös-Rényi model—is sometimes known as 'classical' random graphs, and several of their mathematical properties can be found elsewhere (Bollobás, 2001; Durrett, 2007). In this model we begin with n isolated nodes as illustrated in the first snapshot in Figure 12.1. Then, with probability $p > 0$ each pair of nodes is connected by a link. Consequently, in this model the network is determined only by the number of nodes and links, and usually an Erdös-Rényi random graph is written as $G(n, m)$ or $G(n, p)$. In Figure 12.1 we illustrate some examples of Erdös–Rényi random graphs with the same number of nodes and different linking probabilities.

It is easy to realise that if we repeat the process for the same number of nodes and the same probability, we will not necessarily obtain the same network. Therefore, the correct way of referring to these graphs is by considering the whole ensemble of networks that can be formed for a given n and p. We designate these ensembles as $G_{ER}(n, m)$ or $G_{ER}(n, p)$. In this case we can see that the mean number of links $\overline{m}$ is directly related to the number of nodes and the probability by

$$\overline{m} = \frac{n(n-1)p}{2} \tag{12.1}$$

The expected value (or mean) for the node degree of $G_{ER}(n, p)$ is also determined by the number of nodes and the probability only:

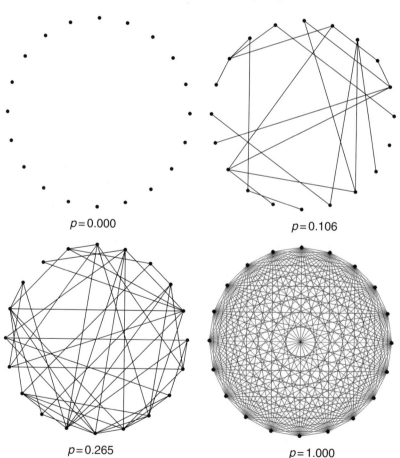

$p=0.000$

$p=0.106$

$p=0.265$

$p=1.000$

Fig. 12.1
Erdös–Rényi random networks.
Erdös–Rényi random networks with twenty nodes and different probabilities of connecting a pair of nodes.

$$\bar{k} = (n-1)p \qquad (12.2)$$

We recall from Chapter 8 that the density of a network is defined as

$$\delta(G) = \frac{\bar{k}}{(n-1)} \qquad (12.3)$$

which means that $p = \delta(G)$.

For instance, if we consider one particular random realisation of $G_{ER}(1000, 0.04)$ and plot the probability $p(k)$ of finding a node of degree k, versus the degree, we obtain Figure 12.2, where it can be seen that the maximum of the distribution is about the value $\bar{k} = (n-1)p = 39$. Obviously, the probability $p(k)$ follows a binomial distribution of the form:

$$p(k) = \binom{n-1}{k} p^k (1-p)^{n-1-k} \qquad (12.4)$$

For large values of n, expression (12.4) becomes

$$p(k) = \frac{e^{-\bar{k}}\bar{k}^k}{k!} \qquad (12.5)$$

which is the Poisson distribution found in Chapter 2. In Figure 12.2 we also display the heterogeneity plot for $G_{ER}(1000, 0.04)$, where two characteristic features of ER networks are observed. The first is a typical dispersion of the points around the value $x = 0$, and the second is the very small value of $\rho(G)$, which in this case is 0.0066.

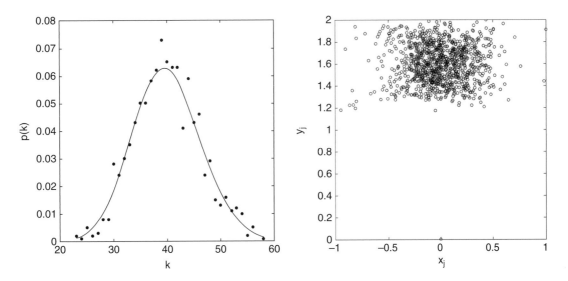

Fig. 12.2

Heterogeneity of Erdös–Rényi random networks. A typical Poissonian degree distribution of an Erdös–Rényi random network with 1,000 nodes $p = 0.04$ (left), and the characteristic heterogeneity plot for the same network.

In the case of a large random network $H \in G_{ER}(n,\ p)$, the average path length is given by

$$\bar{l}(H) = \frac{\ln n - \gamma}{\ln(pn)} + \frac{1}{2} \qquad (12.6)$$

where $\gamma \approx 0.577$ is the Euler–Mascheroni constant. This means that Erdös–Rényi random graphs display a very small average path length. This is a characteristic of most large networks, as analysed in Chapter 3.

On the other hand, the average clustering coefficient of $G_{ER}(n,\ p)$ is given by

$$\overline{C} = p = \delta(G) \qquad (12.7)$$

which means that for sparse Erdös–Rényi random networks the clustering coefficient is very small—in general, much smaller than that for real world networks with the same density.

One interesting discovery already reported by Erdös and Rényi is that when increasing p, most nodes tend to be clustered in one giant component, while the rest of the nodes are isolated in very small components. This means that if we plot the number of nodes in the giant component versus the probability that two nodes are linked in an Erdös-Rényi network, we obtain a plot similar to that illustrated in Figure 12.3.

Note that in the plot shown in Figure 12.3, a jump takes place around the value $p \approx 10^{-3}$. As this network has 1,000 nodes, the jump is a value of about $pn \approx 1$. In fact, this is a fundamental result found by Erdös and Rényi in their paper of 1960 (Erdös and Rényi, 1960), and some more recent results on the same topic can be found in the literature (Achlioptas et al., 2009; Spencer, 2010). According to their analytical results, the structure of $G_{ER}(n,\ p)$ changes as a function of $p = \bar{k}/(n-1)$, giving rise to the following three stages:

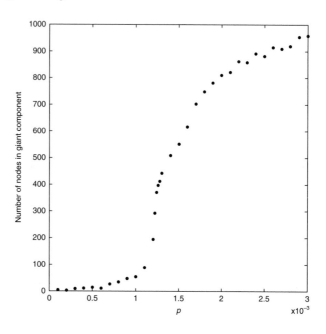

Fig. 12.3
Connectivity of Erdös–Rényi random networks. Change of the size of the main connected component in an Erdös–Rényi random network with 1,000 nodes as a function of the linking probability.

1. Subcritical $\bar{k} < 1$, where all components are simple and very small. The size of the largest component is $s = O(\ln n)$.
2. Critical $\bar{k} = 1$, where the size of the largest component is $S = \Theta(n^{2/3})$.
3. Supercritical $\bar{k} > 1$, where the probability that $(f - \varepsilon)n < S < (f + \varepsilon)n$ is 1 when $n \to \infty$ for $\varepsilon > 0$, where $f = f(\bar{k})$ is the positive solution of the equation: $e^{-\bar{k}f} = 1 - f$. The rest of the components are very small—the second largest being about $\ln n$.

The largest component in the supercritical stage is termed the *giant component*. The value of $\bar{k} = 1$ for the critical states has been further analysed by finding a critical window for this transition. The reader is directed to more specialised literature for the details. In Figure 12.4 we illustrate a snapshot of Erdös–Rényi random networks with $n = 100$ for the subcritical, critical, and supercritical states.

The largest eigenvalue of the adjacency matrix in an ER network grows proportionally to n (Janson, 2005): $\lim_{n\to\infty}(\lambda_1(\mathbf{A})/n) = p$, where p is the probability that each pair of vertices is connected by a link. The second largest eigenvalue grows more slowly than λ_1—$\lim_{n\to\infty}(\lambda_2(\mathbf{A})/n^\varepsilon) = 0$ for every $\varepsilon > 0.5$ —and the smallest eigenvalue also grows with a similar relation to $\lambda_2(\mathbf{A})$: $\lim_{n\to\infty}(\lambda_n(\mathbf{A})/n^\varepsilon) = 0$ for every $\varepsilon > 0.5$. The spectral density of ER random networks displays a typical shape, which is well characterised mathematically (Arnold 1967; 1971; Chung et al., 2003; Farkas et al., 2001; Juhász, 1978). The reader is directed to Chapter 2 for an introduction to spectral density. In the case of random graphs it is convenient to define the following normalisation factor:

$$r = \sqrt{np(1 - p)} \qquad (12.8)$$

The spectral density of an ER random network follows Wigner's semicircle law (Wigner, 1955), which by using the previously defined normalisation factor is simply written as

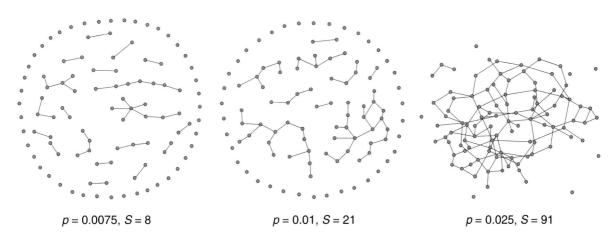

$p = 0.0075,\ S = 8$ $p = 0.01,\ S = 21$ $p = 0.025,\ S = 91$

Fig. 12.4
Components of Erdös–Rényi random networks. Size of the main connected component s for an Erdös–Rényi random network with 100 nodes and different linking probabilities. The nodes in the largest connected component are drawn darker.

$$\rho(\lambda) = \begin{cases} \dfrac{\sqrt{4-\lambda^2}}{2\pi} & -2 \leq \lambda/r \leq 2 \\ 0 & \text{otherwise} \end{cases}$$

12.2 Small-world random networks

One of the cornerstones of network theory is an experiment carried out by Stanley Milgram in 1967 (Milgram, 1967). The experiment consisted in selecting at random several people in the cities of Omaha (Nebraska) and Wichita (Kansas), located at the centre of the USA, and asking those people to send a letter to a target person living in Boston (Massachusetts), on the east coast. The individuals at the starting points were asked to send letters to someone they knew on a first-name basis. In those cases in which the letter arrived at its target, the researcher had the opportunity of following the trajectory that the letter followed through the US. If you were selected as one of the starting points, you must first think whether you were know personally by the target, in which case you simply directed the letter to him/her. If not, then you must think about someone you knew personally, and you think that he/she has a large probability of knowing the target personally. If we consider that the starting points and the target are separated by more than 2,000 km, it is not strange to think that the number of steps that the letter needs to take is very large. The results produced the following conclusions:

4. The average number of steps for a letter to arrive at its target was around 5.5 or 6.
5. There was large group interconnection, resulting in an acquaintance of one individual feeding back into his/her own circle, normally eliminating new contacts.

Let us translate these two findings into the language of networks with which we are already acquainted. The one indicates that the average path length in the network was $\bar{l} = 6$, and the second indicates that the clustering coefficient of every individual was relatively large. The first characteristic is popularly known as 'six degrees of separation'—perpetuating the myth that every pair of individuals in the world is separated by another by only six steps. The second is understood as the transitivity of social relations. If A knows B and B knows C, there is a large chance that A also knows C, closing a triangle of relations. From a network structural point of view, what is interesting here is that these 'small-world' networks share two properties: small average path length, and high clustering coefficient, which are not simultaneously found in the random network model studied in the previous section. That is, an Erdös–Rényi random network displays a very small average path length as the 'small-world' network, but fails in reproducing a large clustering coefficient, which, as we have seen, is only $\overline{C} = \delta(G)$ for Erdös–Rényi networks.

In 1998, Watts and Strogatz (1998) proposed a model which reproduces the two previously mentioned properties in a simple way. Let n be the number of nodes, and let k be an even number. The Watt–Strogatz model starts by using the following construction. Place all nodes in a circle, and connect every node to its first $k/2$ clockwise nearest neighbours as well as to its $k/2$ counterclockwise nearest neighbours. This will create a ring, which for $k > 2$ is full of triangles and consequently has a large clustering coefficient. This situation is illustrated for $n = 20, k = 6$ in Figure 12.5 (top left). In fact, it has been proved that the average clustering coefficient for these networks is given (Barrat and Weigt, 2000) by

$$\overline{C} = \frac{3(k-2)}{4(k-1)} \tag{12.9}$$

which means that the clustering coefficient of these networks is independent of network size, and tends to the value $\overline{C} = 0.75$ for very large values of k.

As can be seen in Figure 12.5 (top left), the shortest path distance between any pair of nodes which are opposite to each other in the network is relatively large. This distance is, in fact, equal to $\lceil \frac{n}{k} \rceil$. This situation is traduced into a relatively large average shortest path length, which is represented by the following expression:

$$\overline{l} \approx \frac{(n-1)(n+k-1)}{2kn} \tag{12.10}$$

This means that when the network is sparse, $n \gg k$, as the network size grows the average path length tends to $n/2$, which is far larger than the value required for a network to be a small-world; that is, $\overline{l} \sim \ln n$. Watts and Strogatz cleverly solved this situation by considering a probability for rewiring the links in the ring, so that the average path length decreases very fast while the clustering coefficient still remains high. The rewiring can be considered as the process through which, with probability p, we replace each link r, s with a link r, t, where t is a randomly chosen node different from r and s. If r, t is already contained in the modified network, no action is considered. This process is illustrated in Figure 12.5. Obviously, as $p \to 1$ the network tends to a completely random graph of the type studied in the previous section. Therefore, a Watts–Strogatz network can be written as a three-parameter graph: $WS(n, k, p)$.

In Figure 12.6 we illustrate what happens to the two parameters that we are considering here—clustering and average path length—as the rewiring probability changes from 0 to 1 in a network with $n = 400$ and $k = 12$. As can be seen, the average path length $\overline{l}$ decreases very rapidly as p increases, while $\overline{C}$ decreases more slowly. As a consequence there is a region of p values for which the networks display relatively large $\overline{C}$ and small $\overline{l}$, which match very well with the characteristics expected from some real-world networks.

If we consider the heterogeneity index $\rho(G)$ discussed in Chapter 8 (see (8.13)) for the Watts–Strogatz model for different rewiring probabilities, we observe that all networks in the range $0 \leq p \leq 1$ have $\rho(G) \approx 0$, indicating that they are very homogeneous. In Figure 12.7 we plot the values of $\rho(G)$ versus probability for the WS networks previously built. Of course, $\rho(G) = 0$ is obtained for the regular network given by $WS(400, 12, 0)$, and the value only increases to 10^2 for $WS(400, 12, 1)$.

In Figure 12.8 we illustrate the heterogeneity plots for WS networks with different rewiring probabilities. In the case of $p = 0.00$ the plot is the typical

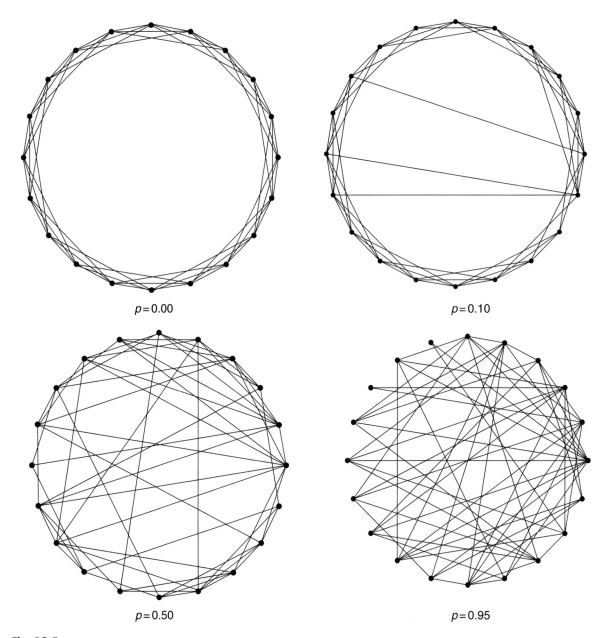

$p = 0.00$ $p = 0.10$

$p = 0.50$ $p = 0.95$

Fig. 12.5
Watts–Strogatz random networks. The rewiring process, which is the basis of the Watts–Strogatz model for small-world networks. Starting from a regular network with $n = 20$, $k = 6$, some links are rewired with probability p.

one for regular networks, and they tend to that of an ER network as the rewiring probability tends to 1.

The fact that the Watts–Strogatz model finds the networks with the so-called 'small-world' property in some point in which the rewiring probability is neither zero nor one, has given rise to the myth that complex networks

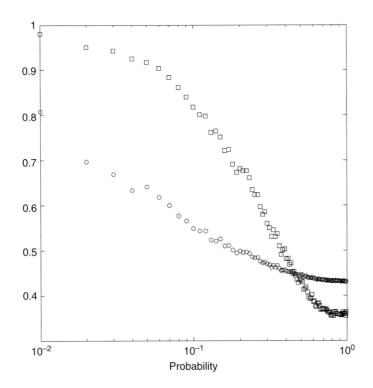

Fig. 12.6
Structural evolution of the Watts–Strogatz model. The variation of the average path length and clustering coefficient with the change of the rewiring probability for a network having $n = 400$ and $k = 12$. The values of the average path length and clustering coefficient are normalised by dividing them by the respective values obtained for $WS(400,12,0)$.

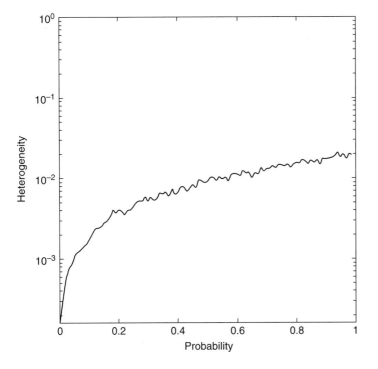

Fig. 12.7
Heterogeneity of the Watts–Strogatz networks. The variation of the heterogeneity of a Watts–Strogatz network having $n = 400$ and $k = 12$, with the change of the rewiring probability.

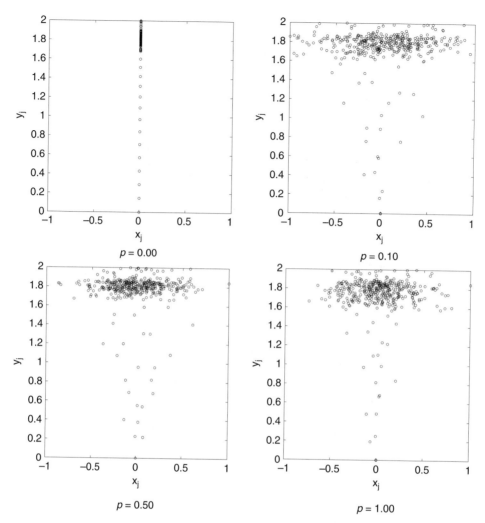

Fig. 12.8

Heterogeneity plots for Watts–Strogatz networks. Characteristic plots of the heterogeneity for Watts–Strogatz networks having $n = 400$ and $k = 12$, with different rewiring probability.

are 'between regularity and randomness.' Some clarification is needed at this point to try to avoid the propagation of this misconception. First, the starting network in the Watts–Strogatz model is not any regular network, but a very particular one. These networks correspond to the subclass of *circulant graphs* $G_n(L)$, which are defined as the network of n nodes in which the ith node is adjacent to the $(i + j)$th and $(i - j)$th nodes, each j in the list L. In particular, the circulant graphs used in the Watts–Strogatz model are those of the form $G_n(1, 2, \ldots, k)$ for $2 \leq k < \lfloor n/2 \rfloor$. Note that the networks used in the previous paragraphs correspond to $G_n(1, 2)$, while $G_n(1) = C_n$ and $G_n(1, 2, \ldots, k)$ for $k = \lfloor n/2 \rfloor$ is the complete network, K_n.

Therefore, it is incorrect to claim that 'small-world' networks and real-world networks are somewhere between regular and random. If we are referring in particular to the properties of having relatively small average path length—say $\bar{l} \sim \log n$—and having relatively large clustering coefficient, then we can find regular networks fulfilling both conditions. In fact, regular networks can encompass the entire spectrum of clustering from zero to one. In addition, they can have average path length as small as a random graph—for example, random regular graphs. Consequently, what is correct is that 'small-world' networks are between circulant graphs of the type $G_n(1, 2, \ldots, k)$ for $2 \leq k < \lfloor n/2 \rfloor$ and random networks. The reasons why such kinds of circulant graph were selected are obvious: they display large average path length and maximum clustering. But are they the least random networks among all regular ones? It is probable, but then a measurement of randomness should be required in order to quantify this property and to prove such statement.

12.3 'Scale-free' networks

As we have seen in the previous two sections, both Erdös–Rényi and Watts–Strogatz networks display Poissonian degree distributions. However, we have seen in Chapter 2 that there are real-world networks that display a very different degree distribution, which is characterized by a few nodes of high degree and a large proportion of nodes with relatively low degree. Although such real-world networks can display a large variety of 'fat-tail' degree distributions (see Chapter 2), the easiest way of conceptualising such topological characteristics is to consider a model in which $p(k) \sim k^{-\gamma}$: that is, a model in which the probability of finding a node with degree k decreases as a power-law of its degree. The most popular of these models is the one introduced by Barabási and Albert (1999), which is described below.

In the Barabási–Albert model a network is created by using the following procedure. Begin with a small number, m_0, of nodes. At each step, add a new node u to the network, and connect it to $m \leq m_0$ of the existing nodes $v \in V$ with probability

$$p_u = \frac{k_v}{\sum\limits_{w} k_w} \tag{12.11}$$

The resulting network is known as the Barabási–Albert network, and is designated as $BA(n, n_0, d)$.

In the model originally proposed by Barabási and Albert there is no specification concerning how the initial m_0 nodes are connected among them. Bollobás (2003) have remarked that this model is 'rather imprecise', but the situation can be saved in several ways. For instance, we can consider that we start from a connected random network of the Erdös–Rényi type with m_0 nodes, $G_{ER} = (V, E)$. In this case the BA process can be understood as a process in which small inhomogeneities in the degree distribution of the ER network grow with time. Another option is that developed by Bollobás and Riordan, in which it is first assumed that $d = 1$ and that the ith node is attached to the jth one with probability:

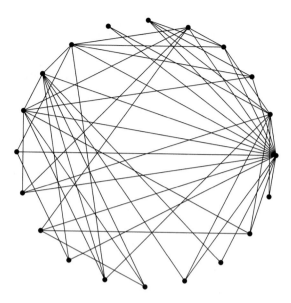

Fig. 12.9

Barabási–Albert networks. Example of a random network obtained with the preferential attachment method of Barabási and Albert, with $n = 20$ and $d = 4$.

$$p_i = \begin{cases} \dfrac{k_j}{1 + \sum\limits_{j=0}^{i-1} k_j} & \text{if } j < i \\[4ex] \dfrac{1}{1 + \sum\limits_{j=0}^{i-1} k_j} & \text{if } j = i \end{cases}$$

Then, for $d > 1$ the network grows as if $d = 1$ until nd nodes have been created, and the size is reduced to n by contracting groups of d consecutive nodes into one. The network is now specified by two parameters as $BA(n, d)$. Multiple links and self-loops are created during this process, and they can be simply eliminated if we need a simple network. For instance, the network shown in Figure 12.9 corresponds to $BA(20, 4)$.

A characteristic of BA networks is that the probability that a node has degree $k \geq d$ is given by

$$p(k) = \frac{2d(d - 1)}{k(k + 1)(k + 2)} \sim k^{-3} \tag{12.12}$$

which immediately implies that the cumulative degree distribution is given by

$$P(k) \sim k^{-2} \tag{12.13}$$

In Figure 12.10 we illustrate the cumulative degree distribution for a random BA network with 1,000 nodes and $d = 4$ in a log–log scale, where it can be seen that the slope of the straight line is about 2. In the same figure we also show the heterogeneity plot for the same network, where it can be seen as the characteristic triangular shape of the plot (see Chapter 8).

However, there is a fundamental difference between the degree distribution of BA networks and their heterogeneity as measured by $\rho(G)$ for different

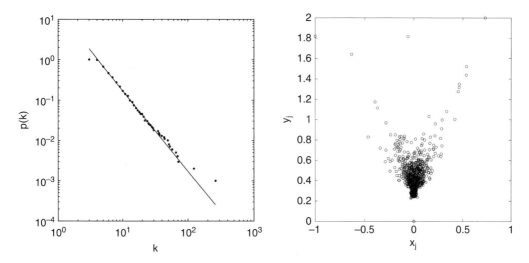

Fig. 12.10
Degree distribution of Barabási–Albert networks. The power-law cumulative degree distribution (left) and heterogeneity plot (right) for a random network obtained with the Barabási–Albert model with 1,000 nodes and $1 \leq d \leq 30$. The cumulative degree distribution is plotted in a log–log scale.

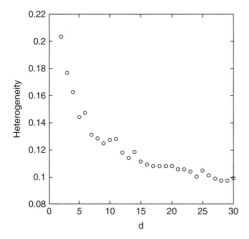

Fig. 12.11
Heterogeneity of Barabási-Albert networks. Change of heterogeneity measure in Barabási–Albert random networks with 1,000 nodes and $1 \leq d \leq 30$.

values of d. First, we note that the average degree of the network is $\bar{k} = 2d$. Then, as d increases a BA network tends to a saturation point which is exhibited by the complete graph. This means that we would expect that the heterogeneity of a BA network would decrease as the average degree increases. This is exactly what we observe in Figure 12.11, when we plot $\rho(G)$ versus d for a BA network with 1,000 nodes and $1 \leq d \leq 30$. However, the exponent of the power-law degree distribution does not reflect this change in heterogeneity of the networks, as it remains constant at the value of 3 for any value of d.

It is also interesting to note that the typical triangular shape of the networks with power-law degree distributions does not change when the values of d changes, as can be seen in Figure 12.12.

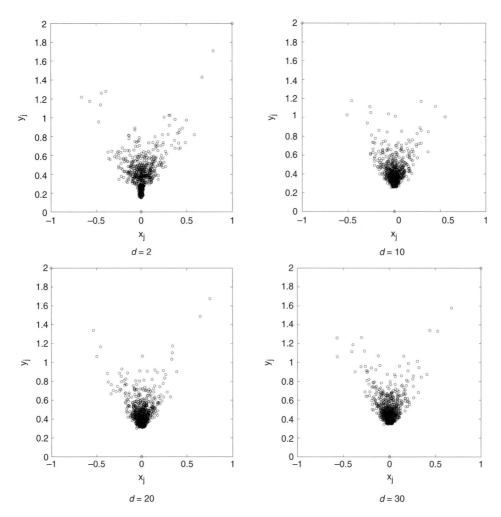

Fig. 12.12

Typical heterogeneity plots of scale-free networks. Heterogeneity plots for several random networks obtained with the Barabási–Albert model with 1,000 nodes and different values of d.

In Figure 12.13 (top) we illustrate the spectral scaling (see Chapter 9) for two BA networks with 1,000 nodes having $d = 2$ and $d = 4$, respectively. As can be seen, the network with $d = 2$ displays a typical pattern characteristic of networks of class IV, which are those having a core-periphery structure in which the periphery is characterised by several well-connected clusters. However, the network with $d = 4$ is typically a good expansion or Class I network, characterised by a relatively high topological homogeneity. As illustrated in Figure 12.13 (bottom), networks created by using the BA preferential attachment method display a transition from Class IV to very topologically homogeneous (Class I) networks as the average degree increases. In fact, Gkantsidis et al. (2003), using arguments from max-flow min-cut theory, have shown that networks obeying power-law degree distribution have good expansion

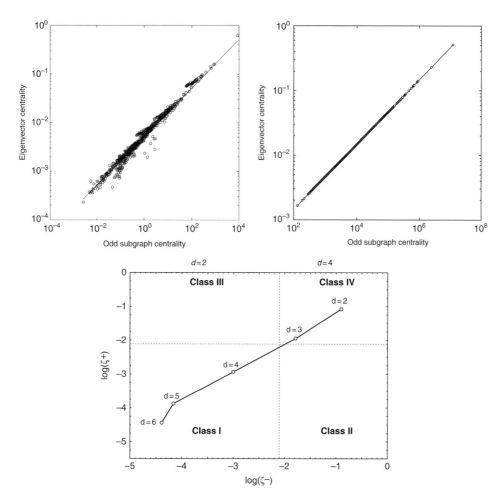

Fig. 12.13
Structural classes of scale-free networks. Spectral scaling plots for two BA networks with 1,000 nodes having $d = 2$ and $d = 4$ (top, left, and right). The first plot is characteristic of Class IV networks (see Chapter 9), and the second of Class I networks. Changes of structure in the networks obtained with the Barabási–Albert model for different values of d. As d increases, BA networks change from Class IV to Class I without passing through Classes II or III.

properties, as they allow routing with $O(N \log^2 N)$ congestion, which is close to the optimal value of $O(N \log N)$ achieved by regular expanders.

It is obvious from inspection of Figure 12.13 (bottom) that the BA networks do not reproduce the characteristics of two important structural classes of networks: namely, Classes II and III. The same has been observed for the Erdös–Rényi model of random networks.

For fixed values $d \geq 1$, Bollobás (2003) has proved that the expected value for the clustering coefficient $\overline{C}$ is given by

$$\overline{C} \sim \frac{d - 1}{8} \frac{\log^2 n}{n} \qquad (12.14)$$

for $n \to \infty$, which is very different from the value $\overline{C} \sim n^{-0.75}$ reported by Barabási and Albert (1999) for $d = 2$.

On the other hand, the average path length has been estimated for the BA networks to be as follows (Bollobás and Riordan, 2004):

$$\bar{l} = \frac{\ln n - \ln(d/2) - 1 - \gamma}{\ln \ln n + \ln(d/2)} + \frac{3}{2} \qquad (12.15)$$

where γ is the Euler–Mascheroni constant. This means that for the same number of nodes and average degree, BA networks have smaller average path length than their ER analogues. In other words, BA networks are smaller small-worlds than ER random networks. Some authors have referred to BA networks as 'super small-worlds.' In Figure 12.14 we illustrate the change of the average path length as a function of network size for ER and BA networks.

Another characteristic feature of BA networks and other random graphs having power-law degree distribution is that the spectral density follows a triangular distribution (Farkas et al., 2001, 2002; Goh et al., 2001a), which when using the normalisation factor given for ER networks is simply written as

$$p(\lambda) = \begin{cases} (\lambda + 2)/4 & -2 \leq \lambda/r \leq 0 \\ (2 - \lambda)/4 & 0 \leq \lambda/r \leq 2 \\ 0 & \text{otherwise} \end{cases}$$

In Figure 12.15 we illustrate the spectral densities for random ER and BA networks. If we consult Chapter 2 we can see that many real-world networks deviate from these two patterns in their spectral densities, indicating that although they can show Poisson or power-law degree distributions, there are other structural factors which influence their spectral properties.

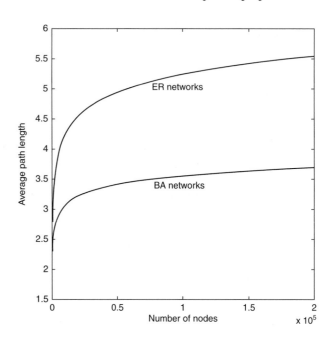

Fig. 12.14

Small-worldness of BA networks. Change in the average path length of random networks created with the Barabási–Albert and Erdös–Rényi models, with the change in the network size.

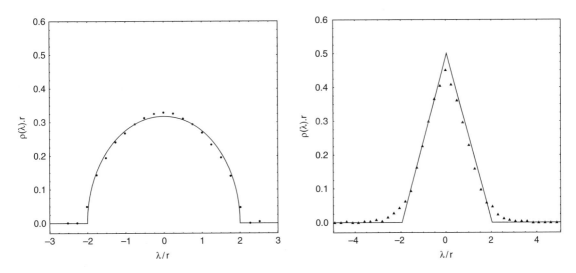

Fig. 12.15
Spectral density of random networks. Plot of the spectral density of a random graph with Poisson (left) and power-law degree distribution. In both cases the plots are the average of 100 realizations for networks having 1,000 nodes and 8,000 links.

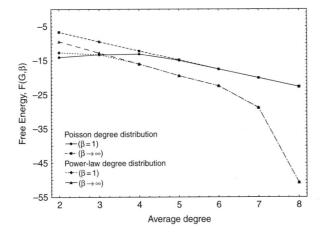

Fig. 12.16
Free energy of random networks. Values of free energy for random networks with Poisson and power-law degree distributions for $\beta = 1$ and $\beta \to \infty$.

In Figure 12.16 we illustrate the trends of the free energy (see Chapter 5) for random networks displaying Poisson and power-law degree distribution. The networks are generated with the Erdös–Rényi and Barabási–Albert models respectively. As can be seen, the networks generated from the preferential attachment BA mechanism display very fast convergence towards the minimal free energy as the average degree of the network increases. This convergence is lower in the ER networks. Consequently, networks with large average degree having power-law degree distributions are much more stable than their analogues with Poisson degree distributions. They are characterized by very low entropy, which indicates a clear localization of the networks in the ground state.

Several extensions and variations of the Barabási–Albert model have been proposed in the literature in order to make the model more flexible

(Caldarelli, 2007). In one such model, Dorogovtsev and Mendes (2003) proposal is to start in a similar way as for the BA model in which at each step s a new node is added to the set V_{s-1} and is attached to the existing ones with probability which is proportional to the degrees of the nodes in V_{s-1}. They then consider a constant $\chi \geq 0$, and add χ^d links between the nodes in V_{s-1} with probability $p_{u,v} \sim k_u k_v$, with the condition that the link $\{u, v\}$ does not yet exist. The process generates networks with the power-law degree distribution

$$p(k) \sim k^{-\left(2 + \frac{1}{1+2\chi}\right)}$$

which allows for power-law degree distributions with exponents between 2 and 3.

In another modification of the BA model, Holme and Kim (2002) proposed a model allowing for generating networks with scale-free degree distribution with scaling exponent equal to 3, and different values of the clustering coefficient which can reach higher values than in the classical BA model.

12.4 Random geometric networks

In Chapter 3 we saw that there are networks which can growth under certain geometrical constraints—such as a network of streets in a city, or a network of ant galleries or termite nests. Therefore, it is interesting to analyse some methods that create networks based on certain geometrical criteria for joining pairs of nodes. Such networks are referred to as *random geometric networks*.

A random geometric network (Penrose, 2003) of size n is built by placing n points randomly distributed in the unit square in which node i has coordinates (x_i, y_i). The coordinates are independent, and are identically distributed with uniform $[0, 1)^2$ distribution. For an illustration see Figure 12.17 (left), where 100 points are randomly distributed in the unit square. The second step consists in joining those pairs of points placed below certain specified Euclidean distance. Let r be some specified radius, while points i and j are linked if and only if

$$(x_i - x_j)^2 + (y_i - y_j)^2 \leq r^2 \qquad (12.16)$$

In Figure 12.17 (right) we illustrate the use of $r = 0.15$ for the set of points shown at left in this figure. The topological information contained in the Figure 12.17 (right) represents a network, in which we are interested only in the adjacency relationships between pairs of nodes (points), and not in the geometric information used in building such graphs.

A characteristic feature of random geometric networks is that there is a phase transition from disconnected networks to a connected network for a certain value of the radius. That is, for a given value r_C, a network generated with the procedure previously described will be connected with high probability. The critical radius is given by

$$r_C = \frac{\sqrt{\ln n + O(1)}}{\pi n} \qquad (12.17)$$

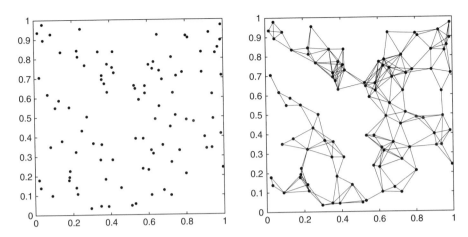

Fig. 12.17
Random geometric networks. In the formation of a random geometric network, 100 nodes are first placed randomly and independently in the unit square (left). Then, pairs of nodes are linked according to their Euclidean distance, using a cut-off radius of $r = 0.15$ (right).

In Figure 12.18 (left) we illustrate a plot of the size of the main connected component versus the radius for a random geometric network with 100 nodes. Note that the critical radius for a network with 100 nodes is $r_C \approx 0.14$. The figure shows a network created for $r < r_C$ (centre), and another for $r > r_C$ (right).

A generalisation of this model has been proposed by Avin (2008) in the form of *random distance graphs*. In this case, n nodes are placed uniformly at random in the unit disk. Then, a link is placed between each pair of nodes i, j separated at distance d_{ij}. Such links are placed independently of all other links with probability $p_{ij} = g(d_{ij})$. It is then easy to consider a step function with a threshold r, which creates links with probability $\alpha = \alpha(n)$ for nodes at distance $d_{ij} \leq r$, and with probability $\beta = \beta(n)$ for nodes at distance $d_{ij} > r$.

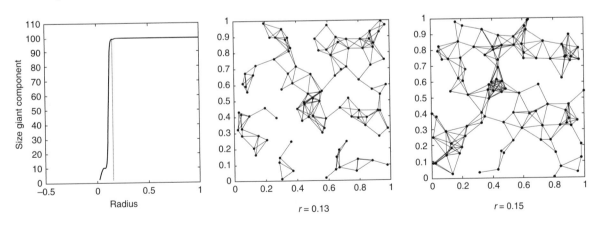

Fig. 12.18
Connectivity of random geometric networks. Change of the size of the main connected component versus the radius for a random geometric network with 100 nodes (left). Two examples of networks obtained with this model which are below (centre) and after (right) the critical radius.

It is desirable to choose $\beta(\alpha)$ as a function of α and

$$(1 - \alpha)\pi r^2 = \beta(1 - \pi r^2) \tag{12.18}$$

in order to keep the average degree of the network invariant with respect to α. Then, for $\pi r^2 \le \alpha \le 1$ we have

$$g_r^\alpha = \begin{cases} \alpha & \text{if } d \le r \\ \beta = \dfrac{(1-\alpha)\pi r^2}{1 - \pi r^2} & \text{if } d > r. \end{cases}$$

These random networks are denoted as $D(n, g_r^\alpha)$, and have the following connectivity characteristics. Let

$$\pi r^2 = \frac{\log n + c_n}{n} \tag{12.19}$$

The network is connected with high probability if $c_n \to +\infty$ and disconnected if $c_n \to -\infty$.

For connected random distance networks, the following properties have been found with high probability:

$$\bar{k} = \Theta(\log n) \tag{12.20}$$

$$\overline{C} = 0.5865\,\alpha + o(1) \text{ for } \pi r^2 \le \alpha \le 1 \tag{12.21}$$

$$l_{\max} = \Theta\left(\frac{\log n}{\log \log n}\right) \text{ for } \pi r^2 \le \alpha \le 1 - \varepsilon, \ \varepsilon > 0 \tag{12.22}$$

An important result proved by Avin (2008) is that with high probability there is a random distance network with $\bar{k} = \Theta(\log n)$, high clustering, small diameter, and short local routing. Local routing consists of sending a message from a source to a destination using only local information. That is, random distance networks display the important characteristics of 'small-world' networks that have been found in the real world.

12.5 Other random models

Several variations and generalisations of the previous models for generating random networks have been described and analysed in the literature. Other models have been inspired in real-world networks, and are designed to account for certain characteristics of such networks. One model that represents a variation of the Watts and Strogatz (1998) model was introduced by Kleinberg (2000a). It starts by considering an $n \times n$ lattice as illustrated in Figure 12.19 (left). Then, a node is connected to all its neighbours, which are at a distance of at most d, where the Manhattan distance is used. In addition, each node is connected to q other nodes which are at a distance larger than d. These nodes are selected independently and at random, with probability proportional to $r^{-\alpha}$, where $r > d$ is the distance between the two nodes, and $\alpha \ge 0$ is a fixed clustering exponent. The most general Kleinberg graph can then be defined by the following set of parameters: $G_k(n, d, q, \alpha)$. An example of such networks is shown in Figure 12.19 (right).

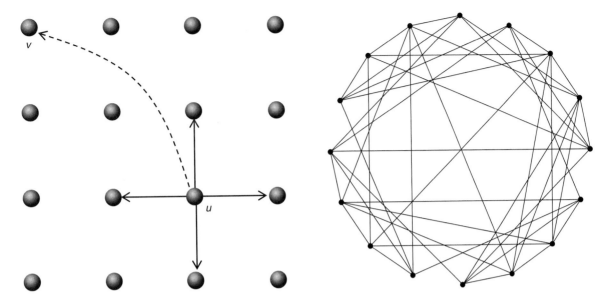

Fig. 12.19

Kleinberg model of random networks. The method used by Kleinberg to generate random networks by using both short and long-range connections between nodes (left). An example of a network generated with this model (right) has the following parameters: $G_K(20, 1, 1, 2)$.

In the previous section we discussed the concept of local routing of a message in a network. This refers to sending a message from a source to a target with local information only, as in the case of the Milgram experiment described in the small-world section. One way of studying the transmission of such messages is by using decentralised algorithms. Kleinberg (2000a) has used these algorithms in order to pass a message from the holder across one of its short-range or long-range connections. As before, the current holder does not know the long-range connections of nodes which have not touched the message. The efficiency of such algorithms is measured via the *delivery time*, t, which is the expected number of steps required to forward a message from a random source to a target. Kleinberg (2000b) has found that for G_k the delivery time is bounded as follows:

$$t \geq cn^\xi \tag{12.23}$$

where

$$\xi = \begin{cases} (2 - \alpha)/3 & \text{for } 0 \leq \alpha \leq 2 \\ (\alpha - 2)/(\alpha - 1) & \text{for } \alpha > 2 \end{cases}$$

which is represented graphically in Figure 12.19 (left). As can be easily inferred from this expression and plot, the value of $\xi = 2$ is critical for the delivery time. This means—as Kleinberg (2000b) has remarked—that 'efficient navigability is a fundamental property of only some small-world structures.' In particular, when long-range connections follow an inverse-square distribution ($\xi = 2$) the delivery of a message carried out by a decentralized algorithm is very fast. In Figure 12.20 we also illustrate the change of the

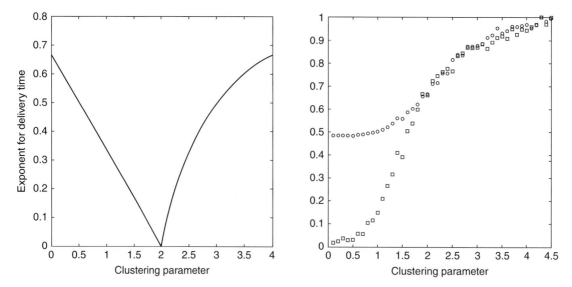

Fig. 12.20

Message delivery in the Kleinberg random networks. Change in the exponent of the delivery time with the change in the clustering parameter (left) in Kleinberg networks (see text for an explanation). Change of the average clustering coefficient and path length with the change of clustering parameter for a Kleinberg network with the following parameters: $G_K(900, 1, 1, \alpha)$ (right). In this plot, the squares represent the clustering coefficient and the circles represent the average path length.

clustering coefficient and average path length for Kleinberg networks with the variation of the clustering parameter α. As can be seen for $\alpha = 2$—the value at which a fast delivery of messages is possible—the network displays characteristics of small-world systems. However, for values of $\alpha < 2$ the clustering coefficient decays very fast, and for $\alpha > 2$ the average path length grows quickly. The value $\alpha = 2$ then appears as critical for which all desired properties of these networks are observed.

Some models of random networks have been inspired by the structure of biological networks. For instance, Grindrod (2002) has proposed a *range-dependent random network* model inspired by protein–protein interaction networks. In this model, the nodes are ordered in a natural linear way: $i = 1, 2, \ldots, n$. Then, a link between nodes i and j is created independently of the other links, with probability $\alpha \eta^{|j-i|-1}$, where $\alpha > 0$ and $\eta \in (0, 1)$ are fixed parameters. When $\alpha = 1$, all nearest neighbours are connected, and the network contains a Hamiltonian path connecting all nodes independently of η. Gridrod has determined that these networks have the following properties:

$$\bar{k} = \frac{2\alpha}{1 - \eta} \tag{12.24}$$

$$\overline{C} = \frac{3\alpha\eta}{(1 + \eta)(1 + 3\eta)} \tag{12.25}$$

which means that for $\alpha = 1$, $\overline{C} \to 3/8$ as $\eta \to 1$. The clustering coefficient in these networks displays a maximum which is located at $\eta = 3^{-0.5}$. If we consider that $\alpha = \bar{k}(1 - \eta)/2$, we can write the clustering coefficient as

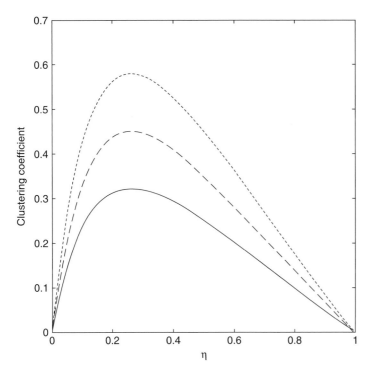

Fig. 12.21
Clustering coefficient in the Grindrod random network. Change of the average clustering coefficient in range-dependent random networks developed by Grindrod, with the change in the parameter η for three different average degrees: continuous line ($\bar{k} = 2.5$), discontinuous line ($\bar{k} = 3.5$), and dotted line ($\bar{k} = 4.5$).

$$\overline{C} = \frac{3\bar{k}(1 - \eta)\eta}{2(1 + \eta)(1 + 3\eta)} \tag{12.26}$$

which has a maximum at $\eta = (\sqrt{8} - 1)/7 \approx 0.261$, independent of the average degree. This is illustrated in Figure 12.21, in which we plot the clustering coefficient as a function of η for three values of the average degree.

Other models inspired by protein–protein interaction networks are the *lock-and-key* and *stickiness* models (Morrison et al., 2006; Pržulj and Higham, 2006). The first model is based on the paradigm that two proteins interact because they have some physically complementary parts—one of them considered as a key, and the other as a lock (see Chapter 15 for a discussion of this topic). The existence of various types of key is accommodated in this model by using different colours for the nodes, and for each type of key there is a matching lock of the same colour. In the lock-and-key model, each node has the same chance of having each colour of lock and each colour of key. That is, for a given number of colours C, for each node we independently assign to it each possible lock and key with some fixed probability p. The links of the network are created by considering that two nodes are connected if and only if one has a key and the other has a lock of the same colour. In Figure 12.22 (left) we illustrate a lock-and-key network with twenty nodes, $c = 2$ colours, and $p = 0.6$.

The stickiness model is a modification of the previous one which allows for different degree distributions. In this case, a non-negative vector $\widehat{\mathbf{k}}$ whose ith entry $\widehat{k_i} = k_i/\sqrt{\sum_{i=1}^{n} k_i}$ is a normalised degree of a target node in a given real

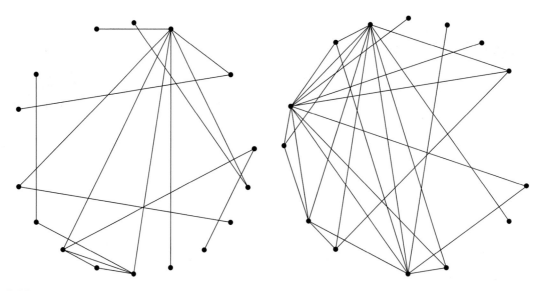

Fig. 12.22

Examples of lock-and-key networks. Two examples of random networks created with the lock-and-key model with twenty nodes, $c = 2$, and $p = 0.6$ (left), and with the stickiness model with twenty nodes and degree distribution $p(k) \sim k^{-2.5}$ (right).

network. Then, a link between two nodes is created with probability $p_{ij} = \widehat{k_i}\widehat{k_j}$, and in this way the expected degrees in the random model match the degrees in the real network. In Figure 12.22 (right) we illustrate a network created with the stickiness model having twenty nodes and matching a degree sequence of a network with a power-law degree distribution of the type $p(k) \sim k^{-2.5}$.

The reader needs to consider that this chapter is not an exhaustive compilation of all existing models for generating random networks. In fact, several variations of the existing models have been proposed in the literature in order to modulate some of the most important structural parameters that have been observed in real-world networks. The reader is therefore referred to the more specialised literature to obtain more information about these models and their characteristics.

Applications

I regarded as quite useless the reading of large treatises of pure analysis: too large a number of methods pass at once before the eyes. It is in the works of application that one must study them; one judges their utility there and appraises the manner of making use of them.

Joseph Louis Lagrange

Genetic networks

13

> Successive interactions of differentiated regions and the calling into play
> of additional genes may lead to any degree of complexity of pattern in the
> organism as a largely self-contained system.
>
> Sewall G. Wright, *Cold Harbor Symp. Quant. Boil.* **2** (1934), 142

According to the central dogma of molecular biology (Crick 1958; 1970), DNA passes information to itself and to RNA through replication and transcription respectively, and RNA passes information to proteins through translation. This is illustrated schematically in Figure 13.1. There are networks of interactions at the different levels of organisation of this general scheme of biology (Barabási and Oltvai, 2003; de los Rios and Vendruscolo, 2010; Koyutürk, 2010), including gene–gene relations, protein–protein interactions, and metabolic reactions, among others. Gene networks represent a starting point for the understanding of these relations, and they are the first to be studied in this part of the book. These networks account for the relationships established between genes, which are based on the observation of how a change in the activity of one gene affects the activity of the others.

13.1 A primer on gene networks

A basic paradigm in the construction of gene networks—known as 'guilt by association' (Oliver, 2000)—is that if two genes show similar expression profiles they are supposed to have similar functions. In this way, *gene coexpression networks* are constructuted by computing a similarity score for each pair of genes. The genes are then represented by the nodes of the networks, and two genes are connected if their similarity is above a certain threshold. There are some approaches for correcting the false positives that arise from this general principle of guilt by association—such as the use of refinements based on comparing orthologs across divergent species. *Gene regulatory networks*, also known as *gene transcriptional regulatory networks*, is another type of gene network in which the connections are between transcription factors and the genes that they regulate. These genes are of two possible types: regulatory genes, and target genes. The first are genes coding for transcription factors, and the

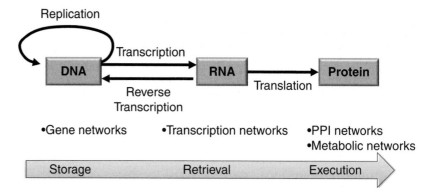

Fig. 13.1

Central dogma of molecular biology. The central dogma of molecular biology in which DNA passes information to itself and to RNA, and then RNA passes it to proteins.

second are genes regulated by transcription factors. It has to be remarked that a regulatory gene can also play the role of a target gene, as in auto-regulation. This type of network provides insights about the gene function, as well as on regulatory processes in the cell, which is an extra value in comparison with gene coexpression data. *Synthetic lethal networks* constitute another type of gene network, in which pairs of nodes representing genes are connected if the simultaneous deletion of both genes causes lethality. General information about these different types of gene network can be found in the literature (Janga and Babu, 2010; Brilli et al., 2010; Markowetz and Spang, 2007; Pisabarro et al., 2008; Sharom et al., 2004).

The information needed for constructing gene networks can be obtained from DNA microarray experiments. Most of the functional genomic data available today comes from gene expression microarray experiments, due to the relatively low cost and accessibility of this technology. Microarray analysis can also be used for identification of coregulated groups of genes. In this case the use of chromatin immunoprecipitation microarray (ChIP) allows for identification of direct binding of a specific protein complex to DNA on a whole-genome scale. It can be integrated with the gene expression microarray to identify coregulated groups of genes, their regulators, and the corresponding transcription factor binding sites (Collas, 2010). In order to build synthetic lethal networks, a technique known as 'synthetic lethality analysis by microarray' (SLAM) has been proposed. It uses a transformation-based strategy to create a pool of double mutants which can then be analysed by the barcode microarrays (Boone et al., 2007; Pisabarro et al., 2008).

DNA microarray technology is based on the capacity of DNA and RNA to form double-stranded molecules by hybridisation. This principle is based on the complementarity of base pairing, which guarantees that adenine (A) forms a base pair only with thymine (T) or uracil (U) in RNA, and guanine (G) forms a base pair only with cytosine (C). These pairings correspond to bases which are in two different DNA (RNA) chains, such that the pairs are formed through non-covalent bonds between both bases—in particular, by using hydrogen bonds. Then, if a sequence of a DNA chain is written as ATTGAGCTCTAG, the only complementary chain that can be constructed from it is TAACTCGAGATC. If we mark the first chain with a fluorescent or radioactive tag, it can be used to isolate the complementary chain from a very complex mixture. The first chain, after being marked with a tag, is known as the *probe*, and the complementary is known as the *target*. Then, if we place in contact the probe and the target, the hybridisation between both chains will take place as illustrated in Figure 13.2.

Briefly, a DNA microarray is a dense arrangement of DNA cells in a miniature support, each of them representing a different gene (Troyanskaya, 2005). The array is usually hybridised with a probe generated by labelling a complex mixture of RNA molecules derived from a particular type of cell. Then, using a method which depends of the nature of the tag used (radioactive or fluorescent), the intensity of each feature on the microarray is obtained. Such intensity represents the relative expression level of the corresponding gene. These microarrays have densities of 5,000 to 1 million features per square centimetre.

Gene expression obtained by using microarrays can be stored in the form of a rectangular matrix. The rows of this matrix represent the genes placed on the cells of the microarray, and the columns represent the experimental conditions applied. Then, if there are N_E experiments and N_G genes, the gene expression matrix is $N_E \times N_G$. The entries in this matrix represent the levels of expressions of the genes in the corresponding experiments. The genes are then grouped together according to similar expression profiles, which are defined by the correlation between the different rows of the gene coexpression matrix. The Pearson linear correlation coefficient is generally used to transform the gene coexpression matrix into a similarity matrix, which is then represented by a dendrogram. The gene coexpression network can be built by dichotomising the similarity matrix into an adjacency matrix, as illustrated in Figure 13.3 (Brazhnik et al., 2002; Pisabarro et al., 2008).

In order to investigate protein–DNA interactions inside the cell, which are necessary to obtain transcriptional regulatory networks, a method known as 'chromatin immunoprecipitation' (ChIP) is currently the technique of choice. The method is based on the formation of reversible cross-links between DNA and proteins by using an agent such as formaldehyde, which has the capacity of reacting with primary amines such as those present in nucleic acids and amino acids. These bonds are formed when the protein and DNA are separated at about 2 Å, which makes it possible to use the method for detecting proteins which interact directly with DNA. Once the cross-linking process is completed, the cells are lysed, and chromatin is fragmented to small pieces having between 200 and 1,000 base pairs by using sonication or other techniques such as enzymatic digestion. The complex between DNA and the protein is then immunoprecipitated by using an antibody that acts against the protein of interest. At this point, after the elimination of possible impurities, the cross-links are reversed by heating, and proteins are digested by using proteinase K. Finally, the DNA is purified and identified by using any of the techniques available. Depending on the identification technique used, several variations of ChIP are known: basically, end-point polymerase chain (PCR), by labelling and hybridising onto a microarray chip (ChIP-on-chip), or by direct high-throughput sequencing (ChIP-seq). In Figure 13.4 we reproduce a scheme of the ChIP assay and some possible methods of analysis (Collas, 2010).

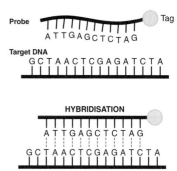

Fig. 13.2

DNA/RNA hybridisation. Schematic representation of the process of DNA or RNA hybridisation by means of which two complementary nucleotic chains are paired through hydrogen bonds.

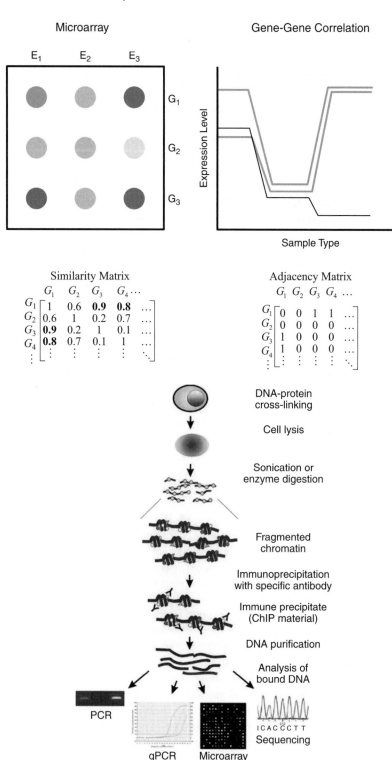

Fig. 13.3

Construction of gene coexpression networks. Schematic representation of the process for building adjacency matrices from the similarity matrices obtained from micro-array experiments. Highly correlated genes, according to the Pearson correlation coefficient r, are marked as bold in the similarity matrix. The similarity matrix is dichotomised to obtain the adjacency matrix. For example, instances of r $\geq$ 0.8 in the similarity matrix are replaced by 1s, and the rest of the values and those in the main diagonal by zeros.

Fig. 13.4

Chromatin immunoprecipitation (ChIP) assay. Schematic representation of the chromatin immunoprecipitation assay and various methods of analysis. From Collas, P. (2010). The current state of chromatin immunoprecipitation. *Mol. Biotech.* **45**, 87–100. (This illustration, and any original (first) copyright notice displayed with material, is reproduced with the permission of Springer Science+Business Media.)

Although these are the most commonly used techniques for creating gene networks, numerous other variations and techniques are also used, including genetical genomics, native ChIP (NChip), DNA adenine methyltransferase Identification (DamID), Methylated DNA Immunoprecipitation (MeDIP), and so on (Collas, 2010; de Koning and Haley, 2005). Several computational strategies—most of them using concepts from network theory—have been proposed and analysed in the literature, including discrete Boolean, Bayesian, and weighted networks (de Jong , 2002; D'haeseleer et al., 2000; Hasty et al., 2001; Zhang and Horvath, 2005).

13.2 Topological properties of gene coexpression networks

We begin here by analysing some of the topological properties described so far for gene coexpression networks. The yeast coexpression network has been analysed from a complex network perspective, showing both 'small-worldness' and 'scale-freeness' (van Noort et al., 2004). The coexpression network studied contained 4,077 genes and 65,430 links, which indicates an average degree of 32. The main connected component is formed by 3,945 nodes, for which the average path length is $\bar{l} = 4$, noting that $\bar{l} < \ln n$ and comparable to that of a random graph of the same size and density, $\bar{l}_{rand} \approx 2.8$. The average Watts–Strogatz clustering coefficient of the whole network is $\bar{c} = 0.6$, which is significantly larger than that of a random network of the same size, $\overline{C}_{rand} = 0.008$. van Noort et al. (2004) also studied the degree distribution for the gene coexpression networks obtained by considering different cut-off values of the correlation coefficient in defining the gene network. They observe that when the Pearson correlation coefficient used is larger than 0.6, the networks obtained have power-law degree distributions of the type $p(k) \sim k^{-1}$, and the clustering coefficient of the network remains practically invariable (see Figure 13.5). However, the consideration of less stringent values of the

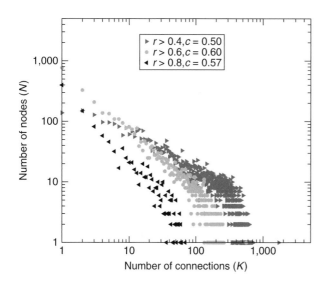

legend:
- $r > 0.4, c = 0.50$
- $r > 0.6, c = 0.60$
- $r > 0.8, c = 0.57$

x-axis: Number of connections (K)
y-axis: Number of nodes (N)

Fig. 13.5
Degree distribution in the yeast coexpression network. Node degree distributions for the gene coexpression network of yeast, according to van Noort et al. (2004), in which coexpression is defined by a correlation in expression pattern higher than 0.4 (right-pointing arrows), 0.6 (circles), or 0.8 (left-pointing arrows). From van Noort, V., Snel, B., and Huynen, M. A. (2004). The yeast coexpression network has a small-world, scale-free architecture and can be explained by a simple mode. *EMBO Rep.* **5**, 1–5. (Reproduced with the permission of Macmillan Publishers Ltd.)

similarity between gene coexpressions—for example, a correlation coefficient of 0.4—gives rise to networks with smaller clustering coefficient and exponential degree distribution, which is indicative of the inclusion of some random connections between the genes.

The gene coexpression network of yeast has also been analysed by Carlson et al. (2006) by dividing it into three separated networks corresponding to three separate microarray datasets studying DNA damage (DD), cell cycle (CC), and environmental response (ER). In addition, they constructed weighted networks instead of simple ones. The weights of the links were assigned by using the correlation coefficients obtained from microarray data and the function $A_{ij} = |corr(x_i, x_j)|^\beta$, where the parameter β is chosen using a scale-free topology criterion. The networks built in this way were observed to display fat-tail degree distributions, which the authors claimed as power-law and exponentially truncated power-law ones. More evidence of power-law degree distributions in gene networks has been obtained, for instance, for human and mouse gene coexpression networks (Tsaparas et al., 2006). Both networks have more than 7,000 nodes, average degree over 40, average Watts–Strogatz clustering coefficient of 0.4, and average shortest path length of 4.7. Figure 13.6 shows the log–log plot for the degree distributions of both coexpression networks.

Degree distribution is one of the topological properties of gene coexpression networks that have received more attention. Bortoluzzi et al. (2003) analysed nineteen different adult human tissues based on the information of 27,924 genes, observing that for all networks the degree distribution follows a power-law of the type $p(k) = ak^{-\gamma}$, with $1.2 \leq \gamma \leq 2.4$ and very high correlation coefficients. The implications of scale-freeness on the robustness and evolvability of genetic regulatory networks have been studied by Greenbury et al. (2010).

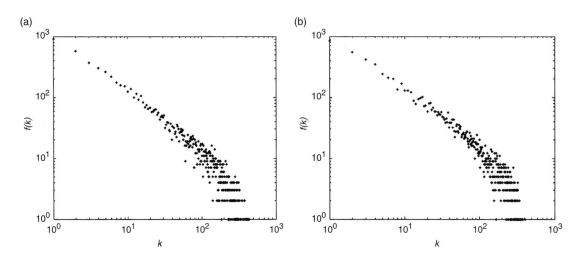

Fig. 13.6
Degree distributions for human and mouse coexpression networks. Node degree distributions for human (a) and mouse (b) coexpression networks studied by Tsaparas et al. (2006). (Reproduced with the permission of Tsaparas et al.)

A simple evolutionary model was originally proposed and tested by van Noort et al. (2004) as a plausible explanation of the fat-tail degree distribution observed for a yeast gene coexpression network. This model is based on transcription factor binding sites (TFBSs) which are present in a gene. The model can be briefly described as follow:

1). Consider an initial genome consisting of a set of genes in which certain numbers of TFBSs are randomly allocated.
2). Let the genome evolve by allowing the following mechanisms:
  i) Gene duplication.
  ii) Gene deletion.
  iii) TFBS duplication.
  iv) TFBS deletion.
3). Construct a network by considering the genes in the evolved genome as the nodes and connecting two nodes if the corresponding genes share TFBSs.

In the model it is assumed that the probability of obtaining a specific TFBS is proportional to its frequency in the genome, and that for a new TFBS is (150 − total number of different TFBSs present)/(150 + total number of TFBSs). Using this model, and assuming that the rate of TFBS duplication and deletion is within the same order of magnitude, both small-worldness and scale-freeness of the resulting networks are obtained.

Another characteristic of gene coexpression networks revealed from the studies carried out so far is their modularity. For instance, Figure 13.7

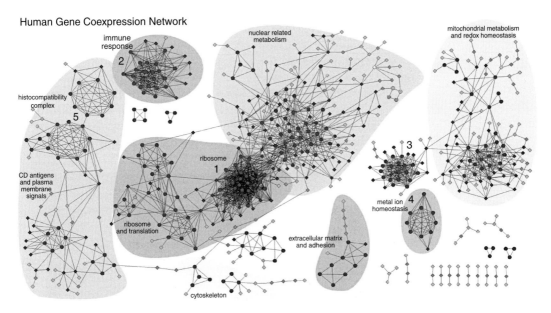

Fig. 13.7

Human gene coexpression network. The human gene coexpression network obtained by Prieto et al. (2008) by the intersection of two different datasets. From Prieto, C., Risueño, A., Fontanillo, C., and De las Rivas, J. (2008). Human gene coexpression landscape: confident network derived from tissue transcriptomic profiles. *Plos One* **3**, e3911. (Reproduced courtesy of J. de la Rivas, by permission of the authors.)

illustrates a network consisting of 615 genes in humans and 2,190 coexpression links (Prieto et al., 2008). This network represents a validated core of a coexpression network of 3,327 nodes and 15,841 links. The validation was carried out by using two independent methods. The nodes are clearly organised into several clusters. The three larger ones are a cluster of genes involved in nuclear activity and nuclear-driven metabolism, a cluster of genes involved in mitochondrial metabolism and redox homeostasis, and a third cluster formed by genes involved in the immune response. Other small clusters are also recognisable as those related to metal ion homeostasis, genes involved in extracellular matrix and cell adhesion, and genes related to the cytoskeleton.

Gene functions on a global scale have been elucidated by using gene coexpression networks obtained from DNA microarrays from humans, flies, worms, and yeast. Stuart et al. (2003) have found 22,163 coexpression relations across organisms, which have been conserved across evolution. They are interpreted as gene pairs conferring selective advantage and functional relationship. Metagenes are defined here as sets of genes across multiple organisms using a reciprocal best blast hits strategy for protein sequences. A network of 3,416 metagenes connected by 22,163 coexpression interactions was then created, from which only 236 interactions were expected by chance using a selection criterion based on $P < 0.05$, where P is a probability measure based on order statistics. In Figure 13.8 we reproduce the terrain map showing metagenes represented in the $x-y$ plane, based on the negative logarithms of the P values, and displaying the density of genes in a given region as the altitude in the z direction. In this clever pictorial representation of the network regions, highly interconnected metagenes are represented as hills or mountains in the terrain map. These hills represent clusters or components formed mainly by metagenes displaying similar biological functions, such as protein degradation, ribosomal function, cell cycles, metabolic pathways, and so on.

How predictive are topological clusters for real biological functions? According to the principle of 'guilt by association', if a gene is linked to many others that have the same biological function, the first is also expected to display that function. Therefore, it is possible to search for genes with unknown function in the hills of Figure 13.8. These genes must have functions similar to those represented in the clusters. This experiment was carried out by

Fig. 13.8

Terrain map of gene coexpression networks. A terrain map of the gene coexpression networks of humans, flies, worms, and yeast, where metagenes with high conserved expression are placed together. The altitudes of the hills in the terrain correspond to the local density of genes. From Stuart, J. M., Segal, E., Koller, D., and Kim, S. K. (2003). A gene-coexpression network for global discovery of conserved genetic modules. *Science* **302**, 249–55. (Reproduced with permission from AAAS.)

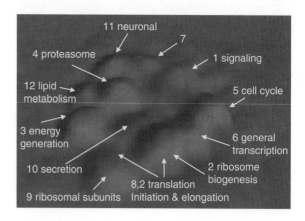

Stuart et al. (2003) by considering five metagenes located in the hill enriched with genes with known involvements in cell proliferation and cell cycle. The original five metagenes were not known to be involved in such functions. The results are very encouraging, as all five genes were shown to be overexpressed in human pancreatic cancers in relation to normal tissue. The observed overexpression is of the same extent as that observed for genes known to be involved in cell proliferation.

13.3 Topological properties of gene regulatory networks

We now turn our attention to another type of gene network: gene transcriptional regulatory networks. We recall that these networks are formed by two types of gene—the regulatory genes (transcription factors) and the target genes—and that the networks are directed pointing from the first to the second ones. In Figure 13.9 we illustrate the transcription regulation network of the bacterium *E. coli* compiled by Shen-Orr et al. (2002) by combining DBregulon information and an extensive search in the literature. It consists of 423 genes—116 of them are transcription factors, and 519 interactions. The network is formed by twenty-nine connected components, the largest one consisting of 328 genes.

A general characteristic observed for transcription networks is an evident asymmetry in the degree distributions for incoming and outgoing links in both prokaryotic and eukaryotic cells. It has been observed that indegree distributions of transcription networks are exponential-like. This type of exponential degree distribution is explained by the fact that there is no such thing as hub target genes. However, the outdegree distribution is fat-tailed, and can be described by using a power-law with exponent between 1 and 2. This clearly points to the fact that there is a small number of hub TFs that regulate a large number of genes. For instance, in the regulation network of *E. coli* illustrated

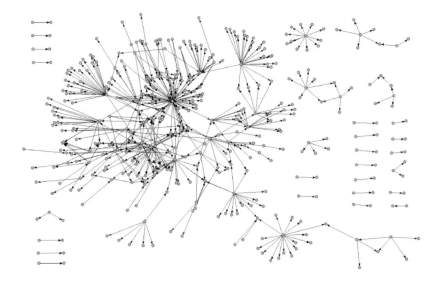

Fig. 13.9
Transcription regulation network of *E. coli*. Representation of the transcription network of *E. coli* as a directed network.

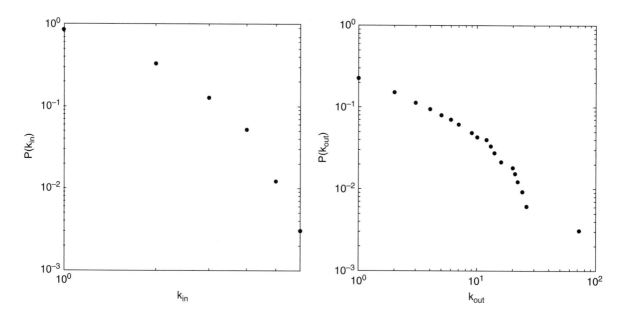

Fig. 13.10

Degree distributions in the transcription network of *E. coli*. Distributions of the indegrees (left) and outdegrees (right) in the gene transcription networks of *E. coli* represented in Figure 13.9.

in Figure 13.9 there is one TF that regulates seventy-two other genes. This characteristic asymmetric degree distribution is illustrated in Figure 13.10 for the transcription network of *E. coli* by using the cumulative degree distribution.

However, a different analysis of the gene regulatory network of yeast carried out by Lee and Rieger (2007) showed that its indegree distribution is broader than that of *E. coli*, suggesting a power-law with exponent 2.7. The discrepancy with the previous results was attributed to the analysis of different datasets of different sizes, which supports our claims in Chapter 2 about the necessity for different characterisation of network heterogeneity. The results of Lee and Rieger (2007) are shown in Figure 13.11.

The 'mesoscopic' analysis of transcriptional regulatory networks has provided evidence for the existence of subgraphs that appear more frequently than expected at random in these networks. These subgraphs are known as 'network motifs', and were studied in Chapter 4. The most abundant of these motifs in regulatory networks is the 'feed-forward loop' (FFL). This motif consists of one transcription factor X, which regulates another transcription factor Y. These two TFs jointly regulate a third or more of the genes, giving rise to the motif illustrated in Figure 13.12. This motif appears forty and seventy times in the transcriptional networks of *E. coli* and yeast, respectively (Yeger et al., 2004). The number of times this fragment is expected in random networks with the same characteristics as the two regulatory networks are seven and eleven, respectively.

Another motif frequently found in transcriptional networks is the 'single input module' (SIM). In this motif a single transcription factor X regulates a series of operons, which are not regulated by other TFs (Yeger et al., 2004).

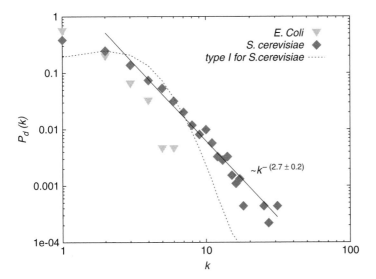

Fig. 13.11

Degree distributions in transcriptional networks. Indegree distributions in the transcription networks of *E. coli* and yeast, as well as its randomised network. From Lee and Rieger (2007). Comparative study of the transcriptional regulatory networks of *E. coli* and yeast: Structural characteristics leading to marginal dynamic stability. *J. Theor. Biol.*, **248**, 618–26. (Reproduced with the permission of Elsevier, ©2007.)

Usually, the TF is autoregulatory, which means that it has a self-loop as illustrated in Figure 13.13.

A third motif found in regulatory networks is the dense overlapping regulons (DORs), in which a set of TFs regulate a set of operons (see Figure 13.14). In this case, one operon can be regulated by more than one TF, and one TF can regulate more than one operon. The bi-fan motif (discussed in Chapter 4) is one particular case of DOR in which there are two TFs that control two operons (Yeger et al., 2004).

All thirteen directed subgraphs with three nodes have been studied for the transcription networks of *E. coli*, yeast, and mammals. In this case the transcription network of mammals represents 279 genes in human, mouse, and rat, and 657 interactions (Goemann et al., 2009). In Figure 13.15 the results are plotted in terms of the Z-score defined in Chapter 4. As can be seen, only the FFL appears over-represented in the three networks, despite its not being among those with the largest score for the mammalian network. The fragments 2 and 4 appear significantly under-represented in the three networks, and can be considered as anti-motifs—patterns that appear less frequent in a network than expected at random. Fragments 6, 11, and 12 are the most over-represented ones in the mammalian network, and those with numbers 3, 7, and 10 appear very much under-represented in this network. Goemann et al. (2009) have termed the under-represented fragments 'non-motifs', and have used them to compare the structural significance of overrepresented fragments; that is,

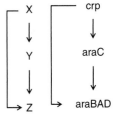

Fig. 13.12

Feed-forward loop in the transcription network of *E. coli*. Schematic representation of a feed-forward loop (left) and the L-arabinose utilisation feed-forward loop (right) found by Shen-Orr et al. (2002) in the transcription network of *E. coli*.

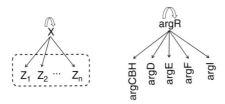

Fig. 13.13

Single input module (SIM) in the transcription network of *E. coli*. Schematic representation of a single input module (SIM) (left) and the arginine biosynthesis SIM (right) found by Shen-Orr et al. (2002) in the transcription network of *E. coli*.

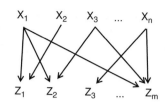

Fig. 13.14

Dense overlapping regulon (DOR) in the transcription network of *E. coli*. Schematic representation of a dense overlapping regulon (DOR) (left) and the stationary phase response DOR (right) found by Shen-Orr et al. (2002) in the transcription network of *E. coli*.

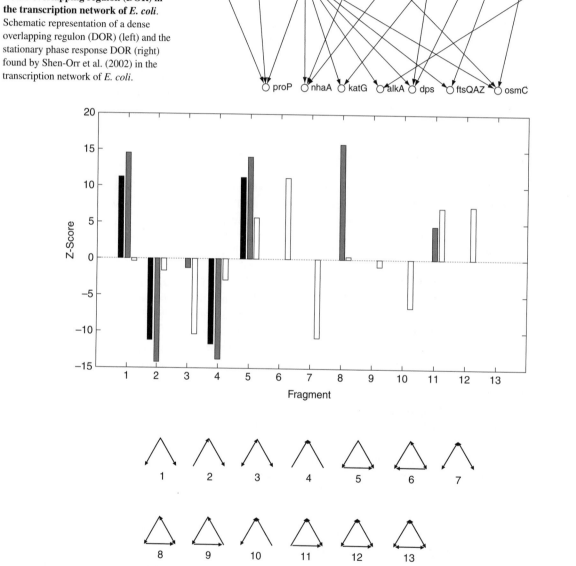

Fig. 13.15

Motifs and anti-motifs in transcription networks. Representation of the Z-scores (see Chapter 4) for the thirteen subgraphs with three nodes in the transcription networks of *E. coli* (black), yeast (grey), and mammals (white).

motifs. It would be preferable to use fragments for which the frequency in the transcription network is not statistically different from that of a random network. If there is any biological reason why some fragments appear more frequently in a transcription network than expected at random, a similar reasoning can be applied for those fragments that were 'avoided' during evolution with a frequency significantly larger than in a random graph. We would then prefer to consider these fragments as anti-motifs rather than non-motifs. However, it is interesting that from the results obtained by Goemann et al. (2009) there are only very few fragments whose removal from the network breaks many of the pairwise connections between nodes. In general, the elimination of motifs was not observed to significantly change the structure of the networks studied. But should we expect such changes? We think it is wrong to expect that a motif—which is an important fragment for the dynamical processes taking place in the transcription network—can also play a fundamental role as a structural skeleton for the network. Let us present a simple example. If we remove all the elevators from a skyscraper we render the building completely non-functional, even though the structure of the building is not damaged in any way.

A complete mammalian transcription network consisting of 121 genes and 212 interactions was extracted from TRANSPATH® database by using a unipartite graph at ortholog abstraction level (Potapov et al., 2005). A subset of this network centred at the $p53$ gene and consisting of 44 genes and 80 interactions was also studied by Potapov et al. (2005). Both networks displayed average shortest path lengths of 2.2 and 2.4 respectively, and diameters of 5 for both networks. The authors reported scale-free degree distributions for both the indegree and outdegree, but the correlation coefficient of the fits for the indegree were significantly lower than that of the outdegree. The average clustering coefficient for each network was 0.134 and 0.241 respectively—about five and three times larger than those obtained for similar random graphs. The log–log plots of the clustering coefficients versus the degree in both cases indicate the existence of certain power-law scaling for the clustering: $C(k) \sim k^{-\omega}$, with exponent close to 1. This scaling of the clustering coefficient can be indicative of a certain 'hierarchy' in the network, but an alternative topological explanation is provided in Chapter 4.

A local-level analysis of the mammalian transcription network illustrates the relevance of certain centrality measures in identifying relevant genes (Potapov et al., 2005). For instance, the p53 gene is identified as the node with the highest betweenness centrality among all genes, which indicates that this TF-gene is an important intermediary in this regulatory network. This gene also displays the highest outdegree, $k_{out} = 10$, among all genes in the transcription network. Most of the genes in the top-ten ranking according to the betweenness centrality are known to be involved in the regulation of cell proliferation, and some of them have features of tumour-suppressors or proto-oncogenes. Some of these genes, with their degree and betweenness centrality, are listed in Table 13.1. Note that although the tumour-suppressing gene *WT1* has small values of degree centrality, it is ranked among the top according to the betweenness.

Centrality measures have also been used to study the transcription regulatory network of *E. coli*. In this case a series of centrality measures was

Table 13.1 Centrality in mammalian transcription network. Values of the indegree (k_{in}) and outdegree (k_{out}) as well as of the betweenness centrality (BC) for some relevant genes in a complete mammalian transcription network (Potapov et al., 2005).

Gene	k_{in}	k_{out}	BC
p53	8	10	0.0188
c-fos	24	3	0.0139
Egr1	3	6	0.0047
c-jun	8	5	0.0046
WT1	2	1	0.0029

calculated for all nodes in this regulatory network, and the nodes were then ranked according to the values of the centralities in decreasing order. By selecting the top twenty-five genes according to every ranking, Koschützki and Schreiber (2008) analysed how many of these genes are global regulators which have been defined on the basis of their pleiotropic phenotype and the gene ability to regulate operons that belong to different metabolic pathways. Martínez-Antonio and Collado-Vides (2003) identified eighteen global regulators in the transcriptional regulatory network of *E. coli*, which were then used as the targets of the centrality measures analysis by Koschützki and Schreiber (2008). These centrality measures are (1) outdegree, (2) PageRank of the reverse network, (3) Katz index for the reverse network, (4) betweenness centrality, (5) radiality, (6–8) motif centrality indices based on chains, dominant node of the FFL, and the total FFL motif (all of which are defined and analysed in Chapter 7). In general, most of these centrality measures are intercorrelated for the transcription network of *E. coli*, with the exception of the betweenness and motif centralities based on FFL motif, which are poorly correlated with the rest and are among them (see Figure 13.16).

Fig. 13.16

Intercorrelation between centrality measures. A plot of the correlation matrix for the eight centrality indices studied by Koschützki and Schreiber (2008) in the transcription network of *E. coli*.

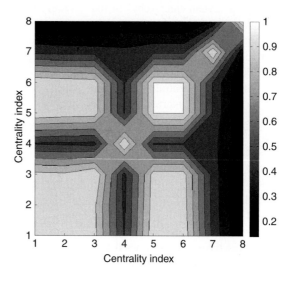

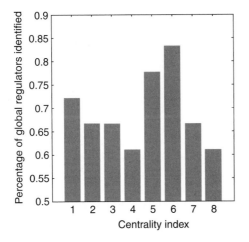

Fig. 13.17
Centrality-based identification of
global regulators. Performance of the
eight centrality measures studied by
Koschützki and Schreiber (2008) for the
identification of global regulators in the
transcription network of *E. coli*.

In Figure 13.17 we illustrate the percentages of global regulators identified among the top twenty-five genes ranked by eight different centrality measures. As can be seen in this figure, all centralities perform relatively well in identifying global regulators. For instance, there are global regulators in the top five of most of the rankings, except the one with the betweenness. The motif centrality based on chains produces the best results, identifying 83% of global regulators in the network.

The reason why the chain motif centrality provides the best ranking of the genes in the transcription network of *E. coli* for identifying global regulators could be related to the role played by *transcriptional cascades* in these networks. A transcriptional cascade is a sequence of transcriptional factors that regulate each other in a sequential basis. That is, they are linear chains of TFs. It has been shown that long cascades are rare in most transcriptional networks (see Figure 13.18), because they have been found to produce long response delays (Rosenfeld and Alon, 2003; Shen-Orr et al., 2002; Nikoloski et al., 2010). However, although in *E. coli* most of the cascades are of one step only, the results shown by Koschützki and Schreiber (2008) may be indicative

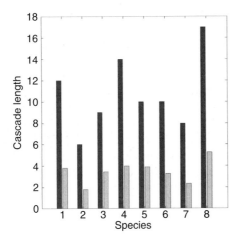

Fig. 13.18
Cascades in transcription networks.
Length of the longest cascade (dark grey)
and average cascade length (light grey)
in the transcription networks of *E. coli*
(1), *Saccharomyces cerevisiae* (2), sea
urchin ectoderm (3) and endomesoderm
(4), *Drosophila melanogaster* (5), *Mus
musculus* (6), *Ratus norvegicus* (7), and
Homo sapiens (8). The results are taken
from Nikoloski et al. (2010).

that those scarce long cascades play some fundamental role in the network, as they involve global regulators with high frequency, and also provide a method for identifying then in a transcriptional network—at least in that of *E. coli*.

An important area of research in the study of gene networks is the relation between network structure and genomic function. It has been proposed—by Brazhnik et al. (2002) for instance—that gene networks can be used to describe function unequivocally, and that they can be used for genome functional annotation. Using network theory approaches, Schlitt et al. (2003) have identified functional relationships for several genes in yeast, including the assignation of functions for seven genes of yeast which were previously uncharacterised. Accordingly, the genes KIN3 and YMR269W were identified as having possible involvement in biological processes related to cell growth and/or maintenance, and IES6, YEL008W, YEL033W, YHL029C, YMR010W, and YMR031W, probably have metabolic functions. More recent interest in this area of structure–function relationships for gene networks is related to human diseases, such as cancer, immunological diseases, and schizophrenia (Nacu et al., 2007; Potkin et al., 2010; Ratkaj et al., 2010).

13.4   Other gene networks

A different source of gene–gene networks is provided by bipartite networks in which one group of nodes represents genes and the other group represents effects produced by genes or molecules that interact with them. For instance, suppose that we consider all genes involved in hereditary diseases in humans as one of the bipartitions of the network, and the hereditary diseases as the second group. In this case we have a *diseasome network* relating hereditary diseases with the genes which are involved in them (Goh et al., 2007; Loscalzo et al., 2007). One of the projections of this network is a gene–gene network representing connections between pairs of genes which are involved in the same disorder. This projection corresponds to a disease gene network as illustrated in Figure 13.19.

This network consists of 1,377 genes, from which the giant connected component has 903 genes. The average size of the giant component generated by 10,000 random realizations is $1,087 \pm 20$ genes, which is significantly larger than the actual size of the giant size in the real network. A plausible explanation for this divergence is that in the real network, most of genes prefer to be linked to others belonging to the same disorder class. That is, in the gene–gene network there is a large assortativity by disorder class of the genes. This assortativity is also reflected in the topological structure of this network. In fact, the Newman assortativity coefficient for this network is positive and high: $r = 0.619$. We recall that a network is assortative if $|P_{2/1}| < |P_{3/2}| + C$, where $|P_{r/s}|$ is the ratio of paths of length r and s, and c is the clustering coefficient. In this gene–gene network, $|P_{2/1}| = 24.844$ and $|P_{3/2}| = 28.8221$, which clearly explains the assortativity of this network. In addition, the clustering coefficient is very high—$C = 0.619$—which indicates a large transitivity in the gene–gene relationships. The degree distribution in this network is of exponential type, and there is no hierarchical modularity, as can be seen in Figure 13.20. However, it is evident from Figure 13.19 that

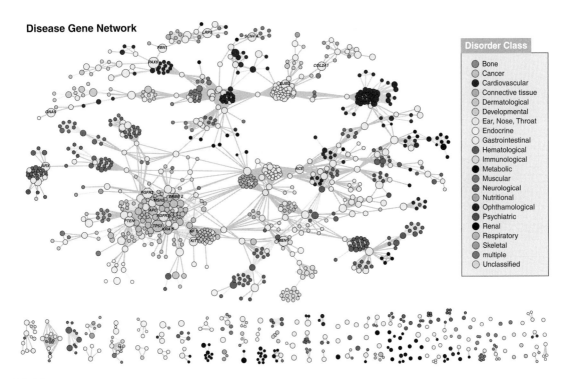

Fig. 13.19

Disease gene network. Projection of the human disease network into the gene–gene network, where nodes represent genes, and two nodes are connected if they are related to the same disease. From Goh, K-II., Cusick, M. E., Valle, D., Childs, B., Vidal, M., and Barabási, A-L. (2007). The Human disease network. *Proc. Natl. Acad. Sci. USA* **104**, 8685–90. (National Academy of Sciences, USA, ©2007.)

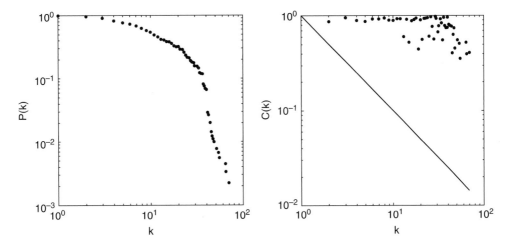

Fig. 13.20

Properties of the disease gene network. Degree distribution (left) and decay of the clustering as a function of the degree (right) for the nodes in the gene–gene projection of the gene-disease network studied by Goh et al. (2007).

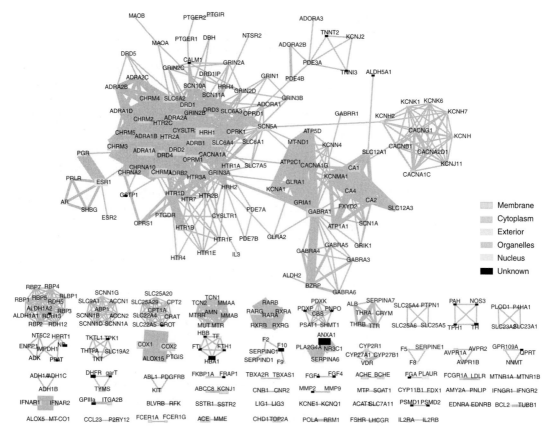

Fig. 13.21

Target-protein network. Representation of the protein–protein projection of the drug-protein interaction network, where nodes represent proteins and two nodes are connected if they are targeted by the same drug. From Yildirim, M. A., Goh, K-II., Cusick, M. E., Barabási, A-L., and Vidal, M. (2007). Drug-target network. *Nature Biotech.* **25**, 1119–26. (Reproduced with the permission of Macmillan Publishers Ltd., ©2007.)

genes are organised in a modular way in this network, in which modules clearly correspond to the disorders in which they are involved.

Another gene–gene network that has been studied in the literature is the one obtained from the interaction of drugs and their targets (Yildirim et al., 2007). This bipartite drug-target network can be projected into the target–target space, giving rise to a network in which nodes represent genes targeted by drugs, and in which two genes are connected if there is at least one drug interacting with them. In Figure 13.21 we reproduce the target network. It has been observed, by analysing the networks derived from this study, that a large number of drugs target the same 'already targeted' genes. However, new trends have been observed concerning newly developed drugs to more functionally diverse targets.

Protein residue networks

14

From the intensity of the spots near the centre, we can infer that the protein molecules are relatively dense globular bodies, perhaps joined together by valency bridges, but in any event separated by relatively large spaces which contain water.

Dorothy C. Hodgkin (1934)

The study of interatomic networks can be traced back to the nineteenth century and the works of several chemists and mathematicians—among the most notable of them being Arthur Cayley and James J. Sylvester. We recall that the origin of the word 'graph' arose in this context in a letter published in *Nature* in which Sylvester(1878) analysed the connection between chemistry and algebra (see Chapter 1). An interatomic network is a simple pictorial representation of a molecule in which atoms, or atomic groups, are represented as the nodes of the network and the interatomic interactions are represented by the links of the network. These networks are known in the literature as 'molecular graphs'—in particular when the interactomic interactions are the covalent bonds keeping together atoms in a molecule. However, as we will see, these interactomic interactions can also refer to any kind of non-covalent physical interactions among atoms. Examples of these networks in which links represent non-covalent physical interactions are the protein residue networks (Gursoy et al., 2008).

14.1 A primer on protein structure

A protein is a polymer in which the monomeric units are amino acids. An amino acid is an organic compound having the groups $-NH_2$ (amino) and $-COOH$ (carboxylic acid) in the same structure. For proteins found in nature, both groups are bonded to the same carbon atom, which is denoted as alpha-carbon, C_α. A third chemical group is attached to this C_α, which we represent generically in Figure 14.1 (left) as R_1 (Lesk, 2001). In nature there are twenty different R_1 groups, which form the natural amino acids. For instance, if $R_1 = H$, the amino acid is glycine (Gly, G); if $R_1 = CH_3$, the amino acid is alanine (Ala, A); if $R_1 = CH(CH_3)_2$, the amino acid is Valine (Val, V); and so forth. The list of amino acids and their three-letter and one-letter codes are given in Table 14.1. The carbon atom of R_1 which is directly bonded to the

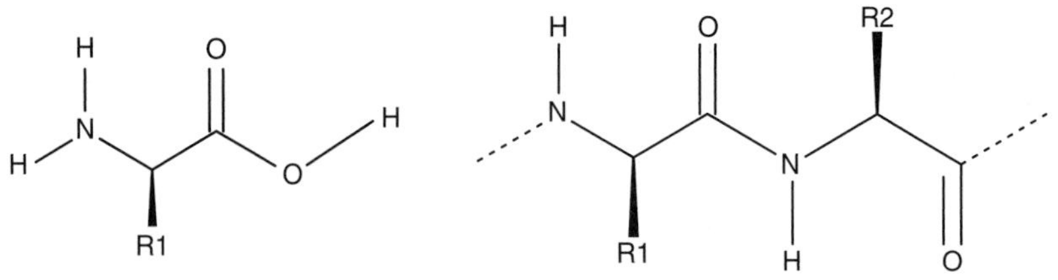

The Structure of Complex Networks

Fig. 14.1

Molecular formulae of amino acids and peptide. Representation of the molecular formula of one α-amino acid (left) in which R_1 represents the side chain. The structure of a peptide is schematically represented at right.

alpha-carbon is named the beta-carbon, C_β. Obviously, glycine does not have any C_β, and it is the only natural amino acid in this situation.

Two amino acids (not necessarily different) can react to form a new molecule (dipeptide), which is characterized by the presence of a peptide bond, $C(=O)$—N, represented in Figure 14.1 (right). The condensation of many amino acids in this way gives rise to a protein. Therefore, if we represent every amino acid by a single letter, like those shown in Table 14.1, we can immediately realise that a protein is a 'word' of n letters (amino acids). This sequence of amino acids is known as the primary structure of the protein.

From a network perspective, representing the primary structure of a protein is rather mundane, as it corresponds to a path P_n. However, the three-dimensional structure of proteins is very far from being boring. The linear chain of the protein backbone is folded to a well-determined 3-D structure, in which some regions of the protein form helices, sheets, or loops, as illustrated in Figure 14.2.

Information about the 3-D structure of a protein is obtained experimentally by using X-ray crystallography, or nuclear magnetic resonance (NMR) spectroscopy. This information is then deposited in the Protein Databank (PDB), in which every protein has a unique PDB identification code (Westhead et al., 2002). This database is freely accessible at http://www.pdb.org. The structure of PDB files is standard, and begins with a general heading containing information about the protein, how its structure was determined, who determined it, and where it was originally published. An example is given in Figure 14.3 for the protein with PDB code 1ash.

The information about the 3-D structure of the protein starts with the head 'ATOM', and consists on the following in columns: number of the atom in

Table 14.1 Natural amino acids. Names and codes of the twenty naturally occurring amino acids.

Amino acid	3-letter	1-letter	Amino acid	3-letter	1-letter
Alanine	Ala	A	Leucine	Leu	L
Arginine	Arg	R	Lysine	Lys	K
Asparagine	Asn	N	Methyonine	Met	M
Aspartic acid	Asp	D	Phenylalanine	Phe	F
Cysteine	Cys	C	Proline	Pro	P
Glutamic acid	Glu	E	Serine	Ser	S
Glutamine	Gln	Q	Threonine	Thr	T
Glycine	Gly	G	Tryptophan	Trp	W
Histidine	His	H	Tyrosine	Tyr	Y
Isoleucine	Ile	I	Valine	Val	V

Fig. 14.2
Protein backbone. Representation of
the three-dimensional structure of the
backbone of a folded protein.

```
HEADER    OXYGEN STORAGE                          06-JAN-95   1ASH
TITLE     THE STRUCTURE OF ASCARIS HEMOGLOBIN DOMAIN I AT 2.2
TITLE    2 ANGSTROMS RESOLUTION: MOLECULAR FEATURES OF OXYGEN AVIDITY
COMPND    MOL_ID: 1;
COMPND   2 MOLECULE: HEMOGLOBIN (OXY);
COMPND   3 CHAIN: A;
COMPND   4 ENGINEERED: YES
SOURCE    MOL_ID: 1;
SOURCE   2 ORGANISM_SCIENTIFIC: ASCARIS SUUM;
SOURCE   3 ORGANISM_COMMON: PIG ROUNDWORM;
SOURCE   4 ORGANISM_TAXID: 6253;
SOURCE   5 EXPRESSION_SYSTEM_PLASMID: PET-8C
KEYWDS    OXYGEN STORAGE
EXPDTA    X-RAY DIFFRACTION
AUTHOR    J.YANG,F.S.MATHEWS,A.P.KLOEK,D.E.GOLDBERG
REVDAT   2   24-FEB-09 1ASH    1          VERSN
REVDAT   1   27-FEB-95 1ASH    0
JRNL        AUTH   J.YANG,A.P.KLOEK,D.E.GOLDBERG,F.S.MATHEWS
JRNL        TITL   THE STRUCTURE OF ASCARIS HEMOGLOBIN DOMAIN I AT
JRNL        TITL 2 2.2 A RESOLUTION: MOLECULAR FEATURES OF OXYGEN
JRNL        TITL 3 AVIDITY.
JRNL        REF    PROC.NATL.ACAD.SCI.USA        V.  92  4224 1995
JRNL        REFN                   ISSN 0027-8424
JRNL        PMID   7753786
JRNL        DOI    10.1073/PNAS.92.10.4224
```

Fig. 14.3
Heading of PDB files. Section of the heading of the PDB file for the protein with PDB code 1ash.

the sequence, label for the atom, name of the residue, one-letter code for the
residue, number of residues, orthogonal coordinates x, y and z for the corresponding atom, occupancy, temperature factor, element symbol, and charge if
it exists. In Figure 14.4 an example is given for the first two residues of the protein with PDB code 1ash, where we have highlighted the rows corresponding
to the beta-carbons, which are denoted by CB in the third column.

ATOM	1	N	ALA	A	0	11.081	1.847	9.557	1.00	64.40	N
ATOM	2	CA	ALA	A	0	10.369	0.997	10.519	1.00	60.28	C
ATOM	3	C	ALA	A	0	9.181	0.318	9.833	1.00	56.39	C
ATOM	4	O	ALA	A	0	9.344	-0.374	8.829	1.00	52.67	O
ATOM	5	N	ASN	A	1	7.992	0.669	10.335	1.00	55.08	N
ATOM	6	CA	ASN	A	1	6.691	0.239	9.830	1.00	53.00	C
ATOM	7	C	ASN	A	1	6.406	0.688	8.391	1.00	47.79	C
ATOM	8	O	ASN	A	1	5.757	-0.071	7.678	1.00	46.55	O
ATOM	9	CB	ASN	A	1	5.615	0.752	10.808	1.00	60.06	C
ATOM	10	CG	ASN	A	1	4.159	0.416	10.450	1.00	65.81	C
ATOM	11	OD1	ASN	A	1	3.751	-0.731	10.250	1.00	68.33	O
ATOM	12	ND2	ASN	A	1	3.303	1.433	10.367	1.00	68.59	N
ATOM	13	N	LYS	A	2	6.837	1.833	7.838	1.00	43.12	N
ATOM	14	CA	LYS	A	2	6.677	1.983	6.389	1.00	38.31	C
ATOM	15	C	LYS	A	2	7.612	1.040	5.632	1.00	31.24	C
ATOM	16	O	LYS	A	2	7.173	0.454	4.658	1.00	35.55	O
ATOM	17	CB	LYS	A	2	6.951	3.406	5.907	1.00	43.90	C
ATOM	18	CG	LYS	A	2	6.088	3.566	4.640	1.00	50.06	C
ATOM	19	CD	LYS	A	2	6.084	4.892	3.921	1.00	54.79	C
ATOM	20	CE	LYS	A	2	7.505	5.186	3.441	1.00	59.95	C
ATOM	21	NZ	LYS	A	2	7.488	6.172	2.371	1.00	65.93	N

Fig. 14.4
Atomic coordinates in a PDB file. Section of the PDB file for the protein with PDB code 1ash, with the coordinates of every atom in the protein.

14.2 Protein residue networks

A consequence of the 3-D folding of a protein is that some amino acids are close to each other even though they can be far separated in the linear sequence. As a result, the 3-D structure of a protein contains a large number of non-covalent interactions between amino acids. These interactions can be of different natures, such as electrostatic, hydrophobic, van der Waals forces, ionic interactions, hydrogen bonds, and so on (Daune, 1999). In Figure 14.5

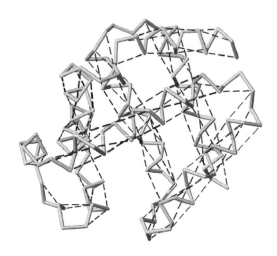

Fig. 14.5
Hydrogen bonds in a protein.
Hydrogen bonds (dotted lines) between amino acids in the protein represented in Figure 14.2.

we illustrate, as dotted lines, the hydrogen bonds between amino acids in a protein as an example of non-covalent interactions.

Non-covalent interactions are very weak in comparison with covalent bonds. In fact, they decay very quickly with the distance separating the interacting atoms. For instance, the energy of charge–charge interactions decays with $1/r$, where r is the distance separating the interacting species. However, dispersion and van der Waals forces decay with $1/r^6$ and $1/r^{12}$, respectively. Therefore, although the distance at which a pair of residues in a protein interact changes according to the type of interaction taken into account, it is almost certain that such interactions do not exist for residues separated at distances larger than 10 Å (Daune, 1999). The relevance of these observations for network representation of proteins is fundamental. This means, for instance, that we can select a certain cut-off distance r_C and consider that any pair of residues i and j separated at a distance $r_{ij} \leq r_C$ have some kind of non-covalent interaction. Consequently, we can consider that they are connected in our network representation of this protein. However, before proceeding with the construction of these networks, we still need to clarify how we can define the distance between two residues. An amino acid (residue) is formed by several atoms, and we can define the distance between both residues in several different ways. Many studies have considered, for example, the distances between C_α of both residues. These methods, however, do not take into account the proximity effects between atoms in the side chains of the residues. A step forward was provided by Miyazawa and Jernigan (1996), who considered the distances between the centres of residues. This method takes into account the proximity effects between side chains, but is more time-consuming than those based on C_α. An intermediate compromise has been proposed by considering the distances between C_β of the residues, while C_α is considered for glycine. We then have the necessary ingredients to transform the information contained in the 3-D structure of a protein into a residue network.

The nodes of the network represent the amino acid residues of the protein, centred at their C_β atoms—with the exception of glycine, for which C_α is used. In order to connect the nodes of the network we consider a cut-off radius r_C, which represents an upper limit for the separation between two residues in contact. The distance between two residues is measured by taking the distance between C_β atoms of both residues. The distance between residues i and j is represented by

$$r_{ij} = \sqrt{(x_i - x_j)^2 + (y_i - y_j)^2 + (z_i - z_j)^2} \qquad (14.1)$$

where x_i, y_i, z_i are the coordinates for the C_β of the residue i. For instance, for the residues represented in the PDB file previously shown, we have a distance of 5.73 Å.

Then, when the inter-residue distance is equal to or less than r_c, both residues are considered to be interacting. In this case, the corresponding nodes in the residue network are connected. The elements of adjacency matrix of the residue network are obtained by

$$A_{ij} = \begin{cases} H\left(r_C - r_{ij}\right) & i \neq j \\ 0 & i = j, \end{cases}$$

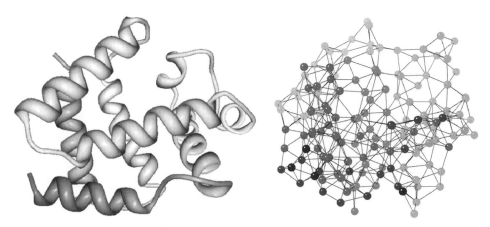

Fig. 14.6
Protein residue network. A protein residue network (right) obtained for the protein with PDB code 1ASH (left).

where $H(q > 0) = 1$ and $H(q \leq 0) = 0$. Then, a protein is represented by the network $G = (V, E)$, where V represents the set of amino acids centred at their beta-carbons, and E represents the set of interactions between them. Note that the links here represent both covalent and non-covalent interactions in the protein.

The remaining topic in building residue networks is the selection of the cut-off distance r_C. Several values have been proposed and used in the literature, ranging from 3.8 Å to 9.0 Å, and their influence of certain topological properties of the networks obtained have been analysed (Brinda and Vishveshwara, 2005). While the first allows only strong non-covalent interactions between residues, such as hydrogen bonds, the second is very permissive, allowing all kinds of interaction, but probably including amino acids which are not physically interacting between them. In a recent study, da Silveira et al. (2009) have proposed that the cut-off value at about 7.0 Å is the most appropriate one. At this distance, all contacts between amino acids are complete and legitimate. In Figure 14.6 we illustrate the residue network for the protein with PDB code 1ash, built by using a cut-off value of 7.0 Å.

14.3 Small-worldness of residue networks

One of the most investigated topological features of protein residue networks is their small-world characteristics. Vendruscolo *et al.* (2002) studied 978 representative proteins structures from the PDB having sizes between 50 and 1021 amino acids. They found that the average value in the distribution for the average shortest path length L is 4.1, and that for the Watts–Strogatz clustering coefficient C is 0.58. These results indicate that proteins are between random and circulant networks (see Chapter 12) when plotted in an L–C plot, as illustrated in Figure 14.7. According to the Kolmogorov–Smirnov test, the probability of observing these differences by chance is close to zero. However, Vendruscolo *et al.* (2002) do not find significant differences in L and C between homopolymers, clusters, and proteins, which according to them

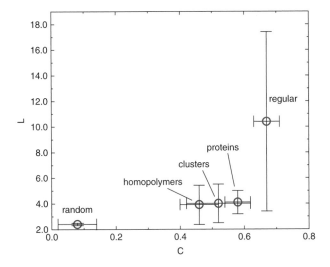

Fig. 14.7
Small-worldness of protein residue networks. Representation of protein residue networks in the clustering-average path length space. From Vendruscolo, M., Dokholyan, N. V., Paci, E., and Karplus, M. (2002). Small-world view of the amino acids that play a key role in protein folding. *Phys. Rev. E* **65**, 061910. (Reproduced with the permission of the American Physical Society, ©2002.)

may be due to the fact that somewhat different energy functions were used to model the various systems.

Similar results were obtained by Bagler and Sinha (2005) by using a cut-off value of 7.0 Å for eighty proteins with 73 to 2,359 amino acids, which were classified in the following structural classes according to their secondary structures: α, β, $\alpha + \beta$ and α/β.[1] In this case the average values of l and c for these proteins is 6.88 and 0.553, respectively. These values are significantly different from those for comparable random graphs, which were 2.791 and 0.031, respectively, as well as for those of comparable circulant networks, which were 29 and 0.643, respectively. As shown in Figure 14.8a, these results are independent of the structural classes of proteins. A closer look at the $L - C$ plot of α and β proteins indicates that there are statistically significant differences in the clustering coefficient of both types of protein, which have mean clustering of 0.588 and 0.538 respectively (see Figure 14.5b). This is mainly due to the fact that the amino acids are more densely packed in helical structures than in the flat β sheets. However, those classes composed by mixtures of both kind of structure, $\alpha + \beta$ and α/β, do not show any significant differences in their clustering coefficients.

A different kind of analysis was carried out by Greene and Higman (2003), who built residue networks by distinguishing two types of interaction between residues. Accordingly, they define short-range interactions as those in which the interacting residues are relatively close in the sequence of a protein and consequently in their 3-D structures. On the other hand, long-range interactions are those occurring between amino acids which are far in the protein sequence but close in the protein 3-D structure. They determined that the clustering coefficient for residue networks having both short-range and long-range interactions is considerably larger than those for networks when only long-range

---

[1] The so-called α and β proteins are those having mainly helices and sheets in their structures, respectively. $\alpha + \beta$ and α/β proteins have combinations of both types of secondary structure.

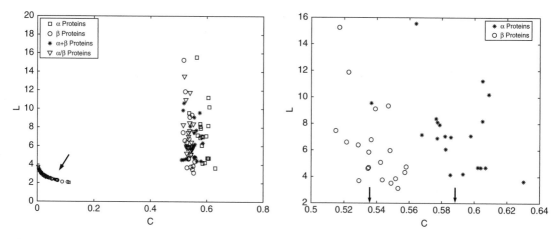

Fig. 14.8

'Small-worldness' and protein classes. Plot of the average path length (L) versus the average Watts–Strogatz clustering coefficient for proteins in the four structural classes (left), and a closer view of the same for proteins in classes α and β only. From Bagler, G., and Sinha, S. (2005). Network properties of protein structures. *Physica A* **346**, 27–33. (Reproduced with the permission of Elsevier, ©2007.)

interactions are considered, as illustrated in Figure 14.9. We will return to this point later in this chapter.

According to Bartoli et al. (2007), 'the characteristic path length and clustering coefficient are not useful quantities for "protein fingerprinting",' because they can be reproduced by using random networks in which constraints similar to those induced by the backbone connectivity are imposed. It has been claimed that these results explain why the networks generated by using only long-range interactions are indistinguishable from random graphs (see Figure 14.9). The procedure followed by Bartoli et al. (2007) can be summarized as follows:

i) Assign 1s to the first two diagonals (up and down the main diagonal) of the adjacency matrix in order to define the backbone contacts, which are equal to any protein folding type.

Fig. 14.9

'Small-worldness' plot of proteins. Plot of the average path length (L) versus the average Watts–Strogatz clustering coefficient for proteins studied by Greene and Higman (2003). The triangles and circles at extreme left represent random networks with the same number of nodes and average connectivity, and the triangles at inner left and circles at right represent real proteins. Triangles depict long-range interaction networks, and circles depict long-range and short-range networks. From Greene, L. H., and Higman, V. A. (2003). Uncovering network systems within protein structures, *J. Mol. Biol.* **334**, 781–91. (Reproduced with the permission of Elsevier, ©2003.)

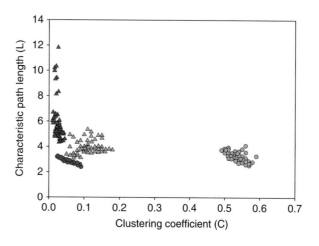

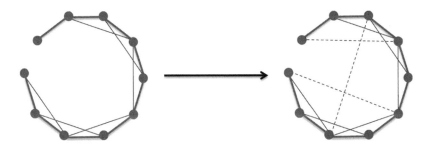

Fig. 14.10
Constrained random residue
networks. Schematic representation of
the process developed by Bartoli
et al., (2007) for the construction
of random protein networks.

ii) Randomly select a pair of residues i and j with a probability that decreases linearly with the distance separating these residues in the protein sequence.

iii) Assign 1s to the entries of the adjacency matrix corresponding to all nine residue pairs generated by the Cartesian product of $\{i - 1, i, i + 1\} \times \{j - 1, j, j, +1\}$.

iv) Iterate the last procedure until the number of links in the random graph is close to those of the real protein.

The process is schematically represented by Figure 14.10.

Although it is evident that the second step of this algorithm introduces randomness into the definition of the residue network, it includes some deterministic elements which should be taken into account. The first step in this algorithm simulates the generation of the secondary structure of a protein. Here this process is not random but deterministic. The secondary structure of the protein is known to depend on the amino acid sequence, which is the basis of many methods for building the secondary structure of proteins. Therefore, this step implicitly assumes that we also have a known amino acid sequence. The third step of the algorithm assumes that the probability that a pair of residues interact through non-covalent long-range interactions decays linearly with their separation in the primary sequence. It is known that the entire 3-D structure of a protein depends very much on the primary and secondary structures of the protein. Therefore, because the method assumes that the sequence and the secondary structure of the protein are determined, it is not a completely random process of linking pairs of nodes selected according to their separation in the sequence. The message we want to transmit here is the following. If many physical constraints are imposed on the process used to generate a network, then very little is left to the random part of the process. Consequently, because of these physical constraints the final network obtained with this process might appear like the real one, which could be primarily responsible for the structure of the real networks. In other words, Nature, in its non-randomness, is being imitated.

14.4 Degree distributions of residue networks

Another topic that has been widely studied for residue networks is their degree distribution. By studying the whole residue networks, Atilgan et al. (2004) found Gaussian-like degree distribution when the networks were built with a cut-off value of 7 Å. These results were obtained for residues located at the

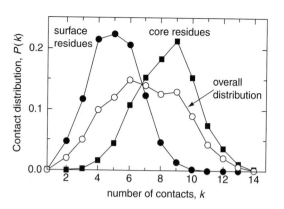

Fig. 14.11

Degree distribution in protein residue networks. Distribution of the number of contacts (degree) in residue networks as well as their core and surfaces. From Atilgan, A. R. et al. (2004). Small-world communication of residues and significance for protein dynamics, *J. Biophys. J.* **86**, 85–91. (Reproduced with the permission of Elsevier, ©2004.)

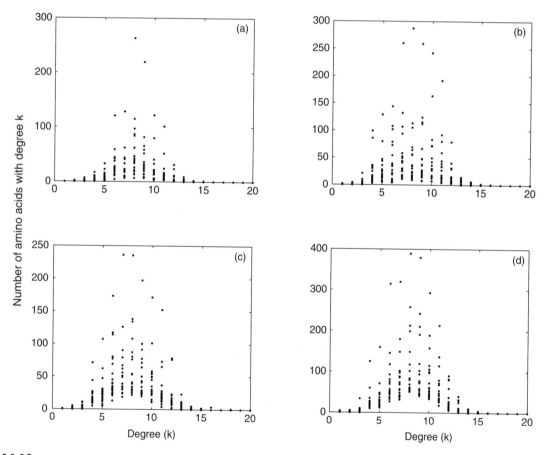

Fig. 14.12

Degree distribution and protein classes. Degree distribution of α (a), β (b), $\alpha + \beta$ (c), and α/β. From Bagler, G., and Sinha, S. (2005). Network properties of protein structures. *Physica A* **346**, 27–33. (Reproduced with the permission of Elsevier, ©2005.)

surface as well as in the core of the protein, as can be seen in Figure 14.11. Bagler and Sinha (2005) and Kundu (2005) also observed bell-shaped distributions for the protein residue networks which they studied independently of their structural classes, as can be seen in Figure 14.12.

The analysis of the degree distribution for the networks constructed using short-range and long-range interactions, as proposed by Greene and Higman (2003), displays some interesting results. First of all, when both short-range and long-range interactions are considered in the network, the degree distributions for nine protein folds display bell-shaped, Poisson-like distributions. However, if only long-range interactions are considered, the degree distributions of the same nine representative folds display typical fat-tailed shapes, as can be seen in Figure 14.13. According to Greene and Higman (2003), although these distributions have an underlying scale-free behaviour they are dominated by an exponential term.

One of the principal consequences of this observation is that the residue networks based on long-range interactions resemble other networks in their resilience to random and targeted attacks. That is, the removal of the most highly connected (in long-range terms) amino acids has a dramatic effect on the integrity of the protein structure in contrast to a random removal of residues, which practically does not affect the protein structure. These findings allow the design of new molecules that can interact with such 'hub' residues in order to affect protein structure and function.

14.5 Centrality in residue networks

The analysis of node centrality of residue networks has important implications for understanding protein structures and functions. Alves and Martinez (2007) studied 160 protein structures represented as residue networks and representing structural classes of α, β, $\alpha + \beta$, and α/β proteins. The degree of a given amino acid is then calculated for all 160 structures studied, and the average is taken as a characteristic of that residue. The four most connected residues are valine, isoleucine, leucine, and phenylalanine, and the least connected are glutamine, glutamic acid, aspartic acid, lysine, and arginine. The plot in Figure 14.14 illustrates these results. It is straightforward to realise that the most connected amino acids are among the most hydrophobic ones, while those poorly connected are among the most polar ones. In fact, the four amino acids with highest degree have hydrophobic side chains, while among the least connected ones, glutamic acid and aspartic acid are polar and negatively charged, lysine and arginine are polar and positively charged, and glutamine is polar and uncharged (see Figure 14.14). In fact, Alves and Martinez (2007) found a strong correlation between these two parameters: degree centrality and hydrophobicity.

In another analysis, Atilgan et al. (2004) found that the average path length of a given residue correlates very well with the thermal fluctuation of this residue in the protein, as illustrated in Figure 14.15. We recall that the closeness centrality is related to the inverse of the sum of all distances from one node to the rest of nodes in the network. Then, the smaller the average path length for a given residue, the larger its closeness centrality. The authors claimed that

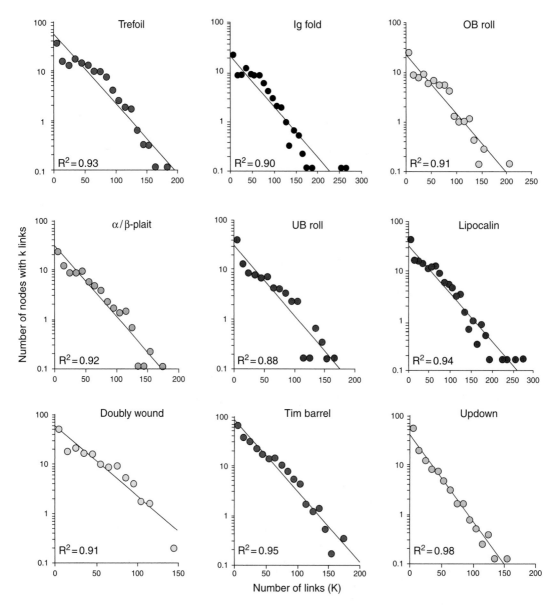

Fig. 14.13

Degree distribution in long-range residue networks. Degree distribution for residue networks built from long-range interactions only for proteins having different characteristics folds. The plots are on a semi-log scale, and the squared correlation coefficients for the best linear fits are displayed. From Greene, L. H., and Higman, V. A. (2003). Uncovering network systems within protein structures. *J. Mol. Biol.* **334**, 781–91. (Reproduced with the permission of Elsevier, ©2003.)

there is 'an intriguing balance between these two measurable.' However, the results are quite predictable from the following perspective. Residues located at the core of the protein are closer to the centre of gravity of the protein. They will therefore have a relatively smaller average shortest path length than those which are on the surface. The last ones are far away from those which are

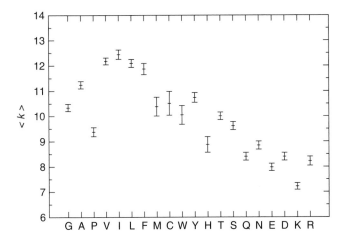

Fig. 14.14
**Average degree of amino acids in
proteins**. Representation of the average
degree and their standard deviations for
each of the twenty natural amino acids
occurring in proteins. From Alves, N. A.,
and Martinez, A. S. (2007). Inferring
topological features of proteins from
amino acid residue networks. *Physica A*
375, 336–344. (Reproduced with the
permission of Elsevier, ©2007.)

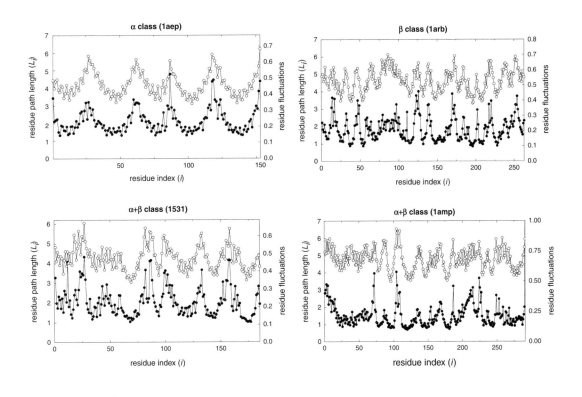

Fig. 14.15
Path length and residue fluctuations. Representation of the average shortest path length for every amino acid (empty circles) and residue
thermal fluctuations (full circles) revealed from X-ray crystallography of four proteins belonging to different structural classes. From Atilgan,
A. R. et al. (2004). Small-world communication of residues and significance for protein dynamics, *J. Biophys. J.* **86**, 85–91. (Reproduced with
the permission of Elsevier, ©2004.)

located on the surface in an opposite hemisphere. Residues in the core are very packed, and will be very little affected by thermal fluctuations. However, those at the surface are allowed to oscillate more freely and are then more affected by thermal factors. Consequently, a linear correlation should be expected between average shortest path per residue and its thermal fluctuations.

The importance of these results is evident from the fact that in a protein, the regions having large oscillations are usually more flexible and functionally important. We can therefore use the values of the vibrational node displacements obtained in Chapter 7 as a more appropriate measure for quantifying these oscillations. We recall that the node displacement has been defined as

$$(\Delta x_i)^2 = \frac{1}{\beta k} (\mathbf{L}^+)_{ii} \tag{14.2}$$

where $\mathbf{L}^+$ is the Moore–Penrose generalised inverse of the Laplacian matrix (see Chapter 7). These atomic displacements have been used previously by Bahar et al. (1997) to describe thermal fluctuations in proteins measured experimentally by the B-factors in X-ray diffraction experiments. These authors used the 3-D structure of the protein to make their calculations. Here, however, we instead use the protein residue network to study seven proteins with structures that have been revealed by X-ray crystallography (http://www.pdb.org). In Table 14.2 we give the correlation coefficients between the experimental B-factors and the topological atomic displacements.

In Chapter 7 we discussed a new relationship between the node displacement Δx_i and the resistance distance R_i for a given atom, which allows a new interpretation of the correlations found between the atomic displacement and the B-factor. We recall that this relationship is as follows (see Chapter 7):

$$R_i = n \left[(\Delta x_i)^2 + \overline{(\Delta x)^2} \right] \tag{14.3}$$

Table 14.2 Topological atomic displacement and thermal fluctuations. Correlation coefficients (r) between the topological atomic displacement (equation (15.2)), and the experimental B-factors (thermal fluctuations) of residues in eleven proteins. The PDB code, number of residues, and atomic resolution at which the protein structure was resolved, are also included.

PDB	Residues	Resol. (Å^2)	r
2pii	112	1.90	0.52
1cpq	129	1.72	0.60
1vid	213	2.00	0.68
1akz	233	1.57	0.65
1fnc	296	2.00	0.56
1tca	317	1.55	0.74
1pmi	440	1.70	0.64

The resistance distance is a graph metric, which is identical to the shortest path distance in the case of acyclic networks only. The relationships obtained by Atilgan et al. (2004), shown in Figure 14.15, can therefore be understood as an approximation to the expected correlation between atomic displacements, which are related to the resistance distance, and the thermal fluctuations in proteins. The resistance distance allows for a better understanding of what is happening in a protein, using the following reasoning. If a cycle exists in the path between any two atoms in a molecule, the resistance distance is smaller than in the case when there is only a single path. In this case the transmission of any oscillation/vibration from one atom to another is attenuated along every path. Consequently, the most rigid atoms in a molecule are those taking part in a large number of cycles. On the other hand, the most flexible atoms are those taking part in the least cyclic structures in the molecule. As a consequence, it is expected that in a protein represented by its residue network, as described previously, the most flexible residues are those on the surface of the protein, as they take part is fewer cycles than those in the core, where more cliquishness is expected due to the larger number of non-covalent interactions.

In Figure 14.16 (top) we reproduce the plot of Δx_i versus the experimental B-factors for the protein with PDB code 1tca. It can be seen that the regions of the maximal flexibility and the maximal rigidity identified by the topological atomic displacements perfectly match those identified by the experimental values of the B-factors (Atilgan et al., 2004; Atilgan and Baysal, 2004; Estrada and Hatano, 2010b). As can be seen in Figure 14.16 (bottom), the residues having the largest values of the topological atomic displacements, which are the most flexible ones, are placed on the surface of the protein—most of them in protein loops. The residues with the lowest values of Δx_i are located at the core of the protein, which corresponds to the most rigid region of the protein. This corroborates our previous explanation of why good correlations with the closeness centrality are observed.

The betweenness centrality was used by Vendruscolo *et al.*, (2002) as a measure for comparing the native state of proteins and their transition state ensembles, which were obtained by a Monte Carlo sampling procedure. They have found that certain 'key residues'—which are critical for forming the nucleus that encodes the overall native structure—can be identified by using the betweenness centrality. In fact, in the transition states there are only two to four residues displaying large values of the betweenness, and in general they correspond to those key residues (see Figure 14.17).

One important structural characteristic of proteins is the existence of binding sites. These are specific regions of the protein which allow other molecules to attach the protein by forming molecular complexes responsible for certain functions in the cell. In particular, enzymes have 'active sites' in which such interactions are responsible for the catalytic activity of these proteins. The identification of these active sites is a very difficult task—particularly if only the 3-D structure of the protein is used. Once the binding sites have been identified, networks based on their similarity can be built (Zhang and Grigorov, 2006), allowing the construction of novel binding site classifications and for new molecular design strategies. Amitai et al. (2004) have used a network

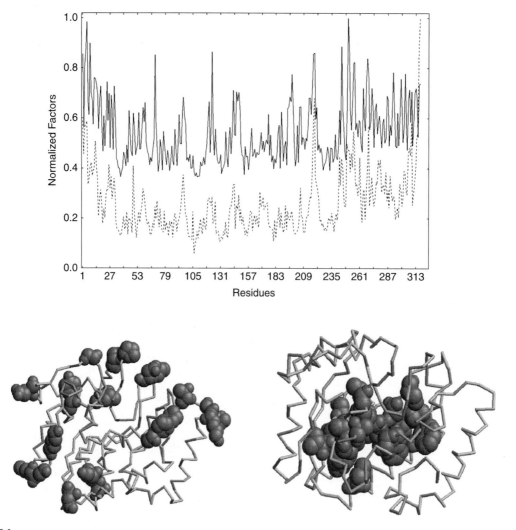

Fig. 14.16

Topological atomic displacement, thermal fluctuations, and protein packing. Profiles (top) of the experimental B-factors (solid line) and topological atomic displacements (broken line) for 1tca. (Bottom) Illustration of the residues with the maximal (left) and minimal (right) topological atomic displacements in the protein 1tca, which clearly correspond to protein regions with minimal and maximal packing respectively.

representation of 178 protein chains of enzymes with known structure and uniformly defined active sites. They found that those residues which form the binding sites of these proteins display significantly larger closeness centrality than the remainder of the residues, including those in the protein core. These results are illustrated in Figure 14.18.

These results have been confirmed by del Sol et al. (2006) by studying 46 protein families, including twenty-nine enzyme and seventeen non-enzyme families. They carried out protein family alignment to find those amino acids which are conserved in a given family, calculated the closeness centrality for each of these amino acids in the residue network, and selected those having

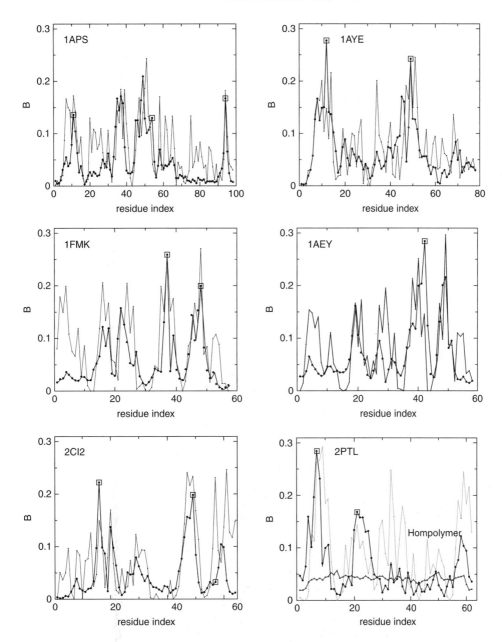

Fig. 14.17

Residue betweenness in proteins. Betweenness centrality for the residues in the transition state (thick line) and in the native states (thin lines) in six proteins. For the protein with PDB code 2ptl, the B profile for a homopolymer of the same length is also displayed. From Vendruscolo, M., Dokholyan, N. V., Paci, E., and Karplus, M. (2002). Small-world view of the amino acids that play a key role in protein folding. *Phys. Rev. E* **65**, 061910. (Reproduced with the permission of the American Physical Society, ©2002.)

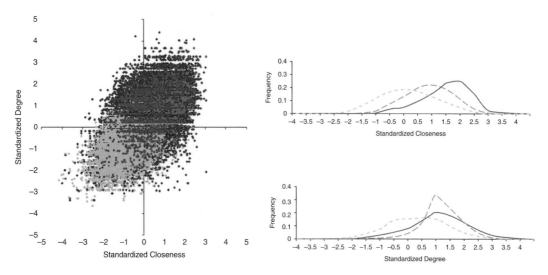

Fig. 14.18

Residue closeness and degree in enzymes. Representation of the standardised closeness centrality for the residues with different expositions to the solvent (left). The distributions of the standardised closeness and degree (right) represent highly exposed residues (dotted lines), core residues (broken lines) and active site residues (solid line). From Amitai, G., Shemesh, A., Sitbon, E., Shklar, M., Netanely, D., Venger, I., and Pietrokovski, S. (2004). Network analysis of protein structures identifies functional residues, *J. Mol. Biol.* **344**, 1135–46. (Reproduced with the permission of Elsevier, ©2004.)

Fig. 14.19

Closeness centrality of residues. Closeness centrality of the residues in the subtilisin DY protease (PDB code 1bh6). The most central residues are those located in the binding sites of this protein, illustrated with a synthetic inhibitor (left) and with a sodium atom (right). From Amitai, G., Shemesh, A., Sitbon, E., Shklar, M., Netanely, D., Venger, I., and Pietrokovski, S. (2004). Network analysis of protein structures identifies functional residues, *J. Mol. Biol.* **344**, 1135–46. (Reproduced with the permission of Elsevier, ©2004.)

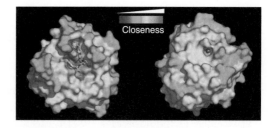

statistically significant closeness centrality ($z - scores \geq 2$) in at least 70% of the structures in such family. These residues are called 'centrally conserved positions'. They found that 80% of these central positions corresponded to active site residues or residues in direct contact with these sites. This percentage increased to 91% for enzyme families, while for non-enzyme it was only 48%. A classification of the binding sites found in these proteins sheds even more light on the capacity of the closeness centrality to identify them. For instance, 70% of the centrally conserved positions are located in catalytic sites of enzymes, which are located mainly in clefts in protein surface, 64% are in binding sites used by hetero-atoms, which are in general deep cavities inside the proteins, and 16% represent protein–protein interfaces, which are more planar configurations. An illustration of two of these types of binding site and the closeness centralities of residues in the subtilisin DY protease, according to Amitai et al. (2004), is shown in Figure 14.19.

14.6 Global topological properties

In this chapter we have analysed some global topological properties of protein residue networks, such as the average path length, clustering coefficient, and degree distributions. Here we study some other properties which have also been studied in the literature, and which can provide insights about the structure of proteins. The first of these properties is the assortativity of residue networks. Alves and Martinez (2007) found that all protein residue networks which they studied display assortativity instead of the characteristic disassortativity found for most of the biological networks. Bagler and Sinha (2007) also found that all protein residue networks studied so far 'exhibit the exceptional topological property of "assortative mixing" that is absent in all other biological and technological networks.' In Chapter 2 we developed a combinatorial expression for the assortativity coefficient in terms of ratios between certain fragments in the network. In Table 14.3 we give the values of these fragments and the assortativity coefficient for the residue networks of ten representative proteins having between 50 and 1,021 amino acids. We recall that the assortativity coefficient can be expressed as

$$r = \frac{|P_2|(|P_{3/2}| + C - |P_{2/1}|)}{3|S_{1,3}| + |P_2|(1 - |P_{2/1}|)} \qquad (14.4)$$

where $|P_k|$ is the number of paths of length k, $|S_{1,3}|$ is the number of star fragments of four nodes, $|P_{r/s}| = |P_r|/|P_s|$, and C is the clustering coefficient defined as the ratio of three times the number of triangles to the number of 2-paths.

As can be seen in the Table 14.3, all protein residue networks are assortative due to the fact that

$$|P_{2/1}| < |P_{3/2}| + C \qquad (14.5)$$

as proved in Chapter 2. More interestingly, in all protein residue networks the ratios $|P_{2/1}|$ and $|P_{3/2}|$ are very similar, which implies that the assortativity of these networks is mainly determined by their clustering coefficients. In

Table 14.3 Assortativity of protein residue networks. Values of the assortativity coefficient (r) for several proteins. The number of residues (n) and the parameters in equation(15.4) are also shown, with the PDB code for each protein.

PDB	n	$|P_{2/1}|$	$|P_{3/2}|$	C	r
1cdr	77	5.455	5.363	0.348	0.290
2aak	150	5.516	5.379	0.329	0.309
1arb	263	6.826	6.742	0.362	0.375
1pedA	351	7.052	7.061	0.360	0.449
1kapP	470	6.282	6.179	0.348	0.353
1occA	514	6.470	6.328	0.352	0.380
1aorA	606	6.998	6.918	0.352	0.417
1oacA	719	5.760	5.670	0.329	0.378
1qba	863	6.389	6.212	0.347	0.292
1bglA	1021	6.013	5.871	0.338	0.329

other words, the assortativity is determined by the fact that the transitivity of interactions between amino acids is very large. This is probably determined by short-range interactions between residues which are close to each other in the primary sequence of the proteins and consequently also in their 3-D structures. That is, in structures such as A–B–C there is a large probability that A and C are separated at a distance less than the cut-off value, producing a triangle in the residue network. We think that if only long-range interactions are considered, as in the case of the study carried out by Greene and Higman (2003), then this assortativity will disappear, because in this case the clustering coefficient is significantly smaller than the combination of both short-range and long-range interactions (see Figure 14.9). Therefore, we think that the assertion of Alves and Martinez (2007)—that 'the positive assortative measured corroborates the marginal stability of proteins'—should be constrained to the case of when only short-range interactions among amino acids are concerned.

Another global property of residue networks that has been investigated is the topological classes to which they belong. In Chapter 9 we studied a method based on the spectral scaling between subgraph and eigenvector centrality to classify complex networks into four universal topological classes. The spectral scaling method has been applied to 595 residue networks built by using a cut-off value of 7 Å between beta-carbons of amino acids. A typical spectral scaling for protein residue networks for the protein with PDB code 1ash is illustrated in Figure 14.20. In this case, all residues deviate negatively from perfect spectral scaling. The value of the mean square error of negative

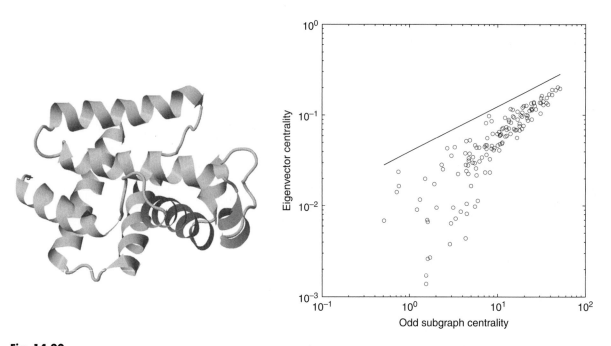

Fig. 14.20

Spectral scaling of a protein residue network. Typical spectral scaling (right) between the eigenvector and subgraph centralities for the protein with PDB code 1ash, whose three-dimensional structure is illustrated (left).

deviations for this protein is $\xi^- = 0.498$, and that for positive deviations is exactly zero. We recall that this type of spectral scaling is typical of networks having structures characterized by the presence of internal holes, for which

$$\Delta \log EC(i) \leq 0 \implies [EC(i)]^2 \sinh(\lambda_1) \leq EE_{odd}(i), \forall i \in V. \qquad (14.6)$$

It was found that 564 out of the 595 non-homologous proteins have ξ^+ exactly equally to zero and $\xi^- \gg 0$. The other thirty-one proteins have small positive deviations from the perfect scaling, having average positive deviations of $\bar{\xi}^+ = 0.0011$ and average negative deviations of $\bar{\xi}^- = 1.550$ (Estrada, 2010b). These results indicate that clearly 95% of all these proteins are in Class II, while 5% are in Class IV according to the classification given in Chapter 9. Therefore, for most of the 595 proteins studied, which have between 54 and 1,021 residues and less than 25% of homology in their sequences, we can model their structures as networks in which several highly connected clusters are separated from each other by forming structural cavities (see Chapter 9).

An analysis of the sizes for the residue networks in Classes II and IV revealed that the average size of protein networks in Class II is 261 residues, which is relatively larger than that for the whole dataset of proteins; that is, 254 residues. However, those protein networks classified in Class IV have an average size of only 126 residues, which is significantly lower than the average size for the whole data set of proteins. In fact, most proteins in Class IV have less than 200 residues (Estrada, 2010b).

This size dependence on the topological classes can be explained by using the following facts about the domain structure of proteins. A domain is a part of the protein which has a compact 3-D structure and can often be independently stable and folded. Most domains found in proteins have between 50 and 150 residues. Consequently, most of the small proteins have only one domain, while larger proteins tend to be combinations of such domains. In a residue network those amino acids which are in the same domain tend to form highly interconnected clusters, while the number of interactions between two different domains in a multi-domain protein is relatively low. Therefore, it is plausible that most of the residue networks in Class II are multi-domain proteins, which explains why practically all residue networks with more than 240 amino acids are clearly in Class II. Those proteins in Class IV are formed by small clusters of secondary structures, such as helices and sheets. These residue networks can then be thought of as being formed by combinations of quasi-cliques and quasi-bipartite structures, which give rise to networks of Class IV, as we saw in Chapter 9.

In Figure 14.21 we show a hole—a chordless cycle—of length 15, which is formed by the residues 27, 30, 33, 40, 43, 59, 60, 62, 64, 67, 71, 92, 95, 96, and 101 in the residue network for the protein with PDB 1ash. That is, the residues form a cycle of length 15, in which no two of them are joined by a link which does not belong to the cycle. A close look at the structure of this protein according to PDBsum reveals that the residues having contacts with the ligand are 30, 33, 40, 43, 44, 60, 64, 67, 68, 71, 95, 96, 101, 103, 108, and 140, which are illustrated in Figure 14.15 (right). This means that there is an overlap

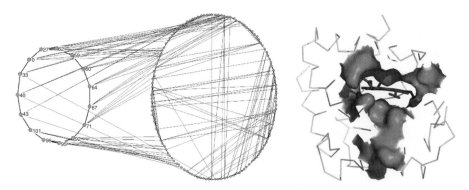

Fig. 14.21

Holes and binding sites in proteins. A hole (chordless cycle) having fourteen nodes (left) in the residue network of the protein with PDB code 1ash. The binding site of this protein, which almost matches the hole found, is illustrated at right (Estrada, 2010b).

of 71% between the list of residues in the binding site and the topological cavity found in the residue network, which illustrates the relationship between topological cavities and potential binding sites in proteins. Unfortunately, the detection of holes in networks is an NP-complete problem, and many problems must be solved before it can be applied as a standard method for detecting cavities in proteins (Nikolopoulos and Palios, 2004; Reitsma and Engel, 2004; Spinrad, 1991). One of them is simply the large number of holes that exist in a given residue network, which makes the analysis of possible binding sites very difficult, if not intractable. Therefore, the use of biological information to reduce the extent of this search should be the best way for the application of these techniques to practical binding site detection algorithms.

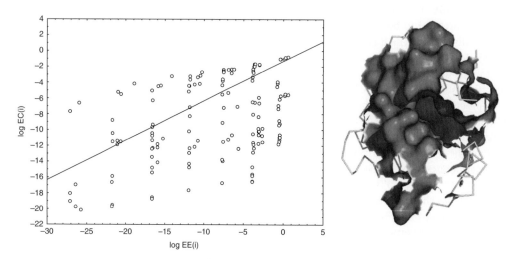

Fig. 14.22

Spectral scaling at high resolution. Spectral scaling (left) for the residue network of the protein with PDB code 1ash, produced by using a small cut-off distance to define the connectivity of the network. In this case the resultant network has large 'rugosities', as illustrated by the very many cavities in the surface of the protein (right).

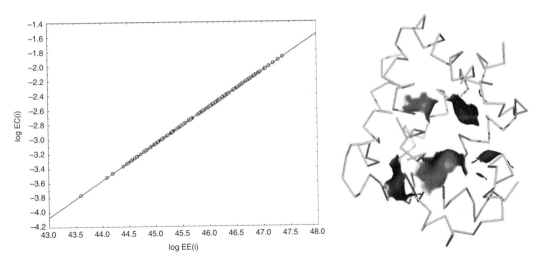

Fig. 14.23

Spectral scaling at low resolution. Spectral scaling (left) for the residue network of the protein with PDB code 1ash, produced by using a very large cut-off distance to define the connectivity of the network. In this case the resultant network has a very smooth surface without small cavities or 'rugosities,' and only very large holes are detectable (right).

Finally, we would like to remark that the topological structural classes admirably reflect the level of coarse graining with which the protein structures have been transferred to the residue network. For instance, the use of a relatively small cut-off value for the non-covalent interaction produces a network which includes only 146 links connecting nearest-neighbour residues in the protein backbone plus forty other inter-residue interactions. This is the equivalent of considering very small probe spheres to explore the protein. That is, if we use a sphere of small radius which is rolling through the surface of the protein, it will be introduced into all small cavities reporting a large rugosity of the protein surface. For instance, in this case the binding cavity previously found for this protein consists of a 53-node hole formed by residues 32–34, 55–58, 60–78, 85–96, 101–103, and 107–118, which together with the spectral scaling (Estrada, 2010b) is illustrated in Figure 14.22.

At the other extreme is the use of a very large cut-off value, such as 15 Å. In this case we are exploring the surface of the protein with a ball of very large radius, which cannot enter the small cavities of the protein and therefore reports a very smooth surface. In other words, the structure of the residue network is similar to a very homogeneous network of Class I, as illustrated in Figure 14.23. In this case only very large cavities in the surface of the protein are detectable, which in this particular case correspond to a hole formed only by few residues, such as the one formed by residues 40, 60, 62, 71, and 96 (see Figure 14.16). Different cut-off values can be useful for different kinds of exploration of protein structures. If we are interested in studying all rugosities of the protein surface, very small cutoff values—$r_C < 5.0$ Å, for example—are recommended, while the use of very large cut-off values is recommended only when very large cavities need to be detected in the protein (Estrada, 2010b).

Protein–protein interaction networks

15

> To fully understand the cellular machinery, simply listing the proteins is not enough—all the interactions between them need to be delineated as well.
>
> Christian von Mering et al. (2003)

15.1   A primer on protein–protein interactions

In Chapter 14 we studied the networks representing the three-dimensional structure of proteins, which are kept in folded states due mainly to the non-covalent interactions between residues (Daune, 1999). However, non-covalent interactions can also take place between proteins, giving rise to stable complexes, which are fundamental to many biological processes (Kleanthous, 2000). These protein–protein interactions (PPIs) are central to the functioning of the immune system, enzyme regulation, release of neurotransmitters, intercellular communication, and programmed cell death, among other functions. An example of these complexes is illustrated in Figure 15.1, in which the rho GTPase-activating protein forms a signalling complex with the GTP-binding protein RhoA. The first protein is a translation product of the ARHGAP1 gene, and the second is a rho-related small GTPase that is a translation product of the RHOA gene.

As we have seen previously, the forces that stabilise these complexes are of non-covalent nature and include van der Waals attraction and repulsion, electrostatic interactions, hydrogen-bond formation, and so on (Daune, 1999). Despite the fact that multimolecular complexes can be formed and some of them have been detected, the most common type of PPI is the formation of one-to-one complexes, which can be written by the following chemical equilibrium:

$$A + Q \rightleftharpoons A\,Q$$

These complexes can be of a different nature, allowing several classifications of the PPIs (Nooren and Thornton, 2003). The simplest one is to consider the nature of the chains involved in the complex, and distinguishing between *homo-* and *hetero-oligomeric complexes*, depending on the involvement of two identical ($A = Q$) or non-identical ($A \neq Q$) chains respectively. On the other hand, PPIs can also be classified by taking into account the stability of the promoters when isolated. Then, in *obligate PPI*, the promoters are not stable as separate structures on their own, and the complex is necessary for stabilising their structures. *Non-obligate complexes* are formed by promoters which exist independently in stable forms. Finally, complexes can be classified according to their lifetime, such that we distinguish between *permanent* and *transient PPIs*. While the first are stable interactions that allow the complex to exist for life, the second is characterized by the eventual association and dissociation of the complex *in vivo*. Therefore, we can have the combinations of these classes to produce obligate homo- and heterodimers, non-obligate homo- and heterodimers, non-obligate permanent heterodimers, and non-obligate transient heterodimers. For instance, the complex illustrated in Figure 15.1 corresponds to a non-obligate heterodimer.

The PPI complex illustrated in Figure 15.1 was determined by using X-ray crystallography, which together with NMR and a range of other spectroscopic and calorimetric techniques is the traditional method for identifying PPIs. These methods are classified as low-throughput biophysical techniques, as they are able to identify and characterize one complex at a time, in contrast with the high-throughput methods, which can study PPIs at a proteome-wide scale. There are two main high-throughput approaches for identifying PPIs: methods based on *yeast two-hybrid*, and methods based on *mass spectrometry*. For an authoritative review of the many different method used for identifying protein–protein interactions, the reader is directed to the work of Uetz et al. (2010).

The yeast two-hybrid (Y2H) method (Causier, 2004; Uetz et al. 2010; Williamson and Sutcliffet, 2010) is based on the use of a transcription factor (TF)—originally the yeast transcription factor GAL4. In the native TF, a DNA-binding domain (BD) is linked to an activation domain (AD), which activates the expression of a reporter gene. Then, the Y2H is based on splitting the TF and fusing a bait encoding protein X to its BD, while a prey-encoding protein Y is fused to its AD. If the proteins X and Y do not interact, it is impossible to re-establish the BD–AD complex that activates the reporter gene. However, if proteins X and Y physically interact with each other, then the BD and AD

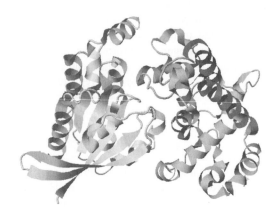

Fig. 15.1

Protein–protein interaction. The three-dimensional structure of the complex formed by the interaction of the Transforming Protein Rhoa (left) and Rho-Gtpase-Activating Protein 1 (right).

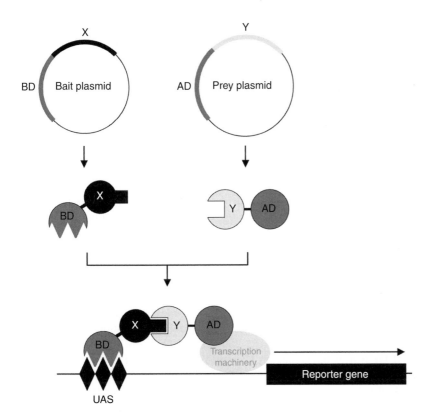

Fig. 15.2
Yeast two-hybrid (Y2H) method. A schematic representation of the Y2H system (explained in the text). From Causier, B. (2004). Studying the interactome with the yeast two-hybrid system and mass spectrometry, *Mass Spectr. Rev.* **23**, 350–67. (Reproduced with the permission of John Wiley and Sons, ©2004.)

are linked together and a functionally active transcription activator complex is re-established. This functionally active TF binds to a region of the DNA in the promoters of the reporter genes known as the 'upstream specific activation sequences' (UAS), which activates their expression. A popular reporter gene is the *E. coli* lacZ, the activation of which can be detected by the blue colour developed by cells that have growth in the presence of the reactant X-Gal. The Y2H method is schematically illustrated in Figure 15.2. The following PPI networks have been studied by using Y2H: Kaposi sarcoma–associated herpesvirus (KSHV) and varicella-zoster virus (VZV), vaccinia virus, hepatitis C virus, *Archaeoglobus fulgidus*, *Bacillus subtilis*, *Escherichia coli*, *Helicobacter pylori*, *Saccharomyces cerevisiae*, *Plasmodium falciparum*, *Caenorhabditis elegans*, *Drosophila melanogaster*, and *Homo sapiens* (see Appendix). There are many variations of the Y2H method, and the reader is directed to the specialised literature for details.

The main characteristics of the Y2H method for detecting PPIs can be summarized as follows:

1) They mainly detect direct binary interactions—particularly those which are transient,
2) They are not suitable for detecting interactions between proteins that cannot be reconstructed in the nucleus, such as membrane proteins.
3) The PPIs detected by this method are not necessarily taking place physiologically, and it then detects a high proportion of 'false positives', which are interactions unlikely to happen *in vivo*.

On the other hand, the application of mass spectrometry (MS) to the detection of PPIs is based on the possibilities that this technique brings for the analysis of the composition of partially purified protein complexes together with a control purification, in which the complex of interest is not enriched (Causier, 2004; Uetz et al., 2010). The method proceeds by the bait expression of the protein in yeast, the subsequent affinity purification of the protein complex, and finally the analysis of the bound proteins. These approaches allow the characterization of protein complexes containing three or more proteins, which is a primary difference of the Y2H method. In the leading two published studies using MS for identifying PPIs, the methods start by creating bait proteins based on attaching tags to many different proteins. Furthermore, DNA molecules encoding these bait proteins are then introduced into yeast cells in order to express the modified proteins in the cells, by which proteins form physiologically relevant complexes with other proteins. Then, each bait protein, together with all proteins in the complex formed, is pulled out of the cell. The protein complexes are then subjected to single immunoaffinity purification. The components of each purified complex are then resolved by utilising one-dimensional denaturing polyacrylamide gel electrophoresis (PAGE). Finally, protein bands are excised from stained gels, and are analysed by using MS. Two approaches which have been used for the identification of the proteins are MALDI-TOF (Gavin et al., 2002) and LC–MS/MS (Ho et al., 2002). MALDI-TOF is the acronym of Matrix-Assisted Laser Desorption/Ionization (MALDI) combined with Time of Flight (TOF) mass spectrometry. The second method is based on the combination of Liquid Chromatography with MS. In Figure 15.3 we illustrate the general procedure for MS-based PPI detection. These methods have been applied to the detection of PPI complexes in yeast (Causier, 2004).

15.2 Global structure of PPI networks

Several PPI networks are available in the literature, and many of their structural properties have been studied (Zhang, 2009). Here we use a collection of these networks to show some of their general structural features and compare them with others found in the literature. The networks considered here (see Appendix) are those of Kaposi sarcoma–associated herpesvirus (KSHV), varicella-zoster virus (VZV), *Archaeoglobus fulgidus*, *Bacillus subtilis*, *Escherichia coli* (network of validated protein complexes), *Helicobacter pylori*, *Saccharomyces cerevisiae*, *Plasmodium falciparum*, *Caenorhabditis elegans*, *Drosophila melanogaster* (higher confidence map), and *Homo sapiens*. Some global structural properties of the main connected components of these networks are presented in Table 15.1.

As can be seen in the Table, all PPI networks are sparse, having densities which are well below 1; that is, $\delta(G) \ll 1$. The median of the average shortest path length for these eleven PPI networks is 4.03, indicating that all of them are 'small-worlds' due to the fact that $\bar{l} \sim \ln n$. The median of the diameter is 9—which is relatively small in comparison with their size. In addition, all these networks display relatively large clustering. For instance, the median of average Watts–Strogatz clustering coefficients is 0.127. According to equation

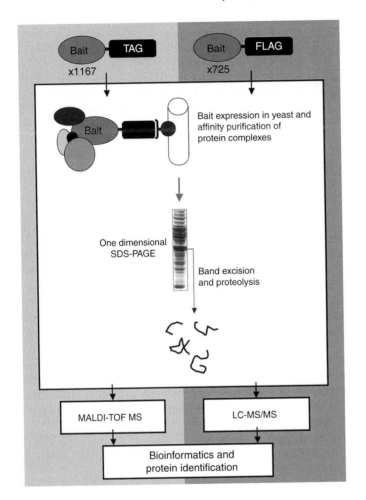

Fig. 15.3
Protein complexes from mass spectrometry (MS). A schematic representation of the large-scale analysis, using MS, of yeast protein complexes. From Causier, B. (2004). Studying the interactome with the yeast two-hybrid system and mass spectrometry, *Mass Spectr. Rev.* **23**, 350–67. (Reproduced with the permission of John Wiley and Sons, ©2004.)

Table 15.1 **Global properties of PPI networks**. Values of the number of nodes n, density $\delta(G)$, maximum degree k_{max}, average path length $\bar{l}$, diameter $diam\ (G)$, and average Watt–Strogatz clustering coefficient $\bar{C}$, for the main connected component of several PPI networks.

No.	Network	n	$\delta(G)$	k_{max}	$\bar{l}$	$diam(G)$	$\bar{C}$
1	*A. fulgidus*	32	0.0746	10	3.60	7	0.127
2	KSHV	50	0.0963	16	2.84	7	0.147
3	VZV	53	0.1117	34	2.36	5	0.397
4	*B. subtilis*	84	0.0299	17	4.03	8	0.080
5	*P. falciparum*	229	0.0231	35	3.38	8	0.184
6	*E. coli*	230	0.0264	36	3.78	11	0.315
7	*C. elegans*	314	0.0074	28	6.69	15	0.063
8	*H. pylori*	710	0.0055	55	4.15	9	0.025
9	*S. cerevisiae*	2224	0.0027	65	4.38	11	0.201
10	*H. sapiens*	2783	0.0016	129	4.84	13	0.109
11	*D. melanogaster*	3039	0.0008	40	9.43	27	0.046

(12.7), the density of an Erdös–Rényi random network is equal to the clustering coefficient, which means that the clustering of these PPI networks is far from being similar to those of this type of random graph.

A comparison of the Watts–Strogatz clustering coefficient with the Newman's transitivity index for these PPI networks shows that in general they are linearly correlated, with a Pearson correlation coefficient of 0.91. However, the non-parametric Kendall tau coefficient is 0.70, indicating that both indices produce different ranking of the global clustering. We have seen in Chapter 4 that the Watts–Strogatz index is better understood as a characterisation of the clustering at a local level, while that of Newman produces a better characterization of the global clustering of the network.

One of the most studied properties of PPI networks is their degree distributions (Li et al., 2004; 2006; Krogan et al., 2006; Stumpf and Ingram, 2005; Thomas et al., 2003; Uetz et al., 2006; Wuchty, 2001; Wuchty et al., 2003). The networks of KSHV and VZV were both found to display scale-free characteristics, with degree distributions of the type $p(k) \sim k^{-\gamma}$, and power-law coefficients γ of 0.95 and 0.78 respectively. The networks of validated interactions in *E. coli*, and those of *C. elegans*, *S. cerevisiae*, and *H. sapiens*, were also reported as scale-free, following power-law degree distributions with different exponents. The confidence PPI network of *D. melanogaster*, however, was reported as displaying a power-law degree distribution with exponential tail; that is, $p(k) \sim k^{-\alpha}e^{-\beta k}$, with $\alpha = 1.26$ and $\beta = 0.27$. As discussed briefly in Chapter 2, Stumpf and Ingram (2005) have reanalysed the degree distributions for several of these PPI networks. They tested six different

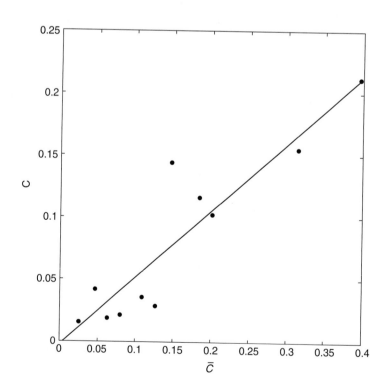

Fig. 15.4

Clustering in PPI networks. Linear correlation between the average Watts–Strogatz clustering coefficient and the Newman transitivity index for the PPI networks given in Table 15.1.

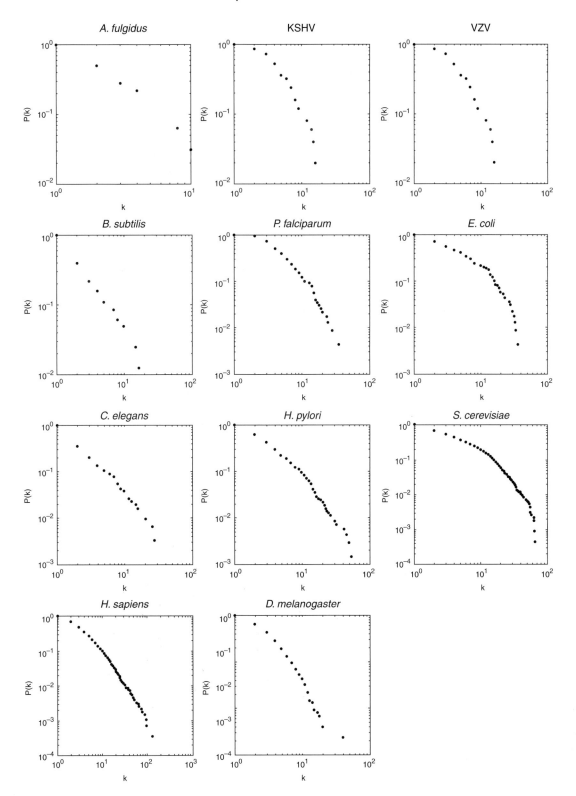

Fig. 15.5
Degree distribution in PPI networks. Cumulative degree distributions for the PPI networks given in Table 15.1.

types of distribution among approximately 200 existing ones. The distributions tested were Poisson, exponential, Gamma, power-law, log-normal, and stretched exponential. They determined that by using the Akaike information criterion (AIC), the PPI networks of *H. pylori*, *S. cerevisiae*, and *E. coli* better fit to log-normal distribution of the type

$$p(k) \sim Ce^{-\ln((k-\theta)/m)^2/(2\sigma^2)} / \left[(k-\theta)\sigma\sqrt{2\pi} \right] \qquad (15.1)$$

while for the PPI networks of *C. elegans* and *D. melanogaster*, the best fit was obtained for stretched exponential distributions of the type

$$p(k) \sim Ce^{-\alpha k/\bar{k}}k^{-\gamma} \qquad (15.2)$$

for $k \geq 0$ in both cases. Figure 15.5 illustrates the plots of the cumulative degree distributions for the main connected components of the eleven PPI networks that we are analysing in this chapter.

In the figure it can be observed that most of the studied PPI networks display some kind of fat-tailed degree distribution characterized by the existence of very few hubs and many nodes of low degree. We have not provided any fit to the distributions shown in this figure due to the following difficulties usually encountered with fitting the 'best' distribution to a given dataset:

1) There are hundreds of distributions to be tested, which requires a considerably extended statistical exercise to find the best fit.

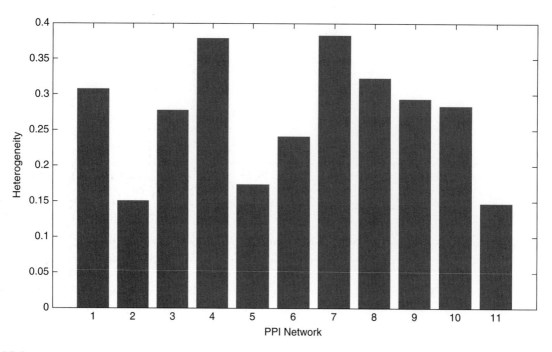

Fig. 15.6

Degree heterogeneity in PPI networks. Values of the degree heterogeneity index $\rho\,(G)$ for the PPI networks given in Table 15.1.

2) It is usually difficult to decide between two or more 'best fits', due to small differences in the statistical parameters of those correlations.

3) For small networks, or for those showing relatively low variations in their degrees, it is difficult to fit the data-points to a given distribution, due to the scarcity of existing data-points.

In order to produce a quantitative characterization of the degree of heterogeneity of PPI networks, we use the degree heterogeneity index studied in Chapter 8. In Figure 15.5 we illustrate the values of this index for the eleven PPI networks given in Table 15.1. The average heterogeneity for these networks is about 0.27, which is well above the values of 0 and 0.12 expected for ER and BA random graphs respectively (see Chapter 12).

The H-plots for some of the PPI networks studied here are illustrated in Figure 15.7, in increasing order of their degree heterogeneity indices.

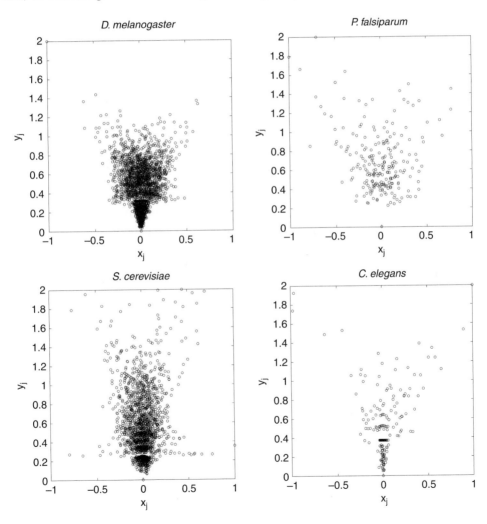

Fig. 15.7
Degree heterogeneity plots (H-plot) for PPI networks. H-plots for four PPI networks displaying general characteristics of fat-tailed degree distributions.

Table 15.2 **Assortativity in PPI networks.** Values of the Newman assortativity coefficient r for PPI networks. Also shown are the values of the path ratios $|P_{2/1}|$ and $|P_{3/2}|$, and the transitivity index C.

No.	Network	$|P_{2/1}|$	$|P_{3/2}|$	C	r
1	*A. fulgidus*	2.889	1.586	0.028	−0.472
2	KSHV	6.210	5.878	0.144	−0.058
3	VZV	9.919	7.269	0.212	−0.296
4	*B. subtilis*	4.630	1.953	0.021	−0.463
5	*P. falciparum*	8.613	7.922	0.116	−0.083
6	*E. coli*	13.291	13.022	0.155	−0.015
7	*C. elegans*	5.942	1.559	0.018	−0.451
8	*H. pylori*	10.719	7.024	0.015	−0.243
9	*S. cerevisiae*	15.656	14.082	0.102	−0.105
10	*H. sapiens*	14.776	10.160	0.035	−0.137
11	*D. melanogaster*	3.241	2.886	0.041	−0.060

The values of these indices are 0.148, 0.174, 0.294, and 0.383, respectively. All plots show characteristic features of the networks with fat-tailed degree distributions, which is reflected by the existence of V-shaped H-plots.

Another global characteristic which can be considered for PPI networks is their degree assortativity. Although this property has not been intensively studied for PPI networks, we show here that it displays some interesting characteristics. In Table 15.2 we give the values of the Newman assortativity coefficient for the degree–degree correlation (see Chapter 2). As can be seen, most PPI networks are strongly disassortative, with some exceptions which are only weakly disassortative. We recall (from Chapter 2) that a network is disassortative $(r < 0)$ if $|P_{2/1}| > |P_{3/2}| + C$, where $|P_{r/s}|$ is the ratio of paths of length r to paths of length s, and C is Newman's clustering or transitivity. The reason why all PPI networks are disassortative is evident from an analysis of the data in Table 15.2, which indicates that in general, $|P_{2/1}| > |P_{3/2}| + 1$—with the exception of the PPI networks of KSHV, *P. falciparum*, *E. coli*, and *D. melanogaster*. Because $C \leq 1$, the condition $|P_{2/1}| > |P_{3/2}| + C$ is immediately satisfied for the majority of them, without the necessity of having a large transitivity. Let us interpret these results. The ratio $|P_{2/1}|$ gives the fraction of complexes of type $A - B$ which grows to complexes of type $A - B - C$ in some evolutionary step of the development of the PPI network. In the next step, the complexes $A - B - C$ can growth in two different ways: $A - B - C - D$ and $A - B(D) - C$, in which D is directly linked to B. The first type of complexes increases the ratio $|P_{3/2}|$, but not the second one, which instead increases the ratio $|P_{2/1}|$. Because we have seen that $|P_{2/1}| > |P_{3/2}| + 1$, we can infer that in the evolutionary process of PPI networks at least one step of preferential attachment in the complexes $A - B - C$ gives rise to the disassortativity observed. This also agrees with the fat-tailed degree distributions observed for these networks.

15.3 Subgraph-based techniques in PPI networks

In Chapter 4 we studied the distribution of the clustering coefficient as a function of the degree of the nodes. The plots of $C(k)$ versus k for all the PPI networks studied here are shown in Figure 15.8.

As can be seen in most PPI networks, a power-law of the type $C(k) \sim k^{-\beta}$ is observed in the plots of $C(k)$ versus k. This has been interpreted as an indication of the existence of certain 'hierarchical' structure in complex networks. In Chapter 4 we provided an alternative explanation for this decay of the clustering with the degree, which is basically that it is very costly for high-degree nodes to have large clustering. The most efficient way for a protein in a PPI network to increase its clustering is to be involved in complexes having cliques. However, for high-degree proteins this means that they need to be involved in clique-like structures with a very large number of other proteins, which can be a very costly process. As a consequence, we do not expect to see very large cliques in PPI networks. In Figure 15.9 we plot the relative abundance of cliques of sizes from 3 to 10 in all the PPI networks analysed here. In order to obtain these plots we have simply divided the number of cliques of size k by the number of 3-cliques. The numbers of 3-cliques for the eleven networks given in the order shown in Table 15.1 are 1, 21, 44, 3, 130, 114, 13, 76, 825, 718, and 141. As can be seen in Figure 15.9, the number of cliques decays very fast with the increase in the size of the cliques, and none of the PPI networks analysed has cliques of length larger than 9. In fact, except the PPIs of yeast and *E. coli*, PPI networks do not have cliques of size larger than 5. Spirin and Mirny (2003), studying a different version of the PPI network of yeast, obtained results showing that the largest clique is of size 14, and that the number of cliques decays very quickly with an increase in the size of the clique, as we have also observed here. These results should be contrasted with the maximum degree for a node observed in these PPI networks, which are given in Table 15.1.

These findings do not imply that there are no clusters of highly connected proteins in a PPI. For instance, the existence of quasi-cliques has been detected by Bu et al., (2003) in the PPI network of yeast, and it was shown there that these quasi-cliques display characteristic functional properties. However, when the density of connections of these quasi-cliques is analysed by comparing the fraction of links in the quasi-clique with the number of links in a proper clique of the same size, it is observed that most of them have values of this density which are well below 1. This topic will be further discussed later in this chapter, when we study the communities in PPI networks.

The cliques of small size, however, can play fundamental functional roles in PPI networks. In fact, this is the basis of the search for functional complexes following several methods proposed in the literature (Spirin and Mirny, 2003; Bu et al., 2003; Pereira-Leal et al., 2004; Yu et al., 2006; Futschik et al., 2007). The idea is that a functional module is formed by a group of proteins which are highly interconnected amongst themselves, but little connected with the rest. That is, they are small 'communities' formed around a clique of proteins in the PPI network. Therefore, by identifying k-cliques, a group of highly

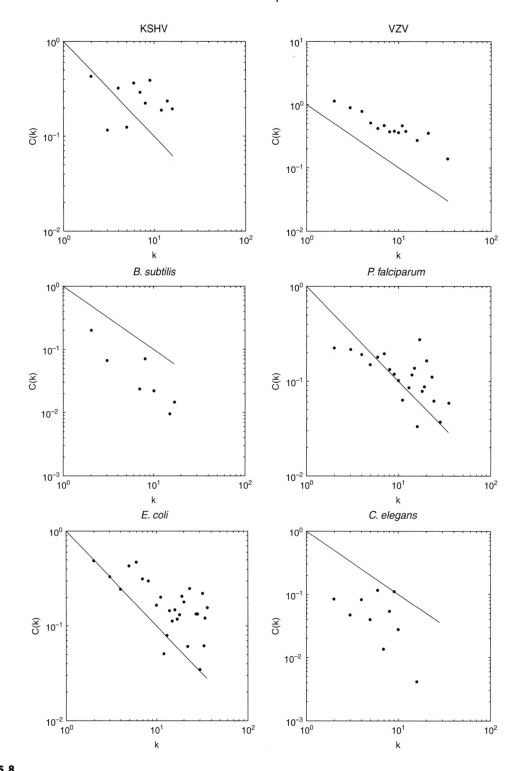

Fig. 15.8
Decay of clustering with node degree in PPI networks. Plot of the Watts–Strogatz clustering coefficient of as a function of the degree of the corresponding node for PPI networks. The solid line corresponds to $C(k) \sim k^{-1}$.

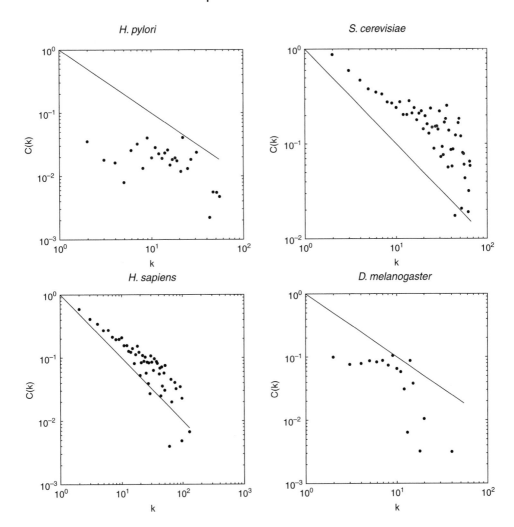

Fig. 15.8
(*continued*)

interconnected proteins can be found, which can then be expanded by considering, for instance, the union of several adjacent k-cliques (Futschik et al. 2007), or by maximising the density of the quasi-clique (Spirin and Mirny, 2003; Bu et al. 2003). The number of k-clique communities found then decreases very quickly with their size, as expected from the previous analysis of the number of cliques of different sizes in PPI networks.

The following are some examples of the k-clique communities found by Futschik et al. (2007) in the human PPI network. Three cliques of sizes 11, 10, and 9, containing proteins related to transcription initiation, chromatin modification, and transmembrane receptor protein tyrosine kinase signalling, respectively. In addition, two quasi-cliques of size 10 and 9 were identified—the first having a core clique of nine proteins, and the second containing a clique of seven proteins.

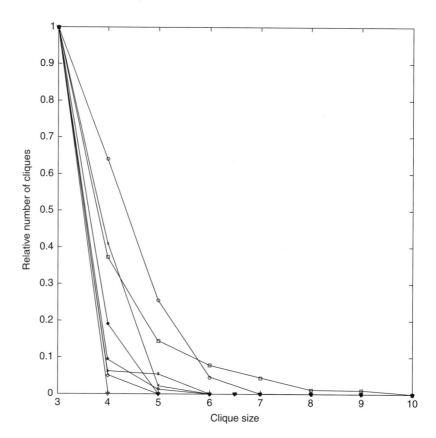

Fig. 15.9

Clique distribution in PPI networks.
Fast decay of the number of cliques as a
function of the clique size in PPI
networks.

Using a k-clique community technique, Josson and Bates, (2006) have
studied human proteins involved in cancer. They have found that cancer proteins have an average degree which is larger than those not documented as
being mutated in cancer. In addition, cancer-involved proteins appear more
frequently as central hubs than in peripheral ones. However, when k-clique
communities are found, the cancer proteins appear very frequently at the
interface of different communities, which indicates that these proteins have
domains with a high propensity for mediating protein interactions. Some of
the communities found by Josson and Bates (2006) are shown in Figure 15.10,
where proteins involved in cancer are represented by triangles.

The importance of motifs in PPI networks has been studied by Wuchty
et al. (2003), who have shown that proteins that participate in certain subgraphs are conserved to a significantly higher degree than those that do not
form such patterns. The conservation was analysed by studying 678 proteins
in *S. cereviciae* (yeast) with an ortholog in each of five higher eukaryotes. The
results were then compared with the random conservation rate defined as the
fraction of subgraphs that is fully conserved for a random ortholog distribution.
In Figure 15.11 we illustrate the results for some of the subgraphs for which
we have given analytic formulae for counting their numbers in a network (see
Chapter 4). These results indicate that some small subgraphs may represent

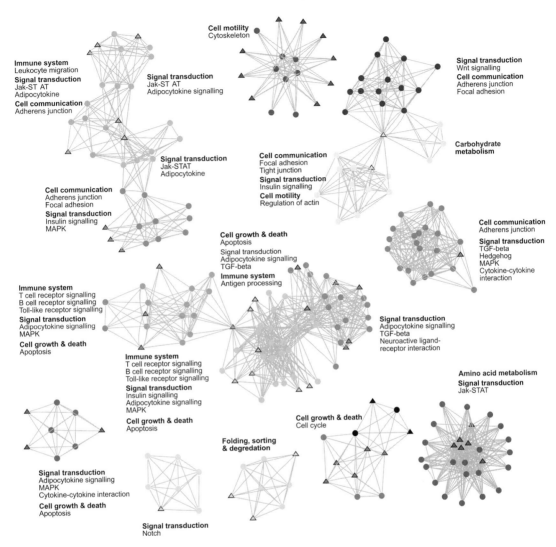

Fig. 15.10

Protein communities in human interactome. Some of the k-clique communities found in the human PPI network. Cancer proteins are shown as triangles. From Josson, P. F., and Bates, P. A. (2006). Global topological features of cancer proteins in the human interactome. *Bioinformatics* **22**, 2291–7. (Reproduced with the permission of Oxford University Press.)

evolutionary conserved units in PPIs. Such subgraphs are formed by groups of proteins which have evolved for certain specific functions in the cell.

The analysis of small subgraphs in PPI networks can provide some interesting indications concerning the structure of these networks. In Chapter 4 we presented analytic expression for calculating the number of nineteen different subgraphs in a given network on the basis of the node adjacency relationships. We use these formulae in order to calculate the abundance of these fragments in the eleven PPI networks studied here. We have normalised the number of fragments in each network by dividing them by the number of the most abundant fragment found. The results are illustrated in Figure 15.12.

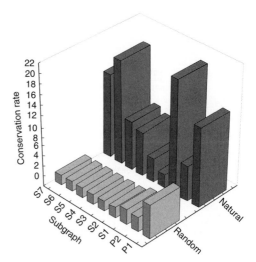

Fig. 15.11

Subgraphs and evolution in PPI networks. Conservation rate of proteins in the yeast PPI network and in analogue random networks, according to Wuchty et al. (2003). The subgraph labels are those shown in Figure 15.12.

According to the plots shown in Figure 15.12, the PPI networks can be separated into two general classes. The first one is characterized by a high abundance of fragments S_2 and S_3, accompanied by a relatively large abundance of P_2. The other group is characterised by a relatively large abundance of fragments S_{11} and S_{15}, and a relatively low abundance of fragments S_2, S_3, and P_2. In the first group we have the PPI networks of *A. fulgidus*, *B. subtilis*, *C. elegans*, and *D. melanogaster*. The second group is formed by the other seven PPI networks. The most salient characteristic of fragments S_2, S_3, and P_2 is that they are acyclic and consequently bipartite, while fragments S_{11} and S_{15} are cyclic and contain an odd-cycle, which makes them non-bipartite fragments. Bipartite and almost-bipartite structures can play a fundamental role in PPI networks, and we shall consider them in the next section of this chapter. A vector representing the number of rooted subgraphs in which a protein takes place—known as a 'graphlet' (see Chapter 4)—has been used to analyse the differences between PPI networks (Pržulj, 2007).

15.4   Bipartite structures in PPI networks

We begin by calculating the values of the global bipartivity of the PPI networks, using the expression $b_S(G) = EE_{even}(G)/EE(G)$ studied in Chapter 11 . We recall that $0.5 \leq b_S(G) \leq 1$, where the lower bound is reached for networks which are not bipartite, and the upper bound is obtained for bipartite networks. Figure 15.14 illustrates the results obtained by using the bipartivity index $b_S(G)$ for the PPI networks studied in this chapter.

It can be seen that the four networks for which we have found a high relative abundance of the fragments S_2, S_3, and P_2 have a bipartivity index close to 1, indicating that they have *almost-bipartite structures*. Network PPI-8—which corresponds to *H. pylori*—also has significant bipartivity, and in Figure 15.13 we can see that it has an intermediate abundance of the fragments S_2 and S_3.

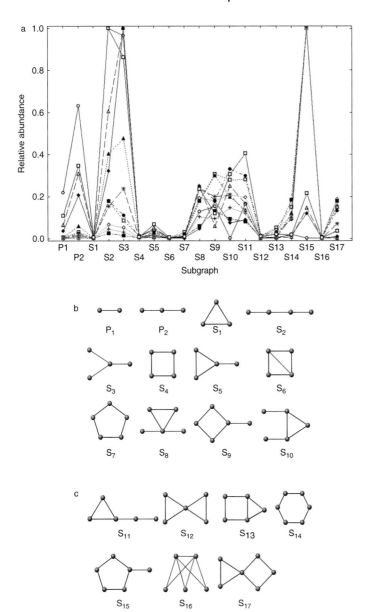

Fig. 15.12
Small subgraphs in PPI networks.
Relative abundance of small subgraphs
in the PPI networks given in Table 15.1.

The other networks have relatively small bipartivity, and in particular the PPI
network of *S. cereviciae* has practically none.

The existence of bipartivity in the structures of PPI networks can be
explained as being the result of the nature of interactions between proteins.
For instance, a classical picture of PPIs is that of a 'lock-and-key' structure,
in which a protein interacts with another which has some complementary
features, such as shapes, or electrostatic or hydrophobic properties. Therefore,
one of the proteins can be considered as a lock in which the other acts as a
key. Clearly, there can be a bipartition of these networks having locks and keys

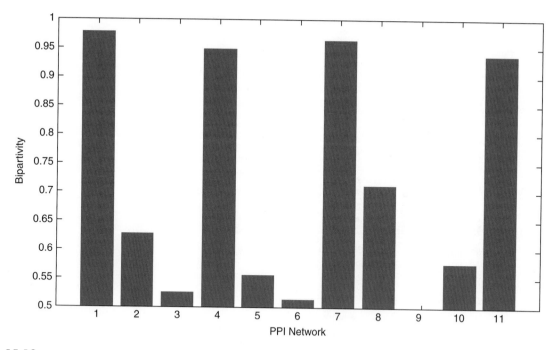

Fig. 15.13

Bipartivity in PPI networks. Values of the spectral bipartivity measure (see Chapter 11) for the PPI networks given in Table 15.1. Values close to 0.5 indicate a lack of bipartivity, and those close to 1.0 indicate a high degree of bipartivity in such networks.

at both sides of each interaction. Note that this does not imply that one of the groups in the bipartition corresponds to all proteins acting as locks or keys—as can be erroneously inferred from some works (Morrison et al., 2006)—but that on an individual basis, every link connects a lock with a key.

Now we can use the method explained in Chapter 11 , based on the communicability at negative absolute temperatures for identifying the bipartitions in PPI networks. For instance, by using exp(–A) we have studied the bipartition of the PPI network of *A. fulgidus*. In this case, the *node-repulsion graph* (see Chapter 11) is formed by two connected components, indicating the existence of two partitions, which, as illustrated in Figure 15.14, are represented by squares and circles for the nodes in each partition.

Note that in this case the network is almost perfectly bipartite, because there is only one interaction between two proteins in the same partition, that is, AF1790/DP-1 interacting with AF1722/DP-2. This bipartivity is already reflected in the value of $b_S(G) = 0.978$ observed for this network, which is very close to 1. However, the situation appears quite different for networks with relatively small bipartivity. For instance, for the PPI network of KSHV the value of $b_S(G) = 0.626$ indicates that there is some bipartivity, but that there are several links which are 'frustrated' as they connect proteins in the same partition. In Figure 15.15 we illustrate this situation by representing the inter-partition links as thin solid lines, and the intra-partition links as thick solid and dotted lines for each partition respectively.

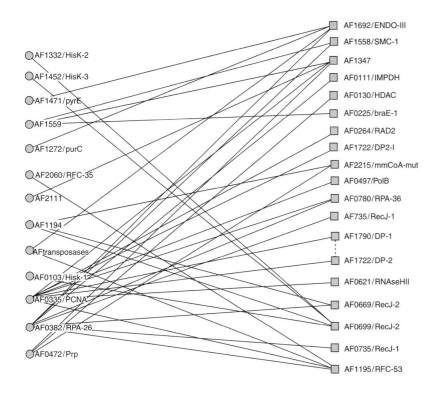

AF1332/HisK-2
AF1452/HisK-3
AF1471/pyrE
AF1559
AF1272/purC
AF2060/RFC-35
AF2111
AF1194
AFtransposases
AF0103/Hisk-1
AF0335/PCNA
AF0382/RPA-26
AF0472/Prp

AF1692/ENDO-III
AF1558/SMC-1
AF1347
AF0111/IMPDH
AF0130/HDAC
AF0225/braE-1
AF0264/RAD2
AF1722/DP2-I
AF2215/mmCoA-mut
AF0497/PolB
AF0780/RPA-36
AF735/RecJ-1
AF1790/DP-1
AF1722/DP-2
AF0621/RNAseHII
AF0669/RecJ-2
AF0699/RecJ-2
AF0735/RecJ-1
AF1195/RFC-53

Fig. 15.14
**Bipartite clustering in the *A. fulgidus*
PPI network**. The best bipartition found
by the communicability function at
negative absolute temperature for the PPI
network of the archae bacterium *A.
fulgidus*. Nodes in one partition are
represented by circles, and those in the
other by squares.

Overall, the results obtained here indicate that only a few PPI networks display some bipartivity. In particular, those in which the number of proteins is relatively large display almost no bipartivity at all. This means that in general the concept of PPI networks as global lock-and-key systems is not very appropriate for their generality. However, this does not deny the existence of bipartite or quasi-bipartite complexes of proteins, which reflect the lock-and-key character of some PPIs (Morrison et al., 2006; Thomas et al., 2003). In order to understand the existence of bipartite or non-bipartite complexes of proteins in PPI networks we should investigate how proteins evolve and how this evolution is reflected in the local topological properties of nodes in PPI networks.

15.5 Local structure of PPI networks

Let us begin this section by analysing how protein evolution can be reflected in the topological properties of the nodes in a PPI network (Pastor-Satorras et al., 2003; Presser et al., 2008). First, we consider that a protein can evolve by means of a duplication-divergence process (Ohno, 1970); that is, the gene encoding a given protein is first duplicated. During this duplication, all links of the original gene are duplicated by its copy. However, in the divergence step of the process some of these links can be removed and some new ones can be created. Let us consider a gene which has interactions with several others that do not share interactions amongst themselves. That is, we are

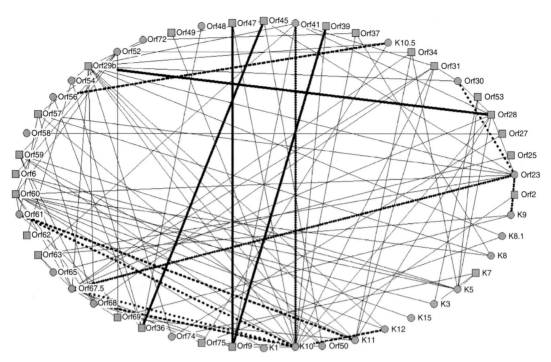

Fig. 15.15
Bipartite clustering in the KSHV PPI network. The best bipartition found by the communicability function at negative absolute temperature for the PPI network of the *Kaposi sarcoma* herpes virus (KSHV). Nodes in one partition are represented by circles, and those in the other by squares. Inter-partition links are represented by thin continuous lines, square–square links by thick continuous lines, and circle–circle links by thick dotted lines.

considering the evolution of a bipartite cluster of genes like that depicted in Figure 15.16. After the duplication of the gene marked in black in the figure, we obtain the pattern which is displayed in the middle. Then, in the divergence stage of the evolutionary process there are two different possibilities: (i) that a large number of new interactions are created, giving rise to the complex at the top right of the figure, or (ii) that very few new links are created and very few old links are removed, producing the complex at the bottom right. It is straightforward to realise that the divergence process (i) gives rise to a non-bipartite complex, while process (ii) gives rise to an almost-bipartite complex. Then, following the 'principle of minimum effort', we can think that process (ii) is the most probable one, as the number of new links that should be created and the number of old ones removed is very small in comparison with those in process (i). As a consequence, the evolution by duplication-divergence of a gene in a bipartite complex should produce an almost-bipartite evolved complex with more chances than a non-bipartite complex (Estrada, 2006a).

The same reasoning can be applied if we begin the process by considering a gene in a non-bipartite complex as the one shown in Figure 15.17. In this case, under similar evolutionary conditions and assumptions the most probable replica of the original complex will be one displaying bipartivity properties similar to its originator's.

EVOLUTION FROM A BIPARTITE TARGET

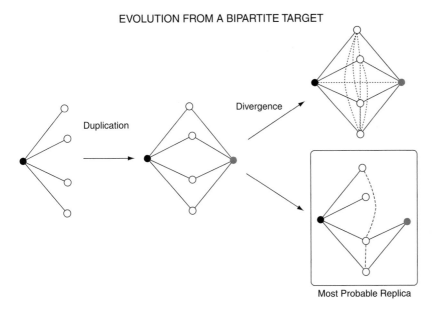

Duplication

Divergence

Most Probable Replica

Fig. 15.16
Evolution of a bipartite protein.
A schematic representation of the
hypothetical evolution by duplication
(centre) and divergence (right) of a
bipartite protein (left). The most
probable replica, following the 'principle
of minimum effort', is represented in
a box.

EVOLUTION FROM A NONBIPARTITE TARGET

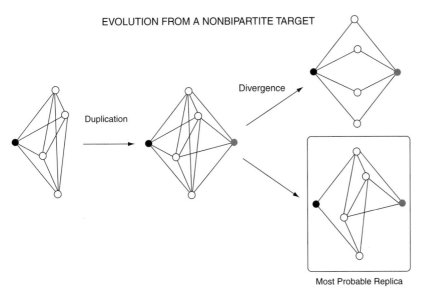

Duplication

Divergence

Most Probable Replica

Fig. 15.17
Evolution of a non-bipartite protein. A
schematic representation of the
hypothetical evolution by duplication
(centre) and divergence (right) of a
non-bipartite protein (left). The most
probable replica, following the 'principle
of minimum effort', is represented in a
box.

This simple model can have dramatic implications for the study of topological properties of nodes in PPI networks, as well as for their biological
interpretations. For instance, it should be expected that when a gene is duplicated it conserves its function, which can then be diversified by means of the
divergence part of the evolutionary process. Consequently, we can expect that
proteins which are grouped together in complexes—like those illustrated in
Figures 15.16 and 15.17—display common functions. For instance, if a protein
with function A in a non-bipartite complex evolves by duplication-divergence,
it should give rise to a non-bipartite complex in which the proteins display the

function A. This is exactly what has been observed by several authors who have found that cliques and quasi-cliques of proteins in PPI networks share identical functions. We maintain that a clique is the structure with the highest non-bipartivity that we can build with a given number of nodes. For instance, Bu et al. have identified forty-eight quasi-cliques and six quasi-bipartite complexes in which proteins share the same function. The abundance of non-bipartite structures with respect to the bipartite structures can be explained by the transitivity of the protein–protein interactions. That is, if proteins A, B, and C share the same function and protein A interacts with B and C, it is very probable that B and C also interact. Transitivity or clustering—reflected here by the presence of triangles—decreases the bipartivity due to the presence of odd-cycles.

If the previous hypothesis is true it can be expected that essential proteins are grouped together into quasi-cliques or quasi-bipartite. An essential protein is one that, when knocked out, renders the cell unviable. We can therefore expect that essential proteins have evolved, via duplication-divergence, from other essential proteins. Consequently, if the first were in non-bipartite (bipartite) complexes we should expect that the last are also found in complexes with bipartivity similar to the original ones. We can expect the same kind of transitivity as explained previously for the interaction of essential proteins in PPI networks, and we should therefore expect a high proportion of essential proteins to be present in the form of non-bipartite complexes, of which cliques are simply an example. In Figure 15.18 we illustrate four protein complexes forming cliques in the PPI network of yeast studied here. Essential proteins are represented by stars, while non-essential proteins are depicted by circles.

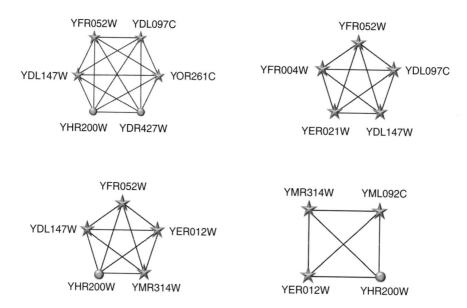

Fig. 15.18
Essential proteins in yeast PPI non-bipartite complexes. A schematic representation of several non-bipartite PPI complexes in yeast. Essential proteins are represented by stars, and non-essential ones by circles.

As can be seen, a very large proportion of the proteins in these complexes are essential. Indeed, we have found that 85.7% of the essential proteins in this PPI network have a bipartivity index below 0.6—that is, $b_S(i) \leq 0.6$— while among proteins with $0.9 \leq b_S(i) \leq 1.0$ only 3.9% are essential, where $b_S(G) = EE_{even}(i)/EE(i)$ is the bipartivity of the node i. These results agree with those of Yu et al. (2004), who reported that the average clustering coefficient of essential proteins in a different version of the yeast PPI network is 0.182, while that of the non-essential ones is 0.095. This again indicates that essential proteins are part of non-bipartite clusters with more preference than to bipartite clusters—at least in the PPI network of yeast.

The question that arises now is whether we can design methods for identifying these essential proteins in PPI networks by using only topological information about the nodes in these networks. This, of course, introduces the field of network centrality; that is, the way in which protein essentiality is reflected in the centrality of the nodes in PPI networks. The object here is to design new molecules which are able to inhibit protein–protein interactions (Arkin and Wells, 2004). For example, in the case of pathogenic organisms these drug candidates should inhibit the interaction, involving at least one essential protein, which will result in the death of such a microorganism. The first breakthrough in this direction was reported by Jeong et al. (2001), who discovered that the fraction of essential proteins with exactly k links increases with their connectivity (degree). They concluded that 'the robustness against mutations in yeast is also derived from the organization of interactions and the topological position of individual proteins'—in other words, that essentiality can be predicted from the centrality of the proteins in the PPI network.

We can therefore design an *in silico* method that allows us to investigate the efficiency of centrality measures in identifying essential proteins in the yeast PPI (Estrada, 2006a; 2006b; Hahn and Kern, 2005). One experiment, for instance, consists in ranking the proteins in the PPI network according to a given centrality measure (Estrada, 2006a; 2006b). Then, by selecting the top fraction of these ranked proteins we can count how many of them are essential to this organism. We can compare our results using centrality by taking random rankings of the proteins in the PPI network and counting the number of essential proteins in the top fraction of these rankings. In this case we have taken the average of 1,000 random rankings. As can be seen in Figure 15.19, the best that can be done by selecting proteins at random identifies only 25% of the essential proteins—which means that of every 100 proteins submit to testing for essentiality, about 75 will be not essential. However, this percentage of success can be improved by using, for instance, the degree centrality, which raises the percentage of hits to an average of about 43%, which is similar to the percentage produced by information centrality (IC). Centrality measures based on shortest paths do not produce any improvement with respect to the degree, as they are able to correctly identify only 38% of essential proteins. The two spectral centrality measures—eigenvector, and subgraph centrality—improve the percentage of hits by detecting, on average, more than 50% of essential proteins. However, the best results are obtained by using the spectral measure of bipartivity (ranked in decreasing numerical order), which identifies, on average, 55% of essential proteins. When analysing the top 30–100 proteins ranked

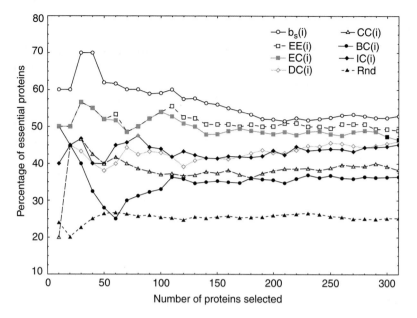

Fig. 15.19

Centrality and essentiality in the yeast PPI network. Percentage of essential proteins among the top-x proteins (number of proteins selected), ranked according to different centrality measures.

by this index we are able to identify around 60–70% of essential proteins. This represents a dramatic improvement with respect to the random selection of proteins, despite the fact that we are using only topological information contained in the PPI network. In an independent study, Hahn and Kern (2005) have shown that the average degree, closeness, and betweenness of essential proteins in yeast, worm, and fly are different from those of the non-essential ones, and that the most central proteins evolve more slowly in these three organisms. These and other results have led to a dramatic increase in the use of centrality measures as a criterion for quantifying the importance of proteins in PPI networks (Lin et al., 2008; Zotenko et al., 2008). Similar approaches have been used to differentiate protein–protein interactions distinguishing malignant tissues from those randomly selected (Platzer et al., 2007).

The structure of reaction networks

16

My main thesis will be that in the study of the intermediate processes of metabolism we have to deal not with complex substances which elude ordinary chemical methods, but with the simple substances undergoing comprehensible reactions.

Frederick G. Hopkins

Chemistry is the science that has shaped our modern lifestyles (Emsley, 1999). Chemical compounds are the basis of all our health-care products, plastics used in the manufacture of industrial products, synthetic dyes and materials used in our clothes, paints, food additives and drugs for human and animal health, and many other applications. Today there are more than 56 million organic and inorganic compounds, of which 44 million are commercially available. This creates a vast universe of molecules. The art and science of transforming one chemical into another is based on the use of chemical reactions which initiate the formation and/or breaking of covalent bonds. Today, there are more than 52 million single-step and multi-step reactions and synthetic preparations shaping the chemical universe. Many of these chemical reactions occur naturally, while many others are the products of the imagination and technical expertise of synthetic chemists. The combination of chemical compounds, represented as nodes, and their chemical reactions transforming some into others, represented by directed links, form the basis of chemical reaction graphs. The initial studies of these reaction graphs— basically, the graphs for individual organic reactions—date back to the work of several mathematical chemists in the 1970s and 1980s, and are compiled in the works of Balaban (1994), Temkin et al. (1996), and Koca et al. (1989). Although much attention has been paid to the study of the kinetic properties of these transformations, this chapter will investigate the information that can be extracted from the topological analysis of reaction networks. It is divided into three sections, encompassing the study of man-made reaction networks, reaction networks occurring in inanimate Nature (atmospheric reactions), and reaction networks in living beings—such as metabolic networks.

16.1 Organic chemical reaction networks

In a chemical reaction, one or more reactants are transformed into one or more products. Both reactants and products are chemical compounds linked by a chemical reaction. In order to build a reaction network we can represent the information about chemical reactions by a bipartite network. In this network there are two types of node—one representing chemicals, and the other representing reactions. In Figure 16.1 we illustrate the bipartite graph for the set of reactions shown at left. In this representation, reactants point towards the reaction nodes, and products point away from the reaction nodes. A reaction network is thus created by considering the reactant-product projection of the bipartite graph. That is, we represent only the nodes representing chemicals, and connect them according to the transformations observed in the bipartite graph. There is a link from chemical i to chemical j if the first is a reactant and the second a product in at least one chemical reaction. In other words, there is a directed link from i to j in the reaction network if there is a directed path of length 2 from i and j in the bipartite graph (see Figure 16.1).

Using the strategy described previously, Fialkowski et al. (2005) and Bishop et al. (2006) have built the reaction network of organic chemistry by using data collected from the Belstein database, starting with 9,550,398 chemicals and 9,293,250 reactions. After pruning to remove catalysts, solvents, substances that do not participate in a reaction, and so on, the reaction network remains in about 6.5 million substances and 6,539,158 chemical reactions. The links in this network were labelled by the year in which the reaction was first published, thus allowing the study of the evolution of the organic chemistry reaction network over time, from 1850 to 2000. The numbers of molecules and reactions have been growing continuously since 1850. However, the rate of growth is different for the period 1850–1900 and later. Starting at the beginning of the twentieth century, there was a deceleration in the growth of chemicals and reactions. For the early period, the average degree of molecules

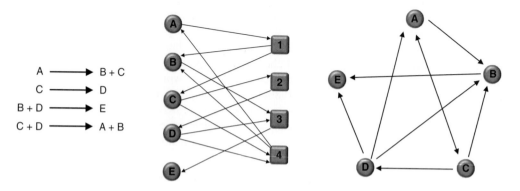

Fig. 16.1

Construction of a reaction network. A hypothetical set of reactions (left) formed by four reactions and five chemical species. These reactions are represented as a bipartite graph (centre) in which chemical species are in one partition and the reactions are in the other. The chemical–chemical reaction network is a projection (right) of the bipartite graph in which nodes are chemical species, and there is a link from one to another if the first is the reactant and the second is a product in at least one chemical reaction.

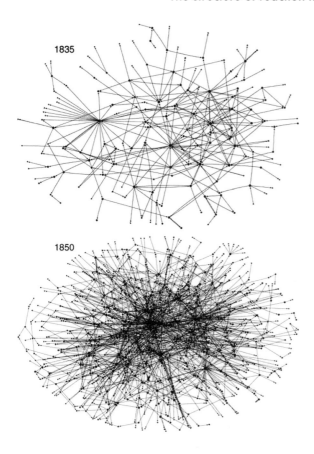

1835

1850

Fig. 16.2
Evolution of organic reaction networks. The main connected component of the network of organic reactions in 1835 (top) and in 1850 (bottom). From Fialkowski, M., Bishop, K. J. M., Chubukov, V. A., Campbell, C. J., and Grzybowski, B. A. (2005). Architecture and evolution of organic chemistry. *Angew. Chem. Int. Ed.* **44**, 7263–9. (Reproduced with the permission of Wiley-VCH Verlag GmbH & Co. KGaA, ©2005.)

in the reaction network increased almost linearly on a yearly basis, reaching a maximum of 3.6 around 1885. Subsequently, the average degree began to decrease year after year, reaching the value of about 2 in the year 2000. This has been interpreted by Fialkowski et al. (2005) as evidence of the existence of two stages in the development of organic reactions, characterized by a first stage in which the wiring of existing molecules was the dominant feature, and a second stage in which the exploration of unknown chemical space dominated the chemical scenario. The reaction networks of 1835 and 1850 are illustrated in Figure 16.2.

Another property of the organic reaction network investigated by Fialkowski et al. (2005) is the degree distribution. It was observed that the indegree and outdegree of the molecules follow power-law distributions of the form $p(k) \sim k^{-\gamma}$. Interestingly, the exponents of the power-law distributions steadily grew from 1850 to 2004, when they reached values of about 2.7 and 2.1 for the indegree and outdegree respectively (see Figure 16.3). This decrease in the power-law exponent shows that the reaction network of organic chemistry is becoming less heterogeneous with time. These power-law distributions demonstrate that those chemicals which are already used in many chemical transformations are more likely to be used in new reactions. This, of course, reflects the fact that these chemicals are useful in many different types of

Fig. 16.3

Degree distributions in organic reaction networks. Indegree and outdegree distributions of molecules in the organic reaction networks existing at four different years. From Fialkowski, M., Bishop, K. J. M., Chubukov, V. A., Campbell, C. J., and Grzybowski, B. A. (2005). Architecture and evolution of organic chemistry. *Angew. Chem. Int. Ed.* **44**, 7263–9. (Reproduced with the permission of Wiley-VCH Verlag GmbH & Co. KGaA, ©2005.)

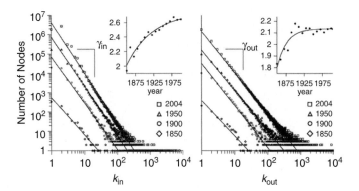

transformations, that they are easy to manipulate, that they are the cheapest, and so on. For instance, Grzybowski et al. (2009) have reported that the price of a substance per mole is inversely proportional to its degree in the reaction network: $s_i \sim k_i^{-1/2}$. In addition, it was observed that the average increase in the indegrees and outdegrees over a given period of time was proportional to their values at the beginning of that period. That is, the chances of a molecule being used as a substrate in chemical reactions (large outdegree) increases with the number of times that it has already been used for such purposes; and the higher the indegree of a substance, the more likely it is that chemists will try to obtain it via a new reaction.

How can these and other findings based on network theory help organic chemists in their arduous labours? Let us consider, for example, the simplest scenario of selecting the most useful set of compounds for diversifying the chemical universe. Grzybowski et al. (2009) have found that the network of organic chemical reactions has a clear core-periphery structure, accompanied by several islands of reactions (see Figure 16.4). The core has been defined as the subset of chemicals in which every pair of molecules is connected by a synthetic path. In this core we find only 4% of all organic compounds. However, what is important is that this diverse set of molecules is involved in more than 35% of all known reactions, and gives rise to more than 78% of known organic compounds existing today—about 5 million compounds. No doubt they are the 'generators' of the chemical universe. More impressive is the fact that a subset of 300 core molecules controls the synthesis of more than

Fig. 16.4

The organic reaction network. Illustration of the topological structure of the organic reaction network in the year 1840 indicating the existence of a main core, a periphery and some islands. From Grzybowski, B. A., Bishop, K. J. M., Kowalczyk, B., and Wilmer, C. E. (2009). The 'wired' universe of organic chemistry. *Nature Chem.* **1**, 31–36. (Reprinted with the permission of Macmillan Publishers Ltd., ©2009.)

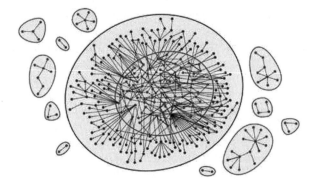

1.5 million other compounds. These are exactly the molecules that you should have in your product line.

The other side of the coin is represented by the islands observed in the organic reaction network. These are small components formed by, on average, less than four molecules, which constitute about 18% of the total number of molecules in the network. Most of these islands, which are disconnected from the core and periphery of the network, are formed by compounds which are difficult to synthesise and for which there is no known procedure to obtain them from simple precursors. These molecules, of course, represent the next challenge for organic chemists, as some of them are known to have interesting pharmacological properties. This is, for instance, the example of the natural product isolated from *Physalis angulata*, which displays antileukemic and antimycobacterial activities, and for which no synthesis from simple precursors has yet been reported.

Other types of support that network theory can bring to organic chemists are the following. The first is related to the optimisation of multiple reactions, in which network-based strategies can help determine which subset of substrates and reaction pathways should be used to minimise the overall cost of the process. The second deals with the set of substances that can be regulated in order to avoid the illegal production of controlled substances and chemical weapons. A better analysis of all possible routes for synthesizing these molecules is needed, as illustrated by the observation made by Grzybowski et al. (2009) that it is still possible to synthesize sarin gas from commercially available and unregulated compounds in just two steps. However, a call of attention is needed here. In this work, a representation of reaction networks in which a compound–compound projection of the bipartite compound-reaction network is used. In this representation, some important information about the real reaction paths connecting reactants and products is missing. Let us consider this with one example in which a series of four reactions is represented as a network, as illustrated in Figure 16.5.

We first consider that compounds A and D are commercially available, and that we are interested in obtaining compound F as our final product. The shortest path analysis of this network indicates that in just two steps we can obtain compound F from a commercially available compound, D; that is, the shortest path connecting a commercially available compound and F is just D → E → F. However, this is nonsense from the chemical point of view, as to obtain compound E you need compound C (reaction 3), which is not commercially available and must be obtained by mean of reactions 1 and 2. Therefore, the shortest paths in the compound–compound projection of a reaction network do not provide the information required to infer real reaction paths. This will

Fig. 16.5
Hypothetical reaction network.
Schematic representation of a hypothetical reaction network (left) for a set of reactions (right). Note that the shortest path D → E → F is chemically meaningless, as the reactant E needs to be produced from A in a sequence of reactions (see text for an explanation).

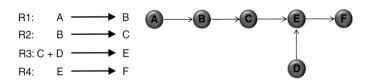

have important implications in the analysis of metabolic networks, discussed in Section 16.3. We need to remark that analysis of the universe of organic reactions is still in its infancy, and many more results are expected from more detailed analysis. We can therefore expect that network analysis will become an important part of organic chemistry in the not-too-distant future.

16.2 Reaction networks in the atmospheres of astronomical bodies

Atmospheres are the gaseous envelopes around planets and satellites, and in our solar system they are present on Earth, Venus, Mars, Jupiter, Saturn, Uranus, Neptune, and Saturn's largest satellite Titan (Wayne, 1985). The Earth's atmosphere is fundamental for the existence of life on our planet, but it has some peculiar characteristics which appear to defy the laws of chemistry and physics. For example, the atmospheres of Venus and Mars have a chemical composition which is basically formed by oxidised compounds such as carbon dioxide, which has a concentration of more than 950,000 parts per million (ppm) by volume. However, the concentration of this gas in the Earth's atmosphere—350 ppm—is almost 3,000 times smaller. By contrast, however, the concentration of oxygen in the Earth's atmosphere is 209,460 ppm—which is 160 times greater than on Mars, and more than half a million times greater than on Venus. Oxygen is a very reactive gas. It reacts with hydrogen, nitrogen, methane, and so on, to form a variety of chemicals, including water, nitrates, carbon dioxide, and many others (Wayne, 1985). Some of these molecules also react with each other, as well as with radiation, to form new molecules, producing a complex network of chemical transformations. Are these differences in the chemistry of the atmospheres in some way reflected in their reaction networks?

The reaction networks of the atmospheres of Earth, Mars, Venus, and Titan have been analysed by Solé and Munteanu (2004). Some of the structural parameters of these networks are shown in Table 16.1. The largest network, and the most sparse, is that of Earth. All networks display small-worldness, in the sense that $\bar{l} \sim \ln n$. They also display a larger Watts–Strogatz clustering coefficient than random analogues; in particular, $\overline{C}$ for the Earth's network is more than ten times larger than that of a random graph. All networks are disassortative, with the Earth's network displaying the largest disassortativity

Table 16.1 Topological properties of atmospheric reaction networks. Values of the number of nodes n, density $\delta(G)$, average path length $\bar{l}$, average Watts–Strogatz clustering coefficient $\overline{C}$ and its ratio to that of a random network $\overline{C}/\overline{C}_{rnd}$, assortativity coefficient r, and modularity in atmospheric reaction networks.

	n	$\delta(G)$	$\bar{l}$	$\overline{C}$	$\overline{C}/\overline{C}_{rnd}$	r	Modular
Earth	248	0.025	2.75	0.31	12.4	−0.31	Yes
Mars	31	0.309	1.89	0.61	1.9	−0.10	No
Venus	42	0.203	1.65	0.59	1.6	−0.14	No
Titan	71	0.159	2.08	0.55	3.4	−0.17	No

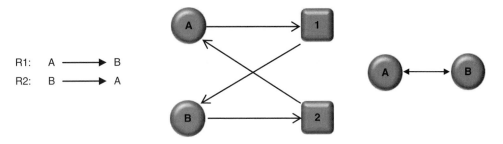

Fig. 16.6
Reversibility in reaction network. A reversible reaction (left), its bipartite graph (centre), and the reaction network (right), which account for the reversibility of the reaction.

among all networks. Solé and Munteanu (2004) have also found that the only atmospheric reaction network that displays modularity is that of the Earth.

A characterisation of small cycles in these four atmospheric reaction networks has been carried out by Gleiss et al. (2001). It was observed that all of them have significantly more triangles and fewer larger cycles than have random graphs. In determining the number of cycles, Gleiss et al. (2001) considered these networks as undirected and unweighted.

We know that some important information can be contained in the directionality of these reactions. The reversibility of the chemical reactions is conserved when we represent them as reaction networks in a compound–compound space derived from the compound-reaction bipartite network. Let us consider, for example, the reversible reactions between A and B as illustrated in Figure 16.6. As can be seen, when the compound–compound projection of the bipartite network is carried out, the reversibility of the reaction is maintained.

We can compare the directed reaction networks of Earth and Mars in order to search for indications which network theory might provide for the analysis of these atmospheres. The first thing to be noted is that the Earth network displays smaller reciprocity, $\rho = 0.193$, than that of Mars, $\rho = 0.411$. Although both networks are reciprocal according to Garlaschelli and Lofredo (2004a) classification, it is evident that the network of reactions in Earth's atmosphere displays a lower degree of reciprocal reactions than that of Mars. Reciprocity in reaction networks can be identified with chemical equilibrium. Consequently, the low reciprocity of Earth's atmosphere is an indication of the large disequilibrium existing in it. This disequilibrium has been identified as evidence of the close interrelation between the atmosphere and the biosphere. For instance, Lovelock and Margulis (1974) have compared the actual and equilibrium concentrations of several constituents of Earth's atmosphere, which are produced by biological sources. These results are partially illustrated in Table 16.2, where it can be seen that the departure from equilibrium expectations is measured in tens of orders of magnitude.

The different levels of equilibrium in the reaction networks of Earth and Mars are also manifested in the degrees of the nodes. First we analyse the most central chemical species according to the indegree and outdegree centralities. The outdegree centrality indicates the number of reactions in which

Table 16.2 **Gases in Earth's atmosphere with origins in the biosphere.** The concentration of different chemical species in Earth's atmosphere, and their expected concentration for an atmosphere of thermodynamic equilibrium. The ratio between both concentrations determined the departure from equilibrium in Earth's atmosphere. Some of the sources for these gases in Earth's atmosphere are included. All data are from Lovelock and Margulis (1974), and are updated according to Wayne (1985).

Species	Present fractional concentration	Expected fractional equilibrium concentration	Departure from equilibrium	Source
N_2	0.78	$< 10^{-10}$	10^{10}	Denitrifying bacteria
CH_4	1.7×10^{-6}	$< 10^{-35}$	10^{29}	Anaerobic fermenting bacteria
N_2O	3.1×10^{-7}	$< 10^{-20}$	10^{13}	Denitrifying bacteria
NH_3	10^{-9}	$< 10^{-35}$	10^{27}	Nearly all organisms
CH_3I	10^{-12}	$< 10^{-35}$	10^{23}	Marine algae

one chemical species takes place as reactant, and in such a way that it is related to the reactivity of the chemical species. In the atmosphere of Earth, the chemical species with largest k_{out} are OH, Cl, and NO_2. In Mars' atmosphere, the species with highest outdegree, such as radiation, O, OH, and HO_2, are also very reactive. However, when the indegree is considered there are significant differences between the two atmospheres. The indegree accounts for the number of reactions that produce a given species. Some of these species are intermediate of other reactions, and others are simply final products, which are in general stable compounds. In the network of the Earth, the species with the highest k_{in} are H_2O, HCl, and O_2—all of which are well-known stable species. In Mars' atmosphere the species with highest indegree are O, OH, NO, and O_2, which, with the exception of molecular oxygen, are very reactive. A criterion of disequilibrium can be obtained by calculating the difference $k_{out}-k_{in}$ for every species in the reaction networks. The highest positive values of $k_{out}-k_{in}$ for Earth's network are for OH (38), radiation (32), Cl (30), and O_3 (23). These are highly reactive species which are more frequently found as reactant than as products. The species with the highest negative values of $k_{out}-k_{in}$ are H_2O (–48) and HCl (–32), which are both well-known stable compounds. For Mars, these disequilibria are not so marked, and the highest positive values of $k_{out}-k_{in}$ are 17 for radiation and 5 for CO_2^+, while the highest negative values are just –8 for NO and –5 for O_2, CO, and O_2^+. The high disequilibrium in Earth's atmosphere does not, however, indicate that it is not highly reactive. As mentioned previously, the high concentrations of oxygen render this atmosphere very oxidant, and this is also reflected in the properties of the reaction network. Several species are created and transformed continuously in Earth's atmosphere. For instance, there are eighteen species for which $k_{out}-k_{in} = 0$, including I, IO, S, Cl_2, FO, FO_2, and $BrNO_2$, among others. However, these species represent only 7.2% of all species in Earth's reaction network. By contrast, 22.6% of species in Mars' atmosphere, including H_2, N_2O, HNO_3, and N_2O_5, among others, have $k_{out}-k_{in} = 0$.

The 'holistic' analysis of reaction networks in atmospheric chemistry can provide useful hints and explanations, as well as allowing new discoveries. The availability of more complete data concerning reactions in planetary and satellite atmospheres will provide network theory with more chances of demonstrating its potential in this field.

16.3 Metabolic networks

The biochemical reactions by which living organisms transform some chemicals into others which are necessary for cellular functions is known as 'metabolism', and the network of the metabolic reactions is known as a *metabolic network* (Lacroix et al., 2008). As in other reaction networks, some of the products of given metabolic reactions are the reactants for others, in such a way that a complex network of metabolic reactions is created. These reactants are usually called *substrates*, and we use the term *metabolites* to refer to the chemicals involved in these reactions. They are small molecules which are imported/exported and/or synthesised/degraded inside the organism. The reactants are usually called *substrates*, and the compounds produced by the biochemical reactions are known as *products*. The other important players in metabolic networks are the *enzymes*—proteins or protein complexes that catalyse some reactions inside the cell. Some biochemical reactions are catalysed by one or more enzymes, while others do not need them. Also, one enzyme can catalyse more than one reaction. Last, but not least, are the *cofactor*s, which are also small molecules that bind enzymes and can enhance or decrease their catalytic activities.

In order to build a metabolic network we need information about the relations between genes, enzymes, and reactions in a given metabolic system. One way of extracting this information is by means of comparative genomics, which starts by determining the catalytic activity of the enzymes which are coded by the genes. It is then necessary to establish the list of reactions which these annotated catalytic activities enable. An important characteristic of metabolic reactions is that of reversibility. That is, under determined physiological conditions, some reactions are *irreversible*; that is, they take place in only one of the two possible directions. The reversibility of metabolic reactions is controlled by the thermodynamics, the kinetics, and the stoichiometry of the reaction. The directionality of metabolic reactions is then added manually *a posteriori*—for example, by the analysis of the stoichiometric matrix of the metabolic network (Yang et al., 2005), or by utilising experimental thermodynamic data such as Gibbs energies of formation and metabolite concentration (Kümmel et al., 2006). Information about metabolism can be found in the following databases:

- KEGG (http://www.genome.jp/kegg/). A collection of databases containing information about genes, biological functions, ligands, and metabolic pathways. The KEGG Pathway database consists of manually drawn metabolic pathways, as well as genetic and environmental information processing, cellular processes, and diseases. An example, for sulphur metabolism, is provided in Figure 16.7.

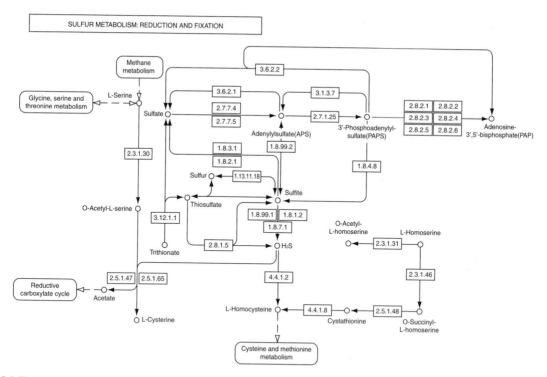

Fig. 16.7

Metabolic network. Schematic representation of the metabolic network for sulphur metabolism, from the KEGG database: http://www.genome.jp/kegg/.

- BioCyc (http://biocyc.org/). A collection of more than 1,000 pathways/genome databases. In each database the genome and metabolism of a single organism is described. It contains one of the most complete existing datasets about *E. coli* metabolism: EcoCyc, which is structured as a Pathway-Genome Database (PGDB). Similar PGDBs exist for other organisms, as well as a cross-species pathway database, MetaCyc, which gathers pathways from more than 1,900 organisms.
- Reactome (http://www.reactome.org). A manually curated and highly reliable database of biological pathways, which includes signalling, innate and acquired immune function, transcriptional regulation, translation, apoptosis, and classical metabolism. It emphasises human metabolism using a cross-species approach. Pathways are curated by experts and are projected on human biology.

Today there are some high-throughput techniques that allow the study of metabolism on a global scale. These techniques are grouped together with the acronym of metabolomics (Kaddurah-Daouk et al., 2008; Hollywood et al., 2006; Verkhedkar et al., 2007). The advantages of using metabolomic techniques are based on the fact that they are able to assay thousands of small organic molecules in cells, tissues, organs, and biological fluids, providing a global biochemical picture of the systems being studied. They also allow for study of the interaction of metabolites with drugs or xenobiotics, which can be of great help in pharmacometabolomic investigations.

16.3.1 Structural analysis

One of the first analyses carried out for metabolic networks was devoted to the study of the degree distributions of metabolites in networks representing the metabolism of different organisms (Jeong et al. 2000; Ma and Zeng, 2003). The networks were built by using annotated genomes for these organisms, combined with data from the biochemical literature. The major finding of this work is that both indegrees and outdegrees of metabolites of metabolic networks for all forty-three different organisms representing archae, bacteria, and eukaryote follow power-laws distributions. In Figure 16.8 we illustrate the degree distributions of the metabolic networks of *Archaeoglobus fulgidus* (archae), *Escherichia coli* (bacterium), and *Saccharomyces cerevisiae* (eukaryote). These results were interpreted on the basis of natural selection for robust and error-tolerant networks, due to the fact that scale-free networks are very resilient to random failures. It has been remarked that although the metabolite–metabolite projection of metabolic networks follows power-law degree distributions, this is not the case for the reaction–reaction projection of the same networks. In this case, binomial distributions are observed—probably as a consequence of the fact that reactions are single events in which the number of participants is restricted to a few reactants (Montañez et al., 2010).

The plots illustrated in Figure 16.8 clearly show that the existence of perfect power-law distributions can be challenged for these networks. The plots also indicate that at least some kind of heavy-tailed degree distribution exists for the majority of these networks. However, the aspect of these findings that has generated the most criticism concerns the biological interpretation of these fat-tailed distributions, the robustness of the networks, and possible reasons for the evolution of the networks into the structures observed so far (Bourguignon et al., 2008). Here, as an example, we present the reasoning of Bourguignon et al. (2008) against the interpretation of hub removals in metabolic networks as a criterion of their robustness. These authors have established that when considering the removal of a hub from these networks, such as water, 'even bacteria with a very small genome already contains hundreds of enzymes that catalyze H_2O-producing reactions. Thus, removing a single hub (the H_2O molecule) from the metabolic network of such an organism would require the deletion of several hundreds of enzymes.' This, of course, will kill the organism long before the last H_2O molecule is removed from the cell.

Another aspect that has generated controversies about the analysis of metabolic networks is the one related to paths and small-worldness. In an early work, Wagner and Fell (2001) studied the metabolite–metabolite projection of the metabolic network of *E. coli*, and observed an average path length of $\bar{l} = 3.88$, which is very close to what should be expected for a random network of the same size. This was reanalysed by Arita (2004), who cleverly used the 'atomic traces' from substrates to products in order to determine biochemically meaningful paths; that is, a biochemical link between two metabolites exists

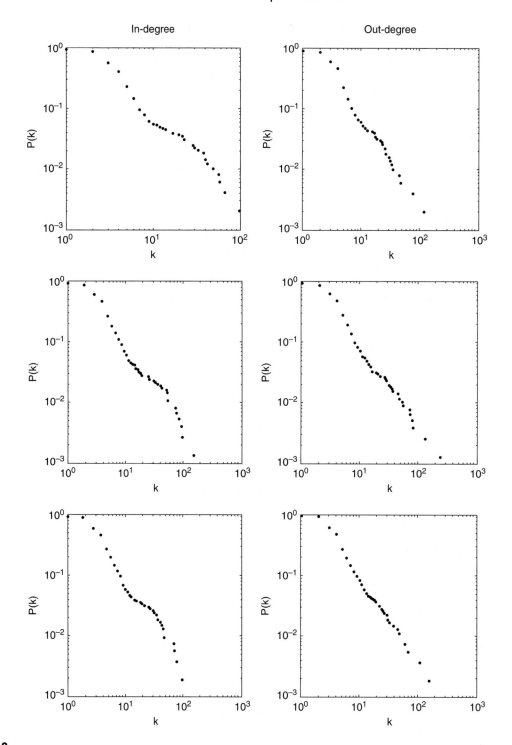

Fig. 16.8
Degree distributions in metabolic networks. Indegree (left) and outdegree (right) distributions for the nodes of the metabolic networks of the archae *Archaeoglobus fulgidus* (top), the bacterium *Escherichia coli* (centre), and the eukaryote *Saccharomyces cerevisiae* (bottom).

only if there is at least one structural moiety which is transferred from one to the other. He then used 'carbon traces'—carbon atoms that can be followed from one metabolite to another in defining a path. In such a way, a path from metabolite X to metabolite Y is defined as a sequence of biochemical reactions through which at least one carbon atom in X reaches Y. As a result of this approach, he found that the average shortest path length is about 8.0 for the directed version of the metabolic network of *E. coli*, which although being larger than that found by Wagner and Fell (2001), is still relatively small compared with the size of the network.

However, a source of additional criticism is the identification of shortest paths in metabolic networks as possible metabolic paths. Here again we return to the problem of representing reaction networks as graphs. We saw in Section 16.1 that a compound–compound projection of a reaction network is not appropriate for selecting chemically meaningful reaction paths. Let us see what happens if we instead use the whole bipartite reaction network, containing both compounds and reactions, as illustrated in Figure 16.9.

We can identify a shortest path from D to F following reactions 3 and 4. However, this path is not at all meaningful for obtaining E, as it is necessary for C to be present to react with D, and C can only be obtained from a sequence of consecutive reactions. This strategy fails even if carbon traces are used, as there will be a sequence of biochemical reactions through which at least one carbon atom in D reaches F. This problem has been largely ignored by researchers in this area, and has produced a considerable amount of criticism of network theory as a tool for studying metabolic networks (Bourguignon et al., 2008). It is not that the network techniques are wrong, but rather, that inappropriate questions have been asked about the representation of such a system.

Some solutions have been proposed relating to the problem of finding biochemically relevant pathways as shortest paths in metabolic networks. For instance, Croes et al. (2006) have used weighted bipartite networks of metabolites and reactions to find metabolic pathways. The approach used by Croes et al. (2006) assigns to every compound in the network a weight equal to its degree, which, due to the bipartite structure used, is equal to the number of reactions in which it takes place. They then searched for minimal paths— those having the minimum sum of the weights between a pair of source and target reactions. The reported results are quite impressive, particularly when compared with those using the raw bipartite graph or a filtered version of it. The raw bipartite network is simply the unweighted version of that used by Croes et al. (2006), while the filtered one is the raw graph from which the thirty-six highest-degree nodes were removed. Using this approach, about 86% of the intermediate reactions and metabolites are correctly inferred, and

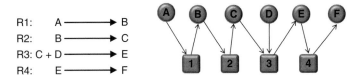

R1: A ⟶ B
R2: B ⟶ C
R3: C + D ⟶ E
R4: E ⟶ F

Fig. 16.9
Hypothetical bipartite reaction network. Schematic representation of a bipartite reaction network (right) for a set of hypothetical reactions (right) (see text for an explanation).

the reactions are positioned in the correct order along the path. To acquire a flavour of these results, let us take the case of the metabolic networks of *E. coli* and *S. cereviciae* extracted from KEGG and aMAZE databases. The raw network displays an average sensitivity of only 31.4% —which is expected, as we have previously analysed the shortest paths in these networks. When this network is filtered, the average sensitivity increases to 68%—a very dramatic improvement of 36.6%. The accuracy also increases from a poor 28.4% up to 65.5%. Before continuing, we need to consider these results very carefully. They tell us, very clearly, that biochemically meaningful pathways are not among the shortest paths in the metabolic network. Instead, they are among paths which display certain average distance which is larger than that of the shortest paths, but not necessarily too large. How can we identify such kinds of path in a network? The answer is easy: simply avoid the hubs. As its name indicates, a hub connects many regions of the network, so that the average shortest path distance among the regions is small. If we avoid them by, for example, eliminating the thirty-six most connected nodes, then we necessarily travel through longer paths—those which are biochemically meaningful. However, the discarding of nodes is ultimately not a good strategy, due to several problems—the most evident being the question of how many should be discarded. Therefore, a better strategy is to weight the nodes with their degree and use the minimal shortest paths, as was carried out by Croes et al. (2006). By using this method they produced a 20% improvement in the average sensitivity and accuracy obtained with the filtered graph. It is clear that network theory has not failed here, because the correct types of question have been asked concerning the representation of the system. But is there any reason why this strategy will work in any metabolic network? The authors suggested that the success of their 'simplistic approach is rooted in the high degree of specificity of the reactions in metabolic pathways, presumably reflecting thermodynamic constraints operating in these pathways.' In the view of the current author, this is a good example of the (sometimes) successful use of phenomenological approaches. If you know *a priori* that the length of the biochemical pathways is longer than the shortest paths of the network, simply avoid those nodes that reduce the topological distance. We consider, however, that the use of more sophisticated network tools, such as reaction hypergraphs, is the only natural way for solving these problems. In other words, the use of reaction graphs for analysing metabolic networks is a reductionist mark inside an holistic approach. For instance, for the hypothetical set of reactions represented in Figure 16.9, the reaction hypernetwork is illustrated in Figure 16.10, where there are six nodes and four hyperedges, which correspond to the existing reactions. There is then only one chemically complete hyperpath, which is the one including all four reactions: $R_1 \rightarrow R_2 \rightarrow R_3 \rightarrow R_4$.

Fig. 16.10

Reaction hypernetwork. Representation of the reaction network illustrated in Figure 16.9 as a directed hypernetwork. Here, R_3 is a hyperlink connecting both C and D with E at the same time.

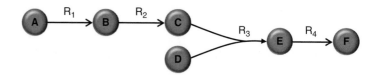

Another structural characteristic of metabolic networks that has attracted much attention in the literature is that of modularity—in particular, 'hierarchical modularity.' According to the findings reported by Ravasz et al. (2002) 'the metabolic networks of forty-three distinct organisms are organized into many small, highly connected topologic modules that combine in an hierarchical manner into larger, less cohesive units, with their number and degree of clustering following a power-law.' In other words, the Watts–Strogatz clustering coefficient decays as a power-law with the degree of the nodes: $C(k) \sim k^{-1}$. Figure 16.11 illustrates such a power-law for the three metabolic networks previously studied in Figure 16.8. The distribution of cliques of different sizes in these metabolic networks is shown at the right of the figure.

In order to understand these power-law decays of the clustering in terms of degree, we have to consider that metabolic networks in general display large clustering coefficients. This has been interpreted as a consequence of the projections of the bipartite metabolite-reaction network (Montañez et al., 2010). In the bipartite network there are no odd cycles and the clustering is zero, but when it is projected to the metabolite–metabolite or reaction–reaction networks, many triangles appear. In order to test this hypothesis, Montañez et al. (2010) have generated random bipartite networks formed by two disjoint sets of nodes V_1 and V_2, and have projected them into V_1–V_1 or V_2–V_2 spaces. The resulting projections display a large clustering coefficient—much larger than random unipartite networks generated by using similar procedures. For instance, the clustering coefficient for the metabolite–metabolite projection of the version of the metabolic network of *E. coli* used by these authors is $\overline{C} = 0.67$. The clustering for a unipartite graph of similar size is only $\overline{C} = 0.01$, but that for a projection of a random bipartite network is $\overline{C} = 0.19$. The bipartite random network generated by these authors is an undirected one, while the bipartite metabolic network is directed. We argue that directionality in the bipartite network can futher increase the clustering of the corresponding projections. More importantly, Montañez et al. (2010) have found that the power-law decay of the clustering coefficient with the degree of the nodes also appears as a consequence of the projection of the bipartite network into metabolite–metabolite spaces. According to the results produced by Montañez et al. (2010), a projection of a random bipartite network produces a similar, though not so pronounced, decay of $C(k)$ versus k. As the authors have stressed, 'such dependence occurs in the ER bipartite null model even without satisfying the scale-free condition and being dramatically distant from the prototype hierarchical one.' As explained in Chapter 4, the presence of such power-law decay is more a consequence of the decay in the number of cliques with the size of the cliques than of the hierarchical architecture of a network. As can be seen in the right-hand side of Figure 16.11, there are no cliques of size larger than 7 in these metabolic networks. In addition, the number of cliques of size k decays very fast with the size of the clique, indicating that the only truly abundant cliques are triangles.

While there is some argument concerning the hierarchical nature of the modularity in metabolic networks (Montañez et al., 2010), there is no doubting their modular structure. For instance, Ravasz et al. (2002) have used a modification

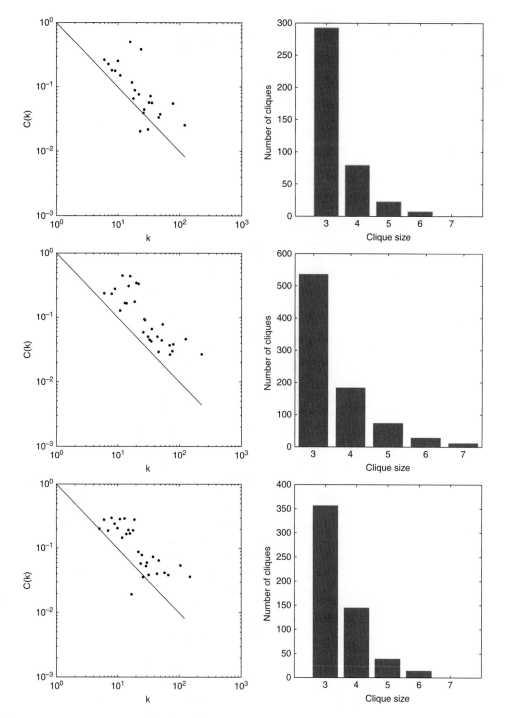

Fig. 16.11

Clustering, degree, and cliques in metabolic networks. Plot of the decay of the Watts–Strogatz clustering coefficient as a function of the degree of the nodes (left) for the metabolic networks of the archae *Archaeoglobus fulgidus* (top), the bacterium *Escherichia coli* (centre), and the eukaryote *Saccharomyces cerevisiae* (bottom). Distribution of the number of cliques as a function of clique size (right) for the same metabolic networks (see text for interpretation of these plots).

of the adjacency matrix of metabolic networks known as the 'overlap matrix' and an hierarchical clustering algorithm for detecting modules in metabolic networks. The overlap matrix O_T has the following entries for nodes i and j:

$$O_T(i, j) = \frac{\sum_{m=1}^{n} A_{im} A_{jm} + A_{ij}}{\min(k_i, k_j) - A_{ij} + 1} \tag{16.1}$$

Using O_T and the unweighted average linkage method, Ravasz et al. (2002) extracted several modules existing in metabolic networks. Then, by reordering the resulting overlap matrix, they grouped metabolites that form tight clusters close to the main diagonal of the matrix. Metabolites are then classified according to their known functional properties, which represent carbohydrate metabolism, nucleated and nucleic acid metabolism, protein, peptide, and amino acid metabolism, lipid metabolism, and so on, as illustrated in Figure 16.12 for the metabolism of *E. coli*.

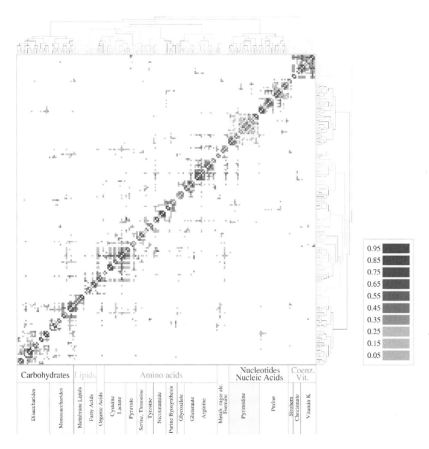

Fig. 16.12

Topological modules in the *E. coli* metabolic network. The topological overlap matrix for the metabolic network of the bacterium *E. coli*. From Ravasz, E., Somera, A. L. Mongru, D. A., Oltvai, Z. N., and Barabási, A.-L. (2002). Hierarchical organization of modularity in metabolic networks. *Science* **297**, 1551–5. (Reprinted with permission from AAAS.)

By using a maximisation of the modularity measure studied in Chapter 10, and with the help of simulated annealing techniques, Guimerá and Amaral (2005) identified modules in metabolic networks. They then classified every group of metabolites into one of the nine existing groups of pathways proposed in the database KEGG: carbohydrate metabolism, energy metabolism, lipid metabolism, nucleotide metabolism, amino acid metabolism, glycan biosynthesis and metabolism, metabolism of cofactors and vitamins, biosynthesis of secondary metabolites, and biodegradation of xenobiotics. It is interesting that in metabolic networks the clusters are very tight, as shown by Guimerá and Amaral (2005), who reported that typically 80% of the nodes display only intracluster links and very few intercluster links. These results are illustrated in Figure 16.13.

The modularity of metabolic networks has been found to be correlated with the variability of the natural environment in which bacteria live. The environmental variability refers to this environment, and includes six classes (Parter et al., 2007):

i) *Obligate bacteria* are those which are obligately associated, intracellularly or extracellularly, with a host.

ii) *Specialised bacteria* live in specialised environments such as marine thermal vents.

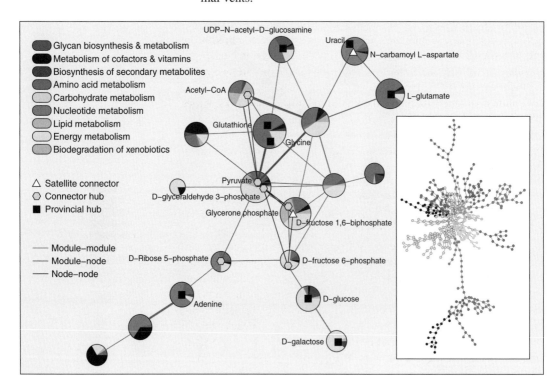

Fig. 16.13

Cartographic representation of the metabolic network of *E. coli*. Representation of the metabolic network of *E. coli* in which each circle represents a module classified according to the KEGG pathway classification. Non-hub connectors (triangles), connector hubs (hexagons), and provincial hubs (squares) are also represented. From Guimerá, R., and Amaral, L. A. N. (2005). Functional cartography of complex metabolic networks. *Nature* **433**, 895–9. (Reprinted with the permission of Macmillan Publishers Ltd., ©2005.)

iii) *Aquatic bacteria* are not associated with any host, and live in freshwater or seawater environments.

iv) *Facultative bacteria* are free living bacteria that often associate with a host.

v) *Multiple bacteria*, such as those with a wide range of hosts, live in several different environments,.

vi) *Terrestrial bacteria* are those that live in the soil.

By considering a series of physical conditions of these environments—such as temperature, osmolarity, oxygen availability, and so on—Parter et al. (2007) have ranked the variability of these six environments in the following order: obligate < specialised ≤ aquatic < facultative < multiple < terrestrial. That is, obligate bacteria have the most constant environments which correspond to biochemically controlled and isolated environments. On the other hand, terrestrial bacteria live in the most variable environment, since soil is highly heterogeneous and has a diverse ecology.

The first interesting observation made by Parter et al. (2007) is that the size of metabolic networks is correlated with environmental variability. That is, bacteria living in more constant environments have metabolic networks smaller than those living in variable environments. This is understandable when considering the number of metabolites as part of the responses of bacteria to the different challenges in their environments. Obligate bacteria, which live in constant environments, are not confronted with many environmental challenges, and can minimise their metabolism to adapt to these circumstances. On the other hand, terrestrial bacteria are constantly encountering challenges from the environment, and different metabolites and pathways are needed to survive in these conditions. However, the total number of nodes produces very little information about the structure and organisation of such networks. Parter et al. (2007) used a normalisation of Newman's modularity based on random graphs which preserve the degree distribution of the real network, and found that the modularity of metabolic networks tends to increase with an increase in environmental variability.

The interpretation of these results is based on the fact that more than one possible route is possible to achieve the same metabolic goal faced by a bacterium. For instance, bacteria need certain amino acids to function properly. If this amino acid exists in the environment the bacterium can import it, but if not then the bacterium must synthesize it. Consequently, both situations can arise for a bacterium living in a variable environment, and the organism has two routes to the amino acid in its metabolism. Conversely, if the bacterium lives in a constant environment, only one of these possibilities exists, and it needs only one of the routes to obtain the amino acid. Consequently, bacteria living in varying environments typically evolve a functional module for each of these routes, while those in constant environments evolve towards fewer modules. One of the main disadvantages of the work of Parter et al. (2007) is that they ignore the directionality of the metabolic reactions. We saw in the previous section that the directionality of chemical reactions, which is related to their reversibility, can play an important role in understanding the equilibrium/out-of-equilibrium nature of the entire system of reactions. Therefore, the use of measures such as network returnability may play an important role in understanding this important relationship between the topology of networks and the environmental variability of different species. This type of study constitutes one of the emerging areas of scientific research in which network theory can play a very important role. It has been termed 'reverse ecology', and has produced the first interesting results about topological signatures of species in metabolic networks, and clear differentiations between organisms in different kingdoms based on their metabolic network data (Borenstein and Feldman, 2009; Borenstein et al., 2008; Zhu and Qin, 2005).

Anatomical networks

17

> The elementary parts of all tissues are formed by cells in an analogous, though very diversified manner, so that it may be asserted, that there is one universal principle of development for the elementary parts of organisms, however different, and that this principle is the formation of cells.
>
> Theodor A. H. Schwann (1847)

By anatomical networks we refer here to the interconnectivity between the physical regions of the whole or parts of an animal or plant. These physical regions can be constituted by the interaction of single cells like those in tissues and neural networks, or can be formed by groups of cells located in definite regions. They can have definite morphologies, such as channels and vessels, in the animal or plant, or can be defined by the connectivity between certain anatomical or functional regions. Examples of anatomical networks are cell-graphs representing tissues, the vascular networks that conduct fluids in organisms, the networks of Havers and Volkmann channels inside the bones of mammals, and the networks interconnecting several regions in the human brain. Here we also consider functional networks, in which attention is paid to the functional connectivity between parts of an organ or organism—that is, the analysis of how patterns of functional relations between areas of an organ appear throughout such organs. In particular, we will be interested in functional networks in the brain.

17.1 Intercellular networks

The study of intercellular networks dates back to the 1970s. The term *cellular sociology* was coined by Rosine Chandebois (1976) in a series of papers published in the 1970s and then formalised by several authors in the 1980s and 1990s (Bigras et al., 1996; Weyn et al., 1999). The term refers to the topological organisation produced by the spatial relationship among cells in different tissues. The general idea of all these methods is to represent the information about the cellular organisation of a tissue in the form of a graph, of which the use of Voronoi's diagrammatic (Voronoi 1907; Aurenhammer, 1991) representation has emerged as the most useful technique. First, images of the stained tissue at the microscopic level are taken by using background correction processing. The images are then segmented, and a Voronoi diagram is constructed as follows. A set of cell markers—such as the centres of gravity of the nuclei—are used, and for any element p of this set Y, the Voronoi polygon $Z(p)$ associated with p is the locus of points r that are closer to p than to any point q in the plane $\mathfrak{R}^2$:

$$Z(p) = \{r \in \mathfrak{R}^2, \forall q \in Y\backslash\{P\}, dist(r, p) < dist(r, q)\} \qquad (17.1)$$

Then, $Z(p)$ makes a partition of the plane which is called the Voronoi diagram, in which every of these partitions represents the zone of influence of the corresponding element. The Voronoi diagram for a group of fourteen cells is illustrated in Figure 17.1.

One of the strategies followed in early studies of cellular sociology considered the Voronoi diagram as the network representation of the tissue. As can be seen in Figure 17.1, the polygons located at the periphery are open to infinity. In order to avoid border effects, Bigras et al. (1996) proposed the use of only triplets of points whose respective circumcircles do not intersect the convex hull of the sampling area. An example of the Voronoi network obtained for a typical carcinoid is reproduced in Figure 17.2, from the work of Bigras et al. (1996).

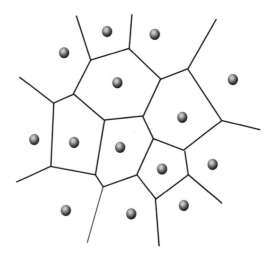

Fig. 17.1
Voronoi diagram. The Voronoi diagram for a set of fourteen nodes placed in a plane.

Fig. 17.2
Voronoi diagram of a typical carcinoi. Example of a typical carcinoid tissue (left) and its Voronoi diagram (right). From Weyn, B., van de Wouwer, G., Kumar-Singh, S., van Daele, A., Scheunders, P., van Marck, E., and Jacob, W. (1999). Computer-assisted differential diagnosis of malignant mesothelioma based on syntactic analysis. *Cytometry* **35**, 23–9. (Reproduced with the pemission of John Wiley and Sons, ©1999.)

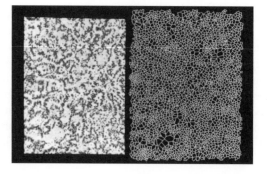

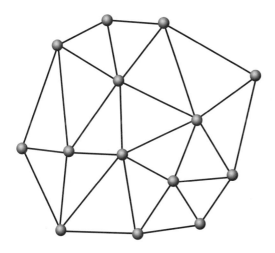

Fig. 17.3
Delaunay triangulation. Delaunay triangulation obtained from the Voronoi diagram illustrated in Figure 17.1.

Other authors developed further transformations of the the Voronoi diagram in order to produce cellular networks. Most of these transformations start by using the Delaunay triangulation. In this triangulation the nodes of the network are the centres of the cells, and two of them are connected if their respective Voronoi polygons have a common edge. In other words, the Delaunay triangulation (Delaunay, 1934; George, 1999) is the dual of the Voronoi diagram in $\Re^2$, which means that there is a natural bijection between the two which reverses the face inclusions. Figure 17.3 illustrates an example of the Delaunay triangulation for the Voronoi diagram in Figure 17.1.

From the Delaunay graph, several types of cellular network have been explored—such as weighted minimal spanning trees, Gabriel graphs, graphs of relative or nearest neighbours, and others. The most used of these graphs are the weighted minimal spanning tree (MST) and the Gabriel graph (GG). The MST is obtained by considering a tree that joins all nuclei in such a way that the sum of the length of the branches is minimal. On the other hand, the GG is obtained from the Voronoi diagram with node set S, in which the points p and q are connected if the closed disc of which the line segment pq is a diameter contains no other points of the set S. An example of a digitized image of a section of a pleural mesotheliome, with its Voronoi diagram, GG, and MST, reproduced from the work of Weyn et al. (1999), is illustrated in Figure 17.4.

In a more recent work, the so-called 'cell-graphs' have been used to study cellular networks in different kinds of tissue (Bilgin et al., 2007, 2009; Demir et al., 2005a, 2005b, 2005c; Gunduz et al., 2004). In order to generate the cell-graphs, they proceeded as usual by carrying out image segmentation, using K-means algorithms in order to cluster the pixels of the microscopic images of tissues, according to their red, green, and blue (RGB) values, into clustering vectors. In this case, the clustering vectors were estimated as maximising an error function defined as the sum of the squares of the difference between the centre of the clusters and the intensity value of the image. Figure 17.5 illustrates an example of the segmentation of a microscopic image of a surgically removed sample of human breast cancer tissue.

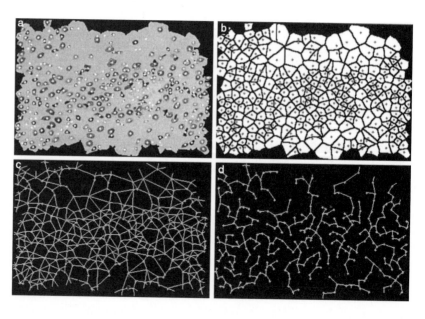

Fig. 17.4

Network representation of a malignant mesothelioma. A digitised image a section of a pleural malignant mesothelioma (A), its Voronoi diagram (B), the Gabriel graph (C), and the minimum spanning tree (D) obtained from them. From Bigras, G., Marcelpoil, R., Brambilla, E., and Brugal, G. (1996). Cellular sociology applied to neuroendocrine tumors of the lung: Quantitative model of neoplastic architecture. *Cytometry* **24**, 74–82. (Reproduced with the permission of John Wiley and Sons, ©1996.)

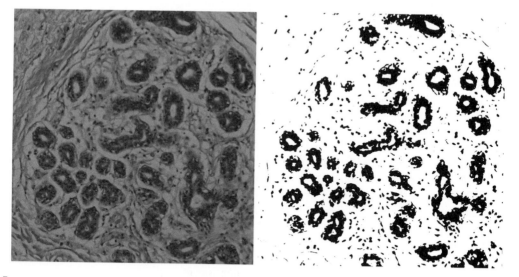

Fig. 17.5

K-means segmentation of breast cancer tissues. Illustration of an invasive tissue extracted from a breast, opened in RGB space (left), and the result of the K-means segmentation of it (right), in which black points are parts of cells and white points are treated as background. (Illustrations courtesy of C. Bilgin.)

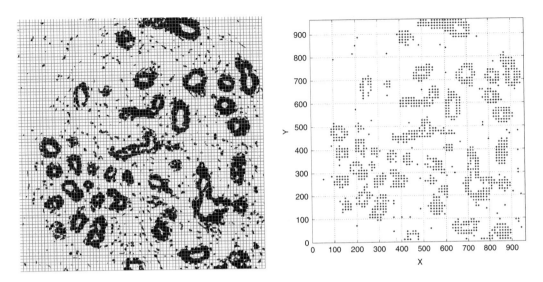

Fig. 17.6
Node identification in segmented breast cancer tissues. The result of the application of a grid (left) to the segmented image of a tissue illustrated in Figure 17.5, and the overall result of node identification (right). (Illustrations courtesy of C. Bilgin.)

In the second step of the generation of cell-graphs, a grid is placed over the segmented image in order to identify the cells. Then, for each grid entry the probability of being a cell is calculated as the ratio of cell pixels to the total number of pixels in the grid. Using thresholding, it is decided whether or not this grid entry is a cell. Figure 17.6 illustrates the results of applying the grid to the image of Figure 17.5, as well as the node identification using thresholding.

At this point, the authors defined three different kinds of cell-graph (Bilgin et al., 2007, 2009; Demir et al., 2005a, 2005b, 2005c; Gunduz et al., 2004). *Simple cell-graphs* are based on considering the Euclidean distance between two cells in order to construct the network. That is, cells are represented as nodes, and two nodes are connected if their Euclidean distance is less than a certain threshold. The second type of network is *probabilistic cell-graphs*, in which a link between two cells p and q is built with probability

$$P(p, q) = d(p, q)^{-\alpha} \qquad (17.2)$$

where $d(p, q)$ is the Euclidean distance between the two cells, and α is a parameter. The third type of network is *hierarchical cell-graphs*, which after the node identification step places a larger grid on top of these cells. Then, by calculating the number of cells in the grid and by dividing this number by the grid size, the probability of being a cluster is calculated for each grid entry. Finally, a threshold is used in order to consider the grid entries with a probability greater than this threshold as a cluster. The simple and hierarchical cell-graphs for the image in Figure 17.5 are illustrated in Figure 17.7.

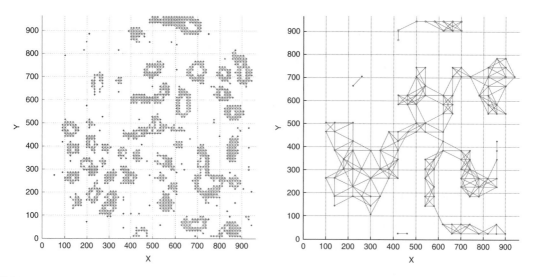

Fig. 17.7

Networks of breast cancer tissues. A simple cell-graph (left) formed on the basis of the location of the cells, and an increase of the grid size (right) in order to capture the cell clusters from which hierarchical graphs are built. (Illustrations courtesy of C. Bilgin.)

17.1.1 Intercellular networks and cancer diagnosis

One of the most interesting applications of intercellular networks is the distinguishing of cancerous cells from healthy and unhealthy but not cancerous cells. This analysis represents an important step forward for the correct diagnosis of cancer. There are many approaches in the automated classification of histopathological images in cancer diagnosis. In general, they use several morphological features of cells, such as area, perimeter and roundness of the nucleus, and/or textural features derived from the co-occurrence matrix. These techniques are based on a large variety of pattern recognition techniques, which include artificial neural networks (ANN), support vector machines, genetic algorithms, and many others. In the following we describe some results obtained by using topological information from cell-graphs in combination with ANN for discriminating cancerous cells from healthy and inflamed cells.

Gunduz et al. (2004) have studied the cell-graphs of a type of brain cancer called 'glioma'. They consist of 1,000–3,000 cells and 2,000–10000 links among them, taken on the basis of the Euclidean distance between pairs of cells. They have also constructed cell-graphs for healthy and inflamed tissues by using exactly the same technique as for the glioma cells. In total, they studied 285 images of 80x consisting of 384 x 384 pixels, taken from twelve different patients. The strategy used here for classifying the cell-graphs in the three classes—cancerous, inflamed, and healthy—is as follows. For each node in a cell-graph, a series of k node measures is calculated. Then, a matrix formed by k columns and S rows is created. The number of rows is given by the sum of the number of nodes in each of the 285 cell-graphs studied. The node measures used in that study were the degree, clustering coefficient, eccentricity—eccentricity 90 being defined as the minimum of the shortest path lengths for

Table 17.1 **Pattern recognition in brain cellular networks.** Percentages of good classification of tissues (healthy, inflamed, and glioma) represented by cell-graphs for training and
prediction sets (patient-dependent and independent), as reported by Gunduz et al. (2004).

	Training set	Patient-dependent set	Patient-independent set
Total	93.58	85.74	94.04
Healthy	98.19	98.39	98.76
Inflamed	86.19	83.05	84.05
Glioma	86.34	86.75	88.93

a node required to reach at least 90% of the reachable nodes around it—and
the closeness centrality. Then, using a multilayer perceptron ANN with five
hidden neurons, they classified every node of 68 images from six patients into
the three classes studied. The accuracy values were then obtained by averaging
over the nodes of the graphs for different runs, which indicates that 93% of
the cell-graphs were correctly classified in this series. Then, two prediction
series—one of 82 images taken from the same six patients used before, and
the other of 135 images taken from the remaining six patients—were used to
analyse how the method makes predictions in patient-dependent and patient-
independent sets. The results are shown in Table 17.1.

In general, it is not very difficult to discriminate between healthy and cancerous cells, and the best accuracy is always reported for healthy tissues. In fact,
simple use of the size of the networks can identify healthy tissues, as most of
these networks have about 500 nodes while the inflamed and cancerous ones
always have more than 1,000 nodes. The difficult task here is to distinguish
cancerous from inflamed tissues. Figure 17.8 shows an example of healthy,
inflamed, and cancerous networks as used in the work of Gunduz et al. (2004).

In another study, Demir et al. (2005b) used global network metrics in order
to classify these cell-graphs into the three groups considered previously. The
global measures used were the average degree, the average Watts–Strogatz
clustering coefficient, the average eccentricity, the fraction of nodes in the giant
connected component, the percentage of nodes with degree 1, the percentage
of isolated nodes, the spectral radius of the adjacency matrix, and the slope of
the sorted plot of the fifty largest eigenvalues of the adjacency matrix in a log–
log plot, called the eigen-exponent. Using these global network indices and
ANN, they analysed 646 images of brain tissue samples from sixty patients. In
this case, the results were quite impressive, showing 99.55% of overall correct
classification in the training set, 95.45% in the prediction set, and accuracies
of 98.15%, 92.50%, and 95.14% for healthy, inflamed, and cancerous cells
respectively. The quality of these results surpasses that of the those obtained
by using a textural approach to classifying the same group of tissues, which
reaches only 89.03% of overall accuracy, as well as 75.21% and 89.04% of
accuracy for inflamed and cancerous tissues respectively.

The study of cell-graphs of different tissues can involve different types of
strategy for building the cellular networks. For instance, the brain tissues previously studied present a type of diffusive structure, which contrasts with the
lobular/glandular architecture exhibited by breast tissues. In order to confront

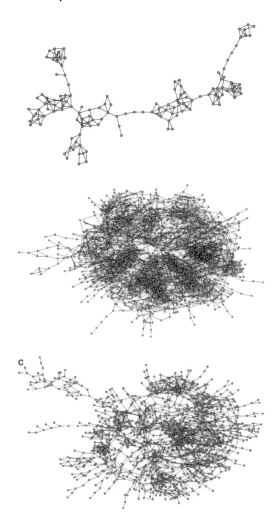

Fig. 17.8

Cellular networks of brain tissues.
Representation of cellular networks of a
healthy (top), an inflamed (centre), and
glioma (bottom) brain tissues.

this structural challenge when studying breast tissues, Bilgin et al. (2007)
used hierarchical cell-graphs characterized by global structural parameters
for classifying breast tissues in three different categories. The first of these
categories is that of healthy tissues, the second consists of tissues with benign
reactive processes, such as hyperplasia, radial scar, or inflammatory changes,
and the third group involves infiltrating carcinomas. The use of a support vector
machine allowed the correct classification of 81.8% of the 446 breast tissue
samples from thirty-six patients. Again, the difficult group is that of the benign
reactive processes, for which 75.6% of tissues were correctly classified, while
83.3% of the invasive tissues and 82.9% of the healthy tissues were correctly
identified.

Other biological information, different from the intercellular proximity, can
be added to the intercellular networks. For instance, Bilgin et al. (2009) have
considered the use of information about the extracellular matrix (EMC), which
plays an important role in the functioning of tissues. After the segmentation

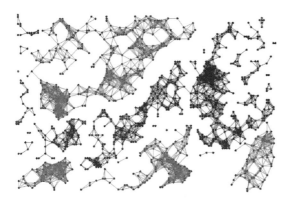

Fig. 17.9
ECM-aware cell network. An ECM-aware cell-graph of fractured bone tissue. From Bilgin, C. C., Bullough, P., Plopper, G. E., and Yener, B. (2009). ECM-aware cell-graph mining for bone tissue modeling and classification. *Data Min. Knowl Discov.*, **20**, 416–38, figure 2(e). (This illustration, and any original (first) copyright notice displayed with material, is reproduced with the permission of Springer Science+Business Media.)

of the microscopic tissue images, a colour, based on the RGB values of its surrounding EMC, is assigned to each cell. Links connect pairs of nodes which are physically in contact, and in addition have the same four colours (red, green, blue, and white), which correspond to the predominant colours of the ECM pixels. The result is a multi-coloured cell-graph, such as example shown in Figure 17.9.

Using a combination of global and spectral invariants, Bilgin et al. (2009) studied twenty images of healthy bone tissues, 39 fractured tissues, and 75 diseased bone tissues. The use of ECM-aware cell-graphs was compared to that of simple cell-graphs and graphs derived from Delaunay triangulation. The three methods were very well able to differentiate healthy from cancerous tissues. However, the differentiation between fractured and cancerous tissues, and between healthy and fractured tissues, are more difficult, and the ECM-aware cell-graphs emerge as the best solution. The analysis of the receiver operator characteristics (ROC) plot, which is a way of comparing sensitivity versus specificity, has an area under the curve of 0.9781 for the ECM-aware graphs, while it is only 0.6667 for the Delaunay graph for fractured versus cancerous tissue classification. For healthy versus fractured, the areas are 1 and 0.9237 for ECM-aware and Delaunay graphs respectively. Overall, the ECM-aware graph is able to identify 93.5% of cancerous tissues, while Delaunay and simple cell-graphs correctly identify 69.17% and 75.7% respectively. The percentages of good classification for fractured tissues are 87.1%, 69.7%, and 79.3% for ECM-aware, Delaunay, and simple graphs respectively. These experiments clearly indicate that the addition of extra information to cell-graphs representing intercellular networks can provide further value for distinguishing tissues in different pathological states.

17.2 Vascular networks

In a vascular network the nodes represent the confluence of channels, and the links represent the connections between them. In one example of this kind of network, da Fontoura and Viana (2006) have constructed the network of Havers and Volkmann channels of two humeri (left and right) of an adult cat. A reproduction of the three-dimensional bone channels and its network

(a) (b)

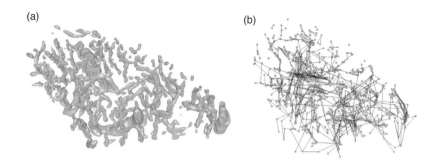

representation is shown in Figure 17.10. Another bone network, consisting of
1,800 nodes and 2,131 links in the right tibia of a pig, has been further studied
by Viana et al. (2010).

The giant connected components of the networks generated for the humerus
consist of 421 nodes and 516 links for the left bone, and 439 nodes and 526
links for the right one. The analysis of them clearly indicates that the node
degrees are distributed according to power-laws of the form $p(k) \sim k^{-\gamma}$, with
$\gamma = 3.8$ and $\gamma = 3.7$ respectively. It was also observed that the highest probability of finding a link in a network occurs for pairs of nodes having degree
equal to 3, which indicates the preferential attachment between bifurcation
points in the bone network. A distinctive characteristic of these bone channel
networks is that they do not display the characteristic branching fractality of
the majority of vascularisation networks in which a node is bifurcated into two
branches, each of them bifurcated into four branches, and so forth. da Fontoura
and Viana (2006) have argued that the presence of this characteristic could
be related to the impossibilities of neovascularisation to provide bypasses to
region irrigation in bones, which necessitates complex and extended reorgani-
sation of Havers and Volkmann channels containing the arteries.

Vascular networks can also be obtained by using Voronoi tessellation of
images obtained from angiographic studies. For instance, the use of centroid
Vororoi tessellation (CVT) of magnetic resonance angiography (MRA) images
has been used by Aylward et al. (2005) to study intracranial vascular networks.
Using MRA data, these authors first identified the location of vessels, their
directions of blood flow, and their branching points. This information was
further processed in order to build the vascular networks using CVT, such as
those illustrated in Figure 17.11.

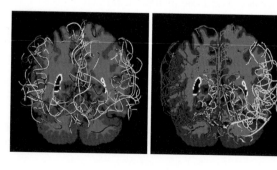

The authors then used several measures, including a variation of the Katz centrality index for directed networks (see Chapter 7). Specifically, they studied nine right-handed male and nine right-handed female subjects with average ages of 34.4 and 35.3 respectively (see Figure 17.11). They were than able to discriminate between males and females based on intracranial vasculature, and further tested the classification method on six subjects not used for training purposes. More challenging, the authors produced a mirror image of the right-handed graphs along the x-axis, which approximates brain lateralisation. Then, using these graphs for right-handed males and mirrored-right-handed males, they classified three testing males known to be left-handed and three others known to be right-handed. Despite the limited amount of data used in these experiments, it is encouraging to see that network theory can be applicable to the extraction of information from intracranial vascular networks.

The combination of network theoretic tools with anatomical or physiological information can provide a better understanding of vascular networks. Espinoza-Valdez et al. (2010) have developed a model for renal arterial branching, which is based on the mechanism of angiogenesis. Using a combination of network theory and physiological information, the authors developed a quantitative model for the vascular tree which considers nodes as the branching points and the links as the vessels. They then considered a growing process in which links are, as usual, assigned to pairs of nodes or can be assigned to one single node to indicate that the vasculature can terminate at this link. Then, by simulating the two processes involved in the angiogenesis—sprouting and splitting—they generated an algorithm that creates networks similar to arterial vascular trees, and specifically studied it for the human kidney. In another study, Reichold et al. (2009) studied the vascular network for the cerebral vasculature by representing it through the use of multigraphs (see Chapter 1 for definitions). In order to incorporate anatomical information, they considered that every link representing a vessel is a straight cylinder of the same length as its real, tortuous counterpart. Also, every node is associated with a blood pressure value, and the conductance of every vessel is assigned on the basis of cross-sectional areas and length of the vessels, as well as the dynamic viscosity and blood density. Then, using a vascular graph model, the authors computed blood pressure, flow, and scalar transport in realistic vascular networks. In addition, these vascular networks are embedded in a computational grid representing brain tissue, with which the authors simulated the oxygen extraction from the vessels. Such modelling allows study of the effects of local vascular changes, such as those occurring during functional hyperaemia, or those taking place when occlusion of an arteriole takes place. These studies demonstrate the capacity of network theory, used in combination with anatomical and physiological information, in studying important problems in vascular networks.

17.3 Brain networks

For, every time a certain portion is destroyed, be it of the brain or of the spinal cord, a function is compelled to cease suddenly, and before the time known beforehand when it would stop naturally, it is certain that this function depends upon the area destroyed.

Julien J. C. Le Gallois (1812)

The study of networks in the brain has become one of the most interesting and exciting areas of application of network theory. This revolution has been conditioned in part by the tremendous development in modern brain-mapping techniques, which include diffusion magnetic resonance imaging (MRI), functional MRI, electroencephalography (EEC), and magnetoencephalography (MEG), among others. On the other hand, this field has benefitted substantially from the concepts and tools developed in modern network theory, as well as in other mathematical and computational methods, giving rise to the new field of 'neuroinformatics'. An excellent compilation of the works in this area is the book *Networks of the Brains*, by Olaf Sporn (Sporn, 2011), which the reader is encouraged to read for a deep account of this field. Brain networks are the result of considering the connectivity of brain regions, in which such connectivity can be determined by anatomical tracts or by functional associations. The nodes of these brain networks are defined by the positions in which electrodes from encephalography or multielectrode-array techniques are placed in the brain, or by regions which are defined anatomically on the basis of histological, MRI, or diffusion tensor imaging techniques (Hagmann et al., 2006). Once the nodes of the network are defined, the links between them must be found. These links are obtained by estimating a continuous measurement of association between the nodes, such as by using spectral coherence or Granger causality measures in MEG, by inter-regional correlations in cortical thickness, by volume MRI measurements in groups of individuals or using probabilistic techniques based on information provided by individual diffusion tensor images, or by other methods. Figure 17.12—reproduced from Bullmore and Sporn (2009)—is a schematic representation of the processes of generating anatomical and functional brain networks.

17.3.1   Anatomical brain networks

In several sections of this book we have studied some topological characteristics of the neural network of the worm *C. elegans*. However, when we consider more complex organisms, the number of neurons and connections increases so dramatically that it is impossible to deal with such networks. For instance, the human neural network is estimated to have 10^{11} neurons and 10^{15} connections, which makes this system untreatable in the near future. In addition, it has been remarked that the alteration in single neurons or synapses have not been shown to have important consequences at the macroscopic level (Sporn et al., 2004). Therefore, because most human cognitive functions depend on the activity and coactivity of large regions of the brain, the analysis of anatomical rather than intercellular networks has been the main choice for studying these systems. However, there is not a unique way of dividing a brain into these regions. It has been considered, for instance, that in the human cerebral cortex there should be about a hundred of these regions and areas. Therefore, it is absolutely

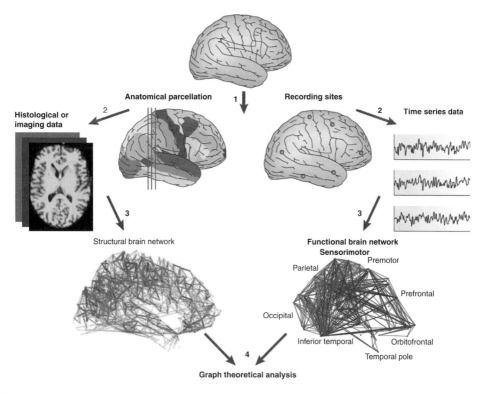

Fig. 17.12

Construction of brain networks. Schematic illustration of the process for building structural and anatomical brain networks. From Bullmore and Sporn (2009). Complex brain networks: graph theoretical analysis of structural and functional system. *Nature Rev. Neurosc.* **10**, 186–98. (Reproduced with the permission of Macmillan Publishers Ltd., ©2009.)

necessary that the nodes and links of the brain networks to be analysed should be carefully selected. The reason is that the neurobiological interpretation of the topological results extracted from such networks depends largely on the selection of nodes and links in the brain.

The first anatomical network of a mammalian cortex was obtained by Felleman and van Essen (1991) from the prior tract-tracing literature. This network consists of thirty-two nodes representing regions of the visual cortex of the macaque monkey, and 305 directed links, which represent the axonal connections between these regions. Networks of the whole cortex were also constructed for the macaque monkey and the cat by Young (1993) and Scannell et al. (1999) respectively. Figure 17.13 represents, after the work of Felleman and van Essen (1991), the network of the visual cortex of the macaque monkey.

There are several versions of these networks in the literature, and the reader should be aware of the differences among them. For instance, Sporn and Kötter (2004) have modified the network of Felleman and van Essen (1991) by consolidating some regions and eliminating a couple of them for which there was lack of connectional information. This resulted in a network having thirty nodes and 311 directed links for the visual cortex of the macaque monkey.

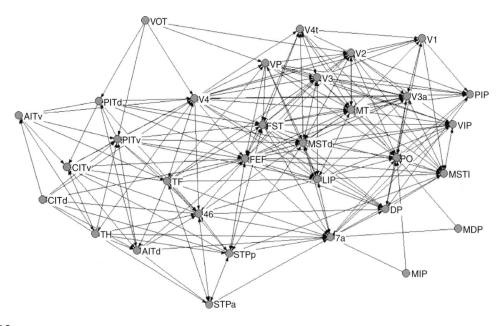

Fig. 17.13

Macaque visual cortex network. Directed network connecting the different anatomical regions of visual cortex of the macaque monkey, according to data from Felleman and van Essen (1991).

They also eliminated the regions of the hippocampus and the amygdale in the original Young (1993) cortex dataset of macaque, resulting in a network having 71 nodes and 746 links. The cat dataset compiled by Scannell et al. (1999) was also modified by Sporn and Kötter (2004) by discarding density information and encoding all pathways as present or absent instead of the gradation given by Scannell et al. (1999). This resulted in a network having 52 nodes and 820 directed links. Finally, in a more recent work, Modha and Singh (2010) have derived a network of the macaque brain by incorporating 410 anatomical tracing studies, consisting of 383 hierarchically organised regions of the cortex, thalamus, and basal ganglia, and 6,602 directed links representing long-distance connections between these regions. A pictorial representation of this network is reproduced in Figure 17.14, where it can be seen that there is a clear discrimination of brain regions.

In Table 17.2 we display some topological properties of the cortex networks of the macaque monkey and the cat, modified by Sporn and Kötter (2004). For the sake of comparison we also include the values of these parameters for the whole network of the cat cortex, as determined by Scannell et al. (1999). The densities of all these networks are relatively high, which indicates a high level of interconnectivity among the different regions of the cortex in these brains. The average path length is relatively small, with values that are not larger than 2 in most of the cases. The clustering coefficient is also high—possibly as a consequence of the high interconnectivity of the brain regions.

The property of these networks of being highly clustered and displaying small average path length—their 'small-worldness'—has received a great deal of attention in the neuroinformatics literature (Bassett and Bullmore, 2006).

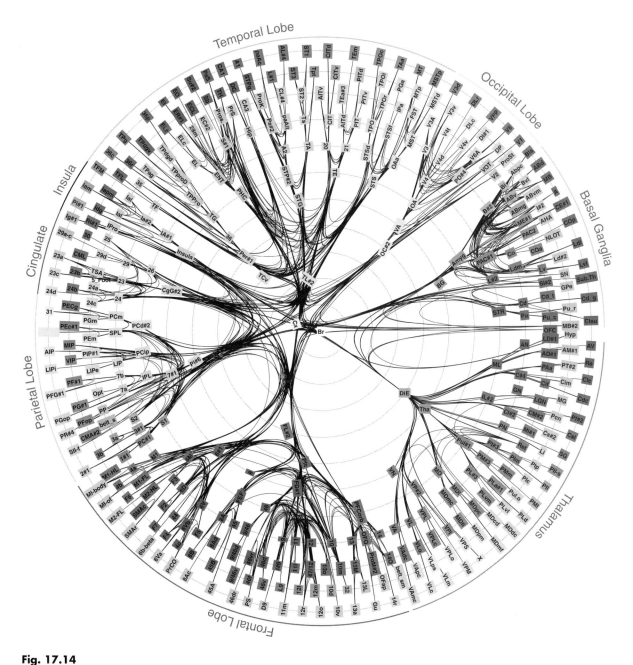

Fig. 17.14

Long-distance network of macaque brain. Representation of the macaque brain network in which nodes correspond to brain regions in an hierarchical brain map, and where two nodes are connected if there is a long-distance connection between the corresponding regions. From Modha and Singh (2010). Network architecture of the long-distance pathways in the macaque brain. *Proc. Natl. Acad. Sci. USA* **107**, 13 485–13 490. (Reproduced with the permission of the authors.)

Table 17.2 **Topological properties of brain networks**. Number of nodes n, global density d, average path length $\bar{l}$, and average Watts–Strogatz clustering coefficient $\bar{C}$, for brain networks.

Network	n	d	$\bar{l}$	$\bar{C}$
Macaque visual cortex	30	0.358	1.716	0.531
Macaque all cortex	71	0.150	2.330	0.461
Cat cortex	52	0.308	1.808	0.552
Cat all cortex	95	0.238	1.923	0.547

The degree distribution of some of these cortex networks has been analysed in a semi-quantitative way by Kaiser et al. (2007), who compared the histograms for node degree of some versions of the macaque monkey and cat cortex networks with those generated at random. Although the networks are directed, they only considered the degree of the undirected versions of these networks. A more quantitative approach to degree distribution was avoided on the basis of three issues: mainly that i) there is a 'sampling problem in that the amount of unknown or not included connections might change the shape of the degree distribution,' ii) due to the small size of the networks 'the degree distribution only consists of two orders of magnitude,' and iii) due to the small size of the networks 'results are unlikely to be precise.' We have seen some of these problems before—particularly the one concerning the best fit when the number of points is relatively small. Having these problems in mind, we can obtain the log–log plots of the cumulative distributions for the indegrees and outdegrees of the networks studied in Table 17.2 (see Figure 17.15). The shape of these distributions does not bear much resemblance to that of power-law degree distributions, and is more similar to exponentially truncated power-law distributions. We will see this type of distribution again, when studying human brain networks. This kind of distribution demonstrates that the removal of most connected nodes causes considerable damage to the structures of these networks, resembling the way in which 'scale-free' networks react to the elimination of hubs.

Analysis of the motifs present in these networks was carried out by Sporn and Kötter (2004), utilising a distinction between 'structural' and 'functional' motifs. Structural motifs are the same those studied in Chapter 4—directed subgraphs appearing more frequently in a network than in a random graph of the same characteristics. According to Sporn and Kötter (2004), functional motifs are formed by the same set of nodes in a structural motif, but using specific combinations of links contained within the structural motif. The authors found that for the modified networks, as displayed in Table 17.2, the number of directed subgraphs is smaller than those in random networks, so we consider them as anti-motifs. However, when the 'functional' subgraphs are analysed, they appear more frequently than in random graphs. Based on these results, the authors concluded that these brain networks maximise the number and diversity of functional motifs, but keep the number of structural motifs small, probably due to geometric constraints. There were, however, several structural subgraphs that appeared more frequently in the brain networks than in the random graphs, and can be considered as motifs according to the classical

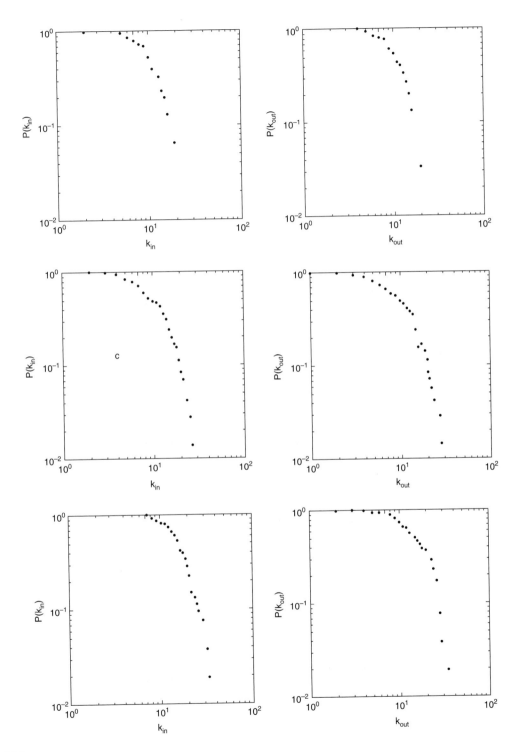

Fig. 17.15
(*continued*)

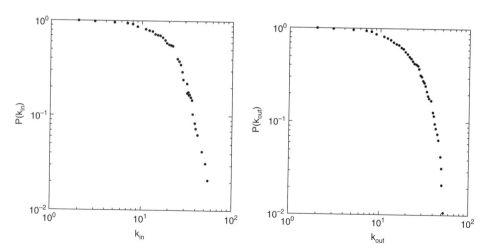

Fig. 17.15

Degree distribution in brain networks. Plot of the cumulative indegree (left) and outdegree (right) distributions for the brain networks of macaque visual cortex (first row), macaque all cortex (second row), cat cortex (third row), and cat all cortex (fourth row).

definition. Five of these motifs appeared very frequently in all three brain networks analysed, and their z-scores are plotted in Figure 17.16.

Sporn and Kötter (2004) have observed that the brain networks' characteristic of having few structural motifs supporting many functional motifs can explain other observed topological properties of these networks. The first of these characteristics is that the structural motifs found must have large reciprocity. A structural motif with reciprocity equal to 1 supports the largest possible number of functional motifs. Observe that the five structural motifs shown in Figure 17.16 display this characteristic. As a consequence, the whole brain network should have large reciprocities, as observed previously in Table 17.3, which includes the Garlaschelli–Lofredo reciprocity index ρ (see Chapter 4).

Another topological characteristic of these brain networks is related to their cyclicity—understanding it as the structural support for the information departing from one node being able to return in a relatively small number of steps. This property can be measured by using the returnability index studied in Chapter 5. In Table 17.3 we give the values of the returnability for the cortex networks previously studied by Sporn and Kötter (2004), where we can see that all of them display relatively high returnability (see Chapter 5). By contrast, the neural network of the worm *C. elegans* is significantly least returnable, as well as least reciprocal, than the anatomical brain networks. The returnability of a network representing an electronic circuit—frequently used as a popular analogue of brain circuitry—is also shown in Table 17.3. Although its returnability is analogous to those of the brain networks, its reciprocity is negative, in contrast with the high positive reciprocity of brain networks. The Roget's Thesaurus network displays both similar reciprocity and returnability, compared with the anatomical cortex networks. In a thesaurus network it is easy to understand the necessity for a certain degree of reciprocal and

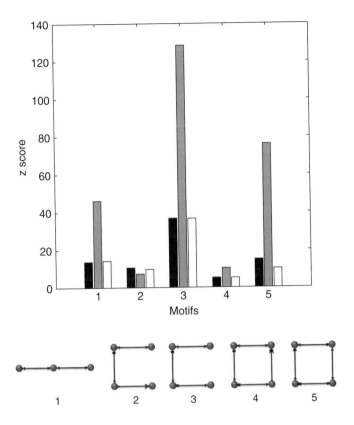

Fig. 17.16
Motifs in brain networks. Plot of the
z-scores (top) for five motifs (bottom)
found in macaque visual cortex (black),
macaque cortex (grey), and cat cortex
(white).

Table 17.3 Reciprocity and returnability in brain networks. Values
of the Garlaschelli–Lofredo reciprocity index ρ and returnability expressed
as pK_r for brain networks in a macaque monkey and a cat. For the sake
of comparison the same parameters are given for the neural network of *C.
elegans*, an electronic circuit, and the Roget's Thesaurus network.

Network	ρ	pK_r
Macaque visual cortex	0.655	1.06
Macaque all cortex	0.795	1.08
Cat cortex	0.625	2.03
Cat all cortex	0.868	1.53
C. elegans neurons	0.158	6.11
Electronic circuit	−0.063	1.84
Thesaurus	0.559	1.45

returnable references between concepts. Then—and at least in this respect—
we can consider that brains are organised in a way similar to a thesaurus.

A further collation of the macaque cortex dataset by Sporn et al. (2007a)
was conducted on a network of 47 nodes and 505 links, which was then
analysed, together with the dataset of the cat cortex having 52 nodes and 818
links, for the identification and classification of hubs. This hub identification
was carried out by considering the individual contribution of nodes, like the
central node of motif 1 in Figure 17.16, as well as by their centrality measures. As centrality measures, the authors considered degree, betweenness,
and closeness, as well as the clustering coefficient. A compilation of several
measures used in the analysis of brain networks has been published by Rubinov
and Sporn (2010), and most of them have been explained in detail in the
first part of this book. The directed version of the betweenness centralities
of the nodes in the macaque monkey and the cat cortices are illustrated in
Figure 17.17.

Using this combination of local indices for the nodes of brain networks,
Sporn et al. (2007a) identified hub regions, many of which have been previously identified as polysensory or multimodal. They then used a classification
for hubs, introduced by Guimerá et al. (2007), into 'provincial' and 'connector'
hubs. A provincial hub is one that has most of its connections within a network
community, and a connector hub is one having most connections between
communities in the network. Using Newman's spectral method for obtaining
network communities based on modularity (see Chapter 10), they found two
communities in the macaque and three in the cat brain networks. Using this
method they found the regions V4 and MT as provincial hubs in the macaque
network, while, for example, the region denoted as 46 is a connector hub.
They argued that V4 appears to mediate information flow between the dorsal
visual stream and the ventral visual stream, while area 46 maintains a diverse
set of projections, including visual somatosensory and motor regions (see
Figure 17.18).

17.3.2 Human brain networks

We have previously mentioned that the human brain is formed by an estimated
10^{11} neurons and 10^{15} connections, and that only the cortex contains approximately 10^{10} neurons and 10^{13} connections. Due to the intractability of these
huge network systems, and the fact that areas of the cerebral cortex appear to
be more influential in brain functionality than individual neurons, the study of
anatomically and functional distinct brain regions is the most feasible route
to the human 'connectome' (Sporn et al., 2004b). Studies of several possible
structural partitions of the human brain into different regions have been based
on the following:

i) Inter-regional covariation of grey-matter volume.
ii) Thickness measurements in structural MRI.
iii) Measurements of white-matter connections between grey-matter regions,
using diffusion tensor imaging (DTI).

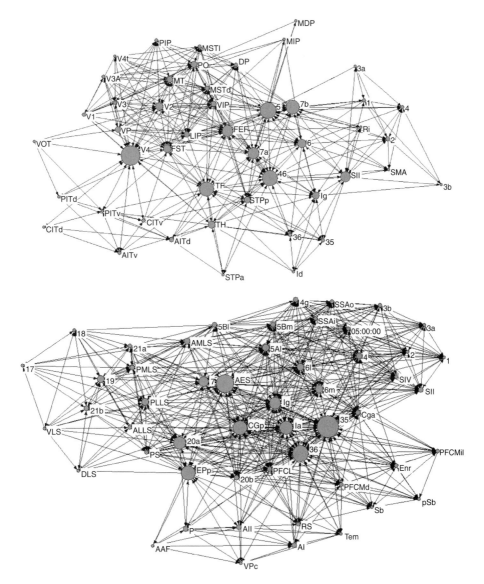

Fig. 17.17

Betweenness in brain networks. Representation of the directed betweenness centrality for the nodes in the brain networks of the macaque monkey (top) and the cat (bottom) cortices. The radii of the nodes is proportional to the betweenness centrality of the corresponding node.

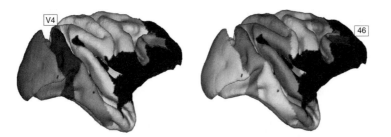

Fig. 17.18

Hubs in brain networks. Surface representation of the areas V4 (left) and 46 (right) in the cortex of a macaque monkey. From Sporn, O., Honey, C. J., and Kötter, R. (2007a). Identification and classification of hubs in brain networks. *PLoS One* 2, e1049, (Reproduced with the permission of the authors.)

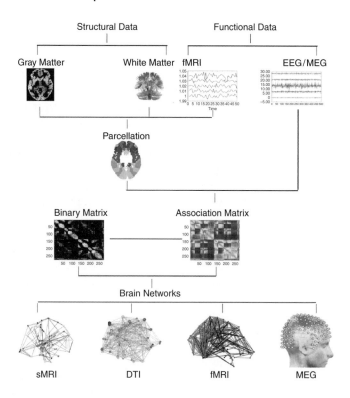

Fig. 17.19

Construction of human brain networks. Schematic representation of the workflow to build human brain networks by starting with either structural or functional data. By dichotomising the association matrix, which is created from pair-wise association between brain regions, the human brain network is finally obtained. From Bassett, H. and Bullmore, E. T. (2009). Human brain networks in health and disease. *Curr. Opin. Neural.* **22**, 340–7. (Reproduced with the permission of the authors.)

A general scheme of the structure of human brain networks is shown in Figure 17.19, from Bassett and Bullmore (2009).

Anatomical networks based on grey-matter volume in MRI data were constructed by Bassett et al. (2008) by analysing inter-regional covariation in 259 healthy volunteers and 203 people with schizophrenia. The networks were constructed by segmenting the structural MRI images to produce maps of grey matter. The brain was divided into 104 regions, and the partial correlation of regional grey matter was estimated for each pair of regions generating a correlation matrix. These regions correspond to 48 cortical regions in each hemisphere, plus the amygdale, the hippocampus, the striatum, and the thalamus bilaterally. The adjacency matrices were then generated from these correlation matrices by using a threshold $0 < \tau < 1$. If a pair of regions had a correlation larger than τ they were considered to be connected in the network. They then studied how the density and connectivity of the networks created change by varying the threshold parameter. The density (see Chapter 8) is referred to as the 'cost'. As $\tau \to 1$ the network tends to a complete graph, and as $\tau \to 0$ it tends to an empty graph. They found by empirical analysis that for $\tau > 0.15$ the network contains only one connected component, which was then further studied. The first property determined so far was the 'small-worldness' of the network—in the sense, defined by Watts and Strogatz (1998), of having both smaller average path length and larger clustering coefficient than analogous random networks. A method of computing this small-worldness is by using the coefficient σ (Humphries et al., 2006):

$$\sigma = \frac{\overline{C}/\overline{C}_R}{\overline{l}/\overline{l}_R} \tag{17.3}$$

where $\sigma \gg 1$ corresponds to small-world networks. Bassett et al. (2008) found that σ decays with the density of the network, and is larger than 2 only when the density is smaller than 0.05. In addition to densities larger than 0.25, the brain networks generated were indistinguishable from random graphs, which indicated that a density between 0.15 and 0.25 is appropriate for having connected small-world brain networks. The ubiquity of small-worldness in these networks is interpreted as a principle of brain topology which supports both high modularity and distributed processing dynamics. When the small-worldness parameters of healthy volunteers were compared with those of people with schizophrenia, Bassett et al. (2008) did not find significant differences, suggesting that key aspects of anatomical brain organisation are highly conserved in the presence of this neuro-developmental disorder. However, they observed that the clustering coefficient of twenty-three regional nodes showed significant differences between healthy and schizophrenic subjects. Most of them (78%) were located in the left hemisphere, and in 61% of the cases the differences were produced by an increase of the clustering in the schizophrenic subjects.

When functional connectivity is studied in healthy and schizophrenic people, a different picture emerges. Micheloyannis et al. (2006) generated networks from the EEG signals from twenty-eight cap electrodes, which constitute the nodes of the network, and synchronisation likelihoods between all pairs of electrodes were used for determining the links of the network. They found that all of the twenty healthy individuals, when studied for the working memory task, had networks with small-world property. Here, as in other areas of the application of complex network theory, a small-world network is understood in the sense of having a similar average path length and a larger clustering coefficient than random analogues. This was observed for the alpha, beta, and gamma bands in which EEGs were recorded. However, for twenty stabilised patients with schizophrenia who were able to work, the small-world property of their functional brain networks were not observed at all. Therefore, in the case of these functional brain networks it appears that schizophrenia produces disturbances in their small-world property, which can be indicative of partial disorganisation in the brain due to this illness. In a further study, Liu et al. (2008) confirmed the disrupted small-worldness of functional brain networks in schizophrenic individuals obtained from functional MRI. In this study, functional connectivity between ninety cortical and subcortical regions were estimated by using partial correlation analysis. Here again, healthy subjects display networks with small-world property, which is disrupted in subjects with schizophrenia. These alterations due to schizophrenia are observed in many brain regions, such as the prefrontal, parietal, and temporal lobes. Stam et al. (2007; 2009) have found that the small-world property is also affected in subjects with Alzheimer's disease (AD). In this case, they studied networks constructed from functional connectivity of beta band-filtered EEG for fifteen AD patients and thirteen control subjects. They observed that in patients with AD the average path length was significantly larger than in the control, with no change in the clustering coefficient. The results have been replicated by Supekar et al. (2008) by considering functional MRI from twenty-one AD patients and eighteen age-matched controls. In this case, however, a significant decrease of the clustering coefficient for the left and right hippocampus was observed for patients with AD with respect to controls. With this single parameter it was possible to distinguish the two groups with 72% sensitivity and 78% specificity. More recently, Sanz-Arigita et al. (2010) found differences in the average path length but not in the clustering coefficient between AD patients and controls when using networks constructed from MRI at 1.5 Tesla. In the light of these results it will be important to investigate the differences between these networks by using more 'sophisticated' network measures which account for more specific topological properties.

Also related to the small-world phenomenon in brain networks are the results that connect some topological properties of anatomical brain networks and intelligence (Li et al., 2009). Here, the intelligence was measured by using IQ tests, such as full-scale IQ (FSIQ), performance IQ (PIQ), and verbal IQ (VIQ). Li et al. (2009) observed that higher intelligence scores correspond to shortest average path lengths and higher global efficiency, which is defined as the sum of inverse distances (see Section 8.4).

The degree distribution has also been analysed for several versions of human brain networks. Among the first to study this property were Eguiluz et al. (2005), who found that functional brain networks obtained from functional MRI display 'scale-free' properties. The networks analysed display a skewed distribution of links with a fat tail that approaches a power-law with exponent around 2. He et al. (2009) have found that functional brain networks based on spontaneous fluctuations of the blood oxygen level dependent (BOLD) signal in functional MRI display exponentially truncated power-law as well as the networks obtained using cortical thickness measurements from MRI (He et al., 2007). A similar type of degree distribution was observed by Bassett et al. (2008) for the anatomical brain networks which they studied. These studies indicate that some kind of fat-tailed distribution exists for the degrees of these networks, which allows for the existence of structural hubs.

Several analyses of hubs have been conducted by using node centrality measures. Betweenness centrality was considered for the nodes of an anatomical brain network of human cerebral cortex. This network was obtained by using diffusion spectrum imaging to map pathways within and across cortical hemispheres in humans. The network consists of 998 regions of interest (ROIs), with an average size of $1.5\,cm^2$, obtained from high-resolution connection matrices. The highest betweenness centrality regions were located in areas of the medial cortex, such as the precuneus and posterior cingulated cortex, and for portions of the medial orbitofrontal cortex, the inferior and superior parietal cortex, and the frontal cortex. In Figure 17.20 this is illustrated for the two cerebral hemispheres, from the work of Hagmann et al. (2008). The measure of efficiency that was proposed by Latora and Marchiori (see Chapter 8) and which is related to the closeness centrality, was also studied for this network, and it was found that the posterior cingulated cortex, the precunus, and the paracentral lobule display the highest efficiency centrality.

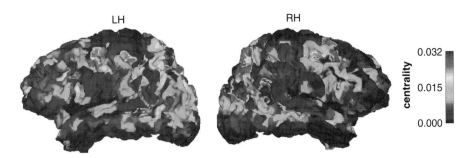

LH RH

centrality

0.032

0.015

0.000

Fig. 17.20
Centrality and efficiency in human brain. Lateral views of the left (LH) and right (RH) cerebral hemispheres showing ROI centrality, with (in the original figure) red representing high centrality and blue representing low centrality. From Hagmann, P., Cammoun, L., Gigandet, X., Meuli, R., Honey, C. J., Wedeen, V. J., and Sporn, O. (2008). Mapping the structural core of human cerebral cortex. *PLos Biol.* **6**, e159. (Reproduced with the permission of the authors.)

It should be remarked that the finding of degree distributions, average path length, clustering coefficients, and efficiency have been found to be independent of the spatial scale which is used to obtain the human brain networks; that is, they are consistent across different parcellation scales at the same resolution (Zalesky et al., 2010). However, there is significant variation in these parameters if the resolution of the networks is changed. For instance, it has been found that small-worldness suffers a dramatic variation if the network is obtained from automated anatomical labelling template, which contains about 100 nodes, in comparison with that obtained from random parcellation, which contains 4,000 nodes (Zalesky et al., 2010).

By using betweenness centrality, He et al. (2009) have identified the most important global hubs in the functional brain networks constructed from BOLD signals in functional MRI. They identified twelve regions having larger betweenness centrality than the mean plus standard deviation. Most of these regions are located at recently evolved associated cortex regions and primitive paralimbic/limbic cortex regions. Five modules or communities in these brain networks were identified. Module I consists of twenty regions, mostly from (pre-)motor, parietal and temporal cortices, module II includes fourteen regions from the occipital lobe, module III is mainly formed by eighteen regions from lateral frontal and parietal cortices, module IV consists of eighteen regions from the medial frontal and parietal cortices, and module V consists of twenty regions from limbic/paralimbic and subcortical systems. A representation of these modules is shown in Figure 17.21.

Using these networks, a provincial/connector type of classification of the hubs was carried out. The authors considered the relative intracluster betweenness centrality and the participation coefficient, which quantifies the extent to which a given node connects different regions in a network. The nodes were then classified as connector hubs, provincial hubs, connector non-hubs, and peripheral non-hubs. The analysis of node removal indicates that the removal of connectors has a dramatic effect on the connectivity of the network. For instance, attacks on connector non-hubs are more significant in destroying the integrity of the network than in attacking provincial hubs. These results show

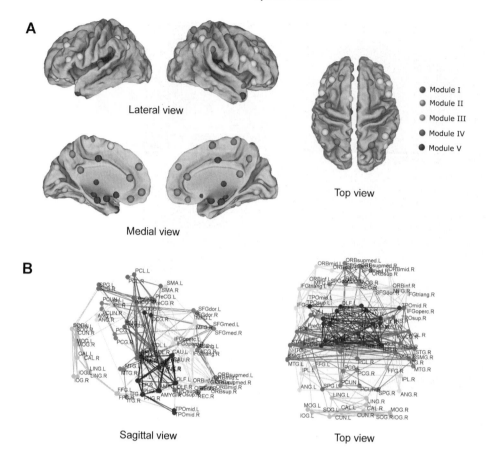

Fig. 17.21

Spontaneous human brain functional network. Sagittal (left) and top (right) views of the spontaneous brain functional network where the nodes and links in the same module are (in the original figure) similarly coloured and the inter-module links are grey. From He, Y., Wang, L., Chen, Z. J., Yan, C., Yang, H., Tang, H., Zhu, C., Gong, Q., Zang, Y., and Evans, A. C. (2009). Uncovering intrinsic modular organization of spontaneous brain activity in humans. *PLos One* **4**, e5226. (Reproduced with the permission of the authors.)

that connector regions in the brain are responsible for retaining the robustness and stability of brain functional networks.

By using anatomical networks of human brains divided into forty-eight cortical and eight subcortical regions, Crofts and Higham (2009) studied the effects of communicability (see Chapter 6) in stroke patients and controls. They generated networks from structural diffusion-weighted imaging data for nine stroke patients at least six months after the first, left-hemisphere, subcortical stroke, and ten age-matched control subjects. Then, using a normalised version of the communicability function (see Chapter 6) they distinguished between stroke patients and controls in an effective way. In further work, Crofts et al. (2011) studied nine chronic stroke patients and eighteen age-matched controls for whom brain networks were built by using diffusion MRI tractography. This time the communicability function was able to differentiate both groups, not only by using information from the lesioned hemisphere but also from the contralesioned one, despite the absence of gross structural

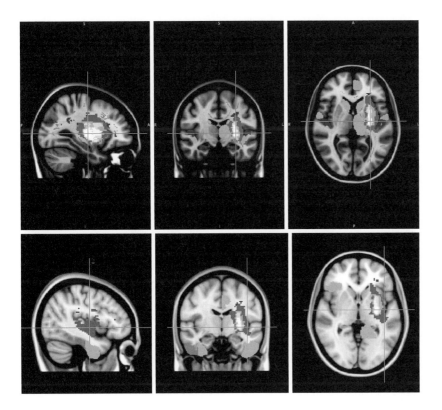

Fig. 17.22
Communicability and stroke lesions.
Stroke lesions in human brains (red in
the original figure), and regions with
reduced (top row) and augmented
(bottom row) communicability. Reduced
communicability is represented in blue
(in the original figure), and regions with
increased communicability are
represented in green (in the original
figure). From Crofts, J. J., Higham, D. J.,
Bosnell, R., Jbabdi, S., Matthews,
Behrens T. E. J., and Johansen-Berg, H.
(2011). Network analysis detects
changes in the contralesional hemisphere
following stroke. *NeuroImage* **54**, 161–9.
(Figure courtesy of J. J. Crofts;
reproduced with the permission of
Elsevier, ©2011.)

pathology in the latter. They found reduced communicability in brain regions
surrounding the lesions in the affected hemisphere as well as in homologous
locations in the contralesional hemisphere for a subset of these regions. In
addition, they identified regions with increased communicability in some brain
regions of the patients—possibly indicating the existence of adaptive, plastic
changes post-stroke. Figure 17.22 illustrates some of these results of decreased
and increased communicability in brain regions as a consequence of stroke
lesions.

Ecological networks

18

> What we understand best about evolution is mostly genetic, and what we
> understand least is mostly ecological. I will go further and suggest that the
> major remaining questions of evolutionary biology are ecological rather
> than genetic in content.
>
> Edward O. Wilson

Complex networks are ubiquitous in the analysis of ecological communities
where individuals interact with each other and with their environment (Proulx
et al., 2005). These interactions can be of very broad type, defined among
conspecific individuals, among populations, among different groups of species,
or among ecosystems. It has been argued by Ings et al. (2009) that these
interactions among organisms comprising ecological networks have played
a central role in the development of ecology as a scientific discipline. The
roots of these ecological networks can be found in the pioneering works of
several scientists at the end of the nineteenth century and the beginning of the
twentieth century. Dunne (2005) has resumed some of these early phases of
research in the field of ecological networks, mentioning the works of Forbes
in 1876 and Camerano in 1880, as well as that of William Dwight Pierce in
1912 (see Chapter 1). Further empirical descriptions of terrestrial and marine
ecological systems appeared, and in 1927 Elton coined the term 'food chain'.
This term has now been superseded by 'food web', defined as 'a network of
feeding interactions among diverse co-occurring species in a particular habitat'
Dunne, (2009). These feeding relations—or 'who-eats-who'—are commonly
referred to as 'trophic' interactions. This chapter discusses some of the main
results obtained from the study of food webs, although it is not limited to
this particular type of ecological network. Ings et al. (2009) have also considered subdivisions of ecological networks: i) traditional food webs, ii) host-
parasitoid webs, and iii) mutualistic webs. These are, to some extent, also
discussed in this chapter, together with ecological landscape networks. Several
challenges for the theoretic understanding of complex ecological networks
have been previously discussed by Green et al. (2005) and here we concentrate
on some of them, using the tools and methods of network theory presented in
the first part of this book.

18.1 Global topological properties

Most of the initial efforts in the quantitative analysis of food webs developed in the 1970s and 1980s (Cohen, 1977; Cohen and Briand, 1984; Cohen and Newman, 1985) concentrated on simple topological parameters, such as the ratio of nodes per links, or the density. In the context of food webs, the density studied in Chapter 8 is known as the 'connectance' of the network, where the number of species (nodes) is designated by S, and the number of trophic links by L. Then, because of the directionality of food webs, the connectance (density) is given by $C = L/S^2$. An initial hypothesis of constant L/S in food webs—known as the 'link-species scaling law' (Cohen, 1977)—was further substituted by the hypothesis of 'constant connectance' (Martinez, 1992). Let us explain what these two hypotheses mean in quantitative terms. According to the 'link-species scaling law',

$$L/S = a \tag{18.1}$$

for a given constant a. This implies that a plot of S versus L for different food webs should produce a linear correlation with slope a; that is $L = aS$. On the other hand, the 'constant connectance' hypothesis implies that

$$L/S^2 = b \tag{18.2}$$

for a given constant b. In this case, a plot of S versus L for different food webs should produce a quadratic correlation with slope b; $L = bS^2$. Both hypotheses can be merged by considering a power-law between the number of species and the number of trophic links:

$$L = \alpha S^{\beta} \tag{18.3}$$

where $\beta = 1$ for the link-species scaling law and $\beta = 2$ for the constant connectance. Montoya and Solé (2003) studied this kind of scaling for several food webs, and developed simulated food webs based on multitrophic assembly models that reproduce the patterns observed. A more recent study of these scaling laws has been carried out by Riede et al. (2010) by considering high quality data of sixty-five taxonomic food webs. It shows that the scaling between the number of species and the number of trophic links follows a power-law with exponent -1.57, which differed significantly from both the 'link-species scaling law' and the constant connectance hypothesis. They also showed that the ratio L/S tends to increase, and the connectance tends to decrease with an increase in the number of species (see Figure 18.1). Here we need to draw attention to Figure 8.3, in which a similar power-law (with different exponent) has been obtained between the number of nodes and the density (connectance) for a series of complex networks in diverse scenarios. We also need to argue that this kind of relationship is a general characteristic of sparse real-world networks, and not a particular characteristic of food webs. In Chapter 8 we argued that for real-world systems there is a very large 'cost' when considerably increasing the number of links as the size of the network increases. Consequently, the density of these systems decays with the increase of the size of the networks. A further analysis carried out by Riede et al. (2010) showed that there are significant differences in the number of links

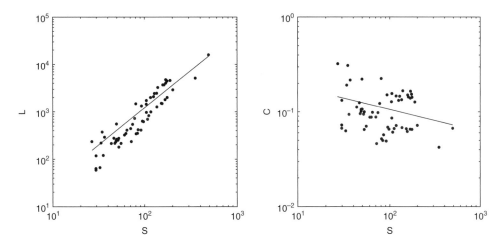

Fig. 18.1
Scaling of the number of species in food webs. Log–log plots of the number of species (S) in sixty-five food webs, compiled by Riede et al. (2010), versus the number of trophic links (left) and the connectance (C) or density (right) of the webs.

and links per node between river and terrestrial ecosystems, but not among other combinations studied (such as estuaries, lakes, and marine). Accordingly, it was suggested that river ecosystems are richer in trophic links and links per species than are terrestrial ecosystems. It was also concluded that there are no significant differences in connectance between any pair of the studied ecosystems, which indicates that the grouping of ecosystem types used in this study does not reveal the pattern observed in Figure 18.1.

Another area of the study of food webs which has captured much attention is the small-world properties of these systems. All studies reported so far (Dunne et al., 2002a; Montoya and Solé, 2002; Camacho et al., 2002; Williams et al., 2002) have found that the average path length of food webs is relatively small, with values around 2 (see Table 18.1). This has given rise to the phrase that species in food webs are just at 'two degrees of separation' (Williams et al., 2002). The average size of the sixty-five food webs studied by Riede et al. (2010) is 101.2 nodes, while for those in Table 18.1 it is just 76.4. Therefore, in these networks, as usual (see Chapter 3), $\bar{l} < \ln n$, and based on this specific criterion they can be considered as 'small-world'.

Here also, however, it is common to consider the clustering coefficient of the networks in relation to random analogues when analysing their small-worldness. This area has generated some controversy because of some contradictory results, published almost at the same time, by Montoya and Solé (2002) and Camacho et al. (2002). Both works reported that the average path length of food webs is very similar to those of random graphs with the same number of nodes and links. In both works, Erdös-Rényi graphs were used as the model of random graphs. However, Montoya and Solé (2002) reported that the average Watts–Strogatz clustering coefficient of four networks was always significantly larger than those obtained for the random analogues. On the other hand, Camacho et al. (2002) reported that for seven food webs the clustering coefficient is in good agreement 'with the asymptotic behaviour

Table 18.1 Topological properties of food webs. Number of species n and of trophic links m, average path length and average Watts–Strogatz clustering coefficient for the real $\bar{l}, \overline{C}$, and random $\bar{l}_{rnd}, \overline{C}_{rnd}$ networks of the same size and density, and the ratio between both clustering coefficients, $\overline{C}/\overline{C}_{rnd}$, for thirteen highly reliable food webs.

	n	m	$\bar{l}$	$\bar{l}_{rnd}$	$\overline{C}$	$\overline{C}_{rnd}$	$\overline{C}/\overline{C}_{rnd}$
Little Rock	181	2327	2.22	1.88	0.34	0.14	2.4
Ythan Estuary 2	92	418	2.25	2.27	0.22	0.10	2.2
El Verde	156	1440	2.30	1.98	0.21	0.12	1.8
Chesapeake Bay	33	71	2.80	2.48	0.20	0.13	1.5
St Martin	44	218	1.93	1.84	0.33	0.23	1.4
St Marks	48	218	2.09	1.95	0.28	0.19	1.5
Reef Small	50	503	1.60	1.59	0.61	0.41	1.5
Coachella Valley	30	241	1.46	1.45	0.71	0.55	1.3
Benguela	29	191	1.62	1.53	0.57	0.47	1.2
Skipwith Pond	35	353	1.42	1.41	0.62	0.59	1.1
Bridge Brook	75	512	2.16	1.85	0.20	0.20	1.0
Canton Creek	108	707	2.35	2.05	0.05	0.12	0.4
Stony Stream	112	830	2.34	1.98	0.07	0.14	0.5

predicted for a random graph.' Then, Dunne et al. (2002a) tried to solve the contradiction by studying a larger pool of food webs, and found that for most of the sixteen networks studied the average clustering coefficient is indistinguishable from those of a random graph. These contradictions are still echoed in the literature. For instance, in a review paper by Ings et al. 2009 it is written that highly resolved webs 'show a higher degree of clustering than their random counterparts,' while in a review by Dunne 2009 in the *Encyclopedia of Complexity and Systems Science* it can be read that 'clustering tends to be quite low in many food webs, closer to the clustering expected on a random networks'—justifying this poor clustering as due to food webs' 'small size compared to most other kind of networks studied.' In the following paragraphs we take a closer look at this problem.

Examination of the values of the clustering coefficient reported by Montoya and Solé (2002) and those reported by Dunne et al. (2002a) shows that they are neither qualitatively nor quantitatively similar for the same networks analysed. For instance, for the network of the Ythan Estuary with parasites, Dunne et al. (2002a) reported a clustering of 0.15 and Montoya and Solé (2002) another of 0.21, and for the network of Little Rock the first group reported 0.25 and the second one reported 0.35. This could be due to the use of different versions of the networks, such as the exclusion of non-trophic species in the work of Dunne et al. (2002a) in comparison with the inclusion of all taxa in Montoya and Solé (2002). In the cases of the random graphs, the values for the clustering coefficients obtained by both groups are surprisingly similar. In view of these differences, we recalculate here the values of the clustering coefficients as well as average path lengths for thirteen food webs, which are among the ones considered by Riede et al. (2010) as highly reliable datasets, and which also excluded parasitoid and parasitic relations. Here we consider all taxa of these food webs. In performing random graph generation we have also taken care to check that the resulting graphs are connected, and as Dunne et al. (2002a) we discarded all realizations ending in a non-connected graph and calculated the average for 200 realisations of connected random graphs with the same number of nodes and links as the food webs. In Table 18.1 we give the values obtained here for the thirteen food webs considered (see Appendix).

As can be seen, the results very much resemble those reported by Dunne et al. (2002a) as there are only five food webs for which the clustering coefficient is more than twice those of the random analogues. The most shocking results are obtained for the clustering of the food webs of Canton Creek and Stony Stream, which are significantly smaller than those of the random networks. Hereafter we will call relative clustering the ratio of the clustering coefficient of the food web and that of the analogues random networks: $\overline{C}_{rel} = \overline{C}/\overline{C}_{rnd}$. There have been two plausible and logic explanations for the small values of the clustering coefficient of food webs:

i) The small relative clustering in food webs is due to their 'small size compared to most other kinds of network' (Dunne, 2009).
ii) The small relative clustering coefficient is due to 'the strict limits on ecologically plausible interactions, e.g., plants do not eat each other [and] the maximal clustering for herbivores is far below 1' (Jordán and Scheuring, 2004).

We have to say that there are poor theoretic bases for arguing about the role of size in the poor relative clustering of these networks, except if we assume that small size implies large connectance (density) in food webs. Large connectance implies that the random network would necessarily be similar to the real ones. For the sake of simplicity, consider a complete network in which we have removed just one link. The density of this network is close to 1, and any random realization for the same number of nodes and links will always produce the same graph, which is exactly the real one. This is just an extreme case used to illustrate that as the density tends to 1 the diversity of the random graphs collapses to zero. However, neither size nor connectance are able to explain the pathological cases of the webs of Canton Creek and Stony Stream, as can be seen in Figure 18.2.

Let us now consider the explanation presented by Jordán and Scheuring (2004). It is true that plants cannot eat each other, and that herbivores do not eat to each other, which means that plants and herbivores will participate in no triangle amongst themselves. Herbivores can participate in triangles only with other predators who eat them, which clearly justifies the argument of Jordán and Scheuring (2004) that they will have maximal clustering far below 1. Canton Creek and Stony Stream are not terrestrial food webs, but they represent rivers and streams. However, we have a way of measuring how much a species is involved in structures of the type described by Jordán and Scheuring (2004). In situations like those described previously, the network necessarily lacks odd-length cycles—that is, they have a high degree of bipartivity. In Chapter 11 we discussed methods for quantifying the degree of bipartivity of a network. In Table 18.2 we give the values of the spectral bipartivity measure for the thirteen food webs studied here. It is true that the networks of Canton Creek and Stony Stream both have high bipartivity, with

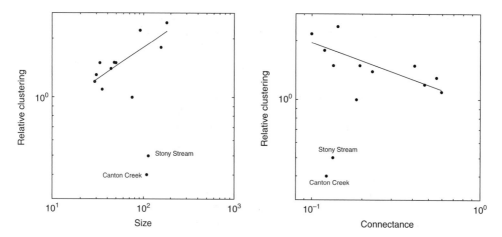

Fig. 18.2
Scaling of clustering in food webs. Log–log plots of the relative clustering coefficient versus the size (number of nodes) (left) and the connectance (density) (right) for the thirteen highly reliable food webs given in Table 18.1. Relative clustering refers to the ratio $\overline{C}/\overline{C}_{rnd}$.

values of $b_S = 0.776$ and $b_S = 0.815$ respectively, which are the largest among all the networks studied. The Erdös–Rényi process is not able to reproduce this characteristic of a network, which may explain why the random graphs are more clustered than the real one. However, the third most bipartite network is that of Chesapeake Bay, with $b_S = 0.710$, which displays a clustering ratio not different from those observed for non-bipartite networks such as St Marks and Reef Small. One point about these structural parameters must be clarified. The fact that a network has small clustering does not imply that the network is bipartite. However, a network with large bipartity necessarily has to have small clustering.

Table 18.2 Bipartivity and spectral scaling in food webs. Values of the spectral bipartivity measure b_s, and the positive ξ^+ and negative ξ^- deviations from the perfect scaling for thirteen highly reliable food webs.

	b_s	ξ^-	ξ^+
Little Rock	0.500	$2.760.10^{-7}$	$3.772.10^{-8}$
Ythan Estuary 2	0.504	$3.028.10^{-4}$	0.0029
El Verde	0.500	0.000	$4.449.10^{-6}$
Chesapeake Bay	0.720	0.085	0.0972
St. Martin	0.503	0.0011	0.0016
St Marks	0.504	0.0046	0.0022
Reef Small	0.500	$1.359.10^{-7}$	$1.187.10^{-7}$
Coachella Valley	0.500	$1.149.10^{-5}$	$2.071.10^{-6}$
Benguela	0.500	$9.502.10^{-4}$	$1.278.10^{-4}$
Skipwith Pond	0.500	0.000	$6.112.10^{-6}$
Bridge Brook	0.509	0.000	0.009
Canton Creek	0.776	0.000	0.183
Stony Stream	0.815	0.000	0.218

It can now be argued that a combination of connectance and bipartivity can solve the puzzle. However, a combination of both structural parameters explains only 62.9% of the variance in the values of the relative clustering for the thirteen food webs analysed with the following equation: $\overline{C}_{rel} = 0.0276.\delta^{-1.5687} + 0.02932.b_s^{-5.3522}$. Therefore, it is evident that even a simple structural parameter like the relative clustering coefficient can reflect the influence of some complex topological structures. We will further explore the topological structures of these food webs in order to search for a plausible explanation of the anomalies in the clustering observed for the webs of Canton Creek and Stony Stream.

We start by considering the analysis of the global topological characteristics of food webs according to the spectral scaling method described in Chapter 9. We have seen that there are four structural classes of network that can be identified by using the spectral scaling method. Most of the food webs studied here are in the Class I, which corresponds to homogeneous or good expansion network. The network representing Chesapeake Bay is in Class IV, which corresponds to networks having a central core connected to several also densely connected clusters which form some structural holes among them. The networks of St Marks and St Martin can be considered as being either in Class IV or in the borderline between Classes I and IV. Finally, the two pathological cases—the networks of Canton Creek and Stony Stream—are in Class III, which corresponds to networks having a central core of highly interconnected nodes and a periphery which has nodes loosely connected to the core and among them. In Table 18.2 we give the values of the negative and positive deviations from perfect spectral scaling (see Chapter 9 for the thirteen food webs studied here, after correcting the values for double precision in the calculations with respect to Estrada (2007b).

The spectral scaling plots for the networks of Canton Creek and Stony Stream food webs are illustrated in Figure 18.3. As we saw in Chapter 12 (see Figure 12.13) the Erdös–Rényi model is unable to reproduce the characteristic topology of networks in Class III, which is a good indication of the existence of the pathological behaviour of those webs when considering the relative clustering. In addition, the web of Bridge Brook also has certain characteristics of Class III networks, and it corresponds to the one in which the relative clustering is exactly 1. Overall, most food webs appear to belong to the classes of homogeneous networks or networks with a central core and interconnected clusters (Class IV), which are well reproduced by the Erdös–Rényi model. However, the existence of food webs of Class III represents a real challenge for the search of a random model, as the existing ones do not reproduce the topological features of this class of network (see Chapter 12). Later in this chapter we will return to this question when we analyse the degree of heterogeneity of these food webs.

Another global topological feature that can provide some important indications about the structural organisation of food webs is their degree distributions. Here again, as usual, there are some contradictions in the literature mainly produced by the problems arising with distributions (see Chapter 2). Montoya and Solé (2002) reported that the networks of Scotch Broom (Silwood Park) and those of Ythan Estuary do not follow degree

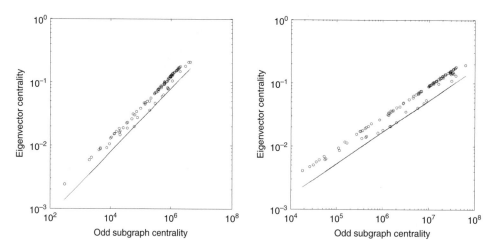

Fig. 18.3

Spectral scaling of core-periphery food webs. Spectral scaling plots for the food webs of Canton Creek (left) and Stony stream (right), which display only positive deviations from the perfect scaling, indicating a topological structure typical of Class III networks (see Chapter 9).

distributions similar to random networks—Poisson distributions—but they 'fit quite well a power-law.' Accordingly, they suggested that 'communities might be self-organised in a non-random fashion.' Almost at the same time, Camacho et al. (2002) found that normalised degree distributions of the food webs of Chesapeake, Bridge Brook, St Martin, Skipwith, Coachella, and Little Rock all display 'universal functional forms.' They showed that the probability densities for the distribution of the number of prey decays exponentially, and that the distribution of the number of predators is a step function with a fast decay. They also found that for Ythan Estuary the results are consistent 'with an exponential distribution of a number of predators.' In another study, Dunne et al. (2002a) considered the degree distributions of both predators and prey links for sixteen food webs, and found that they can be grouped into three global classes:

i) Webs with power-law distributions.
ii) Webs with exponential decay.
iii) Webs with uniform distributions.

In Figure 18.4 we reproduce the results reported by Dunne et al. (2002a) for these sixteen food webs.

In general, it is accepted today that 'empirical food webs display exponential or uniform degree distributions, not power-law distributions' (Dunne, 2009). It is also accepted that there is some kind of universal functional form for normalised degree distributions 'although there is quite a bit scatter when a wide range of data are considered' (Dunne, 2009). It has been shown that the exponential and uniform degree distributions observed in food webs can be recovered from models which include random immigration to local webs from a randomly linked regional set of taxa, and random extinctions in the local webs (Arii and Parrott, 2004).

The previously mentioned analyses of the degree distributions of food webs are far from being a quantitative characterization of the amount of

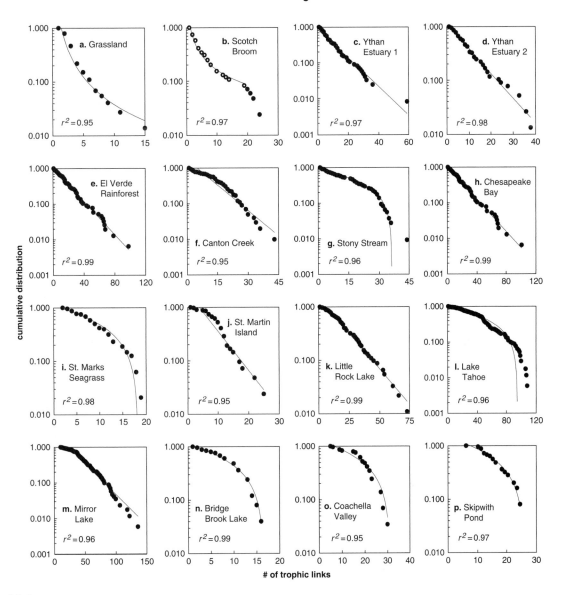

Fig. 18.4

Degree distribution of food webs. Linear–log plots of the cumulative degree distributions for sixteen food webs ordered by increasing density (connectance). Upward curved lines indicate power-law distributions, straight lines are typical of exponential decays, and downward curved lines are typical of uniform distributions. The correlation coefficient for the best fit is also illustrated. From Dunne, J. A., Williams, R. J., and Martinez, N. D. (2002a). Food-web structure and network theory: The role of connectance and size. *Proc. Natl. Acad. Sci. USA* **99**, 12 917–22. (National Academy of Sciences, USA, ©2002.)

heterogeneity of the degrees in these networks, but an analysis can be attempted by using the degree heterogeneity index described in Chapter 8. The results obtained for the thirteen networks studied in that chapter are represented as a bar plot in Figure 18.5. It can be clearly seen that the average heterogeneity of food webs, $\overline{\rho}(G) = 0.162$, is significantly larger than that

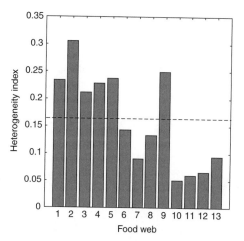

Fig. 18.5

Degree heterogeneity in food webs. Plot of the degree heterogeneity index (see Chapter 8) for the thirteen highly reliable food webs studied in this chapter. The average degree heterogeneity for these webs is represented by a horizontal broken line. The numbers in the x-axis correspond to those in Table 18.1 for the food webs.

expected for random graphs—$\overline{\rho}(ER) \approx 0$. In addition, it can be seen that the Ythan Estuary network is significantly more heterogeneous than the rest, and those of Skipwith, Bridge Brook, and Benguela are hardly distinguishable from random ones; that is, $\rho(G) \approx 0$.

We have seen two different kinds of heterogeneity in the thirteen food webs studied. The first, based on the spectral scaling method, accounts for the global organisation of the nodes and links to several global patterns in the network. When the pattern observed is simply that of self-similarity among the local and global properties, the network is considered to be homogeneous, while the other three classes or heterogeneous organisations are possible (see Chapter 9). The second class of heterogeneity refers only to the way in which the degree of the linked nodes is organised. We can then consider whether the combination of the two classes of heterogeneity of networks considered can explain some of the other observed characteristics of the food webs studied. We can consider, for instance, the ratio $\xi^{+/-} = (\xi^+ + \varepsilon)/(\xi^- + \varepsilon)$ as an indication of the departure of a network from the first type of homogeneity, while $\rho(G)$ is an indication of the departure from the second type of homogeneity. Using both parameters it is possible to reproduce 71.2% of the variance in the relative clustering of the thirteen food webs studied using a simple linear model: $\overline{C}_{rel} = 0.960 - 3.84.10^{-6}\xi^{+/-} + 3.274.\rho(G)$. This model indicates that the relative clustering decreases if the food web has a topological structure which approaches a core-periphery structure, and at the same time has small heterogeneity in the degree distribution. That is, webs like those of Canton Creek and Stony Stream have topologies which clearly display core-periphery structures not reproduced by random networks, although their degree heterogeneity is similar to those of random analogues.

Mutualistic webs have also been analysed for their degree distributions. Jordano et al. (2003) analysed eighty-six mutualistic webs having from seventeen to about 1,000 species, and found that in general they display exponential or truncated power-law distributions of the type $P(k) \sim k^{-\gamma}e^{-k/\xi}$ (see Figure 18.6). The existence of a mechanism such as 'richer-get-richer' that could generate power-law degree distributions in these food webs has been said to

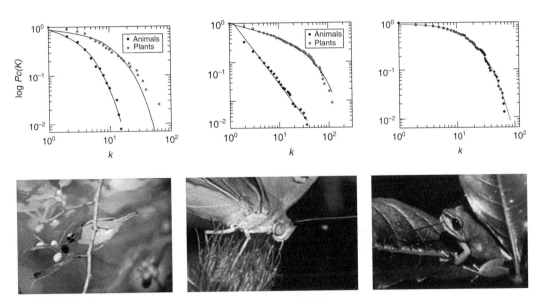

Fig. 18.6
Degree distribution in mutualistic webs. Distribution of the degrees in a frugivore-plant web (a), a pollinator-plant web (b), and in the food web of El Verde rainforest (c), included as a comparison. In (a) and (b) the distributions at right correspond to the numbers of plant species visited by an animal, while those at left represent the number of pollinator species visiting each plant species. Illustrations d–f are examples of these types of web. From Montoya, J. M., Pimm, S. L., and Solé, R. V. (2006). Ecological networks and their fragility. *Nature* **442**, 259–64. (Reproduced with the permission of Macmillan Publishers Ltd., ©2006.)

be at odds with ecological principles: '... as more species of frugivore feed upon a fruit species, then competition for that fruit will increase, and the less likely another frugivore would include it in its diet and the more likely it would feed on other fruits instead.' (Montoya et al., 2006).

18.2 Searching for keystone species

In an ecosystem there are some species 'whose impact on its community or ecosystem is large, and disproportionately large relative to its abundance' (Power et al., 1996). This idea—that all species are not equally important in an ecosystem—has been believed for long time. These species are termed *keystones*, and their investigation is of great importance in conservation ecology (Paine, 1966; Estes and Palmisano, 1974; Mills et al. 1993; Menge et al., 1994; Power et al., 1996; Springer et al., 2003). However, there is no universal way of detecting keystone species in a food web, and several measures for their *a priori* identification have been proposed (Power et al., 1996; Christianou and Ebenman 2005; Jordán et al., 1999; Jordán and Scheuring, 2002; Jordán et al., 2006).

Based on the definition of keystone species given by Power et al. (1996)–as one 'whose impact in a community or ecosystem is large, and disproportionately large relative to its abundance'—it is possible to define the 'community importance' of a given species as a way of identifying keystones in an ecosystem. Power et al. 1996 generalise the concept of community importance (CI) introduced by Mills et al. (1993) and propose to measure it as

$$CI_i = [(t_N - t_D)/t_N](1/p_i) \qquad (18.4)$$

where t_N is a quantitative measure of the trait in the intact community or ecosystem, t_D is the trait when species i is deleted, and p_i is the proportional abundance of species i before it was deleted. A trait refers to a quantitative measure of a community or ecosystem.

At this point it is useful to distinguish between two different types of trait, which hereafter we will refer to as *functional* and *structural* traits. This division is artificial, as it is evident that affecting the structure of an ecosystem will almost surely affect its functioning. However, it will allow us to introduce some topological traits based on network properties for the analysis of keystone species. Among the functional traits we can mention productivity, nutrient richness, or the abundance of one or more functional groups of species or of dominant species. The structural traits are those which can be determined uniquely from the topology of the networks representing the ecosystem. For obvious reasons, we will concentrate on structural keystone species.

The definition of community importance immediately suggests a method for studying the topological influence of one particular species on the global ecosystem. This method consists of removing this species from the network, and analysing the resultant influence on the structure of the network. One important effect of removing a species from a network is that of secondary extinctions (Dunne and Martinez, 2009), which result when there is one or more consumer species that loses all its resource taxa, or when a cannibalistic species loses all its prey except itself. In topological terms this means that the removal of one node changes the outdegree of other(s) from a non-zero value exactly zero. One simple example is provided in Figure 18.7, where the extinction of species 2 leads to the secondary extinction of species 1, as it loses its only resource in this network.

A a primary extinction of a species, however, can trigger a sequence of secondary extinctions in a web. For instance, removal of species 3 in the previous web leaves species 2 with its only resource, which means that it will become secondarily extinct. Once species 2 has disappeared, species 1 and 6 loss their remaining resources in the network and become extinct, resulting in a network of only three nodes (see Figure 18.8).

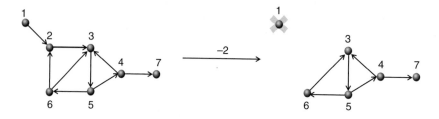

Fig. 18.7

Secondary extinctions in a hypothetical food web. Schematic illustration of a food web (left) and the effect of removing one species (node 2). Note that node 1 has lost all its sources of resources (right) and suffers a 'secondary' extinction.

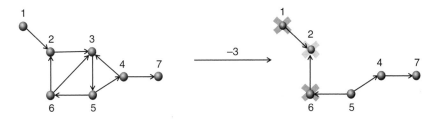

Fig. 18.8

Cascading of secondary extinctions in a hypothetical food web. Schematic illustration of a food web (left) and the effect of removing one species (node 3). After the extinction of species 3, species 2 has lost all its sources of resources and becomes extinct. The extinction of species 2 produces a cascading extinction of species 1 and 6.

Another kind of effect that is frequently analysed when a species is removed from a web is network fragmentation. In this case the web can be considered as undirected, and the analysis consists of quantifying the size of the largest connected component in the web after the removal of one node. This analysis is important, as it can be argued that the fragmentation of the web can give rise to large perturbations, so that some of the species isolated in smaller clusters completely disappear.

The removal of a species for analysing the robustness of the ecosystem can be carried out in different ways. First, the nodes to be removed can be selected at random in order to simulate the effect of random extinctions in an ecosystem. The values of secondary extinctions, as well as web fragmentation due to random extinction of species, can be compared with those obtained by intentional attacks of the most 'central' species of the web. That is, we can use centrality measures to rank the nodes in a food web, and then intentionally remove them. Of course, after a single removal the centrality of the remaining nodes can be changed. Therefore, one possible strategy frequently used (Dunne et al. 2002b) is to determine the centrality of the remaining nodes in the web after each removal. The work of Solé and Montoya (2001) and Dunne et al. (2002b, 2004) showed that the removal of the most connected species (highest-degree nodes) results in significantly higher rates of secondary extinctions than does the random loss of species. The strategy of removing high-degree nodes can be justified by considering, for example, that humans have historically tended to impact higher trophic levels through overfishing and hunting. Dunne et al. 2002b reported the rates of secondary extinctions for sixteen food webs, and measured the robustness as the proportion of species that have to be removed to achieve $\geq 50\%$ total species loss. Using this measure of robustness they determined that it increases approximately logarithmically with an increase in web density (connectance). They also found that protecting basal taxa increases the robustness of the webs. Figure 18.9 shows the effect on secondary extinctions following the removal of primary species.

The analysis of the fragmentation of food webs after the removal of most connected species also revealed interesting structural features. It has been found that the removal of high-degree nodes results in faster fragmentation of the webs. However, in 25% of the food webs analysed (Dunne et al., 2002b), removing species with low degree results in secondary extinctions which are

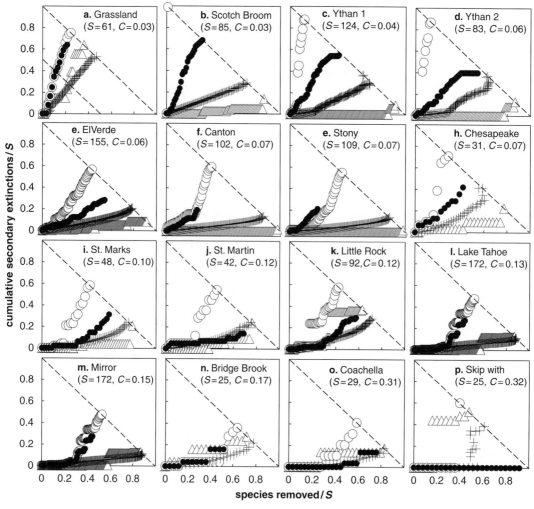

Fig. 18.9

Secondary extinctions in real food webs. Secondary extinctions resulting from the primary extinctions of species in sixteen real food webs. The primary extinctions are produced by removing the most connected species (empty circle), the most connected, no basal species (full circle), randomly selected species (crosses), and the least connected species (up triangles). The number of species (S) and the connectance (C) are also shown, and the webs are ordered by increasing connectance. From Dunne, J. A., Williams, R. J., and Martinez, N. D. (2002). Network structure and biodiversity loss in food webs: Robustness increases with connectance, *Ecol. Lett.* **5**, 558–67. (Reproduced with the permission of John Wiley and Sons, ⓒ2002.)

comparable to, or even greater than, those occurring when high-degree species are removed. Dunne et al. (2002b) have observed that this happens when the connectance of the network is relatively high. These results demonstrate that protecting the most connected species is not sufficient to avoid secondary extinctions in food webs, and that other centrality measures can play an important role in this respect.

In the search for methods for identifying keystone species, Jordán et al. (2006) analysed several 'classical' centrality measures such as betweenness and closeness (see Chapter 7), together with other local measures in the web of Chesapeake Bay. They observed that most of the centrality indices used in their study have high rank correlation coefficients among them. However, this cannot be interpreted as that all centrality measures coincide in identifying the same nodes as the most central ones. In fact, only 46% of centrality measures used identify the same species as the most central one. This situation is repeated for several centrality indices in food webs. For instance, an analysis of the degree, betweenness, closeness, eigenvector, information, and subgraph centrality of sixteen food webs has shown that most of them are highly correlated with each other (Estrada, 2007a). In fact, the use of factor analysis for each individual web showed that only one factor is necessary to account for most of the variance of these six centrality measures (Estrada, 2007a). This factor is a linear combination of the six centrality measures, and it can be used as a unified criterion for identifying keystone species. However, the betweenness centrality scores poorly in this factor for several networks—such as Benguela, Coachella, and Little Rock—which indicates that at least for these networks, betweenness contains information which is not duplicated by other indices. For example, Figure 18.10 illustrates the differences between the ranking obtained by using the degree and the betweenness centrality for the species in the Benguela food web.

Another important aspect of the detection of keystone species is the existence of bottlenecks—structural or functional—in the food webs. A structural bottleneck is a small group of nodes or links which after removal reduces dramatically the size of the network—a clear example being a bridge connecting two main chunks of a network. We have seen (see Chapter 9) that many of the food webs considered so far in the literature are homogeneous, in that what we see locally around a node is similar to what we obtain for the global network. These networks, like those of Little Rock (see Table 18.2), do not have structural bottlenecks. In some of these networks there is also an homogeneous

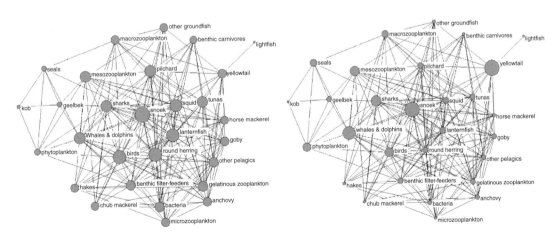

Fig. 18.10
Species centrality in a food web. Representation of the degree (left) and betweenness (right) centrality for the species in the Benguela food web. The radii of the nodes are drawn proportionally to the value of the centrality measure. Note, for instance, that the yellowtail is one of the most central species according to betweenness, but not according to degree.

distribution of the node degrees, as in the networks of Reef Small and Benguela (see Figure 18.4). These homogeneity characteristics of some food webs can explain the observation of Dunne et al. (2002b) that in 25% of the food webs analysed, the removal of species with low degree results in secondary extinctions which are comparable to, or even greater than, those occurring when high-degree species are removed. Conversely, there are food webs in which heterogeneities exist in the global organisation of the nodes, as revealed by spectral scaling (see Table 18.2). We have seen that networks in Classes III and IV are found among the food webs studied. Such networks—particularly those in Class IV—contain structural bottlenecks which are not necessarily among the most central nodes, according to different centrality measures. The elimination of these bottlenecks can therefore produce a dramatic number of secondary extinctions. In fact, small differences in the deviations from spectral scaling can produce significant differences in the robustness of food webs. For instance, in Figure 18.11 we compare the robustness of the St Marks and Chesapeake webs. First, the Chesapeake web displays a larger degree of heterogeneity as measured by the $\rho(G)$ index (see Figure 18.5). It can be seen in Figure 18.11 that Chesapeake is more vulnerable to the loss of hubs than is the web of St Marks. In addition, Chesapeake is a network of Class IV, so it may contain structural bottlenecks. In the figure, this web is even more vulnerable to the loss of these bottlenecks than to the loss of hubs. An analysis

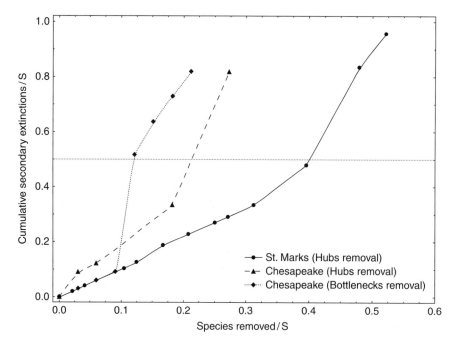

Fig. 18.11

Vulnerability of food webs to 'intentional attacks.' The effects of removing species in two food webs, according to two different 'intentional attack' strategies. The horizontal dotted line represents the loss of 50% of species in the ecosystem. For the St Marks web this point is reached when about 40% of the hub species are removed, but for Chesapeake Bay web it is reached when only 20% of the hubs are eliminated. More critically, the removal of only 10% of the bottleneck nodes in the Chesapeake Bay web leads to the extinction of 50% of all species.

of the effects of both degree distribution and spectral scaling can be found in Estrada (2007b).

Allesina and Bodini (2004) have proposed a method for identifying functional bottlenecks by reducing the topological structure of food webs into linear pathways accounting for the essential chains of energy delivery in a given web. Then, a species is dominant over others if the first passes energy to the others along a chain in the dominator tree. The removal of those species which dominate a large number of other species produces large secondary extinctions in the web. The importance of this approach, and its extension to weighted food webs (Allesina et al., 2006), is that it extends beyond direct interactions in the web to account for indirect interactions.

Finally, it has to be remarked that these studies based on topological keystone species do not necessarily account for all the potential secondary extinctions that can occur in a food web. Some attempts in this direction of 'realistic natural extinction' have been carried out by Srinivasan et al. (2007), and the reader is referred to the original source for details.

18.3 Landscape networks

One of the main consequences of the increased use of land by humans is the loss of natural habitats for other species—a development that is becoming one of the major threats to biodiversity and functional ecosystems (Meffe et al. 2002, Fahrig 2003). An immediate effect of the increased use of land for agriculture and other purposes is fragmentation of the landscape, with consequences for the mobility of the species living in it. If we consider the landscape as consisting of a set of discrete entities that occupy positions in an otherwise undifferentiated space, it is easy to see that human activity is dramatically influencing the landscape connectivity. In this sense, connectivity is the degree to which the spatial pattern of scattered habitat patches in the landscape facilitates or impedes the movement of organisms (Taylor et al. 1993). It is therefore evident that if the connectivity of the landscape is too low, subpopulations become isolated, and, for example, there is a decrease in the possibility of recoveries following local extinction, as successful recolonisation is dependent on the dispersal of species throughout the landscape (Hanski and Ovaskainen 2003, Hanski 1994, Bascompte and Solé 1996). An important point here is not only the size of the patches that remain in a landscape, but also whether the interconnectivity among them is sufficient for providing the necessary mobility of species from patches to patches (Lundberg and Moberg 2003). Obviously, these problems are of major concern for network theory, and some of them will be discussed in this section.

In order to generate landscape networks it is usual to consider the patches as the nodes embedded in a plane, and then to connect them according to certain criteria. One possibility is to consider the functional distance between pairs of patches, which can be defined as the distance between the edge of one patch to the edge of another, or as the centroid–centroid distance. It is also possible to consider the navigability between two patches as a way to obtain 'weighted' inter-patch distances. These distances can be stored in the form of a distance matrix, and the adjacency matrix of the landscape network can be built by considering a threshold distance. Then, two patches are considered as connected if they are at a distance smaller than, or equal to, the threshold. Another possibility for building the adjacency matrix is to consider the dispersal probability between two patches, which is defined as the probability that an individual in patch i will disperse to patch j. Urban and Keitt (2001) have assumed that the dispersal probability decays exponentially with the dispersal distance between the two patches:

$$p_{ij} = e^{\theta d_{ij}} \qquad (18.5)$$

where $\theta < 0$ is a distance-decay coefficient. Again, the adjacency matrix can be constructed by considering a threshold in the probability and connecting pairs of nodes that have values of p_{ij} larger than, or equal to, the threshold. These methods of constructing landscape networks are the same as for the geometrical random graphs studied in Chapter 12, except that in the case of landscape networks the nodes are in specific locations and are not placed at random. These locations are usually determined by using satellite imagery. Figure 18.12 is a Landsat image, obtained in May 2000, of a region of southern Androy, Madagascar, showing forest patches of >1 ha, with a superposed landscape network obtained by Bodin et al. (2006) using a dispersal probability function with distance equal to 1,000 m.

In a similar way as we have seen in Chapter 12 for random geometric networks, here the number of connected components decreases with the increase of the threshold distance used in defining the links of the network. For a given landscape there is a threshold distance at which all patches are connected. Obviously, the diameter of the patches increases as the threshold distance increases. For instance, Bunn et al. (2000) studied an area in North Carolina, USA, consisting of about 580,000 ha and more than 1,400 km of shoreline. They used habitat patches smaller than 100 ha, which resulted in the identification of eighty-three patches, roughly containing 83% of the 53,392 ha of possible habitat. These patches are separated, on average, at 62.7 km. When the threshold distance is reduced to 19 km the landscape is fragmented into many components containing only a few nodes. At the same time, the largest patch has an area of 20 km. Figure 18.13 shows the networks obtained for this landscape at four different distance thresholds.

Fig. 18.12

Landscape connectivity. A Landsat image of an area of southern Androy in Madagascar, and the definition of the connectivity of the landscape according to a network theory model (see text) by setting up the vagility of *Lemur catta* to 1,000 m. From Bodin, Ö., Tengö, M., Norman, A., Lundberg, J., and Elmqvist, T. (2000). The value of small size: loss of forest patches and ecological thresholds in southern Madagascar, *Ecol. Appl.* **16**, 440–51. (Reproduced with the permission of the Ecological Society of America, ©2000.)

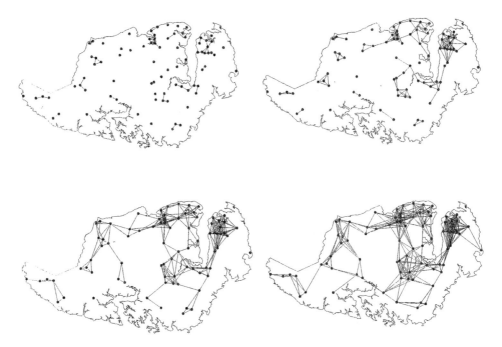

Fig. 18.13
Landscape connectivity in North Carolina, USA. Different connectivity patterns for a landscape in a North Carolina (USA) area in which patches have been connected by using threshold distances of 5, 10, 15, and 20 km respectively. From Bunn, A. G., Urban, D. L., and Keitt, T. H. (2000). Landscape connectivity: A conservation application of graph theory. *J. Environ. Manag.*, **59**, 265–78. (Reproduced with the permission of Elsevier, ©2000.)

One type of important study in landscape networks is the analysis of the influence of node removal on the global properties of the network. Here, node removal simulates habitat removal, and its effects are mainly charactezised by measuring *dispersal flux* and *traversability*. The area-weighted *dispersal flux* is defined as (Bunn et al., 2000).

$$F = \sum_{i=1}^{n} \sum_{j \neq i}^{n} p_{ij} s_i \tag{18.6}$$

where p_{ij} is the dispersal probability considered previously, and s_i is the patch area. Conversely, the *traversability* is the diameter of the largest component in the landscape network after removing a given patch.

Bunn et al. (2000) removed patches from the landscape network which they studied by using three different approaches. The first simulates the random loss of habitat, the second the removal of patches with minimum area, and the third the removal of patches which are end-nodes and also have the smallest area. They observed that both total flux and traversability decrease with the removal of patches, with the removal of end-node patches having the least influence on both parameters. No significant differences are observed in this network when the removal of nodes is carried out at random or by considering the patches of minimal area. This can be indicative of some kind of 'normal' distribution in the size of the patches, which was not studied by Bunn et al. (2000). In another study, Urban and Keitt (2001) studied a landscape network involving the threatened Mexican spotted owl (*Strix occidentalis lucida*) in south-west USA. They show that population persistence of this bird can be maintained despite substantial losses of habitat area, with the condition that the minimum spanning tree of the landscape network is conserved.

An important step forward in the use of network theory for landscape conservation was the comparison between spatially explicit population models (SEPMs) and network theory for selecting important habitat patches. This is attributable to Minor and Urban (2007), who studied a landscape network involving the wood thrush (*Hylocichla mustelina*), and analysed how some network metrics correlate to SEPMs parameters in the search for 'central' patches in a landscape. They used an SEPM model, which is mechanistic, stochastic, and individually based (see Minor and Urban, 2007, for details). The *source strength* and the *persistence* of a patch were then calculated as measures based on SEPM. The first of these accounts for a balance between the size and quality of the patch, and on how well the patch is connected to the rest of the landscape. The second combines the number of birds on each patch after 100 years with the number of years that the patch was occupied in the last ten years. As network parameters, Minor and Urban (2007) used the influx and outflux to and from a patch—defined as the dispersal flux for all links going out of or in a given node—and determined that the outflux correlates very well with the source strength (see Figure 18.14 left), indicating

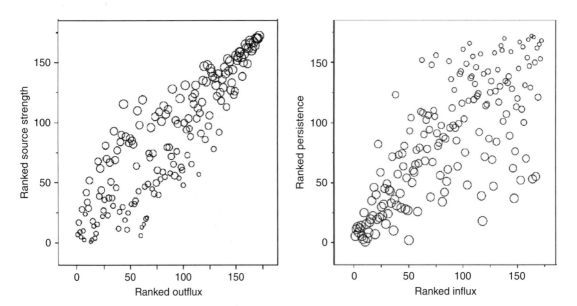

Fig. 18.14
Topological versus mechanistic parameters in landscape networks. Correlation of SEPM metrics versus some topological parameters of landscape networks. In the left plot, the source strength (a SEPM metric) is shown to correlate with the outflow (a topological metric), and in the right plot the persistence (an SEPM metric) is shown to correlate with the influx (a topological metric) for a landscape network. From Minor, E. S., and Urban, D. L. (2000). Graph theory as a proxy for spatially explicit population models in conservation planning, *Ecol. Appl.* **17**, 1771–82. (Reproduced with the permission of the Ecological Society of America, ©2000.)

that the first can be a good indicator for selecting patches with a large number of emigrants (a source patch). Less correlation is observed between the influx and the persistence (see Figure 18.14 right) due to variations in landscape connectivity. Interestingly, they observed that patch clustering may increase patch persistence without affecting influx, and so the analysis of the clustering coefficient of patches is an important objective of landscape studies. Minor and Urban (2007) concluded that network theory is a suitable and possibly preferable alternative to SEPMs for species conservation in heterogeneous landscapes, as it makes similar predictions in most cases, and in others offers valuable insights not available from the SEPMs.

A study of centrality indices in a landscape network of the Androy region in the very south of Madagascar (see Figure 18.12) was carried out by Estrada and Bodin (2008). This landscape consists of hundreds of small forest patches (<1 ha to 95 ha) scattered in an agricultural landscape. These forest patches provide habitat for the ring-tailed lemur (*Lemur catta*), which is an important seed disperser in the area (see Bodin et al., 2006, and references therein). This species rarely moves over distances larger than 1,000 m in search of food. Because the movement of *L. catta* between different forest patches can potentially disperse seeds throughout the landscape, the links of this landscape network were defined using a dispersal probability function with a threshold distance of 1,000 m. The resulting network consists of 259 patches and 1,131 corridors, and is divided into twenty-nine connected components—fourteen of them formed by only one patch, and the very large connected component containing 173 patches.

This study compares the effects of considering a weighted directed network and the unweighted, undirected versions of the same network. Here the weights of the links represent the flux of *L. catta* from one patch to another. Then, an unweighted directed version of the network is obtained by dichotomising the weighted adjacency matrix, and once this dichotomised matrix is symmetrised the simple network is obtained. It was observed that centrality indices depend very much on the version of the network used for defining them. However, even the simplest network representation still provides a coarse-graining assessment of the most central patches in the landscape (Estrada and Bodin, 2008).

In general, there is not a high intercorrelation between the different centrality measures for a given type of network. In the weighted network, the largest correlation coefficient is observed between the subgraph centrality and the indegrees and outdegrees, followed by the correlation between the betweenness and the in-closeness centrality. This situation is also observed for the other two versions of the landscape network. The use of principal component analysis identifies three principal components accounting for 80% of the variance in the centrality measures. The first component was identified as a measure of *the total dispersal of organisms through the closest neighbourhood of the corresponding patch*, as it is formed mainly by contributions from indegree, outdegree, and subgraph centrality. The second principal component measures *the corresponding contribution of a given patch in upholding the large-scale connectivity of the landscape*, as it is formed by the contributions of in-closeness and betweenness centrality (see Figure 18.15).

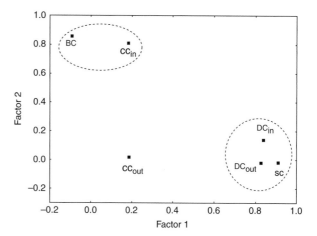

Fig. 18.15

Factor analysis of landscape centrality indices. Representation of indegrees and outdegrees and closeness, betweenness, and subgraph centrality in the space of the two main factors obtained by using principal component analysis (PCA) for the nodes of the Androy region in southern Madagascar. (From Estrada and Bodin, 2008.)

The differences in the two main factors used as generalized centrality indices is shown in Figure 18.16 for the main connected component of the landscape network studied. It can be seen that most of the central patches according to the first principal component are clustered in relatively small areas of the network. We recall that this component has been interpreted as a measure of *the total dispersal of organisms through the closest neighbourhood of the corresponding patch*. Therefore, such high-density clusters of patches imply that species living in them are not able to move as far in the inhospitable matrix. On the other hand, the second principal component, interpreted as a measure of *the corresponding contribution of a given patch in upholding the large-scale connectivity of the landscape*, is mostly located at the centre of the landscape, which reveals the large vulnerability of this system to be split into many islands just as these patches disappear.

The simulation of habitat loss by patch removal in the landscape network revealed dramatic differences in the topology of the network after the removal of the most central nodes according to the two principal components found. For instance, the removal of the most central patches according to the first principal component produces a dramatic reduction of the cliquishness of the network, measured by the clustering coefficient, and by removing about 40% of patches the cliquishness is reduced by 50%. We recall that clustering has been claimed as an important parameter for species mobility in landscape networks. This factor, however, is not affected by removing nodes according to the second principal component. In contrast, the size of the largest component of the landscape is reduced from 173 patches to only 95 when the ten most central patches according to second principal component are removed, but is almost unaffected by removing nodes ranked according to the first component (see Figure 18.17). These two generalised centrality measures account for two important aspects of the landscape topology which dramatically influence the mobility of species in the ecosystem.

Fig. 18.16
Unified centrality measures in landscape networks. Representation of the first (top) and second (bottom) main factors obtained from principal component analysis of the centrality measures shown in Figure 18.15 for the nodes of the Androy region in southern Madagascar. (From Estrada and Bodin, 2008.)

18.4 Other network-based ecological studies

The objective of this chapter is not to present a complete account of all ecological networks, but to provide some illustrations of the use of network theory in this important area of scientific research. However, it is interesting to note that some other important kinds of network are considered in the ecological literature—some of which represent important challenges for network theory. We begin by mentioning the whole area of animal social networks.

Fig. 18.17

Effects of patch removal in landscape networks. Effects of removing patches in the landscape network of the Androy region in Madagascar. The patches are removed according to their values of the two main factors obtained by using PCA, and the effects are analysed according to changes of the Watts–Strogatz clustering coefficient and the size of the main connected component. (From Estrada and Bodin, 2008.)

Fig. 18.18

Temperature and the topology of food webs. Illustration of two Icelandic stream food webs in very near geographical places (less than 2 m away) but at different ambient conditions. At top: a food web in a stream at a temperature of 22°C due to geothermal warming. At bottom: a food web in a stream at a temperature of 5°C. It can be seen that as the temperature increases, the web becomes more 'hierarchical'. and one species (the brown trout) becomes very central. From Woodward, G., Benstead, J. P., Beveridge, O. S., Blanchard, J., Brey, T., Brown, L. E., Cross, W. F., Friberg, N., Ings, T. C., Jacob, U., Jennings, S., Ledger, M. E., Milner, A. M., Montoya, J. M., O'Gorman, E., Olesen, J. M., Petchey, O. L., Pichler, D. E., Reuman, D. C., Thompson, M. S. A., vann Venn, F. J. F., and Yvon-Durocher, G. (2010). Ecological networks in a changing climate. *Adv. Ecol. Res.* **42**, 71–138. (Reproduced with the permission of Elsevier, ©2010.)

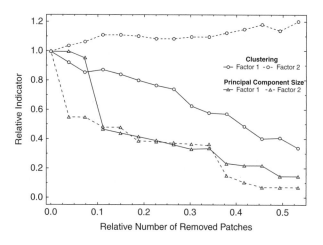

In Chapter 11 we studied a network of bottlenose dolphins in which the links of the network symbolize some kind of social relation between them. This type of animal social network has been studied for a variety of species, including, among others, monkeys, giraffes, deer, and guppies. An excellent account of animal social networks is *Exploring Animal Social Networks* (Croft et al., 2008). Another type of ecological network which has become increasingly relevant in recent years is that of 'individual-based food webs.' In the food webs studied in this chapter, nodes and links represent traits of species or trophic species and their interactions. It has been observed that this practice of aggregating data can lead to the loss of considerable amounts of important biologically meaningful information and give rise to methodological artifacts in the analysis of these networks. The newly proposed approach of using individual-based food webs is expected to improve our understanding of the structure and dynamics of food webs by combining simple metabolic and foraging constraints. An excellent account of this type of approach, with many references, is presented in Woodward et al. (2010). Last but not least, we want to mention the application of network theory in studying the effects of a changing climate on ecological systems. It has been usual to study the effects of climate change on lower levels of biological organisation, such as individuals or populations, instead of considering the global ecosystem in a complex multi-species way. Because 'a picture is worth a thousand words', we show, in Figure 18.18, the influence of temperature on two Icelandic stream food webs. Woodward et al. (2010) have identified a range of approaches and models that are particularly well suited for network studies within the context of climate change, and they also present a good account of previous efforts in this important area.

Social and economic networks

19

> Society is not a mere sum of individuals. Rather, the system formed by their association represents a specific reality which has its own characteristics...The group thinks, feels, and acts quite differently from the way in which its members would were they isolated. If, then, we begin with the individual, we shall be able to understand nothing of what takes place in the group.
>
> Émile Durkheim (1895)

One of the fields in which network theory is more traditional is social network analysis. According to Linton C. Freeman (2004), the origins of social network ideas can be traced back to the nineteenth century and the works of Auguste Comte, who coined the term 'sociology' and defined the main aspects of its study as the investigation of the 'laws of social interconnection', or the 'laws of action and reaction of the different parts of the social system'. Then, several other scientists, including as Durkheim, Simmel, and Morgan, paved the way for the development of more quantitative analyses of social networks. In the 1930s this idea of a quantitative analysis or sociometry was further developed by Jacob Moreno, who saw it as a sort of physics, composed by 'social atoms' and having laws such as 'social gravitation' (Freeman, 2004). The analysis of social networks has given rise to many of the ideas we find today in the study of complex networks as a discipline. The idea of 'small-worldness' arises from the experiment carried out by Stanley Milgram (1967) and the mathematical studies of this phenomenon conducted by de Sola Pool and Kochen in 1978 (see Freeman, 2004, and the references therein). Towards the end of the 1940s, the idea of centrality was introduced by Alex Bavelas (1948, 1950), and in 1949 Luce and Perry formally introduced the notion of 'clique'. Hypergraphs were used by Hobson to represent overlaps among actors in social systems (see Freeman, 2004, and the references therein), and a vast application of algebra and graph theory to solve social network problems can be found in the literature. For an excellent account of the history of social network analysis, the reader is referred to the authoritative review of Freeman (2004).

Another area of the study of complex networks with a long tradition, and of enormous relevance today, is that of financial and economic networks. It is quite remarkable that only twenty-two years after Euler published his solution to the Königsberg bridge problem, François Quesnay (1758) published, in his *Tableau Economique*, a circular flow of financial funds in an economy, utilising network representation. This network is by far more complex than that of the seven bridges of Königsberg, and represents the very first economic and financial network to be studied. Then, others such as Monge, Cournot, Kantorovich, Hitchcock, and Koopmans paved the way for the modern theory of economic networks (see Nagurney, 2003, and the references therein). Many problems on transportation networks were solved in these pioneering works, and the reader is directed to the excellent review by Nagurney (2003) for further reading. In this chapter we will approach these two huge areas of network theory in a modest attempt to account for the most relevant structural properties of these networks.

19.1   Brief on socioeconomic network data

The problem of collecting data in social and economic sciences is not different from those in other disciplines, though it has some peculiarities. When we say that it is 'not different' we mean that it is also subject to the same problems of accuracy, validity, reliability, error, and so on, associated with data collection in physical and biological scenarios. Another similarity is that social data can be obtained 'experimentally' or from archival or historical records. Here we understand 'experimental' as observational techniques, such as questionnaires, interviews, direct observations, or even experiments, such as that of the 'small-world' conducted by Milgram (1967). Wasserman and Faust (1994) dedicated a complete chapter of their book to this topic of data collection, and the reader is referred to their work, as well as other literature, for further details. Some debate about the preference of observation over self-reported data has been raised in the literature (Wasserman and Faust, 1994; Borgatti et al., 2009, Eagle et al., 2009; adams, 2010). There are, of course, obvious cases in which observational data is the preferable, and sometimes the only, possible option. However, in other situations, the questioning of individuals is the only method of obtaining some kind of required relations in building a given network.

In general, socioeconomic network data can be of two different types, according to the 'mode' in which the data are organised. In one-mode networks, all entities represented by the nodes of the network are of the same class—such as individuals, corporations, football teams, and so on. In two-mode networks, however, the entities of the system are of two different natures, such as banks and corporations, cities and commodities, and so on. One possibility is also to have affiliation networks, in which a group of entities is connected to a series of events forming a two-mode network. Multi-modal data can also be considered, but it is more rarely found in the literature. The data obtained for a given set of socioeconomic entities and their relationships can also be of structural or compositional nature (Wasserman and Faust, 1994). The first refers to binary or multiary relationships between pairs or groups of entities, such as friendship, transactions, and so on. The second refers to characteristics of the entities *per se*, such as gender, race, geographical location, number of employees, and so on. One characteristic of social data collection that can be thought as peculiar to these systems is the definition of the boundaries of the population studied. In many cases these boundaries are very well defined, as when all individuals in a small firm, members of a team, or all firms in a given region are studied. In other cases these boundaries are vague, and are difficult (if not impossible) to distinguish. However, the situation is not different from what we have in chemistry or biology, nor even in many areas of physics.

An important aspect of data collection for building socioeconomic networks is that of the nature of the relations which are studied. This determines the very nature of the network we are going to study, and all results analysed so far should be in this particular context. Among these types of relationship are those based on individual evaluations, such as friendship, linking, respect, mutual help, and so on. Another important area includes those that involve transactions or transfer of material resources, such as lending money, commercial transactions, buyer–seller transactions, and others. In our world of communication interchange, the transfer of non-material resources, such as information, is all the more important. Other types of relation can include physical movements, formal roles, kinship, physical interactions, and others. There are several types of study that can be conducted in the context of social network analysis. For example, much attention has been devoted to the study of disease propagation in social networks (Balcan et al., 2009; Holme, 2004; Liljeros et al., 2003; Potterat et al., 2002; Schneberger et al., 2004). Another topic that has been widely treated in the literature is the development of models to explain social network evolution (Carvalho and Iori; Grabowski and Kosiński, 2006; Grönlund and Holme, 2004; Jackson and Wolinsky, 1996; Kossinets and Watts, 2006; Newman et al., 2002; Skyrms and Pemantle, 2000). These studies, however, are mainly concerned with the dynamics of social networks, and the reader is directed to the specialised literature in this field (Barrat et al., 2008). Here, we concentrate on the structural topological organisation of these systems.

19.2   Global topology of social networks

As frequently occurs in multidisciplinary research, different points of view of the same problem arise due to various perspectives according to different disciplines. In social network analysis it has been argued that social and physical scientists tend to have different goals (Borgatti et al., 2009), which has many consequences for the outcome of the results. For instance, Borgatti et al. (2009) consider that physicists tend to concentrate on the discovery of universal properties of social networks—that is, to demonstrate that a series of networks have a certain property which is usually rarely found in random networks. Social scientists, however, use the paradigm that 'different networks (and nodes within them) will have varying network properties, and that these variations account for differences in outcomes for the networks (or nodes)' (Borgatti et al., 2009). The first difference that we can find among the two approaches concerns the very nature of the social networks that both groups tend to study. Physicists, for instance, tend to study large and usually collaborative networks, while social scientists tend to concentrate on smaller networks where a certain type of relation has been 'experimentally' observed. Many of the social networks analysed by physicists are obtained from archival information, and the social relationship among the participants can only indirectly be considered as 'social'. In a network of film actors, for example, the nodes correspond to actors who are connected if they have appeared in the same film. It is easily understandable that many of the actors in these films do not develop any kind of social relation among themselves, and even worse that they probably never meet in the same place at the same time. Co-authorship networks have exactly the same problem. Examples of 'social' networks and some of their topological properties, studied by physicists, are given in Table 19.1.

Examples of the types of social network studied by social scientists are given in Table 19.2, where we can see more variety in the type of social relations between actors. For instance, the networks of 'Galesburg', 'high tech', 'karate club', and 'prison' account for friendship relationships between pairs of individuals in different environments. The 'corporate' network accounts for the co-participation of actors on the boards of different companies; 'geom' is a collaboration network in computational geometry; 'drugs' accounts for the interchange of needles between drug addicts, and 'colo spring' accounts for sexual and injecting drugs partners with HIV infection. (A complete description of these networks is included in the Appendix.)

A comparison of the datasets in Tables 19.1 and 19.2 is self-revealing of the structural differences between the two groups of network. The networks in Table 19.2 are ten times denser than those in Table 19.1, although the average Watts–Strogatz clustering coefficients are similar for both group of network, while the Newman clustering or transitivity is significantly larger for the networks in Table 19.1. More significant is that all networks in Table 19.1 are assortative, while those in Table 19.2 are mainly disassortative, with only three exceptions. The kinds of network in Table 19.1 are those which allowed Newman (2003a) to conclude that almost all networks seem to be disassortatively mixed, except for social networks, which are normally assortative. However, the data in Table 19.2 do not support this conclusion, and as we have seen in Chapter 2 there are many other non-social networks which are also assortative; for example, all protein residue networks. In addition, Hu and Wang (2009)

Table 19.1 Topological properties of collaboration networks. The number of nodes n, density d, average path length $\bar{l}$, average Watts–Strogatz clustering coefficient $\bar{C}$, transitivity index C, and Newman assortativity index r, for some of the collaboration networks usually studied in the physics literature.

Network	n	d	$\bar{l}$	$\bar{C}$	C	r
Jazz[a]	1265	0.048	2.75	0.89	0.77	0.412
physics coauthorship[b]	52 909	0.00017	6.19	0.45	0.56	0.363
math coauthorship[b]	253 339	$1.54.10^{-5}$	7.57	0.15	0.34	0.120
film actors[c]	449 913	0.00025	3.48	0.20	0.78	0.208
biology coauthorship[b]	1 520 251	$1.02.10^{-5}$	4.92	0.088	0.60	0.127

[a]Gleiser and Danon (2003); [b]Newman (2001); Watts and Strogatz (1998), Amaral et al. (2000).

Table 19.2 **Topological properties of social networks**. The number of nodes n, density d, average path length $\bar{l}$, average Watts–Strogatz clustering coefficient $\overline{C}$, transitivity index C, and Newman assortativity index r, for some of the networks usually studied in the social network literature.

	n	d	$\bar{l}$	$\overline{C}$	C	r
galesburg	31	0.144	2.529	0.39	0.28	−0.135
college	32	0.194	2.296	0.33	0.24	−0.119
high tech	33	0.172	2.36	0.52	0.37	−0.087
karate club	34	0.139	2.408	0.59	0.26	−0.476
saw mill	36	0.098	3.138	0.36	0.23	−0.071
prison	67	0.064	3.355	0.33	0.29	0.103
high school	69	0.094	2.955	0.48	0.40	−0.003
colo spring	324	0.007	8.325	0.08	0.038	−0.295
drugs	616	0.011	5.284	0.72	0.37	−0.118
corporate	1586	0.009	3.507	0.50	0.39	0.268
geom	3621	0.001	5.34	0.68	0.22	0.168

have found that most online social networks display disassortative mixing. The size of the networks is not considered as the cause of assortativity, because in Table 19.2 there are some small social networks, such as inmates in prison, which are assortative (see Chapter 14). In the following paragraphs we analyse some models for generating assortative social networks.

It is not difficult to create models that allow for assortativity in networks. The very first consideration is the connectivity of the networks under analysis. In a network formed by several connected components it is relatively easy to be fooled by assortativity questions. For instance, let us consider a network formed by three connected components, each of them a clique of different size (see Figure 19.1). In this case the network can even have a degree–degree correlation coefficient of 1.

Here we concentrate only on the case of connected networks, or in another case, on the main connected component of such networks. Newman was the first to propose a model to construct assortative networks (Newman, 2002a). First, a random network with given degree distribution is built. Then, two links (v_1, w_1) and (v_2, w_2) are selected at random, and the remaining degrees (j_1, k_1) and (j_2, k_2) are calculated. At this point the two links are replaced by two new ones, (v_1, v_2) and (w_1, w_2), with probability $\min(1, (e_{j_1 j_2} e_{k_1 k_2})/(e_{j_1 k_1} e_{j_2 k_2}))$. Different variants of this approach have been studied in the literature. For instance, Catanzaro et al. (2004) developed assortative models for social networks based on a preferential attachment initial graph, which is then modified by adding new links with probabilities

 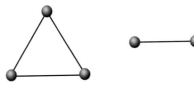

Fig. 19.1
Hypothetical assortative multicomponent network. A network formed by three connected components, which are cliques. This network has an assortativity coefficient of 1.

defined on the basis of the degrees of the nodes being the end-points of such links. Boguñá and Pastor-Satorras (2003) used hidden variables controlling the establishment of links between pairs of nodes, and Xulvi-Brunet and Sokolov (2005) use a link rewiring process satisfying the condition that 'nodes with similar degree connect preferably' to obtain assortative networks. All these models reproduce the assortativity coefficient observed in some of the social networks studied so far. But do they provide any structural information about the organisation of nodes and links in the network? More importantly, do they reveal any information about how social networks are formed?

We have seen previously (in Chapter 12) that random models like those of Erdös–Rényi and Barabási–Albert do not reproduce the structural organisation of many networks, which can be revealed through the use of the spectral scaling method (see Chapter 9). On the other hand, we have seen that assortative (disassortative) networks can have different clumpiness of the highest-degree nodes (see Chapter 8). Let us first investigate these characteristics of social networks, and then return to the problem of assortativity.

The analysis of the spectral scaling of the complex social networks studied in Table 19.2 shows that seven out of eleven of these networks (Galesburg, college, high tech, karate club, saw mill, prison, and colo spring) are of Class IV, while the rest are in Class II. Networks in Class IV (see Chapter 9) are those having a central cluster and several peripheral clusters which are poorly interconnected among themselves and with the central cluster. Class II networks (see Chapter 9) are those having a decentralised structure of clusters poorly connected among themselves. Figure 19.2 illustrates the spectral scaling of two networks in Class IV—one disassortative and the other assortative—and the same for networks in Class II.

The important point is that none of the social networks analysed here are homogeneous, but are of any of the two kinds of networks characterixed by the presence of many clusters or communities. We recall that the topological characteristics of Class II networks are not reproduced by random models, and those of Class IV are only replicated for relatively small density of the networks (see Chapter 12).

There are two other characteristic features of the social networks studied here that need to be considered. The first is the relative clumpiness of the nodes according to their degree (see Chapter 8). In general, social networks appear to have relatively low clumpiness of the nodes according to their degree. That is, the most connected nodes are not clumped together but are spread across the whole network. Even in the case of assortative networks, the relative clumpiness is low, as can be seen in Figure 19.3 (left), where social networks are represented as full dots and other networks are represented as empty circles. We also illustrate in Figure 19.3 (right) the network of friendship among inmates in prison, to show that the 'hubs' of the network are not clumped. Nevertheless, the clumpiness of these real-world networks is significantly larger than that obtained for random graphs (Estrada et al., 2008a).

The second characteristic feature of social networks is their degree heterogeneity. It has been previously observed that many of the social networks analysed so far in the literature are not scale-free (Newman, 2003b). For

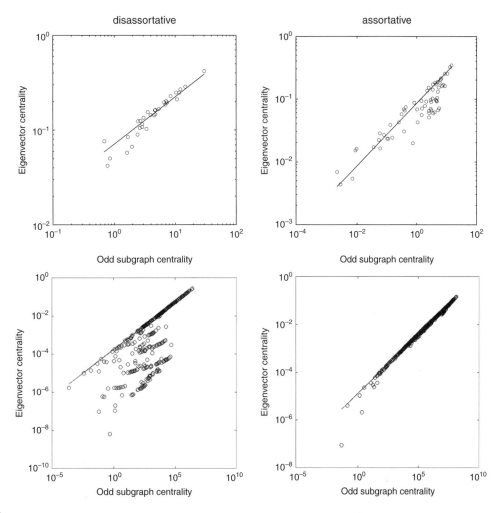

Fig. 19.2
Spectral scaling in social networks. Spectral scaling between the eigenvector and subgraph centralities for the social networks of college (top left), prison (top right), drugs (bottom left), and corporate (bottom right). The first two networks are classified as Class IV, and the other two as Class II.

instance, only 40% of the social networks compiled by Newman in his review paper (Newman, 2003b) are scale-free. We have calculated the degree heterogeneity index (see Chapter 8) for the eleven networks studied here, and the results are plotted in Figure 19.4 (left). The average heterogeneity of these networks is only 0.153, which contrasts with those of the Internet (0.548), the transcription network of yeast (0.448), the USA airport transportation network (0.369), and the ODLIS dictionary (0.342). In many cases—including the networks of inmates in prison, and that of the corporate elite in the USA, which are assortative—the degree heterogeneity is very similar to that of random homogeneous networks. For instance, Figure 19.4 (right) illustrates the H-plot for the network of the corporate elite in the USA, which is very similar to those reported for Erdös–Rényi graphs in Chapter 12.

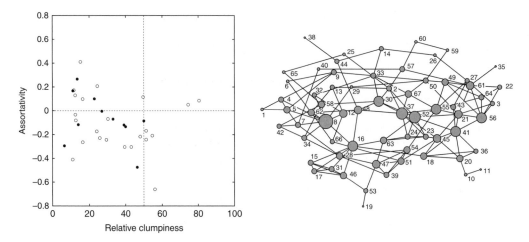

Fig. 19.3
Clumpiness of nodes in social networks. Plot (left) of assortativity versus clumpiness of nodes according to their degree for social (full circles) and other types (empty circles) of network. The degree of the nodes in the network of inmates in prison (right). The radii of the nodes are proportional to their degrees. Note that most central nodes are 'spread' across the network and not clumped together, even though the network is assortative.

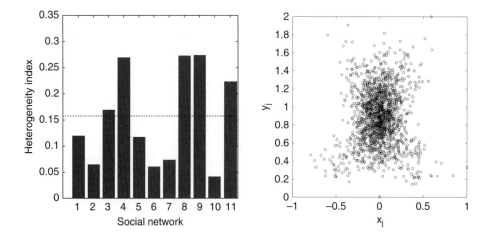

Fig. 19.4
Degree heterogeneity in social networks. Plot of the degree heterogeneity index (left) for the social networks in Table 19.2, and the H-plot for the social network of corporate directors.

Some social networks, therefore, have degree heterogeneity not different from that of random graphs, but have structural characteristics revealed by the spectral scaling, which significantly differ from those of random networks. For instance, the network of inmates in prison and that of the corporate elite have very low-degree heterogeneity similar to those of random networks, but they are more clumped than a random graph, and display characteristics typical of networks in Classes IV and II respectively, which differ from the typical Class I of random graphs. How and why, therefore, are some social networks assortative, having low-degree heterogeneity, topological heterogeneities like those of Classes II and IV, and relatively low clumpiness of the nodes according to their degrees?

We know by the structural interpretation of the assortativity coefficient that there are three structural ingredients contributing to a network assortativity: the clustering coefficient C, the ratio of paths of length 2 to paths of length 1, $|P_{2/1}|$, and the ratio of paths of length 3 to paths of length 2, $|P_{3/2}|$. Then, a network is assortative if, and only if, $|P_{3/2}| + C > |P_{2/1}|$. We can then create assortativity in a network by increasing the number of paths of length 3 without significantly increasing the number of paths of length 2, as well as increasing the clustering. Let us propose a simple model that accounts for these structural requirements and also has some kind of social rationality.

We start by considering k small communities of n_c nodes each. Communities of the same size are used for the sake of simplicity, and the model is perfectly viable when using communities of different sizes. We also simplify the definition of community here by considering that they are cliques. Let us now designate one leader for each community and connect every pair of leaders— as illustrated in Figure 19.5, in which we consider five communities of five nodes each. In this case the network obtained displays assortativity coefficient $r = 0.247$. This simple network of twenty-five nodes and density $d = 0.197$ has degree heterogeneity $\rho = 0.025$ and relative clumpiness $\Phi = 39.01\%$, which are similar to those of real social networks. Note that every pair of non-leaders is connected by a path of length 3, which indicates that the network is also a 'small-world.' In addition, this indicates that as the number of small communities increases, the assortativity coefficient must increase. For instance, for $k = 10$ and $n_C = 5$ the assortativity coefficient is $r = 0.443$. In this case, the network of sixty nodes has the topological properties $d = 0.110$, $\Phi = 35.7\%$, and $\rho = 0.036$, and belongs to the Class II.

This model is, of course, an oversimplification of what could happen in the real-world, and it can be easily relaxed to adapt to more realistic situations. One possibility is to consider that not every pair of leaders is connected,

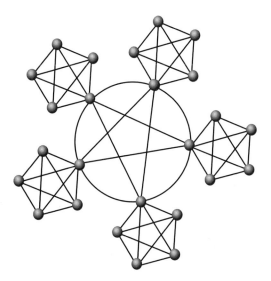

Fig. 19.5
Model for assortative social networks.
A simple model for developing assortative, low-density, topologically heterogeneous, clumped, 'small-world' networks, which very much resembles real-world social networks. Here, a leader from each community connects to the other leaders.

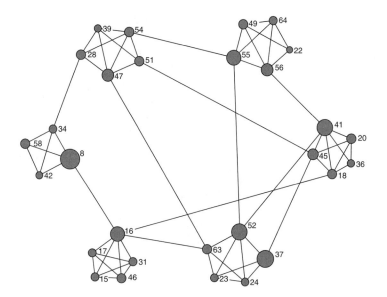

Fig. 19.6

Leader-centre groups in a real social network. Six cliques or quasi-cliques in the network of inmates in prison, displaying the characteristics of the model shown in Figure 19.5. In every small group there are very few leaders—those retaining intergroup connectivity. The radii of the nodes are proportional to their degrees in the whole network.

and to connect them according to some predetermined rule—for example, by random or preferential attachment. The idea behind this model is very easily applicable to that of collaboration networks, where there are some team leaders in each collaboration group who usually collaborate amongst themselves. In the case of the friendship network of inmates in prison, the situation can also be explained by using this analogy. Figure 19.6 shows a group formed by six cliques or quasi-cliques which are found in this network, in which several interconnected leaders are easily identifiable. The assortativity of this core is $r = 0.120$, which is quite close to that of the network as a whole (see Table 19.2).

Overall, a simple model based on the intuition of how we can collaborate in teams and between teams can explain several of the topological characteristics found in assortative social networks. Of course, other models or modifications of this one are necessary to explain the disassortativity found in other social networks, but this will be not discussed here.

19.3 Social network communities

The global topological structure of the social networks studied in the previous section reveals that they are formed by some kind of coarse-grained structure. This structure gives rise to the Classes II and IV observed for these networks, which basically indicates the existence of network communities. These communities are, of course, of a very different nature due to the large variability of these social networks, but in general they account for groups of individuals who are more tightly connected amongst themselves than with the rest of the network. Therefore, the detection of communities in social networks is one of the standard tests for practically every method proposed for finding network communities. An example of this type of community structure organisation is shown in Figure 19.7 for a coauthorship network at

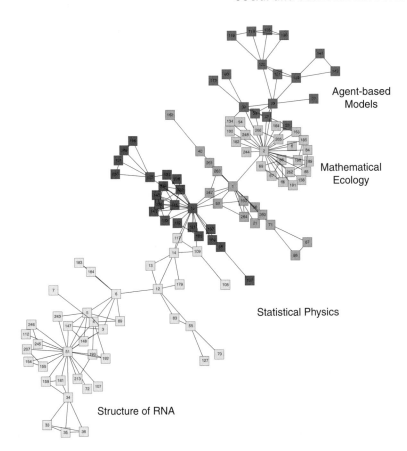

Agent-based
Models

Mathematical
Ecology

Statistical Physics

Structure of RNA

Fig. 19.7
Scientific collaboration network. The largest connected component of the Santa Fe Institute collaboration network. From Girvan, M., and Newman, E. J. (2002). Community structure in social and biological networks. *Proc. Natl. Acad. Sci. USA* **99**, 7821–6. (National Academy of Sciences, USA, ©2002.)

a private research institution (Girvan and Newman, 2002). The patterns of collaboration in different disciplines have been the objective of several studies (Newman, 2001; Newman, 2004c; Guimerá et al., 2005; Palla et al., 2007; Wuchty et al., 2007; Uzzi et al., 2007), and the reader is referred to those works for details.

The topological differences between the communities observed in different kinds of network are evident from the work of Palla et al. (2007), who studied a collaboration network of scientists who together have published papers now gathered in the Los Alamos *cond-mat* archive. This study involved more than 30,000 authors and a publication period of 142 months. The other dataset consists of the complete record of telephone calls between the customers of a mobile phone company, spanning 52 weeks (accumulated over two-week periods) and containing the communication patterns of more than 4 million users. The collaboration network of scientists is quite dense, due to the fact that every team that publishes a paper forms a clique in the network, while in the telephone-call network the communities are less interconnected due to the 'one-to-one' nature of the calls. Figure 19.8 illustrates both networks based on the local community structure at a given time step in the vicinity of a randomly selected node.

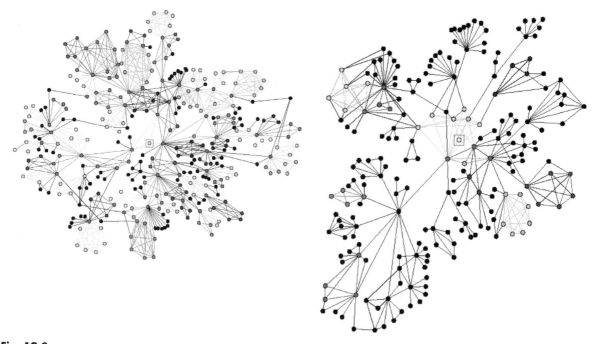

Fig. 19.8

Evolving social networks. A collaboration network among authors who have published papers together (left), and a telephone-call network (right), both of which are dynamically changing over time. From Palla, G., Barabási, A.-L., and Vicsek, T. (2007). Quantifying social group evolution. *Nature* **446**, 664–7. (Reproduced with the permission of Macmillan Publishers Ltd., ©2007.)

One characteristic of the communities studied by Palla et al. (2007) is that they can evolve in time, and as a result, can grow or contract. Groups might merge or split, new communities can be born, and some of the existing communities can disappear. It was found that the size and the age of the communities are positively correlated, indicating that larger communities are, on average, older. The most interesting finding was that the persistence of groups during time depends very much on the group's size. Large groups persist longer if they are capable of changing their membership in a dynamical way. However, small groups tend to be more stable if their composition remains unchanged. In other words, smaller groups are destabilised and can disappear if they drastically change their composition with time, while the large groups increase their adaptability by renewing their composition.

In many social networks there is a natural partition into communities which at the same time possess a natural hierarchy. One of them is the committee assignment network of the US House of Representatives, which has been studied by Porter et al. (2005; 2007). As shown in Figure 19.9, this network is formed by committees (represented by squares) and subcommittees (represented by circles) in the 107th US House of Representatives, and links connecting committees represent the number of common members. The US House of Representatives has an organisation which includes the House floor, groups of committees, committees, groups of subcommittees within larger

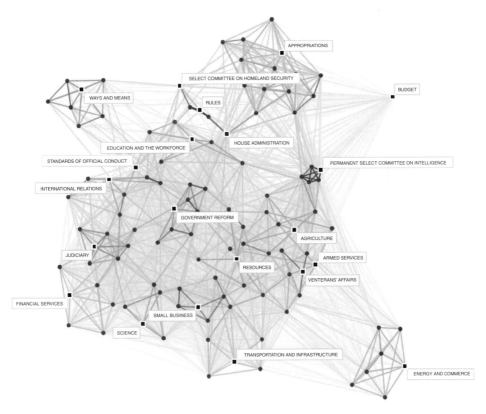

Fig. 19.9

Communities in a hierarchically structured social network. Representation of the communities in the network of committees (squares) and subcommittees (circles) in the 107th US House of Representatives. From Porter, M. A., Mucha, P. J., Newman, M. E. J., and Friend, A. J. (2007). Community structure in the United States House of Representatives. *Physica A* **386**, 414–38. (Reproduced with the permission of Elsevier, ©2007.)

committees, and individual subcommittees. This structure is revealed by the communities found by Porter et al. (2005), in which the different modules inside the House are observed at the same time that small groups of committees belong to larger but less densely connected modules.

There are, however, many real-world situations in which we do not have any knowledge of modular or hierarchical organisation of the social structure in the network. In these cases the analysis of the network communities can reveal the internal structure of the organisations, thereby providing important insights about the way by which they function. An example is the network of injecting drug-users in Colorado Springs. This network has an internal modular organisation that can be revealed by using community detection algorithms. Figure 19.10 shows a partition of this network into twenty different communities, obtained by using the Girvan–Newman algorithm. This partition is the one displaying the maximum modularity among all partitions from two to thirty communities. The discovery and analysis of this kind of modular organisation can provide insights into the best interventional strategies for managing the risk of individuals and groups in such networks.

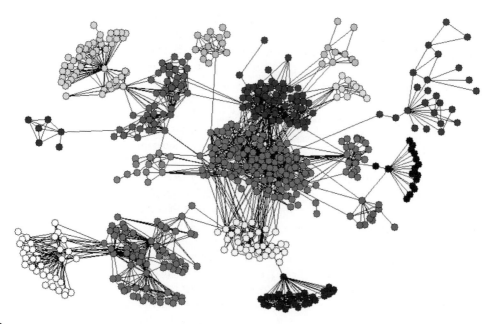

Fig. 19.10

Discovering hidden communities in a social network. The twenty communities identified by the Girvan–Newman algorithm for the network of injecting drug-users in Colorado Springs.

19.4   Centrality in social networks

For some considerable time, the concept of node centrality has been conceptualised and widely applied in social network analysis. The works of Freeman (1979), Friedkin (1991), and Borgatti (2005) can be cited as good examples of the conceptualisation of centrality in social networks, though it should be mentioned that most, if not all, centrality measures defined so far have been introduced in an *ad hoc* way and not derived from any 'first principles'. We have previously analysed (see Chapter 7) some examples of how particular centrality measures are applied in social networks, and a critical review of the role of centrality measures has been published by Landherr et al. (2010). Here, however, we propose a more general approach in which some global characteristics of centrality measures can be extracted.

There have been two main topics of concern in analysing the global properties of centrality measures in social sciences and beyond. The first is related to the 'robustness' of the different centrality measures to incorrect or incomplete data. Bolland (1988) analysed the stability of the degree (DC), closeness (CC), betweenness (BC), and eigenvector (EC) centrality to random and systematic variations of the network structure, and found that BC changes significantly with the variation of the network structure, while DC and CC are usually very stable against such kinds of random and systematic change, with EC being the most stable of all the indices analysed. The BC measure has also been identified by other authors as the most sensible of all these centrality measures to the variations of the structure of the network (Costenbader and Valente, 2003;

Borgatti et al., 2006). Further studies by Borgatti et al. (2006) and Frantz et al. (2009) have extended these investigations into the stability of centrality indices by considering the addition and deletion of nodes and links (Borgatti et al., 2006), as well as by differentiating several types of network topology (Frantz et al., 2009) such as uniform random, small-world, core-periphery, scale-free, and cellular. According to the results of Frantz et al. (2009), DC is almost twice as sensible to structural changes in core-periphery and scale-free networks than in uniform random ones, while in cellular topologies it has about the same robustness as in the random graph. BC is again the most sensible of all the centrality measures analysed, but this sensitivity is specific for core-periphery and scale-free networks, where the effect on accuracy of structural changes is more than double that produced for uniform random graphs. However, in small-world topologies BC is significantly more robust than the random uniform networks. The other two centrality measures display similar behaviour, with larger sensibilities to structural changes when the topologies are core-periphery or scale-free, and smaller for cellular and small-world ones. When analysing the selection of the top 10% of most central nodes, DC appears to be the most robust measure, followed by EC, while BC is the most sensible one. Overall, small-worldness is the most robust of all topologies for any of the centrality measures analysed. Here, small-worldness is understood in the Watts–Strogatz sense of having small average path length and large clustering relative to a random graph. On the other hand, scale-free networks, with their uneven distributions of degrees, display the largest variations in the centrality measures among all the topologies analysed.

The second global property that has been studied for centrality measures is their mutual correlation. The main preoccupation here is that different centrality measures identify different actors as the most central ones in a given network. These kinds of study were conducted by Freeman et al. (1980), Lee (2006), and Kiss and Bichler (2008). The general conclusion of these studies is that different centrality measures in some cases rank the actors of a network in considerably different ways. For instance, among the networks studied by Lee (2006) there are two social networks—one for film actors, and the other a scientific collaboration network. In both cases, BC and DC are significantly correlated with correlation coefficients larger than 0.70. However, BC is poorly correlated with EC and with CC, DC is poorly correlated with CC, and EC is poorly correlated with CC in both social networks. By contrast, when Lee (2006) studied Erdös–Rényi random networks with 1,000, 3,000, and 6,000 nodes and average degree 10, he always observed high correlation among all pairs of centrality measures. In all cases the correlation coefficient was larger than 0.9, with values of up to 0.98 for the correlation of DC versus BC.

In order to further investigate this topic we compare here the degree (DC), closeness (CC), betweenness (BC), subgraph (SC), eigenvector (EC), and information (IC) centrality for the social networks analysed previously in this chapter. We calculate the Kendall τ coefficient (Kendall, 1938) for all pairs of centrality indices as a way of quantifying the similarities in the ranking of the nodes of all these networks. This coefficient represents the probability that in the network the two centralities are of the same order, versus the probability that the two centralities are of different orders. We plot these results in a radial plot in which there are fifteen axes, each of which represents the Kendall coefficient for a pair of centralities. Then, every network is projected into this fifteen-dimensional space as a 15-gon that intersects each axis at the value of its Kendall τ coefficient.

Figure 19.11 illustrates this plot for all the social networks given in Table 19.2. As can be seen in this figure, the highest correlations are obtained for almost every network for the pair SC–EC. This kind of correlation is the basis of the spectral scaling method (see Chapter 9), where even small deviations are important for revealing the global topological organisation of the nodes in a network. However, in some cases, such as the Colorado Springs network, this correlation can be very low, indicating a strong lack of homogeneity between the local and global structures of the network. The poorest correlations are always observed for the pairs EC–BC and SC–BC, indicating that the betweenness contains structural information which is quite far from that represented by the other indices.

We have observed that the correlations between DC–IC, DC–EC, SC–IC, EC–IC, DC–CC, CC–BC, BC–IC, and EC–BC depend very much on the small-worldness properties of the networks involved. In particular, as the average path length of the network decreases, the rank correlation, measured by the Kendall coefficient of the pair of centralities, increases. In other words, there is a negative linear relation between the Kendall coefficient for the previously

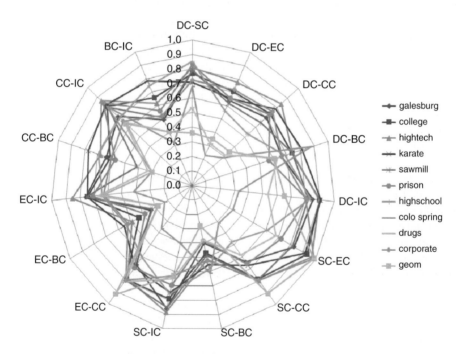

Fig. 19.11

Non-parametric analysis of centrality measures. Radial plot of the Kendall τ index for fifteen pairs of centrality measures analysed in the eleven social networks given in Table 19.2. Every social network is represented by a 15-gon whose vertices are the Kendall indices for the corresponding pair of centralities represented in the given axis.

mentioned pairs of centralities and the average path length of the networks. For example, considering the pair DC–IC, there are four networks which have Kendall coefficients larger than 0.85. These networks have average path length small than 2.6. Four other networks have a Kendall coefficient of about 0.8 and average path length between 3 and 3.5. When the average path length increases to 5.3, the Kendall coefficients decay to 0.6–0.5. Finally, when the average path length is larger than 8, the rank correlation is about 0.33. Similar dependencies are observed for the other pairs of centrality measures, which indicate that the small-worldness of these networks plays a fundamental role in both the robustness of individual centrality measures (Frantz et al., 2009) and the intercorrelation between pairs of centralities. The structural connections between these two types of variable have not yet been investigated.

19.5 Economic networks

There are many challenges in the study of economic networks (Schwitzer et al., 2009)—particularly in a global economy, which is constantly changing—and here we will provide only a few indications about the many roles which network theory can play in the study of economic-related problems. It is straightforward to realise that these kinds of network are intimately related to many of the social systems studied in the previous section, and the transition from social to economic/financial networks is therefore smooth. As a starting point we can consider the network formed by the interlock of members of the boards of directors of US corporations. Although this is a social network, its importance from the economic point of view is evident. It is formed by projecting a bipartite network of corporations and directors into the directors–directors space. However, the corporate–corporate projection produces a network in which nodes are corporations, and links connect pairs of firms which share at least a director on their boards. This network is formed by 789 corporations with 2,993 links, and so the density is 0.0096. The average path length is $\bar{l} = 3.73$, and the clustering coefficient is $\bar{C} = 0.23$. The number of links per corporation decays as an exponential law with a faster tail, as shown in Figure 19.12. The H-plot (see Chapter 8) of this network is more characteristic of an exponential-like degree distribution than of a fat-tailed degree distribution (see the same figure). Finally, the spectral scaling of this corporation network reveals that it belongs to the Class II network (see Chapter 9), and is formed by tightly linked communities separated by structural holes. The analysis of the scaling between the clustering coefficient as a function of node degree is also illustrated in Figure 19.12, where it can be seen that $C(k) \sim k^{-1}$, indicating either the existence of some hierarchical structure or the fact that the number of cliques decays very fast with the size of the cliques. Here again, the second explanation appears more plausible, as there are 535 cliques of size 3, 146 cliques of size 4, forty-four cliques of size 5, eight cliques of size 4, only five cliques of size 7, and no cliques larger than size 7.

Other types of networks of companies have been studied in the literature. Takayasu et al. (2008a) have studied the network of Japanese companies

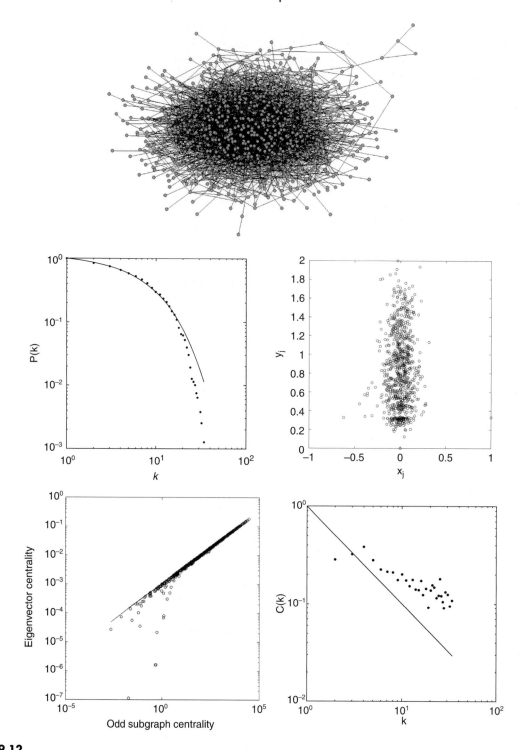

Fig. 19.12

Topological properties of a corporation network. The corporation network of the top US corporations (top), in which nodes represent firms, and two nodes are connected if the corresponding firms have at least one member in common on their boards of directors. Also illustrated here for this network are cumulative degree distribution (centre left), H-plot (centre right), spectral scaling (bottom left), and scaling of clustering as a function of degree (bottom right).

connected by business deals according to the data provided by the Research Institute of Economy, Trade, and Industry (RIETI) of Japan. The network analysed by Takayasu et al. (2008a) consists of 961,318 nodes and 3,667,521 directed links, which represent transactions from customer to supplier. This network needs to be represented by an adjacency matrix having about 10^{12} entries—1 trillion elements. As a comparison, let us consider the Authorised Version of the Bible, which contains 4.017 million characters, and a Paris telephone book, which has about 60 million characters. Takayasu et al. (2008a, 2008b) used the services of the Earth Simulator Centre—one of the most powerful supercomputers in the world. With this huge complex network, they found that the cumulative distribution of the number of links per company follows a power-law distribution with an exponent of about 1.2 for the wide range of degrees between 10 and 1,000. They also analysed the distribution of distances for every pair of nodes in this network, and found that there is about a 45% probability of finding a path between any two nodes selected at random. The eccentricity of the network is just 21, and the average distance is about 5, decaying almost exponentially for distances larger than 8. It can be said that the Japanese intercorporation network is a very small world after all. Ohnishi et al. (2010) have found that the clustering coefficient of this network decays as a power-law with the degree of the nodes: $C(k) \sim k^{-1}$ for large values of the degree. They have interpreted this finding via the classical view of an hierarchical structure of a network, but it must be more plausible to think that in this network the number of cliques decays very fast with the size of the cliques (see Chapter 4), and that most corporations do not prefer to be aggregated into very large clusters.

By considering a subdivision of companies into seventeen job categories, Takayasu et al. (2008a) defined the interaction between two categories by the number of individual links that connect companies in one category to another. From this, a graphical representation of the global industrial interactions in Japan was obtained—as illustrated in Figure 19.13, where every node represents an aggregation of a huge number of companies. For example, in the case of manufacturing there are 143,628 companies, of which about 35% form the strongly-connected component.

The analysis of node centrality in this giant network was carried out by considering the indegree, the outdegree, and the PageRank of every node (Ohnishi et al., 2009; Takayasu et al., 2008b). It was observed that several quantities, such as sale, income, and number of employees, are positively correlated with the centrality of the nodes. However, the correlations are in general very low, indicating more a trend than a strong dependency between the parameters analyzed. The rate of growth was observed to be independent of the centrality of the nodes, indicating that large and small companies have almost similar chances of growing, with very little influence from the network structure—at least in Japan.

A very innovative visualisation analysis of the Japanese corporation network has been carried out by using molecular dynamics simulation (Takayasu et al., 2008b). In this case, every node of the network was treated as a particle having interactions, which are controlled by a Coulomb interaction potential of the following form:

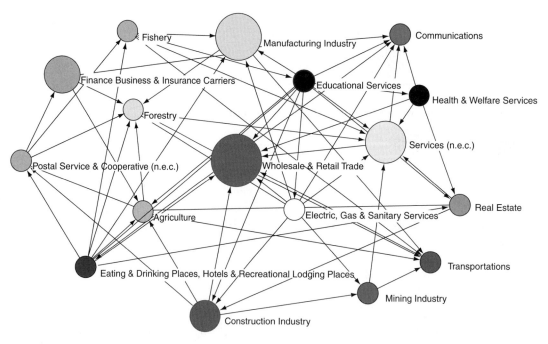

Fig. 19.13

Network of job categories in Japan. Representation of the network of job categories in Japan constructed by aggregating in every node all companies which are in the same sector. The radii of the nodes are proportional to the sum of sales belonging to the category. (Image courtesy M. Takayasu; reproduced from Takayasu et al. (2008a), with the permission of the authors.)

$$\Phi(r_{ij}) = \frac{Q_i Q_i}{r_{ij}} - \frac{1}{2} k_{ij} r_{ij}^2, \qquad (19.1)$$

where Q is the 'electrostatic charge', k_{ij} is a spring constant, and r_{ij} is the interparticle distance. Unit charges, masses, and spring strengths are assigned to every company. It is then assumed that these particles coupled by springs are suspended in three-dimensional viscous fluid. Starting from any initial configuration at high temperature, the particles converge to the most stable configuration after cooling. By using some approximations required to deal with 1 million particles, Takayasu et al. (2008b) decomposed the company configuration by the sectors of industry—as illustrated in Figure 19.14, where it can be seen that each sector has its own characteristic configuration ranging from clustered to more uniform.

An investigation into network motifs in the Japanese corporation network was carried out by Ohnishi et al. (2010), who studied the thirteen three-node connected directed subgraphs. They found that four subgraphs (see Figure 19.15) appear more frequently in this economic network than are expected at random, and they are considered as motifs according to the standard definition given in Chapter 4.

In this network (Ohnishi et al., 2010), the directed links represent flow of money from customer to supplier, and those in the opposite direction show flow of materials and services. Financial transactions, such as loan contracts

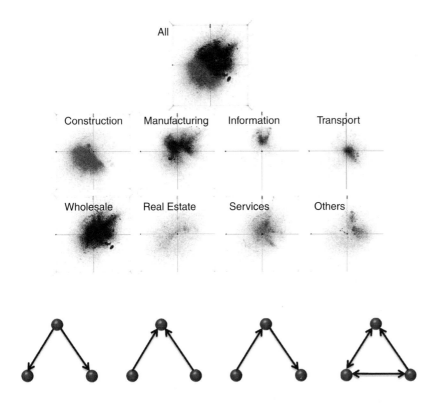

All

Construction   Manufacturing   Information   Transport

Wholesale   Real Estate   Services   Others

Fig. 19.14
Structure of the whole economy in Japan. The optimised structure of the whole economy in Japan, where only nodes (not links) are drawn, and displaying a decomposition into various sectors. (Image courtesy M. Takayasu; reproduced from Takayasu et al. (2008a), with the permission of the authors.)

Fig. 19.15
Motifs in the Japanese corporation network. Four small subgraphs which appear more frequently in the Japanese corporation network than expected at random. Consequently, they are considered as structural motifs in this network.

and purchases of securities, are not considered. A directed link of the type A → B indicates that customer A has transferred money to supplier B, who has transferred some materials/services to the first. The case of bidirected links A ←→ B indicates that both A and B are customers and suppliers in different transactions. The existence of the first two motifs illustrated in Figure 19.15 indicates that there are some 'universal' customers and suppliers in this economy. That is, the central node in the first motif in the figure is a universal customer, as it only transfers money to suppliers who transfer materials/services to it. The central node of the second motif is a universal supplier who always receives money from others to whom it supplies its materials/services. The central node of the third motif is a sort of 'intermediate' who receives money from some, and at the same time transfers money to others. The last motif represents a type of more complicated transaction among triples of companies in which most of them are customer and supplier at the same time.

Three other subgraphs were found significantly less frequently than expected at random, and they are consequently considered as anti-motifs (see Figure 19.16). These are topological structures which corporations avoid during their transactions more frequently than expected at random. The first of them is formed by a universal supplier and a universal customer, who are also interconnected through an intermediate. The second one is the cycle of three intermediates, and the third is very similar to the second except that two companies act as customer and supplier at the same time.

Fig. 19.16

Anti-motifs in the Japanese corporation network. Three small subgraphs which appear less frequently in the Japanese corporation network than expected at random. Consequently, they are considered as structural anti-motifs in this network.

Other types of economic network have also been studied in the literature. They include ownership networks in which nodes represent companies, and links—usually weighted—represent one company owning a portion of another (Bonanno et al., 2003; Garlaschelli and Lofredo, 2004b, 2005; Onnela et al., 2004). Other examples are trade networks, such as the world trade network (Li et al., 2003; Garlaschelli and Lofredo, 2004b), in which every country in the world is represented by a node, and two nodes are connected if they have a trade relationship such as imports or exports. A variation of this type of network is produced by considering the trade of one specific item. For instance, a trade network of miscellaneous manufacture of metal (MMM) among eighty countries during 1994 has been studied (Estrada 2011). The data was compiled by de Nooy et al. (2005) by considering all countries with entries in the paper version of the *Commodity Trade Statistics* published by the United Nations. Countries which were not sovereign at the time of data-compilation—the Faeroe Islands and Greenland, which belong to Denmark, and Macau, which belongs to Portugal—were excluded because additional economic data were not available. Most of the countries omitted are located in central Africa and the Middle East, or belong to the former USSR.

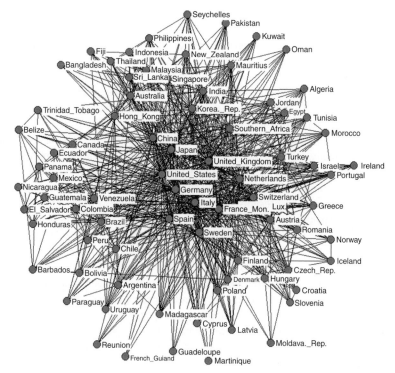

Fig. 19.17

International trade network of metal manufactures. An international trade network of miscellaneous manufacture of metal (MMM), in which nodes represent countries and two nodes are connected if one of them exported MMM to the other in 1994.

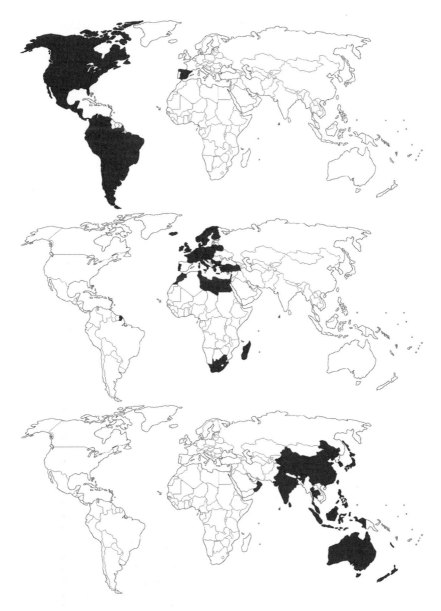

Fig. 19.18
Communities in an international trade network of metal manufactures. Three communities identified by using a communicability-based method for the trade network of miscellaneous metal manufacturers represented in Figure 19.17.

In general, all countries that export MMM also import them from other countries. However, twenty-four countries are importers only, and have no export of MMM at all. They are: Kuwait, Latvia, The Philippines, French Guiana, Bangladesh, Fiji, Reunion, Madagascar, the Seychelles, Martinique, Mauritius, Belize, Morocco, Sri Lanka, Algeria, Nicaragua, Iceland, Oman, Pakistan, Cyprus, Paraguay, Guadalupe, Uruguay, and Jordan. Despite this asymmetry, there are few differences when considering an undirected version of this network rather than a directed version. For instance, the larger exporters are (in order): Germany, the USA, Italy, the UK, China, Japan, France, Belgium/Luxemburg, The Netherlands, and Sweden, and when the undirected version of this network is considered, the same countries appear as those with the highest degree. In addition, there are only slight changes to the order: Germany, the USA, Italy, the UK, Japan, China, France, The Netherlands, Belgium/Luxemburg, and Sweden. On the other hand, the indegree is practically the same for every country. For instance, the average indegree is 12.475, and its standard deviation is only 3.15. The undirected version of this network previously studied is represented in Figure 19.17, where the nodes represent the countries and a link exists between two countries if one of them imports miscellaneous manufacture of metal (MMM) from the other.

We have analyzed the existence of communities in this network by using a method based on the communicability function (Estrada, 2011). Classical methods of community identification—such as the Girvan–Newman algorithm—fail to identify any community structure in this network, due to the large interconnection among the groups existing in it. However, the communicability-based approach clearly identifies three communities, one of which is formed by all Latin American countries plus Canada, the USA, and Spain. The presence of the USA and Canada in this community is justifiable due to their geographical proximity to Central America and South America, while Spain is included because of its long tradition of trade with its former colonies in America. Another community consists of most European countries except Spain, and many of their former colonies or regions of influence in Africa, the Middle East, and South America (French Guiana). The third community represents south-east Asia and oceanic countries (see Figure 19.18).

One characteristic feature of this network is the existence of some global hubs, which is the main reason why algorithms such as the Girvan–Newman fail in identifying these communities. These global hubs are characterized by a large interconnection with members of its community and with other countries in other communities. For example, in the first community the global hub is the USA, which is connected to 72 of the 80 countries in the web. In the second community, Germany, Italy, and the UK are connected to 77, 72, and 63 countries respectively, both inside and outside their community, while Japan and China, which are connected to 58 and 57 countries respectively, are the global hubs in the third community.

Appendix

Datasets cited in this book

A.1 Brain networks

Cortices   The brain networks of macaque visual cortex, macaque cortex, and cat cortex, after the modifications introduced by Sporn and Kötter (2004) (see text). These datasets are available at http://www.indiana.edu/~cortex/CCNL.html.

Neurons   Neuronal synaptic network of the nematode *C. elegans* (White et al., 1986; Milo et al., 2002).

A.2 Ecological networks

Benguela   Marine ecosystem of Benguela, off the south-west coast of South Africa (Yodzis, 2000).

Bridge Brook   Pelagic species from the largest of a set of fifty Adirondack Lake (NY) food webs (Polis, 1991).

Canton Creek   Primarily invertebrates and algae in a tributary, surrounded by pasture, of the Taieri River in the South Island of New Zealand (Townsend et al., 1998).

Chesapeake Bay   The pelagic portion of an eastern US estuary, with an emphasis on larger fish (Christian and Luczkovich, 1999).

Coachella   Wide range of highly aggregated taxa from the Coachella Valley desert in Southern California (Warren, 1989).

Dolphins   Social network of a bottlenose dolphin (Tursiops truncates) population near New Zealand (Lusseau, 2003).

El Verde   Insects, spiders, birds, reptiles, and amphibians in a rainforest in Puerto Rico (Waide and Reagan, 1996).

Little Rock   Pelagic and benthic species, particularly fish, zooplankton, macroinvertebrates, and algae in Little Rock Lake, Wisconsin, USA (Havens, 1992).

Reef Small   Caribbean coral reef ecosystem in the Puerto Rico/Virgin Island shelf complex (Opitz, 1996).

Scotch Broom   Trophic interactions between the herbivores, parasitoids, predators, and pathogens associated with broom, *Cytisus scoparius*, collected in Silwood Park, Berkshire, England (Memmott et al., 2000).

Shelf	Marine ecosystem on the north-east US shelf (Link, 2002).
Skipwith	Invertebrates in an English pond (Yodzis, 1998).
St Marks	Mostly macroinvertebrates, fish, and birds associated with an estuarine seagrass community, *Halodule wrightii*, at the St Marks Refuge, Florida, USA (Goldwasser and Roughgarden, 1993).
St Martin	Birds and predators and arthropod prey of *Anolis* lizards on the island of St Martin in the northern Lesser Antilles (Martinez, 1991).
Stony Stream	Primarily invertebrates and algae in a tributary, surrounded by pasture, in native tussock habitat, of the Taieri River in the South Island of New Zealand (Baird and Ulanowicz, 1989).
Ythan_1	Mostly birds, fish, invertebrates, and metazoan parasites in a Scottish estuary (Huxman et al., 1996).
Ythan_2	Reduced version of Ythan1, without parasites (Hall and Rafaelli, 1991).

A.3 Informational networks

Centrality	Citation network of papers published in the field of Network Centrality (Hummon et al., 1990; de Nooy et al., 2005).
GD	Citation network of papers published in *Proceedings of Graph Drawing* during the period 1994–2000 (Batagelj and Mrvar, 2001).
ODLIS	Vocabulary network of words related by their definitions in the Online Dictionary of Library and Information Science. Two words are connected if one is used in the definition of the other (ODLIS, 2002).
Roget	Vocabulary network of words related by their definitions in Roget's Thesaurus of the English language. Two words are connected if one is used in the definition of the other (Roget, 2002).
Sci_Met	Citation network of papers published in, or citing articles from, *Scientometrics*, for the period 1978–2000 (Batagelj and Mrvar 2006).
Small World	Citation network of papers which cite Milgram's (1967) Psychology Today paper or include Small World in the title (Batagelj and Mrvar 2006).

A.4 PPI networks

PPIs	Protein–protein interaction networks in *D. melanogaster* (fruit fly) (Giot et al., 2003), *Kaposi sarcoma* herpes virus (KSHV) (Uetz et al., 2006), *P. falciparum* (malaria parasite) (LaCount et al., 2005), *Varicella zoster* virus (VZV) (Uetz et al., 2006), human (Rual et al., 2005), *S. cerevisiae* (yeast) (Bu et al., 2003;

von Mering et al., 2002), *A. fulgidus* (Motz et al., 2002), *H. pylori* (Lin et al., 2005; Rain et al., 2001), *C. elegans* (Davy et al., 2001), *E. coli* (Bultland et al., 2005), and *B. subtilis* (Noirot and Noirot-Gross, 2004).

A.5 Protein residue networks

Protein Protein residue networks of 595 proteins compiled by Fariselli and Casadio (1999) (which see for PDB codes of each protein). The protein residue networks were constructed by Atilgan et al. (2004).

A.6 Social and economic networks

Corporate American corporate elite formed by the directors of the 625 largest corporations that reported the compositions of their boards, selected from the *Fortune* 1,000 in 1999 (Davis et al., 2003).

Geom Collaboration network of scientists in the field of computational geometry (Batagelj and Mrvar 2006).

Prison Social network of inmates in prison who chose 'Which fellows on the tier are you closest friends with?' (MacRae, 1960).

Drugs Social network of injecting drug-users (IDUs) who have shared a needle in the last six months (Moody, 2001).

Zachary Social network of friendship between members of the Zachary karate club (Zachary, 1977).

College Social network among college students participating in a course about leadership. The students choose which three members they want to have on a committee (Zeleny, 1950).

ColoSpring The risk network of persons with HIV infection during its early epidemic phase in Colorado Springs, USA, using analysis of community-wide HIV/AIDS contact tracing records (sexual and injecting drugs partners) during 1985–99 (Potterat et al., 2002).

Jazz Collaboration network of jazz musicians, where two musicians are connected if they played in the same band. This includes 198 bands that performed between 1912 and 1940, with most of the bands in the 1920s (Arenas et al., 2004).

Galesburg Friendship ties among 31 physicians (Coleman et al., 1966; Knoke and Burt, 1983; de Nooy et al., 2005).

High_Tech Friendship ties among the employees in a small high-tech computer firm which sells, installs, and maintains computer systems (Krackhardt, 1999; de Nooy et al., 2005).

Saw Mills	Social communication network within a sawmill, where employees were asked to indicate the frequency with which they discussed work matters with each of their colleagues (Michael and Massey, 1997; de Nooy et al., 2005).
MMM	World trade network of miscellaneous manufacture of metals (MMM) in 1994 (Smith and White, 1992; de Nooy et al., 2005).
Heterosexual	Heterosexual contacts, extracted at the Cadham Provincial Laboratory; a six-month block of data from November 1997 to May 1998 (Lind et al., 2005).
Homosexual	Contact tracing study, 1985–99, for HIV tests in Colorado Springs, USA, where most of the registered contacts were homosexual (Lind et al., 2005).

A.7 Technological networks

Electronic	Electronic sequential logic circuits parsed from the ISCAS89 benchmark set, where nodes represent logic gates and flip-flops (Milo et al., 2002).
USAir97	Airport transportation network between airports in the US in 1997 (Batagelj and Mrvar, 2006).
Internet	The Internet at the Autonomous System (AS) level, as of September 1997 and April 1998 (Faloutsos et al., 1999).

A.8 Transcription networks

Trans_E.coli	Direct transcriptional regulation between operons in *Escherichia coli* (Milo et al., 2004a; Shen-Orr et al., 2002).
Trans_sea_urchin	Developmental transcription network for sea urchin endomesoderm development (Milo et al., 2004a; Davidson et al., 2002).
Trans_yeast	Direct transcriptional regulation between genes in *Saccaromyces cerevisae* (Milo et al., 2002; Milo et al., 2004a).

A.9 Software

BGL	Developed by D. Gleiss, and available at http://www.stanford.edu/~dgleich/programs/matlab_bgl/. It consists of several *Matlab*® programs for carrying out elementary operations on graphs, as well as some of the new techniques developed for complex networks available at http://vlado.fmf.uni-lj.si/pub/networks/pajek/.
Combinatorica	An extension of the computer algebra system *Mathematica*® developed by Pemmaraju and Skiena (2003). It performs a large number of operations on graphs and large networks.

CONTEST	A toolbox developed by A. Taylor and D. J. Higham for *Matlab*®, and available at http://www.maths. strath.ac.uk/research/groups/numerical_analysis/contest. It has the implementation of nine models for generating random networks from the classical Erdös–Rényi, Watts–Strogatz, and Barabási–Albert models, and the Range Dependent, Lock-and-Key, and Stickiness models. It also includes the generation of random geometric graphs (Taylor and Higham, 2009).
Pajek	Developed by W. de Nooy, A. Mrvar, and V. Batagelj for the analysis and visualisation of large complex networks. It is freely available at http://vlado.fmf.uni-lj.si/pub/networks/pajek/.
STATISTICA®	Used for statistical analysis, and distributed by Stat-Soft at http://www.statsoft.co.uk/. It provides an array of data analysis, management, visualization, and mining procedures for modelling, classification, and exploratory techniques.
UCINET	Developed by Borgatti, Everett, and Freeman for the analysis of (social) networks. It includes the freeware program *NETDRAW* for visualizing networks, and is distributed by Analytic Technologies at http://www.analytictech.com/ucinet/.

References

Abreu, N. M. M. (2007). Old and new results on algebraic connectivity of graphs. *Lin. Alg. Appl.* **423**, 53–73.

Achlioptas, D., D'Souza, R. M., and Spencer, J., (2009). Explosive percolation in random networks, *Science* **323**, 1453–5.

Adams, J., (2010). Distant friends, close strangers? Inferring friendships from behavior, *Proc. Natl. Acad. Sci. USA* **107**, E29–E30.

Albertson, M. O. (1997). The Irregularity of a Graph. *Ars Combinatorica* **46**, 219–25.

Albert, R., and Barabási, A.-L. (2002). Statistical mechanics of complex networks. *Rev. Mod. Phys.* **74**, 47–97.

Albert, R., Jeong, H., and Barabási, A.-L. (2000). Error and attack tolerance of complex networks. *Nature* **406**, 378–82.

Allesina, S., and Bodini, A. (2004). Who dominates whom in ecosystems? Energy flow bottlenecks and cascading extinctions. *J. Theor. Biol.* **230**, 351–8.

Allesina, S., Bodini, A. and Bondavalli, C. (2006). Secondary extinctions in ecological networks: bottlenecks unveiled. *Ecol. Model.* **194**, 150–61.

Al-Mohy, A. H., and Higham, N. J., (2009). Computing the Fréchet derivative of the matrix exponential, with an applications to condition number estimation. *SIAM J. Matrix Anal. Appl.* **30**, 1639–57.

Alon, N. (1986). Eigenvalues and expanders. *Combinatorica* **6**, 83–96.

Alon, N., and Milman, V. D. (1985). λ_1, isoperimetric inequalities for graphs, and superconcentrators. *J. Combin. Theory Ser. B* **38**, 73–88.

Alon, N., Yuster, R., and Zwick, U. (1997). Finding and counting given length cycles, *Algorithmica* **17**, 209–23.

Alpert, C. J., Kahng, A. B., and Yao, S.-Z. (1999). Spectral partitioning with multiple eigenvectors. *Disc. Appl. Math.* 90, **3**–26.

Alves, N. A., and Martinez, A. S. (2007). Inferring topological features of proteins from amino acid residue networks, *Physica A* **375**, 336–44.

Amaral, L. A. N., Scala, A., Barthelémy, M., and Stanley H. E. (2000). Classes of small-world networks, *Proc. Natl. Acad. Sci.* **97**, 11149–52.

Amitai, G., Shemesh, A., Sitbon, E., Shklar, M., Netanely, D., Venger, I., and Pietrokovski, S. (2004). Network analysis of protein structures identifies functional residues. *J. Mol. Biol.* **344**, 1135–46.

Anderson, J. W. (2005). *Hyperbolic Geometry*. Springer-Verlag, London.

Arenas, A., Danon, L., Díaz-Guilera, A., Gleiser, P. M., and Guimerá, R. (2004). Community analysis in social networks. *Eur. Phys. J. B.* **38**, 373–80.

Arenas, A., Fernández, A., and Gómez, S. (2008). Analysis of the structure of complex networks at different resolution levels. *New J. Phys.* **10**, 053039.

Arii, K., and Parrott, L. (2004). Emergence of non-random structure in local food webs generated from randomly structured regional webs. *J. Theor. Biol.* **227**, 327–33.

Arita, M., (2004). The metabolic world of Escherichia coli is not small. *Proc. Natl. Acad. Sci. USA* **101**, 1543–7.

Arkin, M., and Wells, J. A., (2004). Small-molecule inhibitors of protein–protein interactions: Progressing towards the dream. *Nature Rev. Drug Discov.* **3**, 301–17.

Arnold, L. (1967). On the asymptotic distribution of the eigenvalues of random matrices. *J. Math. Anal. Appl.* **20**, 262–8.

Arnold, L. (1971). On Wigner's semicircle las for the eigenvalues of random matrices. *Z. Wahrsch. Verw. Geb.* **19**, 191–8.

Arora, S., Rao, S., and Vazirani, U. (2009). Expander flows, geometric embeddings and graph partitioning. *J. ACM* **56**, 5.

Artzy-Randrup, Y., Fleishman, S. J., Ben-Tal, N., and Stone, L. (2004). Comment on 'Network motifs: simple building blocks of complex networks' and 'Superfamilies of evolved and designed networks.' *Science* **305**, 1107c.

Asratian, A. S., Denley, T. M. J., and Häggkyist, R. (1998). *Bipartite Graphs and their Applications*. Cambridge University Press, Cambridge, 272 pp.

Assad, A. A. (2007). Leonhard Euler: A Brief Appreciation. *Networks* **49**, 190–8.

Aste, T., Di Matteo, T., and Hyde, S. T. (2005). Complex networks on hyperbolic surfaces. *Physica A* **346**, 20–6.

Atilgan, A. R., Akan, P., and Baysal, C. (2004). Small-world communication of residues and significance for protein dynamics. *Biophys. J.* **86**, 85–91.

Atilgan, A. R., and Baysal, C. (2004). Thermodynamics of residue networks in folded proteins. *WSEAS Trans. Biol. Biomed.* **1**, 205–9.

Aurenhammer, F. (1991). Voronoi Diagrams. A Survey of a Fundamental Geometric Data Structure. *ACM Comput. Surv.* **23**, 345–405.

Avin, C. (2008). Distance graphs: from random geometric graphs to Bernoulli graphs and between. *Proc. 5th Int. Work. Found. Mob. Comput.* New York, pp. 71–8.

Aylward, S. R., Jomier, J., Vivert, C., LeDigarcher, V., and Bullitt, E., (2005). Spatial graphs for intra-cranial vascular network characterization, generation, and discrimination, *Lect. Notes Compt. Sci.* **3749**, 59–66.

Babić, D., Klein, D. J., Lukovits, I., Nicolić, S., and Trinajstić, N. (2002). Resistance-distance matrix: a computational algorithm and its application. *Int. J. Quantum Chem.* **90**, 166–76.

Bagler, G., and Sinha, S. (2005). Network properties of protein structures. *Physica A* **346**, 27–33.

Bagler, G., and Sinha, S. (2007). Assortative mixing in protein contact networks and protein folding kinetics, *Bioinformatics* **23**, 1760–7.

Bahar, I., Atilgan, A. R., Erman, B. (1997). Direct evaluation of thermal fluctuations in proteins using a single-parameter harmonic potential. *Fold. Des.* **2**, 173–81.

Baierlein, R. (1999). *Thermal Physics*. Cambridge University Press, Cambridge, 442 pp.

Baird, D., and Ulanowicz, R. E. (1989). The seasonal dynamics of the Chesapeake Bay ecosystem. *Ecol. Mon.* **59**, 329–64.

Balaban, A. T. (1982). Highly discriminating distance-based topological index. *Chem. Phys. Lett.* **89**, 399–404.

Balaban, A. T. (1994). Reaction graphs, In *Graph Theoretical Approaches to Chemical Reactivity*, Bonchev, D. and Mekenyan, O. (ed.). Kluwer Academic Publishers, Dordrecht, Netherlands, p. 137.

Balaban, A. T. (ed.) (1976). *Chemical Applications of Graph Theory*. Academic Press, London.

Balcan, D., Colizza, V., Goncalves, B., Hu, H., Ramasco, J. J., and Vespignani, A. (2009). Mutiscale mobility networks and the spatial spreading of infectious diseases. *Proc. Natl. Acad. Sci. USA* **106**, 21484–9.

Ball, P. (2009). *Branches. Nature's Patterns: A tapestry in Three Parts*. Oxford University Press, Oxford, p. 44.

Ballantine, C. M., Ramanujan Type graphs and Bigraphs. *Proc. 4th Int. Conf. Dynam. Syst. Diff. Eq.* May 24–27, 2002, Wilmington, NC. pp. 78–82.

Bamdad, H., Ashraf, F., and Gutman, I. (2010). Lower bounds for Estrada index and Laplacian Estrada index. *Appl. Math. Lett.* **23**, 739–42.

Bapat, R. B., Gutman, I., and Xiao, W. (2003). A simple method for computing resistance distance. *Z. Naturforsch.* **58a**, 494–8.

Barabási, A.-L. and Albert, R. (1999). Emergence of scaling in random networks. *Science* **286**, 509–12.

Barabási, A-L., and Oltvai, Z. N. (2003). Network biology: Understanding the cell's functional organization. *Nature Rev. Genet.* **5**, 101–13.

Barber, D., (2008). Clique matrices for statistical graph decomposition and parameterising restricted positive definite matrices, in D. A. Mc Allester and P. Mylgmöki (eds.) *Proc. 24th Ann. Conf. Uncert. Artif. Intel.* UAI PRESS, pp. 26–33.

Barrat, A., and Weigt, M. (2000). On the properties of small-world network models. *Eur. Phys. J. B*. **13**, 547–60.

Barrat, A., Barthélemy, M., and Vespignani, A. (2008). *Dynamical Processes on Complex Networks*. Cambridge University Press, Cambridge, 347 pp.

Barrat, A., Barthélemy, M., Pastor-Satorras, R., and Vespignani, A., (2004). The architecture of complex weighted networks. *Proc. Natl. Acad. Sci.* **101**, 3747–52.

Barrow, G. M. (1962). *Introduction to Molecular Spectroscopy*. McGraw-Hill, New York, 318 pp.

Barthélemy, M., (2004). Betweenness centrality in large complex networks. *Eur. Phys. J. B,* **38**, 163–8.

Barthélemy, M., and Flammini, (2008). Modeling urban street patterns, *Phys. Rev. Lett.* **100**, 138702.

Bartoli, L., Fariselli, P., and Casadio, R. (2007). The effect of backbone on the small-world properties of protein contact maps. *Phys. Biol.* **4**, L1–L5.

Bascompte, J., and Solé, R. V. (1996). Habitat fragmentation and extinction in spatially explicit metapopulation models. *J. Anim. Ecol.* **65**, 465–73.

Bassett, D. S., Bullmore, E., Verchinski, B. A., Mattay, V. A., Mattay, V. S., Weinberger, D. R., and Meyer-Lindenberg, A. (2008). Hierarchical organization of human cortical networks in health and schizophrenia. *J. Neurosci.* **28**, 9239–48.

Bassett, H. and Bullmore, E. T. (2006). Small-World brain networks. *Neoroscientist* **12**, 512–23.

Bassett, H. and Bullmore, E. T. (2009). Human brain networks in health and disease. *Curr. Opin. Neurol.* **22**, 340–7.

Batagelj, V., and Mrvar, A. (2001). *Graph Drawing Contest 2001*. http://vlado.fmf.uni-lj.si/pub/GD/GD01.htm.

Batagelj, V., and Mrvar, A. (2006). *Pajek datasets*. http://vlado.fmf.uni-lj.si/pub/networks/data/.

Bavelas, A. (1948). A mathematical model for group structure. *Human Org.* **7**, 16–30.

Bavelas, A. (1950). Communication patterns in task orientated groups. *J. Acoust. Soc. Am.* **22**, 271–88.

Bell, F. K. (1992). A note on the irregularity of graphs. *Lin. Alg. Appl.* **161**, 45–64.

Benzi, M., and Boito, P. (2010). Quadrature rule-based bounds for functions of adjacency matrices. *Lin. Alg. Appl.* **433**, 637–52.

Bernstein, D. S. (2009). *Matrix Mathematics. Theory, Facts and Formulas*. Princeton Academic Press, Princeton.

Biggs, N. L. (1993). *Algebraic Graph Theory*, 2nd Edition, Cambridge University Press, Cambridge.

Biggs, N. L., Lloyd, E. K., and Wilson, L. (1976). *Graph Theory 1736–1936*. Clarendon Press, Oxford, 239 pp.

Bigras, G., Marcelpoil, R., Brambilla, E., and Brugal, G. (1996). Cellular sociology applied to neuroendocrine tumors of the lung: Quantitative model of neoplastic architecture. *Cytometry* **24**, 74–82.

Bilgin, C. C., Bullough, P., Plopper, G. E., and Yener, B. (2009). ECM-aware cell-graph mining for bone tissue modeling and classification. *Data Min. Knowl Discov.* **20**, 416–38.

Bilgin, C., Demir, C., Nagi, C., and Yener, B. (2007). Cell-graph mining for breast tissue modeling and classification. *29th Annual Int. Conf. IEEE Eng. Med. Bio. Soc.*, Lyon, France.

Bishop, K. J. M., Klajn, R., and Grzybowski, B. A. (2006). The core and most useful molecules in organic chemistry. *Angew. Chem. Int. Ed.* **45**, 5348–54.

Biyikoglu, T., Leydold, J., and Stadler, P. F., (2007). *Laplacian Eigenvectors of Graphs, Perron–Frobenius and Faber–Krahn type Theorems*. Springer, Berlin, 120 pp.

Blanchard, Ph., and Volchenkov, D., (2009). *Mathematical Analysis of Urban Spatial Networks. Understanding Complex Systems*. Springer-Verlag, Berlin, pp. 101–36.

Bodin, Ö., Tengö, M., Norman, A., Lundberg, J., and Elmqvist, T. (2006). The value of small size: loss of forest patches and ecological thresholds in southern Madagascar. *Ecol. Appl.* **16**, 440–51.

Boettcher, S., and Percus, A. G. (2001). Optimization with extremal dynamics. *Phys. Rev. Lett.* **86**, 5211–4.

Boguñá, M. A., and Pastor-Satorras, R. (2003). Class of correlated random networks with hidden variables. *Phys. Rev. E* **68**, 036112.

Boguñá, M., Papadopoulos, F., and Krioukov, D. (2010). Sustaining the Internet with hyperbolic mapping. *Nature Comm.* **1**, 62.

Boguñá, M., Pastor-Satorras, R., and Vespignani, A. (2003). Absence of epidemic threshold in scale-free networks with degree correlations. *Phys. Rev. Lett.* **90**, 028701.

Bolland, J. M. (1988). Sorting out centrality: An analysis of the performance of four centrality models in real and simulated networks. *Social Networks* **10**, 233–53.

Bollobás, B. (2001). *Random Graphs* (2nd edn.). Cambridge University Press, Cambridge.

Bollobás, B., (2003). Mathematical results on scale-free random graphs, In Bornholdt, S., Schuster, H. G. (eds.), *Handbook of Graph and Networks: From the genome to the internet*, Wiley-VCH, Weinheim, pp. 1–32.

Bollobás, B., and Erdös, P. (1998). Graphs of extremal weights. *Ars Combin.* **50**, 225–33.

Bollobás, B., and Riordan, O. (2004). The diameter of a scale-free random graph. *Combinatorica* **24**, 5–34.

Bolshakova, N., and Azuaje, F. (2003). Cluster validation techniques for genome expression data. *Signal Proc.* **83**, 825–33.

Bonacich, P. (1972). Factoring and weighting approaches to status scores and clique identification. *J. Math. Sociol.* **2**, 113–20.

Bonacich, P., (1987). Power and centrality: A family of measures. *Am. J. Soc.* **92**, 1170–82.

Bonanno, G., Caldarelli, G., Lillo, F., and Mantegna, R. N. (2003). Topology of correlation-based minimal spanning trees in real and model markets. *Phys. Rev. E*, **68**, 046130.

Boone, C., Bussey, H., and Andrews, B. J. (2007). Exploring genetic interactions and networks with yeast. *Nature Rev. Gen.* **8**, 437–49.

Borenstein, E., and Feldman, M. W. (2009). Topological signatures of species interactions in metabolic networks. *J. Comput. Biol.* **16**, 191–200.

Borenstein, E., Kupiec, M., Feldman, M. W., and Ruppin, E. (2008). Large-scale reconstruction and phylogenetic analysis of metabolic environments. *Proc. Natl. Acad. Sci. USA* **105**, 14482–7.

Borgatti, S. P. (2005). Centrality and network flow. *Social Networks* **27**, 55–71.

Borgatti, S. P., and Everett, M. G. (2006). A graph-theoretic perspective on centrality. *Social Networks* **28**, 466–84.

Borgatti, S. P., Carley, K. M., Krackhardt, D. (2006). On the robustness of centrality measures under conditions of imperfect data. *Social Networks* **28**, 124–36.

Borgatti, S. P., Mehra, A., Brass, D. J, and Labianca, G. (2009). Network analysis in the social sciences. *Science* **323**, 892–5.

Borowski, E. J., and Borwein, J. M. (1999). *Dictionary of Mathematics*. Collins, Glasgow, 672 pp.

Bortoluzzi, S., Romualdi, C., Bisognin, A., and Danieli, G. A. (2003). Disease genes and intracellular protein networks. *Physiol. Genom.* **15**, 223–7.

Bourguignon, P.-Y., van Helden, J., Ouzounis, C., and Schächter, V. (2008). Computational analysis of metabolic networks. In Frishman, D. and Valencia, A. (Eds.) *Modern Genome Annotation: The BioSapiens Network*. Springer, Berlin, pp. 329–51.

Boutin, F., and Hascoët, M. (2004). Cluster validity indices for graph partitioning. *Proc. Eighth Int. Conf. Inf. Vis.*, 14–16 July, pp. 376–81.

Boyer, J. M., Cortese, P. F., Patrignani, M., and Di Battista, G. (2004). Stop minding your P's and Q's: implementing a fast simple DFS-based planarity testing and embedding algorithm. *Lect. Notes. Comput. Sci.* **2912**, 25–36.

Brand, M., and Huang, K. (2003). A unifying theorem for spectral embedding and clustering, *Proc. 9th Int. Conf. Art. Intell. Stat.*, Keywest, FL.

Brandes, U., Gaertler, M., and Wagner, D. (2003). Experiments on graphs clustering algorithms. *Lect. Not. Comput. Sci.* **2832**, 568–79.

Brandes, U., Gaertler, M., and Wagner, D. (2008). Engineering graph clustering: models and experimental evaluation. *ACM J. Exp. Algor.* **12**, 1.

Braun, J., Kerber, A., Meringer, M., and Rücker, C. (2005). Similarity of molecular descriptiors the equivalence of Zagreb indices and walk counts. *Match Comm. Math. Comput. Chem*, **54**, 74–80.

Brazhnik, P., De la Fuente, A., and Mendes, P. (2002). Gene networks: how to put the function in genomics. *Trends Biotech.* **20**, 467–72

Breiger R., and Pattison P. (1986). Cumulated social roles: The duality of persons and their algebras. *Social Networks* **8**, 215–56.

Brilli, M., Calistri, E., and Lió, P. (2010). Transcription factors and gene regulatory networks. In Buchanan, M., Caldarelli, G., de los Rios, P., Rao, F, and Vendruscolo, M. (eds.). *Networks in Cell Biology*. Cambridge Universisity Press, Cambridge, pp. 36–52.

Brin, S., and Page, L. (1998). The anatomy of a large-scale hypertextual Web search engine. *Comput. Net. ISDN Syst.* **33**, 107–17.

Brinda, K. V., and Vishveshwara, S. (2005). A network representation of protein structures: implications for protein stability. *Biophys J.* **89**, 4159–70.

Britta, R., (2000). Eigenvector-centrality—a node-centrality? *Social Networks* **22**, 357–65.

Bron, C., and Kerbosch, J. (1973). Finding all cliques of an undirected graph. *Comm. ACM*, **16**, 575–7.

Bryan, K., and Leise, T. (2006). The $25,000,000,000 eigenvector: The linear algebra behind Google. *SIAM Rev.* **48**, 569–81.

Bu, D., Zhao, Y., Cai, L., Xue, H., Zhu, X., Lu, H., Zhang, J., Sun, S., Ling, L., Zhang, N., Li, G., and Chen, R. (2003). Topological structure analysis of the protein–protein interaction network in budding yeast. *Nucleic Acids Res.* **31**, 2443–50.

Buchanan, M. (2003). Nexus: Small Worlds and the Groundbreaking Theory of Networks. W. W. Norton & Company, New York.

Buckley, F., and Harary, F. (1990), *Distance in Graphs*. Perseus Books, New York.

Buhl, J., Gautrais, J., Reeves, N., Solé, R. V., Valverde, S. Kuntz, P., and Theraulaz, G. (2006).Topological patterns in street networks of self-organized urban settlements. *Eur. Phys. J. B* **49**, 513–22.

Buhl, J., Gautrais, J., Solé, R. V., Kuntz, P., Valverde, S., Deneubourg, J. L., and Theraulaz, G. (2004). Efficiency and robustness in ant networks of galleries. *Eur. Phys. J.* **B42**, 123–9.

Bullmore, E. and Sporn, O. (2009). Complex brain networks: graph theoretical analysis of structural and functional system. *Nature Rev. Neurosc.* **10**, 186–98.

Bultland, G., Peregrín-Alvarez, J. M., Li, J., Yang, W., Yang, X., Canadien, V., Starostine, A., Richards, D., Beattie, B., Krogan, N., Davey, M., Parkinson, J., Greenblatt, J., and Emili, A. (2005). Interaction network containing conserved and essential protein complexes in Escherichia coli. *Nature* **433**, 531–7.

Bunn, A. G., Urban, D. L., and Keitt, T. H. (2000). Landscape connectivity: A conservation application of graph theory. *J. Environ. Manag.* **59**, 265–78.

Caldarelli, G. (2007). *Scale-Free Networks. Complex Webs in Nature and Technology*. Oxford University Press, Oxford, 309 pp.

Caldarelli, G., Pastor-Satorras, R., and Vespignani, A. (2004). Structure of cycles and local ordering in complex networks. *Eur. Phys. J. B* **38**, 183–6.

Camacho, J., Guimerá, R., and Amaral, L. A. N., (2002). Robust patterns in food web structure, *Phys. Rev. Lett.* **88**, 228102.

Capocci, A., Servedio, V. D. P., Caldarelli, G., and Colaiori, F. (2005). Detecting communities in large networks. *Physica A* **352**, 669–76.

Cardillo, A., Scellato, S., Latora, V., and Porta, S. (2006). Structural properties of planar graphs of urban street patterns. *Phys. Rev. E* **73**, 066107.

Carlson, M. RJ., Zhang, B., Fang, Z., Mischel, P. S., Horvath, S., and Nelson, S, F. (2006). Gene connectivity, functions, and sequence conservation: predictions from modular yeast co-expression networks. *BMC Genom.* **7**, 40.

Carlson, R., (1998). Adjoint and self-adjoint differential operators on graphs. *Elect. J. Diff. Eq.* **1998**, 1–10.

Carvalho, R., and Iori, G. (2008). Socioeconomic networks with long-range interactions, *Phys. Rev. E* **78**, 016110.

Catanzaro, M., Caldarelli, G., and Pietronero, L. (2004). Assortative model for social networks. *Phys. Rev. E* **70**, 037101.

Causier, B. (2004). Studying the interactome with the yeast two-hybrid system and mass spectrometry. *Mass Spectr. Rev.* **23**, 350–67.

Chandebois, R. (1976). Cell Sociology: A way of reconsidering the current concepts of morphogenesis. *Acta Bioth.* **25**, 71–102.

Chartrand, G., and Zhang, P. (2003). Distance in graphs, In Gross, J. L. and Yellen, J, *Handbook of Graph Theory*, CRC Press, Boca Raton, pp. 573–888.

Chartrand, G., Erdös, P., and Oellermann, O. R. (1988). How to define an irregular graph. *College Math. J.* **19**, 36–42.

Chavez, M., Hwang, H.-U., Martinerie, J., Boccaletti, S. (2006). Degree mixing and the enhancement of synchronization in complex weighted networks. *Phys. Rev. E* **74**, 066107.

Chebotarev, P. (2011). A family of graph distances generalizing both the shortest-path and the resistance distances. *Discr. Appl. Math.* **159**, 295–302.

Chebotarev, P., and Shamis, E. (2002). The forest metric for graph vertices. *Electr. Notes Discr. Math.* **11**, 98–107.

Christian, R. R., and Luczkovich, J. J. (1999). Organizing and understanding a winter's seagrass foodweb network through effective trophic levels. *Ecol. Model.* **117**, 99–124.

Christianou, M., and Ebenman, B., (2005). Keystone species and vulnerable species in ecological communities: strong or weak interactors? *J. Theor. Biol.* **235**, 95–103.

Chung, F. R. K. (1996). *Spectral Graph Theory*. American Mathematical Society, Providence, RI, 207 pp.

Chung, F., Lu, L., and Vu, V. (2003). The spectra of random graphs with given expected degrees. *Internet Math.* **1**, 257–75.

Clauset, A., Rohilla Shalizi, C., and Newman, M. E. J. (2010). Power-law distribution in empirical data. *SIAM Rev.* **51**, 661–703.

Cohen, J. E. (1977). Ratio of prey to predators in community food webs. *Nature* **270**, 167.

Cohen, J. E., and Briand, F. (1984). Trophic links of community food webs. *Proc. Natl. Acad. Sci. USA* **81**, 4105–9.

Cohen, J. E., and Newman, C. M. (1985). A stochastic theory of community food webs: I. Models and aggregated data. *Proc. R. Soc. Lond. B* **224**, 421–48.

Cohen, N., Dimitrov, D., Krakovski, R., Skrekovski, R., Vukasinovic, V. (2010). On Wiener index of graphs and their line graphs. *MATCH: Comm. Math. Comput. Chem.* **64**, 683–98.

Coleman, J. S., Katz, E., Menzel, H. (1966). *Medical Innovation. A Diffusion Study*. Bob-Merris, Indianapolis.

Colizza, V., Flammini, A., Serrano, M. A., and Vespignani, A. (2006). Detecting rich-club ordering in complex networks. *Nature Physics* **2**, 110–15.

Collas, P. (2010). The current state of chromatin immunoprecipitation. *Mol. Biotech.* **45**, 87–100.

Collatz, L. and Sinogowitz, U. (1957). Spektren endlicher Grafen. *Abh. Math. Sem. Univ. Hamburg* **21**, 63–77.

Costenbader, E., and Valente, T. W. (2003). The stability of centrality measures when networks are sampled. *Social Networks* **25**, 283–307.

Cottrell, A. H., and Pettifor, D. G. (2000). Models of structure. In Pullman, W., and Bhadeshia, H. *Structure in Science and Art*. Cambridge University Press, Cambridge, pp. 33–47.

Crick, F. (1958). On protein synthesis. *Symp. Soc. Exp. Biol.* **12**, 138–63.

Crick, F. (1970). Central dogma of molecular biology. *Nature* **227**, 561–3.

Croes, D., Couche, F., Wodak, S. J., and van Helden, J. (2006). Inferring meaningful pathways in weighted metabolic networks. *J. Mol. Biol.* **356**, 222–36.

Croft, D. P., James, R., and Krause, J. (2008). Exploring Animal Social Networks. Pricenton University Press, Pricenton, 192 pp.

Crofts, J. J., and Higham, D. J. (2009). A weighted communicability measure applied to complex brain networks. *J. R. Soc. Interface* **6**, 411–14.

Crofts, J. J., Higham, D. J., Bosnell, R., Jbabdi, S., Matthews, Behrens T. E. J., and Johansen-Berg, H. (2011). Network analysis detects changes in the contralesional hemisphere following stroke. *NeuroImage* **54**, 161–9.

Crofts, J.J., Estrada, E., Higham, D. H., and Taylor, A. (2010). Mapping directed networks. *Elec. Trans. Num. Anal.* **37**, 337–50.

Csermely, P., Ágoston, V., and Pongor, S. (2005). The efficiency of multi-target drugs: The network approach might help drug design. *Trends Pharm. Sci.* **26**, 178–82.

Cvetković, D. (2009). Applications of Graph Spectra: An Introduction to the Literature. In, Cvetković, D. and Gutman, I. (eds.) *Applications of Graph Spectra.* Matematiki Institute SANU, Beograd, pp. 7–31.

Cvetković, D., and Rowlinson, P. (1990). The largest eigenvalue of a graph: A survey. *Lin. Multilin. Alg*, **28**, 3–33.

Cvetković, D., and Rowlinson, P., (1988). On connected graphs with maximal index, *Pub. Inst. Math. Beograd* **44**, 29–34.

Cvetković, D., Doob, M., and Sach, H. (1995). *Spectra of Graphs*, 3rd Edition, Joham Ambrosius Barth Verlag, Heidelberg.

Cvetković, D., Rowlinson, P., and Simić, S. (1997). *Eigenspaces of Graphs.* Cambridge University Press, Cambridge.

Cvetković, D., Rowlinson, P., and Simić, S. (2004). *Spectral Generalizations of Line Graphs.* Cambridge University Press, Cambridge.

Cvetković, D., Rowlinson, P., and Simić, S. (2010). *An Introduction to the Theory of Graph Spectra.* Cambridge University Press, Cambridge, 364 pp.

D'haeseleer, P., Liang, S., and Somogy, R. (2000). Genetic network inference: from co-expression clustering to reverse engineering. *Bioinformatics* **16**, 707–26.

da Fontoura, L., and Viana, M. P. (2006). Complex channel networks of bone structure. *Appl. Phys. Lett.* **88**, 033903.

da Silveira, C. H., Pires, D. E. V., Minardi, R. C., Ribeiro, C., Veloso, C. J. M., Lopes, J. C. D., Meira, W. Jr., Neshich, G., Ramos, C. H. I., Habesch, R., and Santoro, M. M. (2009). Protein cutoff scanning: A comparative analysis of cutoff dependent and cutoff free methods for prospecting contacts in proteins. *Proteins* **74**, 727–43.

Daune, M. (1999). *Molecular Biophysics. Structures in Motion.* Oxford University Press, Oxford, 499 pp.

Davidoff, G., Sarnak, P., and Valette, A. (2003). *Elementary Number Theory, Group Theory and Ramanjuan Graphs.* Cambridge University Press, Cambridge.

Davidson, E. H., Rast, J. P. Oliveri, P. Ransick, A., Calestani, C., Yuh, C. H., Minokawa, T., Amore, G., Hinman, V., Arenas-Mena, C., Otim, O., Brown, C. T., Livi, C. B., Lee, P. Y., Revilla, R., Rust, A. G., Pan, Z., Schilstra, M. J., Clarke, P. J., Arnone, M. I., Rowen, L., Cameron, R. A., McClay, D. R., Hood, L., and Bolouri, H. (2002). A genomic regulatory network for development. *Science* **295**, 1669–78.

Davies, D. L., and Bouldin, D. W. (1979). A cluster separation measure. *IEEE Trans. Patt. Recog. Mach. Intel.* **1**, 224–7.

Davis, G. F., Yoo, M., and Baker, W. E. (2003). The Small World of the American Corporate Elite, 1982–2001. *Strategic Organization* **1**, 301–26.

Davy, A., Bello, P., Thierry-Mieg, N., Vaglio, P., Hitti, J, Doucette-Stamm, L., Thierry-Mieg, D., Reboul, J., Boulton, S., Walhout, A. J. M., Coux, O., and Vidal, M. (2001). A protein–protein interaction map of the *Caenorhabditis elegans 26S proteasome.* EMBO Rep. 2, 821–8.

de Aguiar, M. A. M., and Bar-Yam, Y. (2005). Spectral analysis and the dynamic response of complex networks. *Phys. Rev. E* **71**, 016106.

de Jong, H. (2002). Modeling and simulation of genetic regulatory systems: a literature review. *J. Comput. Biol.* **9**, 67–103.

de Koning, D-J., and Haley, C. S. (2005). Genetical genomics and model organisms. *Trends Genet.* **21**, 377–81.

de la Peña, J. A., Gutman, I., and Rada, J. (2007). Estimating the Estrada index. *Lin. Alg. Appl.* **427**, 70–6.

de los Rios, P., and Vendruscolo, M. (2010). Network view of the cell. In Buchanan, M., Caldarelli, G., de los Rios, P., Rao, F, and Vendruscolo, M. (eds.). *Networks in Cell Biology*. Cambridge University Press, Cambridge, pp. 4–13.

de Nooy, W., Mrvar, A., and Batagelj, V. (2005). *Exploratory Social Network Analysis with Pajek*. Cambridge Univ. Press, Cambridge, 362 pp.

Debnath, L. and Mikusiski, P. (1990). *Introduction to Hilbert Spaces with Applications*. Academic Press, Boston.

del Sol, A., Fujihashi, H., Amoros, D., and Nussinov, R. (2006). Residue centrality, functionally important residues, and active site shape: Analysis of enzyme and non-enzyme families. *Protein Sci.* **15**, 2120–8.

Delaunay, B. (1934). Sur la sphère vide. *Iz. Akad. Nauk SSSR, Otdelenie Matemat. Estest. Nauk.* **7**, 793–800.

Demir, C., Gultekin, S. H., and Yener, B. (2005a), Augmented cell-graphs for automated cancer diagnosis. *Bioinformatics* **21** (Suppl. 2), ii7–ii12.

Demir, C., Gultekin, S. H., and Yener, B. (2005b). Learning the topological properties of brain tumors. IEEE/ACM Trans. Comput. Biol. Bioinform. **2**, 1–9.

Demir, C., Gultekin, S. H., and Yener, B. (2005c). Spectral analysis of cell-graphs for automated cancer diagnosis. *4th Conf. Model. Simul. Biol. Med. Biomed. Eng.* Linkoping, Sweden.

Deng, H., Radenković, S., and Gutman, I. (2009). The Estrada index. In Cvetković, D., and Gutman, I. (eds.). *Applications of Graph Spectra*. Matematiki Institute SANU, Beograd, pp. 123–40.

Dixon, S. J., Heinrich, N., Holmboe, M., Schaefer, L., Reed, R. R., Trevejo, J., and Brereton, R. G. (2009). Use of cluster separation indices and the influence of outliers: application of two new separation indices, the modified silhouette index and the overlap coefficient to simulated data and mouse urine metabolomics profiles. *J. Chemom.* **23**, 19–31.

Dobrynin, A. A., and Gutman, I., (1999). The average Wiener index of trees and chemical trees, *J. Chem. Inf. Comput. Sci.* **39**, 679–83.

Dodziuk, J. (1984). Difference equations, isoperimetric inequality and transience of certain random walks. *Trans. Amer. Math. Soc.* **284**, 787–94.

Donetti, L., and Muñoz, M. A. (2005). Improved spectral algorithms for the detection of networks communities. *AIP Conf. Proc.* **779**, 104–107.

Donetti, L., Hurtado, P. I., and Muñoz, M. A. (2005). Entangled networks, super-homogeneity and optimal network topology. *Phys. Rev. Lett.* **95**, 188701.

Donetti, L., Neri, F., and Muñoz, M. A. (2006). Optimal network topologies: expanders, cages, Ramanujan graphs, entangled networks and all that. *J. Stat. Mech.* **P08007**.

Dorogovtsev, S. N., and Mendes, J. F. F. (2003). *Evolution of Networks: From Biological Nets to the Internet and WWW*. Oxford University Press, Oxford, 280 pp.

Doyle, P. G., and Snell, J. L. (1984). *Random Walks and Electric Networks*. Carus Math. Monogr. 22, Mathematical Association of America, Washington DC. http://arxiv.org/abs/math/0001057.

Dunne, J. A. (2005). The network structure of food webs. In Pascual, M., Dunner, J. A. (eds.) *Ecological Networks: Linking Structure to Dynamics in Food Webs*. Oxford University Press, Oxford, pp. 27–86.

Dunne, J. A. (2009). Food webs. In Meyers, R. A. (ed.), *Encyclopedia of Complexity and Systems Science*. Springer, Berlin, pp. 3661–82.

Dunne, J. A., and Martinez, N. D. (2009). Cascading extinctitions and community collapse in model food webs. *Phil. Trans. R. Soc. B* **364**, 1711–23.

Dunne, J. A., Williams, R. J., and Martinez, N. D. (2002a). Food-web structure and network theory: The role of connectance and size. *Proc. Natl. Acad. Sci. USA* **99**, 12917–22.

Dunne, J. A., Williams, R. J., and Martinez, N. D. (2002b). Network structure and biodiversity loss in food webs: Robustness increases with connectance. *Ecol. Lett.* **5**, 558–67.

Dunne, J. A., Williams, R. J., and Martinez, N. D. (2004). Network structure and robustness of marine food webs. *Mar. Ecol. Prog. Ser.* **273**, 291–302.

Durrett, R. (2007). *Random Graph Dynamics*. Cambridge University Press, Cambridge, 212 pp.

Dutt, S. (1993). New faster Kernighan-Lin-type graph-partitioning algorithms. *Proc. 1993 IEEE/ACM Int. Conf. Comp.-Aided Des*. Los Alamitos, CA, pp. 370–7.

Eagle, N., Pentland, A., and Lazer, D. (2009). Inferring social network structure using mobile phone data. *Proc. Natl. Acad. Sci. USA* **106**, 15274–8.

Eguiluz, V. M., Chivaldo, D. R., Cecchi, G. A., Baliki, M., Apkarian, A. V. (2005). Scale-free brain functional networks. *Phys. Rev. Lett.* **94**, 018102.

Ejov, V., Filar, J. A., Lucas, S. K., and Zograf, P. (2007). Clustering of spectra and fractals of regular graphs. *J. Math. Anal. Appl.* **333**, 236–46.

Ejov, V., Friedlan, S. and Nguyen, G. T. (2009). A note on the graph's resolvent and the multifilar structure. *Lin. Alg. Appl.* **431**, 1367–79.

Elsner, U. (1997). Graph partitioning. Tech. Rep. 393, Technische Universität Chemnitz.

Emsley, J. (1999). Molecules at an Exhibition: Portraits of Intriguing Materials in Everyday Life. Oxford Paperbacks, Oxford, 272 pp.

Erdös, P., and Rényi, A. (1959). On random graphs I. *Publ.Math. Debrecen* **5**, 290–7.

Erdös, P., and Rényi, A. (1960). On the evolution of random graphs. *Magyar Tud. Akad. Mat. Kutató Int. Közl.* **5**, 17–61.

Espinoza-Valdez, A., Femat, R., and Ordaz-Salazar, F. C. (2010). A model for renal arterial branching based on graph theory. *Math. Biosci.* **225**, 36–43.

Essam, J. W., and Fisher, M. E. (1970). Some basic definitions in graph theory. *Rev. Mod. Phys.* **42**, 272–88.

Estes, J. A., and Palmisano, J. F. (1974). Sea otters: their role in structuring nearshore communities. *Science* **185**, 1058–60.

Estrada, E. (1995). Edge adjacency relationships and a novel topological index related to molecular volume. *J. Chem. Inf. Comput. Sci.* **35**, 31–3.

Estrada, E. (2000). Characterization of 3D molecular structure. *Chem.Phys. Lett.* **319**, 713–8.

Estrada, E. (2001). Generalization of topological indices. *Chem. Phys. Lett.* **336**, 248–52.

Estrada, E. (2006a). Protein bipartivity and essentiality in the yeast protein–protein interaction network. *J. Proteome Res.* **5**, 2177–84.

Estrada, E. (2006b). Virtual identification of essential proteins within the protein inter-action network of yeast. *Proteomics* **6**, 35–40.

Estrada, E. (2006c). Spectral scaling and good expansion properties in complex net-works. *Europhys. Lett.* **73**, 649–55.

Estrada, E. (2006d). Network robustness. The interplay of expansibility and degree distribution. *Eur. Phys. J.B*, **52**, 563–74.

Estrada, E. (2007a). Characterization of topological keystone species. Local, global and 'meso-scale' centralities in food webs. *Ecol. Complex.* **4**, 48–57.

Estrada, E. (2007b). Food web robustness to biodiversity loss: The roles of connectance, expansibility and degree distribution. *J. Theor. Biol.* **244**, 296–307.

Estrada, E. (2007c). Topological structural classes of complex networks. *Phys. Rev. E*, **75**, 016103.

Estrada, E. (2010a). Quantifying network heterogeneity. *Phys. Rev. E* **82**, 066102.

Estrada, E. (2010b). Universality in protein residue networks. *Biophys. J.* **98**, 890–900.

Estrada, E. (2010c). Generalized walks-based centrality measures for complex biological networks, *J. Theor. Biol.* **263**, 556–65.

Estrada, E. (2011). Community detection based on network communicability. *CHAOS* (in press).

Estrada, E. and Hatano, N. (2010a). A vibrational approach to node centrality and vulnerability in complex networks. *Physica A* **389**, 3648–60.

Estrada, E. and Hatano, N. (2010b). Topological atomic displacements, Kirchhoff and Wiener indices of molecules. *Chem. Phys. Lett.* **486**, 166–70.

Estrada, E., and Bodin, Ö. (2008). Using network centrality measures to manage landscape connectivity. *Ecol. Appl.* **18**, 1810–25.

Estrada, E., and Hatano, N. (2007). Statistical-mechanical approach to subgraph centrality in complex networks. *Chem. Phys. Let.* **439**, 247–51.

Estrada, E., and Hatano, N. (2008). Communicability in complex networks. *Phys. Rev. E* **77**, 036111.

Estrada, E., and Hatano, N. (2009a). Communicability graph and community structures in complex networks. *Appl. Math. Comput.* **214**, 500–11.

Estrada, E., and Hatano, N. (2009b). Returnability in complex directed networks (digraphs). *Lin. Alg. Appl.* **430**, 1886–96.

Estrada, E., and Higham, D. J. (2010). Network properties revealed through matrix functions. *SIAM Rev.* **52**, 696–714.

Estrada, E., and Rodríguez-Velázquez, J. A. (2005a). Subgraph centrality in complex networks. *Phys. Rev. E* **71**, 056103.

Estrada, E., and Rodríguez-Velázquez, J. A. (2005b). Spectral measures of bipartivity in complex networks. *Phys. Rev. E* **72**, 046105.

Estrada, E., Hatano, N., and Gutierrez, A. (2008a). 'Clumpiness' mixing in complex networks. *J. Stat. Mech.* **P03008**.

Estrada, E., Higham, D. J., and Hatano, N. (2008b). Communicability and multipartite structure in complex networks at negative absolute temperatures. *Phys. Rev. E,* **78**, 026102.

Estrada, E., Higham, D. J., Hatano, N. (2009). Communicability betweenness in complex networks. *Physica,* A, **388**, 764–74.

Euler L. (1736). Solutio Problematis ad Geometriam Situs Pertinentis. *Commentarii Academiae Scientiarum Imperialis Petropolitanae,* **8**, 128–40.

Fahrig, L. (2003). Effects of habitat fragmentation on biodiversity. *Ann. Rev. Ecol. Evol. Syst.* **34**, 487–515.

Faloutsos, M., Faloutsos, P., and Faloutsos, C. (1999). On power-law relationships of the internet topology. *Comp. Comm. Rev.* **29**, 251–62.

Fariselli, P., and Casadio, R. (1999). A neural network based predictor of residue contacts in proteins. *Protein Eng.* **12**, 15–21.

Farkas, I. J., Derényi, I., Barabási, A.-L., and Vicsek, T. (2001). Spectra of 'real-world' graphs: Beyond the semicircle law. *Phys. Rev. E,* **64**, 026704.

Farkas, I., Derenyi, I., Jeong, H., Néda, Z., Oltavai, Z. N., Ravasz, E., Schubert, A., Barabási, A.-L., and Vicsek, T. (2002). Networks in life: scaling properties and eigenvalue spectra. *Physica A* **314**, 25–34.

Felleman, D. J., and van Essen, D. C. (1991). Distributed hierarchical processing in the primate cerebral cortex. *Cereb. Cortex* **1**, 1–47.

Fialkowski, M., Bishop, K. J. M., Chubukov, V. A., Campbell, C. J., and Grzybowski, B. A. (2005). Architecture and evolution of organic chemistry. *Angew. Chem. Int. Ed.* **44**, 7263–9.

Fiedler, M. (1973). Algebraic Connectivity of Graphs. *Czech. Math. J.* **23**, 298–305.

Filippone, M., Camastra, F., Masulli, F., and Rovetta, S. (2008). A survey of kernel and spectral methods for clustering. *Patt. Recog.* **41**, 176–90.

Fjällström, P.-O. (1998). Algorithms for graph partitioning: a survey, *Linköping Elect. Art. Comput. Inf. Sci.* **3**, nr 10.

Fortunato, S. (2007). Quality functions in community detection. *SPIE Conf. Ser.* **6601**, 660108.

Fortunato, S. (2010). Community detection in graphs. *Phys. Rep.* **486**, 75–174.

Fortunato, S., and Barthélemy, M., (2007). Resolution limit in community detection. *Proc. Natl. Acad. Sci. USA* **104**, 36–41.

Foss, S., Korshunov, D., and Zachary, S. (2011). *An Introduction to Heavy-Tailed and Subexponential Distributions*. Springer, Berlin, 125 pp.

Frantz, T. L., Cataldo, M., Carley, K. M. (2009). Robustness of centrality measures under uncertainty: Examining the role of network topology. *Comput. Math. Organ. Theor.*, **15**, 303–28.

Freeman, L. C. (1979). Centrality in networks: I. Conceptual clarification. *Social Networks* **1**, 215–39.

Freeman, L. C. (2004). *The Development of Social Network Analysis*. Empirical Press, Vancouver.

Freeman, L. C., Borgatti, S. P., and White, D. R. (1991), Centrality in valued graphs: A measure of betweenness based on network flow. *Social Networks* **13**, 141–54.

Freeman, L. C., Roeder, D., and Mulholland, R. R. (1980). Centrality in social networks. II: Experimental results. *Social Networks* **2**, 119–41.

Friedkin, N. E. (1991). Theoretical foundations for centrality measures. *Am. J. Social*, **96**, 1478–1504.

Friedman, J. (1991). On the Second Eigenvalue and random walks in random d-regular graphs. *Combinatorica* **11**, 331–62.

Friedman, J., and Tillich, J.-P., (2008). Calculus on Graphs, *arxiv:CS/*0408028V1, 63 pp.

Fronczak, A., Hoyst, J. A., Jedynak, M., and Sienkiewicz, J. (2002). Higher order clustering coefficients in Barabási-Albert networks. *Physica A* **316**, 688–94.

Futschik, M. E., Chaurasia, G., Tschaut, A., Russ, J., Babu, M. M., and Herzel, H. (2007). Functional and transcriptional coherency of modules in the human protein interaction network. *J. Integrat. Bioinform.* **4**, 76.

Gallos, L. K., Song, C., and Makse, H. A. (2007). A review of fractality and self-similarity in complex networks. *Physica A*, **386**, 686–91.

Gallos, L. K., Song, C., and Makse, H. A. (2008). Scaling of degree correlations and its influence on diffusion in scale-free networks. *Phys. Rev. Lett.* **100**, 248701.

Garlaschelli, D., and Loffredo, M. I. (2004a). Patterns of link reciprocity in directed networks. *Phys. Rev. Lett.* **93**, 268701.

Garlaschelli, D., and Loffredo, M. I. (2004b). Fitness-Dependent Topological Properties of World Trade Web. *Phys. Rev. Lett.* **93**, 188701.

Garlaschelli, D., Battiston, S., Castri, M., Servedio, V. D. P., Caldarelli, G. (2005). The scale-free topology of market investments. *Physica A* **350**, 491–9.

Gavin, A. C., Bosche, M., Krause, R., Grandi, P., Marzioch, M., Bauer, A., Schultz, J., Rick, J.M., Michon, A.M., Cruciat, C.M., Remor, M., Hofert, C., Schelder, M., Brajenovic, M., Ruffner, H., Merino, A., Klein, K., Hudak, M., Dickson, D., Rudi, T., Gnau, V., Bauch, A., Bastuck, S., Huhse, B., Leutwein, C., Heurtier, M. A., Copley, R. R., Edelmann, A., Querfurth, E., Rybin, V., Drewes, G., Raida, M., Bouwmeester, T., Bork, P., Seraphin, B., Kuster, B., Neubauer, G., and Superti-Furga, G. (2002). Functional organization of the yeast proteome by systematic analysis of protein complexes. *Nature* **415**, 141–7.

George, P.-L. (1999). *Delaunay Triangulations and Meshing*. John Wiley & Sons Inc., 232 pp.

Ghosh, A., Boyd, S., and Saberi, A. (2008). Minimizing effective resistance of a graph. *SIAM Rev.* **50**, 37–66.

Giot, L., Bader, J. S., Brouwer, C., Chaudhuri, A., Kuang, B., Li, Y., Hao, Y. L., Ooi, C. E., Godwin, B., Vitols, E., Vijayadamodar, G., Pochart, P., Machineni, H., Welsh, M., Kong, Y., Zerhusen, B., Malcolm, R., Varrone, Z., Collis, A., Minto, M., Burgess, S., McDaniel, L., Stimpson, E., Spriggs, F., Williams, J., Neurath, K., Ioime, N., Agee, M., Voss, E., Furtak, K., Renzulli, R., Aanensen, N., Carrolla, S., Bickelhaupt, E., Lazovatsky, Y., DaSilva, A., Zhong, J., Stanyon, C. A., Finley Jr., R. L., White, K. P., Braverman, M., Jarvie, T., Gold, S., Leach, M., Knight, J., Shimkets, R. A., McKenna, M. P., Chant, J., and Rothberg, J. M. (2003). A protein interaction map of *Drosophila melanogaster*. *Science* **302**, 1727–36.

Girvan, M., and Newman, E. J. (2002). Community structure in social and biological networks. *Proc. Natl. Acad. Sci. USA* **99**, 7821–6.

Gkantsidis, C., Mihail, M, and Saberi, A. (2003). Conductance and congestion in power law gaphs. *ACM SIGMETRICS Perf. Eval. Rev.* **31**, 148–59.

Gleiser, P., and Danon, L. (2003). Community structure in Jazz. *Adv. Complex Syst.* 6, 565.

Gleiss, P. M., Stadler, P. F., Wagner, A., Fell, D. A. (2001). Relevant cycles in chemical reaction networks. *Adv. Compl. Syst.* **1**.

Goemann, B., Wingender, E., and Potapov, A. P. (2009). An approach to evaluate the topological significance of motifs and other patterns in regulatory networks. *BMC Syst. Biol.* **3**, 53.

Goh, K. I., Kahng, B., and Kim, D., (2001a). Spectra and eigenvectors of scale-free networks, *Phys. Rev. E*, **64**, 051903.

Goh, K.-L., Kahng, B., and Kim, D. (2001b). Universal behavior of load distribution in scale-free networks. *Phys. Rev. Lett.* **87**, 278701.

Goh, K., Oh, E., Jeong, H., Kahng, B., and Kim, D. (2002). Classification of scale-free networks. *Proc. Natl. Acad. Sci.* **99**, 12583–8.

Goh, K-II., Cusick, M. E., Valle, D., Childs, B., Vidal, M., and Barabási, A-L. (2007). The Human disease network. *Proc. Natl. Acad. Sci. USA.* **104**, 8685–90.

Goldwasser, L., and Roughgarden, J. A. (1993). Construction and analysis of a large Caribbean food web. *Ecology* **74**, 1216–33.

Grabowski, A., and Kosiski, R. A. (2006). Evolution of a social network: The role of cultural diversity. *Phys. Rev. E* **73**, 016135.

Grassi, R., Stefani, S., and Torriero, A. (2007). Some new results on the eigenvector centrality. *J. Math. Soc.* **31**, 237–48.

Green, J. L., Hastings, A., Arzberger, P., Ayala, F. J., Cottingham, K. L., Cuddington, K., Davis, F., Dunne, J. A., Fortin, M.-J., Gerber, L., and Neubert, M. (2005). Complexity in ecology and conservation: Mathematical, statistical, and computational challenges. *BioSc*i. **55**, 501–10.

Greenbury, S. F., Johnston, I. G., Smith, M. A., Doye, J. P. K., and Louis, A. A. (2010). The effect of scale-free topology on the robustness and evolvability of genetic regulatory networks. *J. Theor. Biol.* **267**, 48–61.

Greene, L. H., and Higman, V. A. (2003). Uncovering network systems within protein structures. *J. Mol. Biol.* **334**, 781–91.

Gregory, R. D. (2006). *Classical Mechanics*, Cambridge Univ. Press, Cambridge.

Gribkovskaia, I., Halskau Sr., Ø., Laporte, G. (2007). The bridges of Königsberg. A historical perspective. *Networks* **49**, 199–2003.

Grindrod, P. (2002). Range-dependent random graphs and their application to modeling large small-world proteome datasets. *Phys. Rev. E* **66**, 066702.

Grönlund, A., and Holme, P. (2004). Networking the seceder model: Group formation in social and economic systems. *Phys. Rev. E* **70**, 036108.

Gross, J. L. and Tucker, T. W. (1987). *Topological Graph Theory*. Dover Pub., Inc., Mineola, N.Y., 361 pp.

Grzybowski, B. A., Bishop, K. J. M., Kowalczyk, B., and Wilmer, C. E. (2009). The 'wired' universe of organic chemistry. *Nature Chem.* **1**, 31–6.

Güngör, A. D., Çevik, A. S., Karpuz, E. G., Ate, F., and Cangül, I. N. (2010). Generalization for Estrada index. ICNAAM 2010: Int. Conf. Num. Anal. Appl. Math. *AIP Conf. Proc.* **1281**, 1106–10.

Guimerá, R., and Amaral, L. A. N. (2005). Functional cartography of complex metabolic networks. *Nature* **433**, 895–9.

Guimerá, R., Sales-Pardo, M., and Amaral, L. A. N. (2004). Modularity from fluctuations in random graphs and complex networks. *Phys. Rev. E* **70**, 025101.

Guimerá, R., Sales-Pardo, M., and Amaral, L. A. N. (2007). Classes of complex networks defined by role-to-role connectivity profiles. *Nature Phys.* **3**, 63–9.

Guimerà, R., Uzzi, B., Spiro, J., Amaral, L. A. N. (2005). Team assembly mechanisms determine collaboration network structure and team performance. *Science* **308**, 697–702.

Gulbahce, N., and Lehmann, S. (2008). The art of community detection. *Bio Essays* **30**, 934–8.

Gunduz, C., Yenner, B., and Gultekin, S. H. (2004). The cell graphs of cancer. Bioinformatics **20** (Suppl. 1), i145–51.

Gursoy, A., Keskin, O., and Nussinov, R. (2008). Topological properties of protein interaction networks from a structural perspective. *Biochem. Soc. Trans.* **36**, 1386–1403.

Gutman, I., and Graovac, A. (2007). Estrada index of cycles and paths. *Chem. Phys. Lett.* **436**, 294–6.

Gutman, I., and Polanski, O. E. (1987). *Mathematical Concepts in Organic Chemistry*. Springer-Verlag, Berlin, 211 pp.

Gutman, I., and Vidović, D. (2002). Two early branching indices and the relation between them. *Theor. Chem. Acc.* **108**, 98–102.

Gutman, I., and Xiao, O. (2004). Generalized inverse of the Laplacian matrix and some applications. *Bull. Acad. Serb. Sci. Arts.* **29**, 15–23.

Gutman, I., Hansen, P., and Mélot, H. (2005). Variable neighborhood search for extremal graphs. 10. Comparison of irregularity indices for chemical tress, *J. Chem. Inf. Model.* **45**, 222–30.

Gutman, I., Ruscić, B., Trinajstić, N., and Wilcox Jr., C. F. (1975). Graph theory and molecular orbitals. XII. Acyclic polyenes. *J. Chem. Phys.* **62**, 3399–3405.

Györi, E., Kostochka, A. V., uczak, T. (1997). Graphs without short odd cycles are nearly bipartite. *Discr. Math.* **163**, 279–84.

Hage, P., and Harary, F. (1984). *Structural Models in Anthropology*. Cambridge University Press, Cambridge, 220 pp.

Hage, P., and Harary, F. (1995). Eccentricity and centrality in networks. *Social Networks* **17**, 57–63.

Hage, P., and Harary, F. (1996). Island Networks. Communication, Kindship and Classification Structures in Oceania.Cambridge University Press, Cambridge, 318 pp.

Hagmann, P., Cammoun, L., Gigandet, X., Meuli, R., Honey, C. J., Wedeen, V. J., and Sporn, O. (2008). Mapping the structural core of human cerebral cortex. PLos Biol. **6**, e159.

Hagmann, P., Jonasson, L., Maeder, P., Thiran, J-P., Wedeen, V. J., and Meuli, R. (2006). Understanding diffusion MR imaging techniques: From scalar diffusion-weighted imaging to diffusion tensor imaging and beyond. *Radio Graphics* **26**, s205–23.

Hahn, M. W. and Kern, A. D. (2005). Comparative genomics of centrality and essentiality in three eukaryotic protein-iteraction networks. *Mol. Biol. Evol.* **22**, 803–6.

Hall, S. J., and Rafaelli, D. (1991). Food-web patterns - lessons from a species-rich web. *J. Anim. Ecol.* **60**, 823–42.

Hansen, P., and Mélot, H. (2005). Variable neighborhood search for extremal graphs. 9. Bounding the irregularity of a graph. *DIMACS Sec. Disc. Math. Theor. Comput. Sci.* **69**, 254–63.

Hanski, I. (1994). A practical model of metapopulation dynamics. *J. Anim. Ecol.* **63**, 151–62.

Hanski, K., and Ovaskainen, O. (2003). Metapopulation theory for fragmented landscapes. *Theor. Popul. Biol.* **64**, 119–27.

Harary, F. (1967). Graphs and matrices. *SIAM Rev.* **9**, 83–90.

Harary, F. (1969). *Graph Theory*. Addison-Wesley, Reading, MA.

Harary, F. (ed.) (1968). *Graph Theory and Theoretical Physics*. Academic Press Inc., USA, 358 pp.

Harary, F., and Schwenk, A. J. (1979). The spectral approach to determining the number of walks in a graph. *Pacific J. Math.* **80**, 443–9.

Harris, Jr. W. A., Fillmore, J. P., Smith, D. R. (2001). Matrix exponentials. Another approach. *SIAM Rev.* **43**, 694–706.

Hasty, J., McMillen, D., Isaacs, F., and Collins J. J. (2001). Computational Studies of gene regulatory networks: in numero molecular biology. *Nature Rev. Genet.* **2**, 268–79.

Havens, K. (1992). Scale and structure in natural food webs. *Science* **257**, 1107–9.

He, Y., Chen, Z. J., and Evans, A. C., (2007). Small-world anatomical networks in the human brain revealed by cortical thickness from MRI. *Cerebral Cortex* **17**, 2407–19.

He, Y., Wang, J., Wang, L., Chen, Z. J., Yan, C., Yang, H., Tang, H., Zhu, C., Gong, Q., Zang, Y., and Evans, A. C. (2009). Uncovering intrinsic modular organization of spontaneous brain activity in humans. *PLos One* **4**, e5226.

Heyde, C. C., and Kou, S. G. (2004). On the controversy over tailweight of distributions. *Op. Res. Lett.* **32**, 399–408.

Higham, N. J. (2008). *Functions of Matrices. Theory and Computation.* SIAM, Philadelphia, 425 pp.

Ho, Y., Gruhler, A., Heilbut, A., Bader, G. D., Moore, L., Adams, S. L., Millar, A., Taylor, P., Bennett, K., Boutilier, K., Yang, L., Wolting, C., Donaldson, I., Schandorff, S., Shewnarane, J., Vo, M., Taggart, J., Goudreault, M., Muskat, B., Alfarano, C., Dewar, D., Lin, Z., Michalickova, K., Willems, A. R., Sassi, H., Nielsen, P. A., Rasmussen, K. J., Andersen, J. R., Johansen, L. E., Hansen, L. H., Jespersen, H., Podtelejnikov, A., Nielsen, E., Crawford, J., Poulsen, V., Sorensen, B. D., Matthiesen, J., Hendrickson, R. C., Gleeson, F., Pawson, T., Moran, M. F., Durocher, D., Mann, M., Hogue, C. W., Figeys, D., and Tyers, M. (2002). Systematic identification of protein complexes in Saccharomyces cerevisiae by mass spectrometry. *Nature* **415**, 180–3.

Hollywood, K., Brison, D. R., and Goodacre, R. (2006). Metabolomics: current technologies and future trends. *Proteomics* **6**, 4716–23.

Holme, P. (2004). Efficient local strategies for vaccination and network attack. *Europhys. Lett.* **68**, 908–14.

Holme, P., and Kim, B. J. (2002). Growing scale-free networks with tunable clustering. *Phys. Rev. E* **65**, 026107.

Holme, P., and Zhao, J. (2007). Exploring the assortativity-clustering space of a network's degree sequence. *Phys. Rev. E* **75**, 046111.

Holme, P., Liljeros, F., Edling, C. R., and Kim, B. J. (2003). Network bipartivity. *Phys. Rev. E* **68**, 056107.

Hoory, S., Linial, N., and Wigderson, A., (2006). Expander Graphs and their applications, *Bull. Am. Math. Soc.* **43**, 439–561.

Hosoya, H. (1971). Topological index. A newly proposed quantity characterizing the topological nature of structural isomers of saturated hydrocarbons. *Bull. Chem. Soc. Jap.* **44**, 2332–9.

Hu, H.-B., and Wang, X.-F. (2009). Disassortative mixing in online social networks. *Europhys. Lett.* **86**, 18003.

Hubbell, C. H. (1965). An input-output approach to clique identification. *Sociometry,* **28**, 377–99.

Hummon, N. P., Doreian, P., and Freeman, L. C. (1990). Analyzing the structure of the centrality-productivity literature created between 1948 and 1979. *Know.-Creat. Diffus. Util.* **11**, 459–80.

Humphries, M. D., Gurney, K., and Prescott, T. J. (2006). The brainstem reticular formation is a small-world, not scale-free, network. *Proc. Biol. Sci.* **273**, 503–11.

Huxman, M., Beany, S., and Raffaelli, D. (1996). Do parasites reduce the chances of triangulation in a real food web? *Oikos* **76**, 284–300.

Ings, T. C., Montoya, J. M., Bascompte, J., Blüthgen, N., Brown, L., Dormann, C. F., Edwards, F., Figueroa, D., Jacob, U., Jones, J. I., Lauridsen, R. B., Ledger, M. E., Lewis, H. M., Olesen, J. M., vann Veen, F. J. F., Warren, P. H., and Woodward, G. (2009). Ecological networks—beyond food webs. *J. Anim. Ecol.* **78**, 253–69.

Ivanciuc, O., Balaban, T.-S., and Balaban, A. T. (1993). Design of topological indices. Part 4. Reciprocal distance matrix, related local vertex invariants and topological indices. *J. Math. Chem.* **12**, 309–18.

Jackson, M.O., and Wolinsky, A. (1996). A strategic model of social and economic networks. *J. Econ. Theor.* **71**, 44–74.

Jamakovic, A., and Mieghem, P. V. (2008). On the robustness of a complex network by using the algebraic connectivity. *Lect. Notes Comput. Sci.* **4982**, 163–94.

Jamakovic, A., and Uhlig, S. (2007). On the relationship between the algebraic connectivity and graph's robustness to node and link failures, *3rd Euro NGI Conference on Next Generation Internet Networks,* 21–23 May, pp. 96–102.

Jamakovic, A., and van Mieghem, P. (2006). The Laplacian spectrum of complex networks. *Proc. Eur. Conf. Syst.* Oxford, September 25–26.

Janga, S. C., and Babu, M. M. (2010). Transcription regulatory networks. In Buchanan, M., Caldarelli, G., de los Rios, P., Rao, F, and Vendruscolo, M. (eds.). *Networks in Cell Biology.* Cambridge Universisity Press, Cambridge, pp. 14–35.

Janson, S. (2003). The Wiener index of simply generated random trees. *Rand. Struct. Algor.* **22**, 337–58.

Janson, S. (2005). The first eigenvalue of random graphs. *J. Combin. Prob. Comput.* **14**, 815–28.

Jeong, H., Manson, S. P., Barabási, A.-L., and Oltvai, Z. N. (2001). Lethality and centrality in protein networks. *Nature* **411**, 41–2.

Jeong, H., Tombor, B., Albert, R., Oltvai Z. N., and Barabási, A.-L. (2000). The large-scale organization of metabolic networks. *Nature* **407**, 651–4.

Jiang, B. (2007). A topological pattern of urban street networks: Universality and peculiarity. *Physica A* **384**, 647–55.

Jiang, Y., Tang, A., and Hoffmann, R. (1984). Evaluation of moments and their application in Hückel molecular orbital theory. *Theoret. Chim. Acta* **66**, 183–92.

Jordán, F., and Scheuring, I. (2002). Searching for keystones in ecological networks. *Oikos* **99**, 607–12.

Jordán, F., and Scheuring, I. (2004). Network ecology: Topological constraints on ecosystem dynamics. *Phys. Life Rev.* **1**, 139–79.

Jordán, F., Liu, W., and Davis, A. J. (2006). Topological keystone species: measures of positional importance in food webs. *Oikos* **112**, 535–46.

Jordán, F., Takács-Sánta, A., and Molnár, I. (1999). A reliability theoretical quest for keystones. *Oikos* **86**, 453–62.

Jordano, P., Bascompte, J., and Olesen, J. M. (2003). Invariant properties in coevolutionary networks of plant-animal interactions. *Ecol. Lett.* **6**, 69–81.

Josson, P. F., and Bates, P. A. (2006). Global topological features of cancer proteins in the human interactome. *Bioinformatics* **22**, 2291–7.

Juhász, F. (1978). On the spectrum of a random graph. *Coll. Math. Soc. J. Bolyai* **25**, 313–6.

Jungsbluth, M., Burghardt, B., and Hartmann, A. (2007). Fingerprints networks: correlations of local and global network properties. *Physica A* **381**, 444–56.

Kaddurah-Daouk, R., Kristal, B. S., and Weinshilboum, R. M. (2008). Metabolomics: a global biochemical approach to drug response and disease. *Annu. Rev. Pharmacol. Toxicol.* **48**, 653–83.

Kaiser, M., Martin, R., Andras, P., and Young, M. P. (2007). Simulation of robustness against lesions of cortical networks. *Eur. J. Neurosci.* **25**, 3185–92.

Kannan, R., Vempala, S., and Vetta, A. (2004). On clusterings: good, bad and spectral. *J. ACM* **51**, 497–515.

Kashtan, N., Itzkovitz, S., Milo, R., and Alon, U. (2002). *Mfinder Toolguide.* Technical Report, Weizmann Institute of Science, Israel. http://www.weizmann.ac.il/mcb/UriAlon/groupNetworkMotifSW.html.

Katz, L. (1953). A new status index derived from sociometric analysis. *Psychometrica* **18**, 39–43.

Keller, E. F. (2005). Revisiting 'scale-free' networks. *BioEssay* **27**, 1060–8.

Kendall, M. (1938). A new measure of rank correlation. *Biometrika* **30**, 81–9.

Kernighan, B. W., and Lin, S. (1970). An efficient heuristic procedure for partitioning graphs. *Bell Syst. Tech. J.* **49**, 291–307.

Kim, B. J., Hong, H., Choi, M. Y. (2003). Netons: vibrations of complex networks *J. Phys. A: Math. Gen.* **36**, 6329–36

Kim, D., and Kahng, B. (2007). Spectral densities of scale-free networks. *CHAOS* **17**, 026115.

Kim, D.-H., Noh, J. D., and Jeong H. (2004). Scale-free trees: The skeletons of complex networks. *Phys. Rev. E* **70**, 046126.

Kim, J. S., Goh, K.-I., Kahng, B., and Kim, D. (2007). A box-covering algorithm for fractal scaling in scale-free networks. *CHAOS* **17**, 026116.

Kiss, C., and Bichler, M. (2008). Identification of influencers. Measuring influence in customer networks. *Dec. Supp. Syst.* **46**, 233–53.

Kitsak, M., Havlin, S., Paul, G., Riccaboni, M., Pammolli F., and Stanley, H. E. (2007). Betweenness centrality of fractal and non-fractal scale-free model networks and tests on real networks. *Phys. Rev. E* **75**, 056115.

Kleanthous, C. (ed.) (2000). *Protein–protein Recognition*. Frontiers in Molecular Biology. Oxford University Press, Oxford, 314 pp.

Klein, D. J. (1986). Chemical graph-theoretic cluster expansions. *Int. J. Quantum Chem.: Quantum Chem. Symp.* **30**, 153–71.

Klein, D. J., and Ivanciuc, O., (2001). Graph cyclicity, excess conductance, and resistance deficit. *J. Math. Chem.* **30**, 271–87.

Klein, D. J., and Randić, M. (1993). Resistance distance. *J. Math. Chem.* **12**, 81–95.

Kleinberg, J. (2000a) The small-world phenomenon: an algorithmic perspective. *Proc. 32nd ACM Symp. Theor. Comput.*

Kleinberg, J. (2000b). Navigation in a small world. *Nature* **406**, 845.

Knoke, D., and Burt, R. S. (1983). Prominence. In Burt, R. S. and Minor, M. J. (Eds). *Applied Network Analysis. A Methodological Introduction.* Sage Pub., Beverly Hills, pp. 195–222.

Koca, J., Kratochvil, M., Kvasnicka, V., Matyska, L., Pospichal, J. (1989). *Synthon Model of Organic Chemistry and Synthesis Design.* Lecture Notes in Chemistry. Springer, Berlin, 207 pp.

Koschützki, D., and Schreiber, F. (2008). Centrality analysis methods for biological networks and their application to gene regulatory networks. *Gen. Reg. Syst. Biol.* **2**, 193–201.

Kossinets, G., and Watts, D. J. (2006). Empirical analysis of an evolving social network. *Science* **311**, 88–90.

Kosub, S. (2005). Local density. In Brandes, U., and Erlebach, T. (eds.), *Network Analysis. Methadological Foundations.* Springer, Berlin, pp. 113–42.

Kovács, F., Legány, C., and Babos, A. (2006). Cluster validity measurements techniques. *Proc. 5th WSE AS Int. Conf. Artif. Intell., Know Eng. Data Bas.*, Stevens Point, Wisconsin, pp. 372–7.

Koyutürk, M. (2010). Algorithmic and analytical methods in network biology. *WIREs Syst. Biol. Med.* **2**, 277–92.

Krackhardt, D. (1999). The ties that torture: Simmelian tie analysis in organizations. *Res. Sociol. Org.* **16**, 183–210.

Krioukov, D., Papadopoulos, F., Boguñá, M., and Vahdat, A. (2009a). Greedy forwarding in scale-free networks embedded in hyperbolic metric spaces. *ACM SIGMETRICS Perform. Eval. Rev.* **37**, 15–7.

Krioukov, D., Papadopoulos, F., Kitsak, M., Vahdat, A., and Boguñá, M. (2010). Hyperbolic geometry of complex networks. *Phys. Rev. E* **82**, 036106.

Krioukov, D., Papadopoulos, F., Kitsak, M., Vahdat, A., and Boguñá, M. (2009b). On curvature and temperature of complex networks. *Phys. Rev. E* **80**, 035101(R).

Krogan, N.J., Cagney, G., Yu, H., Zhong, G., Guo, X., Ignatchenko, A., Li, J., Pu, S., Datta, N., Tikuisis, A. P., Punna, T., Peregrín-Alvarez, J. M., Shales, M., Zhang, X., Davey, M.,Robinson, M. D., Paccanaro, A., Bray, J. E., Sheung, A., Beattie, B., Richards, D. P., Canadien, V., Lalev, A., Mena, F., Wong, P., Starostine, A., Canete, M. M., Vlasblom, J., Wu, S., Orsi, C., Collins, S. R., Chandran, S., Haw, R., Rilstone, J. J., Gandi, K., Thompson, N. J., Musso, G., St Onge, P., Ghanny, S., Lam, M. H. Y., Butland, G., Altaf-UI, A. M., Kanaya, S., Shilatifard, A., O'Shea, E., Weissman, J. S., Ingles, C. J., Hughes, T. R., Parkinson, J., Gerstein, M., Wodak, S. J., Emili, A., and Greenblatt, J. F. (2006). Global landscape of protein complexes in the yeast *Saccharomyces cerevisiae. Nature* **440**, 637–43.

Kruja, E., Marks, J., Blair, A., Waters, R. (2002). A short note on the history of graph drawing. *LNCS* **2265**, 272–86.

Kümmel, A., Panke, S., and Heinemann, M. (2006). Systematic assignment of thermodynamic constraints in metabolic network models. *BMC Bioinform.* **7**, 512.

Kumpula, J. M., Saramäki, J., Kaski, K., and Kertész, J. (2007). Limited resolution in complex network community detection with Potts model approach. *Eur. Phys. J. B* **56**, 41–5.

Kundu, S. (2005). Amino acid network within protein. *Physica A* **346**, 104–9.

Kuratowski, K. (1930). Sur le problème des courbes gauches en topologie. *Fund. Math.* **15**, 271–83.

LaCount, D., Vignali, M., Chettier, R., Phansalkar, A., Bell, R., Hesselberth, J., Schoenfeld, L., Ota, I., Sahasrabudhe, S., Kurschner, C., Fields, S., Hughes, R. (2005). A protein interaction network of the malaria parasite *Plasmodium falciparum*. *Nature* **438**, 103–7.

Lacroix, V., Cottret, L., Thébault, P., and Sagot, M.-F. (2008). An introduction to metabolic networks and their structural analysis. *IEEE/ACM Trans. Comput. Biol. Bioinform.* **5**, 594–617.

Lancichinetti, A., Fortunato, S., and Radicchi, F. (2008). Benchmark graphs for testing community detection algorithms. *Phys. Rev. E* **78**, 046110.

Lancichinetti, A., Radicchi, F., and Ramasco, J. J. (2010). Statistical significance of communities in networks. *Phys. Rev. E*, **81**, 046110.

Landherr, A., Friedl, B., and Heidermann, J. (2010). A critical review of centrality measures in social networks. *Bus. Inf. Syst. Inf.* **6**, 371–8.

Langville, A. N., and Meyer, C. D. (2005). A survey of eigenvector methods of web information retrieval. *SIAM Rev.* **47**, 135–61.

Langville, A. N., and Meyer, C. D. (2006). *Google's PageRank and Beyond. The Science of Search Engine Rankings*. Princeton University Press, Princeton, 224 pp.

Latora, V., and Marchiori, M. (2007). A measure of centrality based on the network efficiency. *New J. Phys.* **9**, 188.

Leavitt, H. J. (1951). Some effects of certain communication patterns on group performance. *J. Abnor. Soc. Psych.* **46**, 38–50.

Lee, C.-Y., (2006). Correlations among centrality measures in complex networks. *Arxiv: physics*/0605220.

Lee, D-S., and Rieger, H. (2007). Comparative study of the transcriptional regulatory networks of *E. coli* and yeast: Structural characteristics leading to marginal dynamic stability. *J. Theor. Biol.* **248**, 618–26.

Leonard, I. E. (1996). The matrix exponential. *SIAM Rev.* **38**, 507–12.

Lesk, A. M. (2001). *Introduction to Protein Architecture*. Oxford University Press, Oxford, 347 pp.

Li, D., Li, J., Ouyang, S., Wang, J., Wu, S., Wan, P., Zhu, Y., Xu, X., and He, F. (2006). Protein interaction networks of Saccharomyces cerevisiae, Caenorhabditis elegans and Drosophila melanogaster. Large-scale organization and robustness. *Proteomics* **6**, 456–61.

Li, S., Armstrong, C. M., Bertin, N., Ge, H., Milstein, S., Boxem, M., Vadalain, P.-O., Han, J-D. J., Chesneau, A., Hao, T., Coldberg, D. S., Li, N., Martinez, M., Rual, J.-F., Lamesch, P., Xu, L., Tewari, M., Wong, S. L., Zhang, L. V., Berriz, G. F., Jacotot, L., Vaglio, P., Reboul, J., Hirozane-Kishikawa, T., Li, Q., Gabel, H. W., Elewa, A., Baumgartner, B., Rose, D. J., Yu, H., Bosak, S., Sequerra, R., Fraser, A., Mango, S. E., Saxton, W. M., Strome, S., van den Heuvel, S., Piano, F., Vandenhaute, J., Sardet, C., Gerstein, M., Doucette-Stamm, L., Gunsalus, K. C., Harper, J. W., Cusick, M. E., Roth, F. P., Hill, D. E., Vidal, M. (2004). A map of the interactome network of the metazoan *C. elegans*. *Science* **303**, 540–3.

Li, X., and Shi, Y., (2008). A survey on the Randić index. *MATCH Comm. Math. Comput. Sci.* **59**, 127–56.

Li, X., Jin, J. J., and Chen, G. (2003). Complexity and synchronization in the world trade web. *Physica A* **328**, 287–96.

Li, Y., Liu, Y., Li, J., Qin, W., Li, K., Yu, C., and Jiang, T. (2009). Brain anatomical network and intelligence. *PLoS Comp. Biol.* **5**, e1000395.

Liebers, A. (2001). Planarizing graphs. A survey and annotated bibliography. *J. Graph Theor. Alg. Appl.* **5**, 1–74.

Liljeros, F., Edling, C. R., and Amaral, L. A. N. (2003). Sexual networks: implications for the transmission of sexually transmitted infections. *Microb. Infect.* **5**, 189–96.

Lin, C. Y., Chen C. L., Cho, C. S., Wang L. M., Chang C. M., Chen P. Y., Lo, C. Z., and Hsiung, C. A. (2005). hp-DPI: *Helicobacter pylori* database of protein interactomes, A combined experimental and inferring interactions. *Bioinformatics* **21**, 1288–90.

Lin, C.-Y., Chin, C.-H., Wu, H.-H., Chen, S.-H., Ho, C.-W., and Ko, M.-T. (2008). Hubba: hub objects analyzer–a framework of interactome hubs identification for network biology. *Nucl. Acids. Res.* **36**, W438–43.

Lind, P. G., González, M. C., and Herrmann, H. J. (2005). Cycles and clustering in bipartite networks. *Phys. Rev. E.* **72**, 056127.

Link, J. (2002). Does food web theory work for marine ecosystems? *Mar. Ecol. Prog. Ser.* **230**, 1–9.

Liu, J., (2010). Comparative analysis for k-means algorithms in network community detection. *LNCS* **6382**, 158–69.

Liu, Y., Liang, M., Zhou, Y., He, Y., Hao, Y., Song, M., Yu, C., Liu, H., Liu, Z., and Jiang, T. (2008). Disrupted small-world networks in schizophrenia. *Brain* **131**, 945–61.

Loscalzo, J., Kohane, I., and Barabasi, A-L. (2007). Human disease classification in the postgenomic era: a complex systems approach to human pathobiology. *Mol. Syst. Biol.* **3**, 124.

Lovelock, J. E., and Margulis, L. (1974). Atmospheric homeostasis by and for the biosphere: the Gaia hypothesis. *Tellus* **XXVI**, 1–10.

Lubotzky, A., Phillips, R., and Sarnak, P. (1988). Ramanujan graphs. *Combinatorica* **8**, 261–77.

Lundberg, J., and Moberg, F. (2003). Mobile link organisms and ecosystem functioning: Implications for ecosystem resilience and management. *Ecosystems* **6**, 87–98.

Lusseau, D. (2003). The emergent properties of a dolphin social network. *Proc. R. Soc. Lond. B (Suppl.)* **270**, 186–8.

Ma, H.-W., and Zeng, A.-P. (2003). The connectivity structure, giant strong component and centrality of metabolic networks. *Bioinformatics* **19**, 1423–30.

MacRae, D. (1960). Direct factor analysis of *sociometric* data. *Sociometry* **23**, 360–71.

Malarz, K. (2008). Spectral properties of adjacency and distance matrices for various networks. *LNCS* **5102**, 559–67.

Mallion, R. B. (2007). The Six (or Seven) bridges of Kaliningrad: a personal Eulerian walk, 2006. *MATCH: Comm. Math. Comput. Chem.* **58**, 529–56.

Mandelbrot, B. B. (1982). *The Fractal Geometry of Nature.* W.H. Freeman & Co Ltd, 468 pp.

Markowetz, F., and Spang, R. (2007). Inferring cellular networks—a review. *BMC Bioinf.* **8** (Suppl. 6), s5.

Martinez, N. D. (1991). Artifacts or attributes? Effects of resolution on the Little Rock lake food web. *Ecol. Monogr.* **61**, 367–92.

Martinez, N. D. (1992). Constant connectance in community food webs. *Am. Nat.* **139**, 1208–18.

Martínez-Antonio, A., and Collado-Vides, J. (2003). Identifying global regulators in transcriptional regulatory networks in bacteria. *Curr. Op. Microbiol.* **6**, 482–9.

McGraw, P. N., and Menzinger M. (2008). Laplacian spectra as a diagnostic tool for network structure and dynamics. *Phys. Rev. E* **77**, 031102.

McHughes, M. C., and Poshusta, R. D. (1990). Graph-theoretic cluster expansions. Thermochemical properties of alkanes. *J. Math. Chem.* **4**, 227–49.

McLaughlin, M. P. (1999). *A Compendium of Common Probability Distributions.* Appendix A for the Software Regress+. http://www.causascientia.org/math_stat/Dists/Compendium.pdf.

Mead, C. A., and Conway, L. (1980). *Introduction to VLSI systems*. Boston: Addison-Wesley.

Meffe, G. K., Nielsen, L. A., Knight, R. L., and Schenborn, D. A. (2002). *Ecosystem Management: Adaptive, Community-based Conservation*. Island Press, Washington DC.

Meila, M., and Shi, J. (2001). A random walk view of spectral segmentation. Workshop on Artificial Intelligence and Statistics (AISTATS).

Memmott, J., Martinez, N. D., and Cohen, J. E. (2000). Predators, parasites and pathogens: species richness, trophic generality, and body sizes in a natural food web. *J. Animal Ecol.* **69**, 1–15.

Menge, B. A., Berlow, E. L., Blanchette, C. A., Navarrete, S. A., Yamada, S. B. (1994). The keystone species concept: variation in interaction strength in a rocky intertidal habitat. *Ecol. Monogr.* **64**, 249–87.

Merris, R. (1998). Laplacian graph eigenvectors. *Lin. Alg. Appl.* **278**, 221–36.

Michael, J. H., and Massey, J. G. (1997). Modeling the communication network in a sawmill. *Forest Prod. J.* **47**, 25–30.

Michalski, M. (1986). Branching extent and spectra of trees. *Pub. Inst. Math.* **39**, 35–43.

Micheloyannis, S., Pachou, E., Stam, C. J., Breakspear, M., Bitsios, P., Vourkas, M., Erimaki, S., and Zervakis, M. (2006). Small-world networks and disturbed functional connectivity in schizophrenia. *Schizophr. Res.*, **87**, 60–6.

Milgram, S. (1967). The small world problem. *Psychol. Today* 2, 60–7.

Mills, L. S., Soulé, M. E., and Doak, D. F. (1993). The keystone-species concept in ecology and conservation. *Biosc.* **43**, 219–24.

Milo, R., Itzkovitz, S., Kashtan, N., Levitt, R., Shen-Orr, Shai., Ayzenshtat, I., Sheffer, M., and Alon, U. (2004a). Superfamilies of evolved and designed networks. *Science* **303**, 1538–42.

Milo, R., Itzkovitz, S., Kashtan, N., Levitt, R., and Alon, U. (2004b). Response to comment on 'Networks motifs: simple building, blocks of complex networks' and 'Superfamilies of evolved and designed networks.' *Science* **305**, 1107d.

Milo, R., Shen-Orr, S., Itzkovitz, S., Kashtan, N., Chklovskii, D., and Alon, U. (2002). Network motifs: Simple building blocks of complex networks. *Science* **298**, 824–7.

Minor, E. S., and Urban, D. L. (2007). Graph theory as a proxy for spatially explicit population models in conservation planning. *Ecol. Appl.* **17**, 1771–82.

Mitzenmacher, M. (2003). A brief history of generative models for power law and lognormal distributions. *Internet Math.* **1**, 226–51.

Miyazawa, S., and Jernigan, R. L. (1996). Residue-residue potentials with a favourable contact pair term and an unfavourable high packing density terms, for simulation and threading. *J. Mol. Biol.* **256**, 623–44.

Modha, D. S., and Singh, R. (2010). Network architecture of the long-distance pathways in the macaque brain. *Proc. Natl. Acad. Sci. USA* **107**, 13485–90.

Mohar, B., and Pisanski, T. (1988). How to compute the Wiener index of a graph. *J. Math. Chem.* **2**, 267–77.

Moler, C., and van Loan, C. (2003). Nineteen dubious ways to compute the exponential of a matrix, twenty-five years. *SIAM Rev.* **45**, 3–49.

Montañez, R., Medina, M. A., Solé, R. V., and Rodríguez-Caso, C. (2010). When metabolism meets topology: Reconciling metabolite and reaction networks. *BioEssays*, **32**, 246–56.

Montoya, J. M., and Solé, R. V. (2002). Small world patterns in food webs. *J. Theor. Biol.* **214**, 405–12.

Montoya, J. M., and Solé, R. V. (2003). Topological properties of food webs: from real data to community assembly models. *Oikos* **102**, 614–22.

Montoya, J. M., Pimm, S. L., and Solé, R. V. (2006). Ecological networks and their fragility. *Nature* **442**, 259–64.

Moody, J. (2001). Data for this project were provided, in part, by NIH grants DA12831 and HD41877, and copies can be obtained from James Moody (moody.77@sociology.osu.edu).

Morrison, J. L., Breitling, R., Higham, D., and Gilbert, D. R. (2006). A lock-and-key model for protein–protein interactions. *Bioinformatics*, **22**, 2012–19.

Motz, M., Kober, I., Girardot, C., Loeser, E., Bauer, U., Albers, M., Moeckel, G., Minch, E., Voss, H., Kilger, C., and Koegl, M. (2002). Elucidation of an archaeal replication protein network to generate enhanced PCR enzymes. *J. Biol. Chem.* **277**, 16179–88.

Mutzel, P., Odenthal, T., and Scharbrodt, M. (1998). The thickness of graphs: A survey. *Graphs Comb.* **14**, 59–73.

Nacu, Ł., Critchley-Thorne, R., Lee, P., and Holmes, S. (2007). Gene expression network analysis, and applications to immunology. *Bioinformatics* **23**, 850–8.

Nagurney, A. (2003). Financial and Economic Networks: An Overview. In Nagurney, A. (ed.), *Innovations in Financial and Economic Networks*. Edward Elgar Publishing, Cheltenham, England, pp. 1–26.

Newman, M. E. J. (2001). The structure of scientific collaboration networks. *Proc. Natl. Acad. Sci. USA* **98**, 404–9.

Newman, M. E. J. (2002a). Assortative mixing in networks. *Phys. Rev. Lett.* **89**, 208701.

Newman, M. E. J. (2003a). Mixing patterns in networks. *Phys. Rev. E* **67**, 026126.

Newman, M. E. J. (2003b). The structure and function of complex networks. *SIAM Rev.* **45**, 167–256.

Newman, M. E. J. (2004a). Fast algorithm for detecting community structure in networks. *Phys. Rev. E* **69**, 066133.

Newman, M. E. J. (2004b). Analysis of weighted networks. *Phys. Rev. E* **70**, 056131.

Newman, M. E. J. (2005a). A measure of betweenness centrality based on random walks. *Social Networks* **27**, 39–54.

Newman, M. E. J. (2006a). Finding community structure in networks using the eigenvectors of matrices. *Phys. Rev. E* **74**, 036104.

Newman, M. E. J. (2006b). Modularity and community structure in networks, *Proc. Natl. Acad. Sci. USA* **103**, 8577–82.

Newman, M. E. J., (2002b). Random graphs as models of networks, In Bornholdt, S., Schuster, H. G. (eds.) *Handbook of Graph and Networks: From the Genome to the Internet*, Wiley-VCH, Weinheim, pp. 35–65.

Newman, M. E. J., (2004c). Coauthorship networks and patterns of scientific collaboration, *Proc. Natl. Acad. Sci. USA* **101**, 5200–5.

Newman, M. E. J., and Girvan, M. (2003). Mixing patterns and community structure in networks. *Lect. Notes Phys.* **625**, 66–87.

Newman, M. E. J., and Girvan, M. (2004). Finding and evaluating community structure in networks. *Phys. Rev. E* **69**, 026113.

Newman, M. E. J., Forrest, S., and Balthrop, J. (2002). Email networks and the spread of computer viruses. *Phys. Rev. E* **66**, 035101(R).

Newman, M. E. J., Strogatz, S. H., and Watts, D. J. (2001). Random graphs with arbitrary degree distributions and their applications. *Phys. Rev. E* **64**, 026118.

Newman, M. E. J., Watts, D. J., and Strogatz, S. H. (2002). Random graph models of social networks. *Proc. Natl. Acad. Sci. USA* **99**, 2566–72.

Newman, M. J. E. (2005b). Power laws, Pareto distributions and Zipf's laws. *Comtemp. Phys.* **46**, 323–51.

Newman, M. J. E. (2010). *Networks. An Introduction*. Oxford University Press, Oxford, 772 pp.

Ng, A. Y., Jordan, M. I., and Weiss, Y. (2001). On Spectral Clustering: Analysis and an algorithm. *Advances in Neural Information Processing Systems*, **14**.

Nikolić, S., Kovacević, G., Milicević, A., and Trinajstić, N. (2003). The Zagreb indices 30 years after. *Croat. Chem. Acta* **76**, 113–24.

Nikolopoulos, S. D., Palios, L. (2004) Hole and antihole detection in graphs. *SODA 104 Proc. 15th Ann. ACM-SIAM Symp. Discr. Algor.* Philadelphia, PA, pp. 850–9.

Nikoloski, Z., May, P., and Selbig, J. (2010). Algebraic connectivity may explain the evolution of gene regulatory networks, *J. Theor. Biol.* **267**, 7–14.

Noh, J. D. (2007). Percolation transition in networks with degree-degree correlation. *Phys. Rev. E* **76**, 026116.

Noirot, P., and Noirot-Gross, N. F. (2004). Protein interaction networks in bacteria. *Curr. Op. Microb.* **7**, 505–12.

Nooren, I. M. A. and Thornton, J. M. (2003). Diversity of protein–protein interactions. EMBO J. **22**, 3486–92.

ODLIS (2002). *Online Dictionary of Library and Information Science.* http://vax.wcsu.edu/library/odlis.html.

Ohnishi, T., Takayasu, H., and Takayasu, M. (2009). Hubs and authorities on Japanese inter-firm network: Characterization of nodes in very large directed networks. *Prog. Theor. Phys. Suppl.* **179**, 157–66.

Ohnishi, T., Takayasu, H., and Takayasu, M. (2010). Network motifs in an inter-firm network. *J. Econ. Interact. Coord.* **5**, 171–80.

Ohno, S. (1970). *Evolution by Gene Duplication.* Springer, New York.

Oliver, S. (2000). Guilt-by-association goes global. *Nature* **403**, 601–3.

Onnela, J.-P., Kaski, K., and Kertész, J. (2004). Clustering and information in correlation based financial networks. *Eur. Phys. J. B* **38**, 353–62.

Opitz, S. (1996). Trophic interactions in Caribbean coral reefs. *ICLARM Tech. Rep.* **43**, Manila, Philippines, 341 pp.

Oxford English Dictionary (2010). OED Online. Oxford University Press. http://www.oed.com.

Paine, R. T. (1966). Food web complexity and species diversity. *Am. Nat.* **100**, 65–75.

Palla, G., Barabási, A.-L., and Vicsek, T. (2007). Quantifying social group evolution. *Nature* **446**, 664–7.

Palla, G., Derényi, I., Farkas, I., and Vicsek, T. (2005). Uncovering the overlapping community structure of complex networks in nature and society. *Nature* **435**, 814–18.

Papendieck, B., and Recht, P. (2000). On maximal entries in the principal eigenvector of graphs. *Lin. Alg. Appl.* **310**, 129–38.

Parter, M., Kashtan, N., and Alon, U. (2007). Environmental variability and modularity of bacterial metabolic networks. *BMC Evol. Biol.* **7**, 169.

Pastor-Satorras, R., Smith, E., and Solé, R. V. (2003). Evolving protein interaction networks through gene duplication. *J. Theor. Biol.* **222**, 199–210.

Pemmaraju, S., and Skiena, S. (2003). *Computational Discrete Mathematics. Combinatorics and Graph Theory with Mathematica®*. Cambridge University Press, Cambridge, 480 pp.

Penrose, M. (2003). *Random Geometric Graphs.* Oxford University Press, Oxford, 344 pp.

Pereira-Leal, J. B., Enright, A. J., and Ouzounis, C. A. (2004). Detection of functional modules from protein interaction networks. *Proteins: Struct., Funct., Bionform.* **54**, 49–57.

Perna, A., Valverde, S., Gautrais, J., Jost, C., Solé, R. V., Kuntz, P., and Theraulaz, G. (2008). Topological efficiency in the three-dimensional gallery networks of termite nests. *Physica A* **387**, 6235–44.

Perra, N., Zlatic, V., Chessa, A., Conti, C., Donato, D., and Caldarelli, G. (2009). PageRank equation and localization in the WWW. *Europhys. Lett.* **88**, 48002.

Pierce, W. D., Cushman, R. A., Hood, C. E., and Hunter, W. H. (1912). *The Insect Enemies of the Cotton Boll Weevil.* Government Printing Office, Washington DC.

Pisabarro, A. G., Pérez, G., Lavín, J. L. and Ramirez, L. (2008). Genetic networks for the functional study of genomes. *Brief. Funct. Gen. Proteom.* **7**, 249–63.

Pisanski, T., Potočnik, P. (2003). Graphs in surfaces. In Gross, J. L. and Yellen, J. (eds.), *Handbook of Graph Theory*, CRC Press, Boca Raton, pp. 611–24.

Platzer, A., Perco, P., Lukas, A., and Mayer, B. (2007). Characterization of protein-interaction networks in tumors. *BMC Bioinform.* **8**, 224.

Plavšić, D., Nikolić, S., Trinajstić, N., and Mihalić, Z. (1993). On the Harary index for the characterization of chemical graphs. *J. Math. Chem.* **12**, 235–50.

Poe, E. A. (1984). *Complete Stories and Poems of Edgar Allan Poe.* Doubleday; Book Club Edition, 832 pp.

Polis, G. A. (1991). Complex trophic interactions in deserts: an empirical critique of food-web theory. *Am. Nat.* **138**, 123–55.

Porter, M. A., Mucha, P. J., Newman, M. E. J., and Friend, A. J. (2007). Community structure in the United States House of Representatives. *Physica A* **386**, 414–38.

Porter, M. A., Mucha, P. J., Newman, M. E. J., and Warmbrand, C. M. (2005). A network analysis of committees in the U.S. House of Representatives. *Proc. Natl. Acad. Sci. USA* **102**, 7057–62.

Porter, M. A., Onnela, J.-P., and Mucha, P. J. (2009). Communities in networks. *Notices Am. Math. Soc.* **56**, 1082–97 & 1164–6.

Potapov, A. P., Voss, N., Sasse, N., Wingender, E. (2005). Topology of mammalian transcription networks. *Genom. Infor.* **16**, 270–8.

Potkin, S. G., Macciardi, F., Guffanti, G., Fallon, J. H., Wang, Q., Turner, J. A., Lakatos, A., Miles, M. F., Lander, A., Vawter, M. P., Xiaohui, X. (2010). Identifying gene regulatory networks in schizophrenia. *Neuroimage* **53**, 839–47.

Potterat, J. J., Philips-Plummer, L., Muth, S. Q., Rothenberg, R. B., Woodhouse, D. E., Maldonado-Long, T. S., Zimmerman, H. P., Muth, J. B. (2002). Risk network structure in the early epidemic phase of HIV transmission in Colorado Springs. *Sex. Transm. Infect.* **78**, i159–63.

Power, M. E., Tilman, D., Estes, J. A., Menge, B. A., Bond, W. J., Mills, L. S., Daily, G., Castilla, J. C., Lubchenco, J., and Paine, R. T. (1996). Challenges in the quest for keystones. *BioSci.* **46**, 609–20.

Powers, D. L. (1988). Graph partitioning by eigenvectors. *Lin. Alg. Appl.* **101**, 121–33.

Presser, A., Elowitz, M. B., Kellis, M., and Kishony, R. (2008). The evolutionary dynamics of the Saccharomyces cerevisiae protein interaction network after duplication. *Proc. Natl. Acad. Sci. USA* **105**, 950–4.

Prieto, C., Risueño, A., Fontanillo, C., and De las Rivas, J. (2008). Human gene coexpression landscape: confident network derived from tissue transcriptomic profiles. *Plos One* **3**, e3911.

Proulx, S. R., Promislow, D. E. L., and Phillips, P. C. (2005). Network thinking in ecology and evolution. *Trends Ecol. Evol.* **20**, 345–53.

Pržulj, N. (2007). Biological network comparison using graphlet degree distribution, Bioinformatics **23**, e177–83.

Pržulj, N., and Higham, D. J. (2006). Modelling protein–protein interaction networks via a stickiness Index. *J. Roy. Soc. Interf.* **3**, 711–16.

Pržulj, N., Corneil, D. G., and Jurisica, I. (2004). Modeling interactome: scale-free or geometric? *Bioinformatics* **20**, 3508–15.

Qiu, H., and Hancock, E. R. (2004). Graph matching and clustering using spectral partitions. *Patt. Recog.* **39**, 22–34.

Quesnay, F. (1758), *Tableau Economique*. Reproduced in facsimile, with an introduction by H. Higgs. British Economic Society, 1895.

Radicchi, F., Castellano, C., Cecconi, F., Loreto, V., and Parisi, D. (2004). Defining and identifying communities in networks. *Proc. Nalt. Sci. USA* **101**, 2658–2663.

Rain, J.-C., Selig, L., De Reuse, H., Battaglia, V., Reverdy, C., Simon, S., Lenzen, G., Petel, F., Wojcik, J., Schächter, V., Chemama, Y., Labigne, A., and Legrain, P. (2001). The protein–protein interaction map of *Helicobacter pylori*. *Nature* **409**, 211–15.

Ram Murty, M. (2003). Ramanujan graphs. *J. Ramanujan Math. Soc.* **18**, 1–20.

Ramsey, N. F. (1956). Thermodynamics and statistical mechanics at negative absolute temperatures. *Phys. Rev.* **103**, 20–8.

Randić, M. (1975). On characterization of molecular branching. *J. Am. Chem. Soc.* **97**, 6609–15.

Ranjan, D. (2005). *Frank Harary, 1921–2005*. http://www1.cs.columbia.edu/~sanders/graphtheory/harary.html.

Ratkaj, I., Stajduhar, E., Vucinic, S., Spaventi, S., Bosnjak, H., Pavelic, K., and Pavelic, S. K. (2010). Integrated gene networks in breast cancer development. *Funct. Integr. Genom.* **10**, 11–19.

Ravasz, E., Somera, A. L., Mongru, D. A., Oltvai, Z. N., and Barabási, A.-L. (2002). Hierarchical organization of modularity in metabolic networks. *Science* **297**, 1551–5.

Ravasz, E.; Barabási, A.-L. (2003). Hierarchical organization in complex networks. *Phys. Rev. E* **67**, 026112.

Reichold, J., Stampanoni, M., Keller, A. L., Buck, A., Jenny, P., and Weber, B. (2009). Vascular graph model to simulate the cerebral blood flow in realistic vascular networks. *J. Cereb. Blood Flow Metab.* **29**, 1429–43.

Reitsma, F., and Engel, S. (2004). Searching for 2D spatial networks holes. *LNCS* **3044**, 1069–78.

Riede, J. O., Rall, B. C., Banasek-Richter, C., Navarrete, S. A., Wieters, E. A., Emmerson, M. C., Ute, J., and Brose, U. (2010). Scaling of food-web properties with diversity and complexity across ecosystems. *Adv. Ecol. Res.* **42**, 139–70.

Rodríguez, J. A., Estrada, E., Gutiérrez A. (2007). Functional centrality in graphs. *Lin. Multilin. Alg.* **55**, 293–302.

Roget's Thesaurus of English Words and Phrases (2002). Project Gutenberg. http://www.gutenberg.org/etext/22

Rong, Z., Li, X., Wang, X. (2007). Roles of mixing patterns in cooperation on a scale-free networked game. *Phys. Rev. E* **76**, 027101.

Rosenfeld, N., and Alon, U. (2003). Response delays and the structure of transcription networks. *J. Mol. Biol.* **329**, 645–54.

Rousseeuw, P. J. (1987). Silhouettes: a graphical aid to the interpretation and validation of cluster analysis. *J. Comput. Appl. Math.* **20**, 53–65.

Rowlinson, P., (2007). The main eigenvalues of a graph: A survey. *Appl. Anal. Discr. Math.* **1**, 445–71.

Rual, J.-F., Venkatesan, K., Hao, T., Hirozane-kishikawa, T., Dricot, A., Ning, L., Berriz, G. F., Gibbons, F. D., Dreze, M., Ayivi-Guedehoussou, N., Klitgord, N., Simon, C., Boxem, M., Milstein, Stuart., Rosenberg, J., Goldberg, D. S., Zhang, L. V., Wong, S. L., Franklin, G., Li, S., Albala, J. S., Lim, J., Fraughton, C., Llamosas, E., Cevik, S., Bex, C., Lamesch, P., Sikorski, R. S., Vandenhaute, J., Zoghbi, H. Y., Smolyar, A., Bosak, S., Sequerra, R., Doucette-Stamm, L., Cusick, M. E., Hill, D. E., Roth, F. P., and Vidal, M. (2005). Towards a proteome-scale map of the human protein–protein interaction network. *Nature* **437**, 1173–8.

Rubinov, M. and Sporn, O. (2010). Complex network measures of brain connectivity: Uses and interpretations. *NeuroImage* **52**, 1059–69.

Ruhnau, B. (2000). Eigenvector-centrality—a node-centrality? *Social Networks* **22**, 357–65.

Sanz-Arigita, E. J., Schoonheim, M. M., Damoiseaux, J. S., Rombouts, S. A. R. B., Maris, E., Barkhof, F., Scheltens, P., and Stam, C., (2010). Loss of 'Small-World' network in Alzheimer's disease: Graph analysis of fMRI resting-stage functional connectivity. *PLos One* **5**, e13788.

Scannell, J. W., Burns, G. A. P. C., Hilgetag, C. C., O'Neil, M. A., Young, M. P. (1999). The connectional organization of the cortico-thalamic systemof the cat. *Cereb. Cortex* **9**, 277–99.

Schaeffer, S. E. (2007). Graph clustering. *Comput. Sci. Rev.* **1**, 27–64.

Schlitt, T., Palin, K., Rung, J., Dietmann, S., Lappe, M., Ukkonen, E., and Brazma, A. (2003). From gene networks to gene function. *Genome Res.* **13**, 2568–76.

Schneberger, A., Mercer, C. H., Gregson, S. A. J., Ferguson, N. M., Nyamukapa, C. A., Anderson R. M., Johnson, A. M., and Garnett, G. P. (2004). Scale-free networks and sexually transmitted diseases. A description of observed patterns of sexual contacts in Britain and Zimbabwe. *Sex. Trans. Dis.* **31**, 380–7.

Schwitzer, F., Fagiolo, G., Sornette, D., Vega-Redondo, F., Vespignani, A., White, D. R. (2009). Economic networks: The new challenges. *Science* **325**, 422–5.

Seary, A. J., and Richards, W. D. (1995). Partitioning networks by eigenvectors. *Proc. Int. Conf. Soc. Net.* **1**, 47–58.

Seary, A. J., and Richards, W. D. (2003). Spectral methods for analyzing and visualizing networks: an introduction. *Dyn. Soc. Net. Mod. Anal. Workshop Summary and Papers*. The National Academies Press, pp. 209–228.

Serrano, M. A., and Boguñá, M. (2003). Topology of the world trade network. *Phys. Rev. E* **68**, 015101(R).

Serrano, M. A., Krioukov, D., and Boguñá, M. (2008). Self-similarity of complex networks and hidden metric spaces. *Phys. Rev. Lett.* **100**, 078701.

Sharom, J. R., Bellows, D. S., and Tyers, M. (2004). From large networks to small molecules. *Curr. Op. Chem. Biol.* **8**, 81–90.

Shen-Orr, S. S., Milo, R., Mangan, S., and Alon, U. (2002). Network motifs in the transcriptional regulation network of *Escherichia coli*. *Nature Gen.* **31**, 64–8.

Shi, J., and Malik, J. (2000). Normalized cuts and image segmentation. *IEEE Trans. Patt. Anal. Mach. Int.* **22**, 888–901.

Skyrms, B., and Pemantle, R. (2000). A dynamic model of social network formation. *Proc. Natl. Acad. Sci. USA* **97**, 9340–6.

Smith, D. A., and White, D. R. (1992). Structure and dynamics of the global economy - Network analysis of international-trade 1965–1980. *Social Forces* **70**, 857–93.

Smolenskii, E. A. (1964). Application of the theory of graphs to calculations of the additive structural properties of hydrocarbons. *Russ. J. Phys. Chem.* **38**, 700–2.

Snijders, T. A. B. (1981). The degree variance: An index of graph heterogeneity. *Social Networks* **3**, 163–74.

Soffer, S. N., and Vázquez, A. (2005). Clustering coefficient without degree correlations biases. *Phys. Rev. E.* **71**, 057101.

Solé, R. V., and Montoya, J. M. (2001). Complexity and fragility in ecological networks. *Proc. R. Soc. B* **268**, 2039–45.

Solé, R. V., and Munteanu, A. (2004). The large-scale organization of chemical reaction networks in astrophysics. *Europhys. Lett.* **68**, 170.

Solomonoff, R., and Rapoport, A. (1951). Connectivity of random nets. *Bull. Math. Biophys.* **13**, 107–17.

Solow, A. R. (2005). Power laws without complexity. *Ecol. Lett.* **8**, 361–3.

Song, C., Havlin, S., and Makse, H. A. (2005). Self-similarity of complex networks. *Nature* **433**, 392–5.

Song, C., Havlin, S., and Makse, H. A. (2006). Origins of fractality in the growth of complex networks. *Nature Phys.* **2**, 275–81.

Spencer, J. (2010). The giant component: the golden anniversary. *Notices ACM* **57**, 720–4.

Spinrad, J. P. (1991). Finding large holes. *Inf. Proc. Lett.* **39**, 227–9.

Spirin, V., and Mirny, L. A. (2003). Protein complexes and functional modules in molecular networks. *Proc. Natl. Acad. Sci. USA* **100**, 12123–8.

Sporn, O. (2011). *Networks of the Brain*. MIT Press, Cambridge, MA, 412 pp.

Sporn, O., and Kötter, R. (2004). Motifs in brain betworks. *PLoS Biol.* **2**, e369.

Sporn, O., Chialvo, D. R., Kaiser, M. and Hilgetag, C. (2004). Organization, development and function of complex brain networks. *Trends Cogn. Sci.* **8**, 418–25.

Sporn, O., Honey, C. J., and Kötter, R. (2007a). Identification and classification of hubs in brain networks. *PLoS One* **2**, e1049.

Sporn, O., Tononi, G., and Kötter, R. (2007b). The human connectome: A structural description of the human brain. *PLos Compt. Biol.* **1**, e42.

Springer, A. M., Estes, J. A., van Vliet, G. B., Williams, T. M., Doak, D. F., Danner, E. M., Forney, K. A., and Pfister, B. (2003). Sequential megafaunal collapse in the North Pacific Ocean: an ongoing legacy of industrial whaling? *Proc. Natl. Acad. Sci. USA* **100**, 1223–8.

Srinivasan, U. T., Dunne, J. A., Harte, J., and Martinez, N. D. (2007). Response of complex food webs to realistic extinction sequences. *Ecology* **88**, 671–82.

Stam, C. J., de Haan, W., Daffertshofer, A., Jones, B. F., Manshanden, I., van Cappellen van Walsum, A. M., Motenz, T., Verbunt, J. P. A., de Munck, J. C., van Dijk, B. W., Berendse, H. W., and Scheltens, P. (2009). Graph theoretical analysis of magnetoencephalographic functional connectivity in Alzheimer's disease. *Brain* **132**, 213–24.

Stam, C. J., Jones, B. F., Nolte, G., Breakspear, M., and Scheltens, Ph. (2007). Small-world networks and functional connectivity in Alzheimer's disease. *Cerebral Cortex* **17**, 92–9.

Stephenson, K., and Zelen, M. (1989). Rethinking centrality: methods and examples. *Social Networks* **11**, 1–37.

Stuart, J. M., Segal, E., Koller, D., and Kim, S. K. (2003). A gene-coexpression network for global discovery of conserved genetic modules. *Science* **302**, 249–55.

Stumpf, M. P. H., and Ingram, P.J. (2005). Probability models for degree distributions of protein interaction networks. *Europhys. Lett.* **71**, 152–8.

Sugizaki, K. (1992). Mega-city pyramid TRY2004. *Sruct. Eng. Intl.* **4**, 287–9.

Supekar, K., Menon, V., Rubin, D., Musen, M., and Greicius, M. D., (2008). Network analysis of intrinsic functional brain connectivity in Alzheimer's disease. *PLoS Comput. Biol.* **4**, e1000100.

Sylvester, J. J. (1878). Chemistry and algebra. *Nature* **17**, 284–5.

Takayasu, M., Sameshima, S., Ohnishi, T., Iyetomi, H., Ikeda, Y., Takayasu, H., and Watanabe, K. (2008a). Massive economics data analysis by econophysics methods—The case of companies' network structure II. *Ann. Rep. Earth Simul. Cent.* April 2007–March 2008, pp. 237–41.

Takayasu, M., Sameshima, S., Watanabe, H., Ohnishi, T., Iyetomi, H., Iino, T., Kobayashi, Y., Kamehama, K., Ikeda, Y., Takayasu, H., and Watanabe, K. (2008b). Massive economics data analysis by econophysics methods—The case of companies' network structure. *Ann. Rep. Earth Simul. Cent.* April 2007–March 2008, pp. 263–7.

Taylor, P. D., L. Fahrig, K. Henein, and G. Merriam. (1993). Connectivity is a vital element of landscape structure. *Oikos* **68**, 571–3.

Taylor, A., and Higham, D. J. (2009). CONTEST: A Controllable Test Matrix Toolbox for MATLAB. *ACM Trans. Math. Soft.* **35**, 1–26:17.

Temkin, O. N., Zeigarnik, A. V., and Bonchev, D. G. (1996). *Chemical Reaction Networks: A Graph-Theoretical Approach*. CRC Press, Boca Raton, FL, 304 pp.

Thomas, A., Cannings, R., Monk, N. A. M., and Canning, C. (2003). On the structure of protein–protein interaction networks. *Biochem. Soc. Trans.* **31**, 1491–6.

Thurman, B. (1979). In the office: Networks and coalitions. *Social Networks* **2**, 47–63.

Todeschini, R., and Consonni, V. (2000). *Handbook of Molecular Descriptors*. Wiley-VCH, Weinhein, 667 pp.

Townsend, C., Thompson, R. M., McIntosh, A. R., Kilroy, C., Edwards, E., and Scarsbrook, M. R. (1998). Disturbance, resource supply, and food-web architecture in streams. *Ecol. Lett.* **1**, 200.

Trinajstić, N. (1992). *Chemical Graph Theory*. CRC Press, Boca Raton, FL.

Troyanskaya, O. G. (2005). Putting microarrays in a context: Integrated analysis of diverse biological data. *Brief. Bioinf.* **6**, 34–43.

Tsaparas, P., Mariño-Ramírez, L., Bodenreider, O., Koonin, E. V., and Jordan, I. K. (2006). Global similarity and local divergence in human and mouse gene coexpression networks. *BMC Evol. Biol.* **6**, 70.

Uetz P., Dong, Y. A., Zeretzke, C., Atzler, C., Baiker, A., Berger, B., Rajagopala, S. V., Roupelieva, M., Rose, D., Fossum, E., and Haas, J. (2006). Herpesviral protein networks and their interaction with the human proteome. *Science* **311**, 239–42.

Uetz, P., Titz, B., Rajagopala, S. V., and Cagney, G. (2010). Experimental methods for protein interaction identification. In Buchanan, M., Caldarelli, G., de los Rios, P., Rao, F, and Vendruscolo, M. (eds.). *Networks in Cell Biology*. Cambridge Universisity Press, Cambridge, pp. 53–82.

Urban, D., and Keitt, T. (2001). Landscape connectivity: A graph-theoretic perspective. *Ecology* **82**, 1205–18.

Uzzi, B., Amaral, L. A. N., and Reed-Tsochas, F. (2007). Small-world networks and management science research: a review. *Eur. Manag. Rev.* **4**, 77–91.

Valente, T. W., and Foreman, R. K. (1998). Integration and radiality: measuring the extent of an individual's connectedness and reachability in a network. *Social Networks* **20**, 89–105.

van Dam, E. R., and Haemers, W. H. (2003). Which graphs are determined by their spectrum? *Lin. Alg. Appl.* **373**, 241–72.

van Dongen, S. (2000). Performance criteria for graph clustering and Markov cluster experiments. Technical Report INS-R0012, National Research Institute for Mathematics and Computer Science in the Netherlands, Amsterdam.

van Hauwermeiren, M. and Vose, D. (2009). *A Compendium of Distributions*. Ebook. Vose Software, Ghent, Belgium. http://www.vosesoftware.com.

van Noort, V., Snel, B., and Huynen, M. A. (2004). The yeast coexpression network has a small-world, scale-free architecture and can be explained by a simple model. *EMBO Rep.* **5**, 1–5.

van Steen, M., *Graph Theory and Complex Networks: An Introduction*. http://www.distributed-systems.net/gtcn/.

Vázquez, A. (2006). Spreading dynamics on small-world networks with connectivity fluctuations and correlations. *Phys. Rev. E* **74**, 056101.

Vendruscolo, M., Dokholyan, N. V., Paci, E., and Karplus, M. (2002). Small-world view of the amino acids that play a key role in protein folding. *Phys. Rev. E* **65**, 061910.

Verkhedkar, K. D., Raman, K., Chandra, N. R., Vishveshwara, S. (2007). Metabolome based reaction graphs of *M. tuberculosis* and *M. leprae*: A comparative network analysis. *PLoS One* **2**, e881.

Verma, D., and Meila, M. (2005). A comparison of spectral clustering algorithms. *Univ. Washington Dept. Comput. Sci. Tech. Rep.* 03–05–01.

Viana, M. P., Travençolo, B. A. N., Tank, E., and Costa, L. da F. (2010). Characterizing the diversity of dynamics in complex networks without border effects. Physica A **389**, 1771–8.

von Luxburg, U. (2007). A tutorial on spectral clustering. *Stat. Comput.* **17**, 395–416.

von Luxburg, U., Radl, A., and Hein, M. (2011). Getting lost in space: Large sample analysis of the resistance distance. *24th Ann. Conf. Neur. Infor. Proc. Syst. 2010*, 1–9, ed. J. Lafferty, C. K. I. Williams, J. Shawe-Taylor, R. S. Zemel, and A. Culotta. Curran, Red Hook, NY.

von Mering, C., Krause, R., Snel, B., Cornell, M., Oliver, S. G., Fields, S., and Bork, P. (2002). Comparative assessment of large-scale data sets of protein–protein interactions. *Nature* **417**, 399–403.

Voronoi, G. (1907). Nouvelles applications des paramètres continus à la théorie des formes quadratiques. *J. Reine Angew. Math.* **133**, 97–178.

Wagner, A., and Fell, D. A. (2001). The small world inside large metabolic networks. *Proc. R. Soc. Lond.* B **268**, 1803–10.

Waide, R. B., and Reagan, W. B. (eds.) (1996). *The Food Web of a Tropical Rainforest*. University Chicago Press, Chicago.

Walter, E., and Woodford, K. (2005). *Cambridge Advanced Learner's Dictionary*. Cambridge University Press, Cambridge.

Warren, P. H. (1989). Spatial and temporal variation in the structure of a fresh-water food web. *Oikos* **55**, 299–311.

Wasserman, S., and Faust, K. (1994). *Social Network Analysis*. Cambridge University Press, Cambridge.

Watts, D. J. (1999). Small Worlds: The Dynamics of Networks Between Order and Randomness. Princeton University Press, Princeton.

Watts, D. J., and Strogatz, S. H. (1998). Collective dynamic of 'small-world' networks. *Nature* **393**, 440–2.

Wayne, R. P. (1985). *Chemistry of Atmospheres*. Oxford University Press, Oxford, 774 pp.

Wei, T. H. (1952). *The Algebraic Foundations of Ranking Theory*. PhD thesis, University of Cambridge, UK.

Westhead, D. R., Parish, J. H., and Twyman, R. M. (2002). *Bioinformatics. Instant Notes*. BIOS Sci. Pub. Ltd., Oxford, 257 pp.

Weyn, B., van de Wouwer, G., Kumar-Singh, S., van Daele, A., Scheunders, P., van Marck, E., and Jacob, W. (1999). Computer-assisted differential diagnosis of malignant mesothelioma based on syntactic analysis. *Cytometry* **35**, 23–9.

White, J., Southgate, E., Thompson, J., and Brenner, S. (1986). The structure of the nervous system of the nematode *Caenorhabditis elegans*. *Phil. Trans. R. Soc. B* **314**, 1–340.

Wiener, H. (1947). Structural determination of paraffin boiling points. *J. Am. Chem. Soc.* **69**, 17–20.

Wigner, E. P. (1955). Characteristic vectors of bordered matrices with infinite dimensions. *Ann Math.* **62**, 548–64.

Willett, P. (1987). *Similarity and Clustering in Chemical Information Systems*. Research Studies Press, Letchworth, 245 pp.

Willett, P. (2006). Similarity-based virtual screening using 2D fingerprints. *Drug Disc. Today* **11**, 1046–53.

Williams, R. J., Berlow, E. L., Dunne, J. A., Barabási, A.-L., and Martinez, N. D. (2002). Two degrees of separation in complex food webs. *Proc. Natl. Acad. Sci. USA* **99**, 12913–16.

Williamson, M. P. and Sutcliffet, M. J. (2010). Protein–protein interactions. Biochem. Soc. Trans. **38**, 875–8.

Woodward, G., Benstead, J. P., Beveridge, O. S., Blanchard, J., Brey, T., Brown, L. E., Cross, W. F., Friberg, N., Ings, T. C., Jacob, U., Jennings, S., Ledger, M. E., Milner, A. M., Montoya, J. M., O'Gorman, E., Olesen, J. M., Petchey, O. L., Pichler, D. E., Reuman, D. C., Thompson, M. S. A., vann Venn, F. J. F., and Yvon-Durocher, G. (2010). Ecological networks in a changing climate. *Adv. Ecol. Res.* **42**, 71–138.

Wuchty, S. (2001). Scale-free behavior in protein domain networks. *Mol. Biol. Evol.* **18**, 1694–1702.

Wuchty, S., Jones, B. J., and Uzzi, B. (2007). The increasing dominance of teams in production of knowledge. *Science* **316**, 1036–9.

Wuchty, S., Oltvai, Z. N. and Barabási, A.-L. (2003). Evolutionary conservation of motif constituents in the yeast protein interaction network. *Nature Genet.* **35**, 176–9.

Xiao, W., and Gutman, I., (2003). Resistance distance and Laplacian spectrum. *Theor. Chem. Acc.* **110**, 284–9.

Xiao, W.-K., Ren, J., Qi, F., Song, Z.-W., Zhu, M.-X., Yang, H.-F., Jin, H.-Y., Wang, B.-H., and Zhou, T. (2007). Empirical study on clique-degree distribution of networks. *Phys. Rev. E* **76**, 037102.

Xu, X.-K., Zhang, J., Sun, J., and Small, M. (2009). Revising the simple measures of assortativity in complex networks. *Phys. Rev. E* **80**, 056106.

Xulvi-Brunet, R., and Sokolov, I. M (2005). Changing correlations in networks: Assortativity and dissortativity. *Acta Phys. Pol. B.* **36**, 1431–55.

Xulvi-Brunet, R., and Sokolov, I. M. (2004). Reshuffling scale-free networks: From random to assortative. *Phys. Rev. E* **70**, 066102.

Yang, F., Qian, H., and Beard, D. A. (2005). Ab initio prediction of thermodynamically feasible reaction directions from biochemical network stoichiometry. *Metab. Eng.* **7**, 251–9.

Yang, Y., and Zhang, H. (2008). Some rules on resistance distance with applications. *J. Phys. A: Math. Theor.* **41**, 445203.

Yang, Y., Li, X., and Rong, Z. (2010). Assortative degree-mixing patterns inhibit behavioral diversity of a scale-free structures population in high-mutation situations. *Europhys. Lett.* **89**, 18006.

Yeger-Lotem, E., Sattath, S., Kashtan, N., Itzkovitz, S., Milo, R., Pinter, R. Y., Alon, U. and Margalit, H. (2004). Network motifs in integrated cellular networks of transcription-regulation and protein–protein interaction. *Proc. Natl. Acad. Sci. USA* **101**, 5934–9.

Yildirim, M. A., Goh, K-II., Cusick, M. E., Barabási, A-L., and Vidal, M. (2007). Drug-target network, *Nature Biotech.* **25**, 1119–26.

Yodzis, P. (1998). Local trophodynamics and the interaction of marine mammals and fisheries in the Benguela ecosystem. *J. Anim. Ecol.* **67**, 635–58.

Yodzis, P. (2000). Diffuse effects in food webs. *Ecology* **81**, 261–6.

Young, M. P. (1993). The organization of neural systems in the primate cerebral cortex. *Proc. Biol. Sci.* **252**, 13–18.

Yu, H., Greenbaum, D., Lu, H. X., Zhu, X., and Gerstein, M. (2004). Genomic analysis of essentiality within protein networks. *Trends Gen.* **20**, 227–31.

Yu, H., Paccanaro, A., Trifonov, V. and Gerstein, M. (2006). Predicting interactions in protein networks by completing defective cliques. *Bioinformatics* **22**, 823–9.

Zachary, W. (1977).An information flow model for conflict and fission in small groups. *J. Anthropol. Res.* **33**, 452–73.

Zalesky, A., Fornito, A., Harding, I. H., Cocchi, L., Yücel, M., Pantelis, C., and Bullmore, E. T. (2010). Whole-Brain Anatomical Networks: Does the choice of nodes matter? NeuroImage **50**, 970–938.

Zeleny, L. D. (1950). Adaptation of research findings in social leadership to college classroom procedures. *Sociometry* **13**, 314–28.

Zhan, C., Chen, G., Yeung, L. F. (2010). On the distributions of Laplacian eigenvalues versus node degrees in complex networks. *Physica A* **389**, 1779–88.

Zhan, F. B., and Noon, C. E. (1998). Shortest path algorithms: an evaluation using real road networks, *Transport. Sci.* **32**, 65–73.

Zhang, A. (2009). *Protein Interaction Networks. Computational Analysis.* Cambridge University Press, Cambridge, 278 pp.

Zhang, B., and Horvath, S. (2005). A general framework for weighted gene co-expression network analysis. *Stat. Appl. Genet. Mol. Biol.* **4**, 17.

Zhang, J., and Zhang, X.-S. (2010). On resolution limit of the modularity in community detection, Ninth Int. Symp. Op. Res. Appl. (ISORA'10) Chengdu-Jiuzhaigou, China, August 19–23, pp. 492–9.

Zhang, J., Qiu, Y., Zhang, X.-S. (2010). Detecting community structure: from parsimony to weighted parsimony. *J. Syst. Sci. Complex* **23**, 1024–36.

Zhang, Z., and Grigorov, M. G. (2006). Similarity networks of protein binding sites. *Proteins: Struct., Funct., Bioinf.* **62**, 470–8.

Zhou, B., and Trinajstic, N. (2007). On the largest eigenvalue of the distance matrix of a connected graph. *Chem. Phys. Lett.* **447**, 384–7.

Zhou, S. and Mondragon, R. J. (2004). The rich-club phenomenon in the Internet topology. *IEEE Comm. Lett.* **8**, 180–2.

Zhu, D., and Qin, Z. S. (2005). Structural comparison of metabolic networks in selected single cell organisms. *BMC Bioinform.* **6**, 8.

Zotenko, E., Mestre, J., O'Leary, D. P., and Przytycka, T. M. (2008). Why do hubs in yeast protein interaction network tend to be essential: reexamining the connection between the network topology and essentiality. *PLoS Comput. Biol.* **4**, e1000140.

Index